Nonmammalian Animal Models for Biomedical Research

Editor
Avril D. Woodhead, D.Sc.
Associate Biologist
Department of Biology
Brookhaven National Laboratory
Upton, New York

Technical Editor
Katherine Vivirito
Technical Information Division
Brookhaven National Laboratories
Upton, New York

CRC Press, Inc.
Boca Raton, Florida

Library of Congress Cataloging-in-Publication Data

Nonmammalian animal models for biomedical research / editor, Avril D.
Woodhead: technical editor, Katherine Vivirito.
    p. cm.
    Includes bibliographies and index.
    ISBN 0-8493-4763-7
    1. Animal models in research. I. Woodhead, Avril D.
    [DNLM: 1. Disease Models, Animal. 2. Research. W 20.5 N813]
QL55.N66 1989
619—dc19
DNLM/DLC   88-39002
for Library of Congress   CIP

Direct all inquiries to CRC Press, Inc., 2000 Corporate Blvd., N.W., Boca Raton, Florida, 33431.

© 1989 by CRC Press, Inc.

International Standard Book Number 0-8493-4763-7

Library of Congress Card Number 88-39002
Printed in the United States

PREFACE

This volume is about the use of appropriate animal models in biomedical research: it is not about a search for "alternative" animals that may not come under the scope of the institutional Animal Care and Use Committee. All animals used in research deserve our best care and concern. Animal models can encompass the entire animal kingdom, from protozoa that offer unique insights into cellular aging, to apes that afford opportunities for research on language. Our focus in this volume is upon the appropriateness of a particular species for a particular investigation. For some work, mammalian models are indispensible. But in other cases, a rodent model may be used because the animals are readily available, because there is a wealth of information on their physiology and pathology, or because the laboratory is equipped for them and so their use is cost-effective. We all have grown accustomed to the laboratory mouse (or rat), and perhaps our knowledge of comparative physiology of cold-blooded vertebrates and invertebrates has become rusty over the years. It requires determination and a strong motivation to store away the rodent's cages and replace them with aquaria or terreria.

I hope this volume will convince readers that this action may be worthwhile. Although the model finally selected may be far removed from humans on the evolutionary scale, and may look and behave very differently, it may be the best and most relevant animal for the problem. The authors have shown how a variety of nonmammalian animal models have been used to successfully investigate a variety of biomedical questions. The authors' enthusiasms, vividly reflected in their writings, may intrigue researchers to look anew at the wealth of animals in nature. I trust that we also have dealt very well with pragmatic matters, so that no one will feel daunted at the prospect of taking an unusual species into their laboratory. If we have fulfilled our aim, then we may have a new group of researchers joining us in talking of "their fish", "their reptile", "their crustacean", and "their protozoan", as we already do.

THE EDITOR

Avril D. Woodhead is an Associate Biologist in the Department of Biology, and Senior Editor at the Brookhaven National Laboratory, Upton, New York. Dr. Woodhead was a State Scholar at the University of Durham, U. K., and received a First Class Honors in Zoology in 1954; she has a D.Sc. from the University of Newcastle-upon-Tyne, U.K. Her first research was on the migration and endocrinology of the Barents Sea cod at the Fisheries Laboratory, Lowestoft, U.K. She was Ford Foundation Research Fellow at the Zoological Society of London from 1961 to 1964, and then Development Commission Research Fellow until 1966.

Dr. Woodhead worked independently on fish physiology and behavior for several years while living on Heron Island, Great Barrier Reef, Australia, and in Discovery Bay, Jamaica, West Indies. She joined Brookhaven National Laboratory in 1975. She has published 80 papers and has edited 10 books. Her interests are in the mechanisms of aging and carcinogenesis, and her animal models are from amongst the cold-blooded vertebrates, especially fishes.

Dr. Woodhead has an M.A. in Liberal Arts and a New York State Certification in Labor/ Management from the State University of New York at Stony Brook. She is also the Coordinator of Women's Programs for Brookhaven National Laboratory.

CONTRIBUTORS

Stephen L. Adams, M.D.
Associate Chief
Department of Emergency Medicine
Northwestern University Medical School
Chicago, Illinois

Richard G. Andre, Ph.D.
Head
Department of Entomology
Walter Reed Army Institute of Research
Washington, D.C.

David E. Battaglia, Ph.D.
Senior Fellow
Department of Biochemistry
University of Washington
Seattle, Washington

Mark W. Bitensky M.D.
Senior Fellow
Physics and Life Sciences Divisions
Los Alamos National Laboratory
Los Alamos, New Mexico

John A. Brumbaugh, Ph.D.
Professor of Genetics
School of Biological Sciences
University of Nebraska-Lincoln
Lincoln, Nebraska

Gordon M. Burghardt, Ph.D.
Professor of Psychology and Zoology
Department of Psychology
Graduate Program in Ethology
Director
University of Tennessee
Knoxville, Tennessee

David Crews, Ph.D.
Professor of Psychology and Zoology
Department of Zoology
Director
Training Program in Neurobiology and
 Behavior
University of Texas
Austin, Texas

Brian Dale, Ph.D.
Department of Cell Biology
Stazione Zoologica
Naples, Italy

Yesu T. Das, Ph.D.
Senior Research Associate
Department of Entomology
Walter Reed Army Institute of Research
Washington, D.C.

Rebecca Dresser, J.D.
Associate Professor
Center for Biomedical Ethics
School of Medicine
School of Law
Case Western Reserve University
Cleveland, Ohio

Nobuo Egami, D.Sc
Emeritus Proffessor
Zoological Institute
Faculty of Science
University of Tokyo
Bunkyo-ku, Tokyo, Japan

Enrique Font, Ph.D.
Graduate Program in Ethology
University of Tennessee
Knoxville, Tennessee

Neil Greenberg, Ph.D.
Associate Professor
Department of Zoology
Graduate Program in Ethology
University of Tennessee
Knoxville, Tennessee

Jay W. Heinecke, M.D.
Departments of Biochemistry and
 Medicine
University of Washington
Seattle, Washington

Yasuko Hyodo-Taguchi, D.Sc.
Radiation Biologist
Division of Biology
National Institute of Radiological
 Sciences
Anagawa, Chiba-shi, Japan

Paul S. Jones, Ph.D.
Departments of Biochemistry and
 Psychiatry
Health Sciences Center
State University of New York
Stony Brook, New York

Richard E. Jones, Ph.D.
Professor
Department of Environmental,
 Population, and Organismic Biology
University of Colorado
Boulder, Colorado

Arnost Kleinzeller, M.D., Ph.D., D.Sc.
Emeritus Professor of Physiology
Department of Physiology
University of Pennsylvania
Philadelphia, Pennsylvania

Charles M. Lent, Ph.D.
Associate Professor
Department of Biology
Utah State University
Logan, Utah

Carl A. Luer, Ph.D.
Senior Scientist
Mote Marine Laboratory
Sarasota, Florida

David S. Miller Ph.D.
Laboratory of Cellular and Molecular
 Pharmacology
National Instititue of Environmental
 Health Sciences
National Institutes of Health Research
 Triangle Park, North Carolina

David O. Norris, Ph.D.
Professor
Department of Environmental,
 Population, and Organismic Biology
University of Colorado
Boulder, Colorado

William S. Oetting, Ph.D.
Research Associate
Institute of Human Genetics
University of Minnesota
Minneapolis, Minnesota

F. Harvey Pough, Ph.D.
Professor
Section of Ecology and Systematics
Cornell University
Ithaca, New York

Wolfgang Quitschke, Ph.D.
Departments of Biochemistry and
 Psychiatry
Health Sciences Center
State University of New York
Stony Brook, New York

Evaldo Reischl, D.Sc.
Associate Professor
Department of Fisiologia, Farmacologia,
 and Biofisica
UFRGS
Porto Alegre, RS, Brazil

Nisson Schechter, Ph.D.
Associate Professor
Departments of Biochemistry and
 Psychiatry
Health Sciences Center
State University of New York
Stony Brook, New York

Akihiro Shima, D.Sc.
Professor
Zoological Institute
Faculty of Science
University of Tokyo
Bunkyo-ku, Tokyo, Japan

Albert C. Smith, Ph.D., M.D.
Chief
Clinical Pathology
Veterans Administration Medical Center
Bay Pines, Florida
Courtesy Professor
Department of Marine Science
University of South Florida
St. Petersburg, Florida

Joan Smith-Sonneborn, Ph.D.
Professor
Departments of Zoology and Physiology
University of Wyoming
Laramie, Wyoming

Gerald Vaughan
Professor
Department of Zoology
Director
Graduate Program in Physiology
University of Tennessee
Knoxville, Tennessee

Edwin W. Taylor, Ph.D.
Reader in Animal hysiology School of
 Biological Sciences
University of Birmingham
Birmingham, England, U.K.

Katherine Vivirito
Technical Information Division
Brookhaven National Laboratory
Upton, New York

Paul Tesser, Ph.D.
Research Assistant
Department of Psychiatry
State University of New York
Stony Brook, New York

Robert A. Wirtz, Ph.D.
Research Entomologist
Department of Entomology
Walter Reed Army Institute of Research
Washington, D.C.

Avril D. Woodhead, D.Sc.
Associate Biologist and Senior Editor
Biology Department
Brookhaven National Laboratory
Upton, New York

TABLE OF CONTENTS

Chapter 1

THE CHOICE OF NONMAMMALIAN MODELS IN BIOMEDICAL STUDIES

Arnost Kleinzeller

TABLE OF CONTENTS

I. INTRODUCTION

The use of nonmammalian models in biomedical research stems from the belief that these are (eventually) applicable to human physiology and pathology. Such a view implies the assumption that basic mechanisms of physiological phenomena are similar in a broad variety of phyla; other, less efficient mechanisms would have been eliminated in the course of evolution.

This belief in the basic unity of physiology may be traced to Newton's[1] statement in 1686; *"Natura enim simplex est"* (Nature is pleased with simplicity). In the preface to his *Philosophiae Naturalis Principia Mathematica,* Newton[2] implied that the mathematical principles used to study the movement of celestial bodies, and the described forces of nature may also apply to other phenomena: "I wish we could derive the rest of the phenomena of nature by the same kind of reasoning from mechanical principles; for I am induced by many reasons to suspect that they may all depend upon certain forces by which particles of bodies, by some causes hitherto unknown, are either mutually impelled towards each other, or are repelled and recede from each other . . . but I hope that the principles here laid down will afford some light either to that or some truer method of philosophy." These statements imply the universal validity of physical laws to all nature.

Bernard[3] proceeded in this line of thought further when analyzing the aims and conceptual tools of physiology; his views reflect the spectacular progress made in biomedical sciences particularly in the 19th century.

Physiology is the science which studies phenomena characterizing living beings; however, thus defined, this science is still too broad, and should be subdivided into physiology *general* and *descriptive* (*specific* or *comparative*). Each phenomenon of life is always determined by the physical conditions which . . . become the conditions or material causes, immediate or secondary.

General physiology offers an insight into the general conditions of life which are shared by all living beings. Here we are studying the vital process* in itself, independently from its form and special mechanisms by which it is manifested. On the other hand, *descriptive physiology* offers an insight into the forms and particular mechanisms which life employs in order to be expressed in a specific living being. If we now wish to compare the form of these diverse mechanisms, which are of an infinite variety, in order to deduce their laws, this is the task of *comparative physiology*. This approach is of great interest for us since it shows us the infinite variety of life, based upon the unity of its conditions. This unity is revealed to us by *general physiology*. We always have to return to this approach if we wish to understand the driving forces of life.

Bernard's unity of the physicochemical conditions in physiology was then extended in 1926 by the concept of unity in biochemistry (Kluyver and Donker[4]), postulating that basic metabolic processes are shared by a broad spectrum of living cells; differences reflect the properties of the studied object (physiology of the organism and specific functions of its cells), but not different mechanisms of biochemical reactions.

Prosser[5] specifies three reasons for experimentation on various animals: (1) to understand the functional biology of unusual and uniquely adapted animals; (2) to make biological generalizations by comparison of many kinds of animals; and (3) to seek animal models from which, by extrapolation, some contribution can be made to medicine, human or veterinary, or to agriculture. The selection by the biologist of the experimental material for an envisaged project then represents the choice of the basic investigational tool for obtaining answers to specific questions. Such choice requires an evaluation of several points.

The biological properties of the considered material — Does the proposed material display the desired properties in sufficiently specific form so as to represent a distinct advantage over other possible materials? Thus, when studying problems of osmoregulation, the choice of material will depend on the more specific questions asked, reflecting known differences in the osmoregulatory capacity (and mechanisms) of terrestrial animals, animals

---

* Reflecting the relationship between the organism and its environment.

adapted to life in fresh- or seawater, and euryhaline animals capable of rapid adaptation to major changes in the actual salinity of their environment. Furthermore, the tissues involved in the actual osmoregulatory process will have to be considered.

The availability of the considered material — While nowadays the scientist may purchase a considerable spectrum of animal species thousands of miles from the normal habitat of the given animal, the physiological properties of the experimental material may often impose on the investigator the need (temporarily) to locate his/her laboratory close to the model. Hence, to a considerable extent, the attraction of specialized research facilities such as marine biological stations.

The ease of experimental manipulation — Adhering to the osmoregulatory phenomena as an example, in terrestrial animals the kidney (and, more specifically, the distal portion of the nephron) is the crucial tissue dealing with osmoregulation. When studying the whole organ, conventional clearance studies will reveal some aspects. More detailed information may be obtained only by sophisticated techniques involving the perfusion of isolated portions of the nephron.[6] In many freshwater lower vertebrates and in marine teleosts, the crucial osmoregulatory function resides in the gills, and the perfused gill preparation[7] may be used. For more detailed studies, this preparation is cumbersome. However, the discovery of the secretory function of the operculum of the killifish[8,9] opened the door to a convenient experimental approach, permitting an analysis of the transcellular ionic fluxes. Another tissue exposed to major osmotic gradients in freshwater animals and in marine teleosts is the cornea, which can be handled with relative ease for the study of electrolyte and water fluxes.[10] In elasmobranchs, one of the important osmoregulatory organs is the rectal gland, which can be readily studied on the level of the perfused tissue *in vitro*,[11] perfused isolated tubules,[12] or tissue preparations, such as slices[11,13] or isolated cells.[14]

Some other factors — Occasionally the selection of the biological model is affected by other considerations which may eventually prove to be decisive for progress in science. The discovery of the citric acid cycle by Krebs[15] illustrates this point. The crucial experiments of Krebs demonstrated the catalytic effect of C_4-dicarboxylic and C_6-tricarboxylic acids on oxygen uptake by minced pigeon breast muscle. This preparation was chosen after Szent-Gyorgyi showed its activity in the metabolism of C_4-dicarboxylic acids; moreover, pigeons were plentiful in Sheffield. Kreb's results and interpretation were questioned by Breusch because he could not find the catalytic effect of citric acid on the O_2 uptake of minced cat muscle. Subsequently, Krebs's results were verified, and the existence of the cycle was firmly established. In retrospect, the difference between both muscle preparations is based on the fact that pigeon breast muscle is particularly rich in mitochondria, as opposed to the white cat muscle. When I met Professor Breusch in 1961, he offered a cogent reason for his not-so-lucky choice of cat muscle. On the one hand, the budgetary position of his department at the Istanbul University did not permit the purchase of pigeons. On the other, stray cats were plentiful in Istanbul: animals caught in the evening by the laboratory assistant therefore served as welcome experimental material. Obviously, the use of cats would have been out of the question in pet-loving England. These sociological and budgetary aspects thus made a difference in one of the great discoveries in biology.

II. THE USEFULNESS OF NONMAMMALIAN MODELS IN THE STUDY OF OSMOREGULATORY PROCESSES

An analysis of the development of our understanding of osmoregulatory phenomena provides cogent examples of the usefulness of nonmammalian models (and thus of general and comparative physiology) for the study of major questions in human physiology. While at first investigators were challenged by the diversity of responses of animals to their various environments, the implication of the component mechanisms for mammalian physiology eventually became apparent, although at first view the mammal does not face the osmoregulatory problems encountered by animals living in aqueous environments. After 20 years of research at the Mt. Desert Island Biological Laboratory, this student gained a deeper insight into the emergence of new concepts in this field.

Frédericq, the founder of the Liège school of chemical zoology, was the first to demonstrate in 1884 the existence of osmoregulatory phenomena (cf. Florkin[16]). Guided by taste as the only available analytical tool, he found that:

1. The salinity of the blood of marine invertebrates (crab, lobster, octopus) equals that of the sea in which they live.
2. The blood of marine fish (stingray, sole, haddock) is less salty than the seawater; he also commented that their blood did not appear to contain more soluble salts than that of freshwater fish.
3. The blood of crabs living in brackish water tasted less salty than that of crabs living in the sea.

Thus, Frédericq discovered that in marine invertebrates the salinity of the internal and external environment is equal, while in marine teleosts the internal environment is isolated from the external seawater. In subsequent years, having adopted the more objective method of measuring the depression of the freezing point, he extended his studies and emphasized the role of the gills in the maintenance of osmotic phenomena. Eventually, in 1904, he schematically summarized his results[17] (Figure 1), differentiating between different levels of development of aquatic animals:

Stage A — In marine invertebrates the salinity of the external environment does not differ substantially from that of the tissues.

Stage B — At a higher developmental level, the blood is still at stage A, whereas the tissues of crustaceans and marine mollusks contain, in addition to salts, major amounts of organic solutes; however, isotonic conditions pertain between the internal and external environments. In marine elasmobranchs, the blood tonicity is slightly higher than that of seawater; the lower salt level in the blood *and* tissues is compensated by organic solutes.

Stage C — In marine teleosts, the blood and the tissues have a salinity well below that of seawater. Freshwater fish and invertebrates are also at stage C in that the salinity of their blood and tissues is independent of that of the external environment; however, here the osmotic gradient is in the opposite direction.

These levels of phylogenetic development ,in the direction of greater independence of the external environment were related to differences in gill permeability.

Frédericq's observations provided the conceptual framework for the integration of previously scattered information and laid the foundation for a century of investigations (cf. Krogh[18]) concerning the tissues and mechanisms by which:

1. Freshwater vertebrates and invertebrates maintain a concentration gradient between their plasma (and tissues) and the external pond water.
2. Marine teleosts maintain a concentration gradient in the direction from the outside to the inside.
3. Organic solutes play a role in osmoregulation.
4. Terrestrial vertebrates concentrate their urine.

A. ABSORPTION OF ELECTROLYTES AND WATER

The frog skin proved to be the most important tissue for the elucidation of the mechanism by which electrolytes and water are absorbed by epithelia.

Du Bois-Reymond[19] first demonstrated in 1848 the existence of an electrical current across the frog skin (from the external to the internal face); this current was found only in living tissues. Subsequently, Engelmann[20] raised the possibility that this electrical current

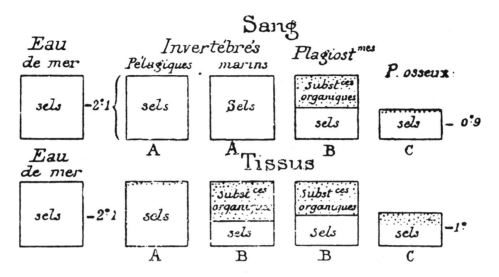

FIGURE 1. Frédericq's[17] scheme: freezing point depression of seawater, blood, and tissues in invertebrates, elasmobranchs, and teleosts.

generated by the skin glands might drive electrolytes (and water) across the skin by electroosmosis.

The role of the frog skin in osmoregulatory phenomena was indicated by Bert[21] by showing that frogs lost water through their skin when placed in hypertonic NaCl (seawater) solutions; in teleosts, the loss of water occurred primarily through the gills. The convincing experiments of Reid[22] in 1892 then demonstrated the absorption of fluid across the frog skin and related it to absorption by the cat intestine; this process was brought about by "the presence in the skin . . . of a vital absorptive force dependent on protoplasmic activity"; on some experimental basis, he dismissed the possibility of an electroosmotic fluid transfer. Later, Reid[23] also showed the secretion of fluid through the skin of frogs in the breeding season, and he related this to the function of secretory skin glands. The next major step in our understanding of osmoregulation in frogs was made by Overton,[24] who clearly demonstrated (in a "preliminary" communication of 19 pages) the permeability of the skin for water in both directions; he also indicated the kidney as the site of formation of hypotonic urine, and the role of the bladder in contributing to the water economy of the frogs. Incidentally, these phenomena also contributed to the water economy of human beings. Darwin[25] reported that South American Indians used the sweet-tasting urine of frogs to quench their thirst.

The next crucial step was provided by Huf[26] in 1936 by demonstrating an "active" (i.e., metabolically dependent) transcellular flux of electrolytes (Cl^- and Na^+) against their concentration gradients from the pond side across the frog skin epithelium to the serosal side of the tissue. He related this process to the osmoregulatory capacity of various aquatic animals.

It took some 15 more years until Ussing and Zerahn[27] elucidated the absorptive process by defining the active sodium transport system as the mechanism by which Na^+ (Cl^- and water) is moved across the frog skin against its electrochemical gradient; the ingenious experiments and conceptual insight thus integrated existing electrophysiological and metabolic information. This study then characterized the first major mechanism in all osmoregulatory phenomena.

B. SECRETION OF ELECTROLYTES AND WATER

In 1930, our understanding of osmoregulation in aquatic animals was still poor, particularly as to specific tissues and cells. For such function in marine teleosts, with the intracellular fluids (approximately 150 mM for Cl$^-$) hypotonic to seawater (540 mM Cl$^-$), Schlieper[28] excluded the possibility that the excess of salts taken in could be extruded by the kidney and considered three views: (1) an excretory role of another organ, (2) the dilution of urine by water produced by the oxidation of nutrients, and (3) an active transport of water from the sea across the fish skin. The nature of the actual regulated mechanism, i.e., osmotic pressure per se, or the water and salt content of the body, was not approached. Several studies in the following 2 years pinpointed the tissue(s) responsible for this phenomenon. Smith[29] in New York provided evidence that of the seawater swallowed by marine teleosts (goosefish, eel, sculpin) only some electrolytes (Ca^{2+}, Mg^{2+}) were excreted by the intestine and the kidney, while the bulk of absorbed Na$^+$ and Cl$^-$ is excreted by an extrarenal route; the gills were suggested as the excretory tissue, performing osmotic work. In Krogh's laboratory in Copenhagen, the technique of a double perfused heart-gill preparation of the eel was developed by Keys;[7] this author then showed[30] that the gills actually extrude a chloride solution into seawater against a major osmotic gradient. Subsequently, Keys and Wilmer[31] suggested that the observed "chloride cells" of the gills may be responsible for the excretory process. At the same time, Schlieper[32] in the Copenhagen laboratory provided conclusive evidence for the excretory role of the gills by demonstrating "active" chloride transport under conditions where osmotic gradients across the gill cells were eliminated. Thus, simple physicochemical processes were excluded as possible mechanisms. By 1937, Krogh[33] realized that in intact animals two different mechanisms are involved in salt absorption and secretion: a cation mechanism and an anion transport system.

The next major step concerning the primary ionic species involved in the secretory process took place some 30 years later. In tissues involved in osmoregulatory phenomena, Zadunaisky demonstrated active transport in frog skin[34] and cornea;[10] in the latter tissue, an involvement of Na$^+$ in the Cl$^-$ transport was found. The discovery[8] that the operculum of the euryhaline *Fundulus* represents a tissue with 40 to 70% chloride secretory cells (as opposed to the gills, where the chloride cells are a minor component) made it possible to use a rigorous biophysical approach to the study of chloride secretion,[9] and the active nature of the process was established. The Cl$^-$ secretion was inhibited *inter alia* by the diuretic furosemide, which has been previously shown to block Cl$^-$ transport in the nephron[35] (ascending limb of Henle's loop). The role of the chloride cells in osmoregulatory phenomena was then clinched when comparing the secretory process in seawater- and freshwater-adapted fish.[36] It remains to be seen whether the same chloride cells secrete NaCl into seawater and absorb NaCl under freshwater conditions. Such a dual role has been ascribed to the chloride cells in the gills.[37]

The analysis of the actual transport mechanism represented the next crucial step. The Na$^+$ dependence of the Cl$^-$ secretion in the opercular epithelium (as well as the inhibition of ouabain, blocking the operation of the Na$^+$ pump at the serosal side, without effect on the passive influx of Na$^+$) suggested[38] a coupling between Na$^+$ and Cl$^-$ influx into the cells. A similar coupling has been postulated for the electroneutral flux of NaCl in intestine and fish gall bladder[39] and was later extended for a variety of epithelial cells.[40] In this concept, the NaCl cotransport proceeds in accordance with the electrochemical gradients of the respective ionic species, and the actual driving force for the secretory (or absorptive) chloride transport is provided by the operation of the Na$^+$ pump. At the present stage of our knowledge, the electroneutral, secondarily active coupled influx of Cl$^-$ into the cells is in effect a process involving Na$^+$, K$^+$, and Cl$^-$ with a stoichiometry of 1:1:2 (Geck et al.[41]). The recognition of the properties of this system then permitted the demonstration of its involvement in a variety of cells capable of secreting or absorbing Cl$^-$, e.g., in the dogfish shark rectal gland,[12] the thick ascending limb of Henle's loop in the nephron,[42] the fish and mammalian intestine,[43,39] the tracheal epithelial,[44] and the cornea.[45]

Thus, the actual mechanism of the secondary active chloride transport now appears to be clarified. The direction of the transport process, i.e., secretion or absorption, is determined by the cellular localization of the Na^+, K^+, Cl^- cotransport system at the basolateral, or luminal membrane, respectively (see Figure 2). The permeability of the cell membranes to water determines whether the excreted fluid is iso- or anisotonic with regard to the plasma or the environment.

C. THE ROLE OF ORGANIC SOLUTES IN OSMOREGULATION

Frédericq's recognition of organic solutes as major osmotic components of the plasma and tissues of marine invertebrates and elasmobranchs (Figure 1) was based on evidence first provided in 1858 by Staedeler and Frerichs[46] that the blood of sharks contains considerable amounts of urea, of the acidic amino acid taurine, and of a new substance called scyllit, correctly suspected to be close to inositol. In 1909, Suwa added betaine[47] and trimethylamine oxide[48] as additional components. The pioneering studies of Florkin[49] and collaborators, Schoffeniels[50] and Gilles[51] then emphasized the important role of a broad spectrum of amino acids as osmotic effectors in invertebrates and vertebrates. By now (cf. Yancey et al.[52]), several classes of organic solutes have been recognized as "osmolytes", i.e., urea, organic amines, amino acids, and polyols.

In marine invertebrates, osmolytes are osmotic effectors in the tissues, while the hemolymph differs in its composition only little from the seawater. In marine elasmobranchs, osmolytes (particularly urea) may represent more than one third of the osmotic components of both plasma and tissues; as to urea, its high concentration is maintained by the relative impermeability of the skin, the gills, and the oral epithelium for this solute, and by its reabsorption by the kidney; as the result, the urine is hypotonic compared with the plasma (Smith[53]). The high urea concentration in the plasma of elasmobranchs prompted a comparison with uremia in human beings.

Evidence for the osmoregulatory role of osmolytes stems primarily from experiments showing that changes of external tonicity considerably affect the plasma (and tissue) levels of osmotic effectors. This point has been clearly established by Smith[53] for urea in elasmobranchs, then for trimethylamine oxide,[54] and more recently for amino acids (Forster and Goldstein[55,56]) and for *myo*-inositol.[57] On the strength of such experiments, Smith[53] concluded in 1931 that, as opposed to previous views, the elasmobranch is indifferent to the urea content (i.e., to osmotic pressure) of the blood, except that osmotic pressure subserves the role of water absorption. The concept thus developed that for the maintenance of the steady state of Bernard's *milieu interieur* the crucial factor is not osmotic pressure per se, but rather the tissue and cell levels of water and specific salts. This point also clearly emerges from the role of osmolytes in the maintenance of cell volume,[58] an osmotic coupling between intracellular osmolytes and electrolytes appears to be the mechanism by which osmotic effectors affect the diffusion gradient of electrolytes.[59]

At first sight, the foregoing observations may be considered to be the results of studies designed to clarify the osmoregulatory processes in fish, particularly in elasmobranchs, from the point of view of the survival value of the phenomenon for the animals. Only very recently was the relevance of these studies for human physiology revealed when it was discovered that osmolytes also play a major role in osmoregulatory processes in the renal medulla of mammals. Thus, inositol[60] and later a broad spectrum of organic osmolytes[61,62] have been identified as components of renal medullary cells, and their role for urinary concentration was indicated.

D. MECHANISMS INVOLVED IN THE CONCENTRATION OF URINE

Mammals concentrate urine against major osmotic gradients. Our knowledge on this topic has been summarized in a symposium[63] honoring Berliner. In essence, the concentrating mechanisms are based on (1) the structural arrangement of nephrons and the renal vasculature in the kidney (permitting the operation of the countercurrent system), (2) the absorption of an isotonic solution of salts (mainly NaCl) from the proximal tubular lumen, (3) the sodium

ABSORPTIVE EPITHELIAL CELL

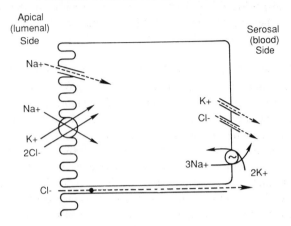

SECRETORY EPITHELIAL CELL

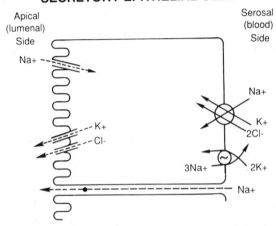

FIGURE 2.   Simplified scheme of ionic fluxes in absorptive and secretory epithelial cells. Solid lines: carrier-mediated fluxes; broken lines: conductive pathways; and $\sim$: ATP-driven pump.

chloride transport system in the thick ascending limb of Henle's loop, and (4) the differential action of hormones on the flows of water and salts. The operation of the countercurrent system is made possible by the concentration in the inner medulla of NaCl but also requires the presence of osmolytes (urea, amino acids, organic amines, and polyols). Hence, the basic mechanisms discussed in the previous sections all contribute to the urinary concentration.

III. THE AUGUST KROGH PRINCIPLE

The preceding survey of the development of our understanding of osmoregulatory phenomena presents convincing examples for the justification of the assumption that basic physiological mechanisms are similar, or identical, in a broad variety of phyla, i.e., for Bernard's unity of physiology. From the point of view of the choice of nonmammalian models for biomedical studies, this emphasis of the unity has been expressed in 1929 by August Krogh's Principle. Krogh[64] wrote:

For a large number of problems there will be some animal choice, or a few such animals, on which it can be most conveniently studied. Many years ago when my teacher, Christian Bohr, was interested in the respiratory mechanism of the lung and devised a method of studying the exchange through each lung separately, he found that a certain kind of tortoise possesses trachea dividing into the main bronchi high up in the neck and we used to say as a laboratory joke that this animal had been created expressly for the study of respiration physiology. I have no doubt that there are quite a number of animals which are similarly 'created' for special physiology purposes, but I am afraid that most of them are unknown to the man for whom they were 'created' and we must apply to the zoologist to find them and lay our hands on them.

The essay of Krebs[65] illustrated this concept by further examples showing that the application of Krogh's Principle led to major progress in biological sciences. The point made by Krebs and Krebs[66] is valid in that Krogh's Principle provides a heuristic tool only when exploring the mechanisms (physiological, biochemical, or biophysical) of biological phenomena — and this brings us back to Bernard's views on the role of general and comparative physiology (Section I). While originally investigators of osmoregulatory processes may have been attracted by the phenomenological differences between various species in their relationship to their environment (i.e., by the diversity in nature), the foregoing survey demonstrates the unifying relative simplicity underlying complex physiological processes. A student of transport physiology is impressed by the emerging knowledge that in all species and phyla studied so far, a broad diversity of physiological processes in polar epithelial cells can be accounted for by the operation of the sodium pump (and possibly some other "minor" ionic pumps), a handful of cotransport and exchange processes, and a limited (?) number of conductive pathways (channels). Thus, the absorption and secretion of fluids, the secretion of acids or alkaline solutions, and the concentration or dilution of fluids all operate using the same limited number of basic functional components, and diversity is rather an expression of quantitative aspects and spatial orientation.

Krogh's principle also represents a real challenge to physiologists. He points out the need of consulting with zoologists as to the choice of the "right" species for a given problem. In this respect, Hill[67] was more demanding:

There is no need to stick to rabbits and frogs. If a problem seems insoluble on one muscle, one should try to define it more precisely to see where the difficulty lies. Discussion with a zoologist, or a visit to a Marine Laboratory, may provide material many times better suited to one's needs. I spent many years trying to measure the heat production of nerve: If I had made the experiment on crab's nerves instead of frog's the answer would have come in 1912 instead of 1926. In 1912 it was not possible to define the problem well enough to get a clear direction to non-medullated nerve, but at least one might have taken a chance and not persisted with the frog's sciatic. If one's instruments, or methods, are too slow, one can make them quicker by using slower material — tortoises, toads or even sloths. That means, of course, that biochemists, like biophysicists, must also be biologists (as Meyerhof has always been and as Hopkins was) — but why not?

Given the increasingly narrow specialization of investigators, Hill's challenge certainly applies today also to many physiologists.

ACKNOWLEDGMENTS

The author wishes to record his appreciation of advice received from Drs. M. M. Civan (University of Pennsylvania, Philadelphia), L. Goldstein (Brown University, Providence, RI), and J. A. Zadunaisky (New York University). This study was supported in part by a grant from the Research Foundation of the University of Pennsylvania.

REFERENCES

1. **Newton, I.,** Rules of reasoning in philosophy, in *Philosophiae Naturalis Principia Mathematica*, Vol. 2, H. D. Symonds, London, 1803, 160.
2. **Newton, I.,** *Philosophiae Naturalis Principia Mathematica*, Vol. 1, H. D. Symonds, London, 1803, b2.

3. **Bernard, C.,** *Leçons sur les Phénomènes de la Vie Communs aux Animaux et aux Vegetaux,* Vol. 1, Baillière et Fils, Paris, 1878, 374.
4. **Kluyver, A. J. and Donker, H. J. L.,** Die Einheit in der Biochemie, *Chem. Zelle Gewebe,* 13, 134, 1926.
5. **Prosser, C. L.,** Animal models for biomedical research V — invertebrates, *Fed. Proc. Fed. Am. Soc. Exp. Biol.,* 32, 2177, 1973.
6. **Burg, M. B., Grantham, J., Abramow, M., and Orloff, J.,** Preparation and study of fragments of single rabbit nephron, *Am. J. Physiol.,* 210, 1293, 1966.
7. **Keys, A. B.,** The heart-gill preparation of the eel and its perfusion for the study of a natural membrane *in situ, Z. Vgl. Physiol.,* 15, 352, 1931.
8. **Karnaky, K. J. and Kinter, W. B.,** Killifish opercular skin: a flat epithelium with a high density of chloride cells, *J. Exp. Zool.,* 199, 355, 1977.
9. **Degnan, K. J., Karnaky, K. J., and Zadunaisky, J. A.,** Active chloride transport in the *in vitro* opercular skin of a teleost *(Fundulus heteroclitus),* a gill-like epithelium rich in chloride cells, *J. Physiol. (London),* 271, 155, 1977.
10. **Zadunaisky, J. A.,** Active transport of chloride in frog cornea, *Am. J. Physiol.,* 211, 506, 1966.
11. **Silva, P., Stoff, J., Field, M., Fine, L., Forrest, J. N., and Epstein, F. H.,** Mechanism of active chloride secretion by shark rectal gland: role of Na-K-ATPase in chloride transport, *Am. J. Physiol.,* 233, F298, 1977.
12. **Greger, R. and Schlatter, E.,** Mechanism of NaCl secretion in rectal gland tubules of spiny dogfish *(Squalus acanthias).* I. Experiments in isolated *in vitro* perfused rectal gland tubules, *Pfluegers Arch.,* 402, 63, 1984.
13. **Kleinzeller, A. and Goldstein, J.,** Effect of anisotonic medias on cell volume and electrolyte fluxes in slices of the dogfish *(Squalus acanthias)* rectal gland, *J. Comp. Physiol. B,* 154, 565, 1984.
14. **Silva, P., Epstein, J. A., Stevens, A., Spokes, K., and Epstein, F. H.,** Ouabain binding in rectal gland of *Squalus acanthias, J. Membr. Biol.,* 75, 105, 1983.
15. **Krebs, H. A.,** The intermediary stages in the biological oxidation of carbohydrate, *Adv. Enzymol.,* 3, 191, 1943.
16. **Florkin, M.,** *L'école liègeoise de physiologie et son maître Léon Frédericq (1851 - 1935). Pionnier de la zoologie chimique,* Vailland-Carmanne, S.A., Liège, 1979.
17. **Frédericq, L.,** Sur la concentration moléculaire du sang et des tissus chez les animaux aquatiques, *Arch. Biol. (Liège),* 20, 709, 1904.
18. **Krogh, A.,** *Osmotic Regulation in Aquatic Animals,* Cambridge University Press, London, 1939.
19. **du Bois-Reymond, E.,** *Untersuchungen über thierische Electricität,* Vol. 2, Sect. 2, G. Reiner, Berlin, 1848, 17ff.
20. **Engelmann, Th. W.,** Die Hautdrüsen des Frosches, *Pfluegers Arch.* 6, 97, 1872.
21. **Bert, P.,** Sur les phenomènes et les causes de la mort des animaux d'eau douce que l'on plonge dans l'eau de mer, *C. R. Acad. Sci. (Paris),* 78, 382, 1871.
22. **Reid, W. E.,** Report on experiments upon "Absorption without osmosis", *Br. Med. J.,* 2, 323, 1892.
23. **Reid, W. E.,** Transport of fluid by certain epithelia, *J. Physiol. (London),* 26, 436, 1901.
24. **Overton, E.,** Neun und dreissig Thesen über die Wasserökonomie der Amphibien und die osmotischen Eigenschaften der Amphibienhaut, *Verh. Phys. Med. Ges. Würzburg,* 36, 277, 1904.
25. **Darwin, C.,** *Journal of Researches into the Natural History & Geology of the Countries Visited during the Voyage of H.M.S. Beagle Round the World Under the Command of Capt. Fitzroy, R.N.,* 2nd ed., D. Appleton, New York, 1871.
26. **Huf, E.,** Über aktiven Wasser- und Salztransport durch die Froschhaut, *Pfluegers Arch.,* 237, 143, 1936.
27. **Ussing, H. H. and Zerahn, K.,** Active transport of sodium as the source of electric current in the short-circuited isolated frog skin, *Acta Physiol. Scand.,* 23, 11, 1951.
28. **Schlieper, C.,** Die Osmoregulation wasserlebender Tiere, *Biol. Rev. Cambridge Philos. Soc.,* 5, 309, 1930.
29. **Smith, H. W.,** The absorption and excretion of water and salts by marine teleosts, *Am. J. Physiol.,* 93, 480, 1930.
30. **Keys, A. B.,** Chloride and water secretion and absorption by the gills of the eel, *Z. Vgl. Physiol.,* 15, 364, 1931.
31. **Keys, A. B. and Wilmer, E. N.,** "Chloride-secreting cells" in the gills of the fish with special reference to the common eel, *J. Physiol. (London),* 76, 368, 1932.
32. **Schlieper, C.,** Über die osmoregulatorische Funktion der Aalkiemen, *Z. Vgl. Physiol.,* 18, 682, 1932.
33. **Krogh, A.,** Osmotic regulation in the frog *(R. esculenta)* by active absorption of chloride ions, *Skand. Arch. Physiol.,* 76, 60, 1937.
34. **Zadunaisky, J. A. and Candia, O. A.,** Active transport of sodium and chloride by the isolated skin of the South American frog *Leptodactylus ocelatus, Nature (London),* 195, 1004, 1962.
35. **Burg, M., Stoner, L., Cardinal, J., and Green, N.,** Furosemide effect in isolated perfused tubules, *Am. J. Physiol.,* 225, 119, 1973.

36. **Mayer-Gostan, N. and Zadunaisky, J. A.**, Inhibition of chloride secretion by prolactin in the isolated opercular epithelium of *Fundulus heteroclitus, Bull. Mt. Desert Isl. Biol. Lab.*, 18, 106, 1978.

37. **Maetz, J.**, Fish gills: mechanism of salt transfer in fresh water and sea water, *Philos. Trans. R. Soc. London Ser. B*, 262, 209, 1971.

38. **Degnan, K. J. and Zadunaisky, J. A.**, Ionic contributions to the potential and current across the opercular epithelium, *Am. J. Physiol.*, 238, R231, 1980.

39. **Diamond, J. M.**, The mechanism of solute transport by the gall-bladder, *J. Physiol. (London)*, 161, 474, 1962.

40. **Frizzell, R. A., Field, M., and Schultz, S. G.**, Sodium-coupled chloride transport by epithelial tissues, *Am. J. Physiol.*, 236, F1, 1979.

41. **Geck, P., Pietrzyk, C., Burckhardt, B. C., Pfeiffer, B., and Heinz, E.**, Electrically silent cotransport of Na^+, K^+, and Cl^- in Ehrlich cells, *Biochim. Biophys. Acta*, 600, 422, 1980.

42. **Greger, R. and Schlatter, E.**, Properties of the basolateral membrane of the cortical thick ascending limb of Henle's loop of rabbit kidney. A model for secondary active chloride transport, *Pfluegers Arch.*, 396, 325, 1984.

43. **Musch, M. W., Orellana, S. A., Kimberg, L. S., Field, M., Halm, D. R., Krasny, J., Jr., and Frizzell, R. A.**, Na^+-K^+-Cl^- co-transport in the intestine of a marine teleost, *Nature (London)*, 300, 351, 1982.

44. **Smith, P. L., Welsh, M. J., Stoff, J. S., and Frizzell, R. A.**, Chloride secretion by canine tracheal epithelium. I. Role of intracellular cAMP levels, *J. Membr. Biol.*, 70, 217, 1982.

45. **Reuss, L., Reinach, P., Weinman, S. A., and Grady, T. P.**, Intracellular ion activities and Cl^- transport mechanisms in bullfrog corneal epithelium, *Am. J. Physiol.*, 244, C336, 1983.

46. **Staedeler, G. and Frerichs, F. T.**, Ueber das Vorkommen von Harnstoff, Taurin und Scyllit in den Organen der Plagiostomen, *J. Prakt. Chem.*, 73, 48, 1858.

47. **Suwa, A.**, Untersuchungen über die Organextrakte der Selachier. I. Mitt. Die Muskelextrakstoffe des Dornhais *(Acanthias vulgarias), Arch. Ges. Physiol.*, 128, 421, 1909.

48. **Suwa, A.**, Untersuchungen über die Organextrakte der Selachier. II. Mitt. Über das aus den Muskelex-trakstoffen des Dornhais gewonnene Trimethylaminoxyd, *Arch. Ges. Physiol.*, 129, 231, 1909.

49. **Camien, M. N., Sarlet, H., Duchâteau, G., and Florkin, M.**, Non-protein amino acids in muscle and blood of marine fresh water *Crustacea, J. Biol. Chem.*, 193, 881, 1951.

50. **Schoffeniels, E.**, Adaptations with respect to salinity, *Biochem. Soc. Symp.*, 41, 179, 1976.

51. **Gilles, R.**, Intracellular organic osmotic effectors, in *Mechanisms of Osmoregulation in Animals*, Gilles, R., Ed., Wiley, Chichester, 1979, chap. 4, p. 111.

52. **Yancey, P. H., Clark, M. E., Hand, S. C., Bowlus, R. D., and Somero, G. N.**, Living with water stress: the evolution of osmolyte systems, *Science*, 217, 1214, 1982.

53. **Smith, H. W.**, The absorption and excretion of water and salts by elasmobranch fishes. II. Marine elasmobranchs, *Am. J. Physiol.*, 98, 296, 1931.

54. **Norris, E. R. and Benoit, G. J.**, Studies on trimethylamine oxide. I. Occurrence of trimethylamine oxide in marine organisms, *J. Biol. Chem.*, 158, 433, 1945.

55. **Forster, R. P. and Goldstein, L.**, Intracellular osmoregulatory role of amino acids and urea in marine elasmobranchs, *Am. J. Physiol.*, 230, 925, 1976.

56. **Boyd, T. A., Cha, C.-J., Forster, R. P., and Goldstein, L.**, Free amino acids in tissues of the skate *Raja erinacea* and the stingray *Dasyatis sabina:* effects of environmental dilution, *J. Exp. Zool.*, 199, 435, 1977.

57. **McGregor, L. C. and Kleinzeller, A.**, Osmolyte function of *myo*-inositol in dogfish *(Squalus acanthias)* rectal gland, *Bull. Mt. Desert Isl. Biol. Lab.*, 26, 168, 1986.

58. **Goldstein, L. and Kleinzeller, A.**, Cell volume regulation in lower vertebrates, *Curr. Top. Membr. Transp.*, 31, 181, 1987.

59. **Kleinzeller, A.**, Trimethylamine oxide and the maintenance of volume of dogfish shark rectal gland cells, *J. Exp. Zool.*, 236, 11, 1985.

60. **Cohen, M. A. H., Hruska, K. A., and Daughaday, W. H.**, Free *myo*-inositol in canine kidney. Selective concentration in renal medulla, *Proc. Soc. Exp. Biol. Med.*, 169, 380, 1982.

61. **Bagnasco, S., Balaban, R. S., Fales, H. M., Yang, Y.-H., and Burg, M.**, Predominant osmotically active organic solutes in rat and rabbit renal medulla, *J. Biol. Chem.*, 261, 5872, 1986.

62. **Balaban, R. S. and Burg, M. B.**, Osmotically active organic solutes in the renal inner medulla, *Kidney Int.*, 31, 562, 1987.

63. **de Roufignac, C. and Jamison, R. L., Eds.**, Symposium on the urinary concentrating mechanism in honor of Robert W. Berliner, *Kidney Int.*, 31, 501, 1987.

64. **Krogh, A.**, Progress in physiology, *Am. J. Physiol.*, 90, 243, 1929.

65. **Krebs, H. A.**, The August Krogh Principle: "for many problems there is an animal on which it can be most conveniently studied", *J. Exp. Zool.*, 194, 221, 1975.

66. **Krebs, H. A. and Krebs, J. R.,** The "August Krogh Principle", *J. Comp. Biochem. Physiol.,* 67B, 379, 1980.
67. **Hill, A. V.,** A challenge to biochemists, *Biochim. Biophys. Acta,* 4, 4, 1950.

Chapter 2

PROTOZOANS AS MODELS OF CELLULAR AGING: THE CILIATES

Joan Smith-Sonneborn

TABLE OF CONTENTS

I. INTRODUCTION

The use of single-celled animals as models of cellular aging began with the pioneering studies of Maupas in 1888[1] and has continued throughout the 20th century. Several ciliates were investigated and showed patterns of age-dependent declining growth, while others exhibited an apparently limitless capacity for cell reproduction.[2]

General interest in cellular aging did not surface because it was believed that aging did not occur at the cellular level in higher organisms. This view was strengthened by the experiments (later found to be flawed) of Carrel, which seemed to indicate indefinite growth of fibroblasts from chick heart cells.[3] The experiments implied that when cells were released from organismal control, their life-span exceeded the life-span of the donor and therefore (1) aging occurred on a supracellular level, and (2) the immortal cultured cells could have no role in the aging phenomenon.[4] The cell biology of human cell aging began with the demonstration of limited *in vitro* life-span of human cells by Hayflick and Moorhead[5] and Hayflick.[6] An understanding of the mechanisms which control the capacity of cells to divide is emerging from studies of human cells in culture. Repressors of cell division have been isolated and identified.[7,8] However, there are some mammalian cells, like skin, hemopoietic systems, and erythropoietic stem cells, which can survive beyond the lifetime of the host.[9,10] Comparisons between cells with widely different life-spans are only beginning.

The cell biology of protozoan aging began in 1954 when Sonneborn[11] found that an apparently immortal line of *Paramecium* was undergoing periodic episodes of self-fertilization and continuously initiating new generations. Yet certain protozoa and lower organisms do not require fertilization for survival, and certain strains of *Tetrahymena* still retain a reputation for unlimited cell division capacity.[12] There are no satisfactory explanations for why certain protozoans or cells do not seem to age, but the ability to exclude or restore damaged parts seems common to the surviving species.[13]

Paramecium became a favored organism for ciliate aging because of its short life-span and the wealth of genetic data, though elegant studies were pursued using *Euplotes, Tokophrya, Tetrahymena, Spathidium, Stylonychia, Amoeba,* and *Volvox*.[14-24]

The present review will focus mainly on ciliates in general and *Paramecium* in particular, since most of the recent data are available in these systems.

II. BACKGROUND

A. THE TECHNOLOGY

Clonal age in ciliates is measured in the number of days or cell divisions (fissions) which have occurred since the origin of the clone (all of the derivatives of a single cell) at fertilization.[1] To count cell divisions, a representative sample of a clone is carried in daily isolations.[11] Fertilized cells are transferred to depressions containing fresh food as single isolates. The following day, the number of cells per depression derived from a single cell is determined, and a single cell from each depression is reisolated and given fresh food. Cells not transferred are permitted to undergo autogamy (self-fertilization) and serve as a source for initiating new progeny lines. The process of counting and reisolation of cells is repeated daily. The $\log_2$ of the number of cells derived from a single cell is the daily fission rate; the sum of the number of daily fissions from the origin of the clone to a given time is the age of the cell line. When an isolated cell dies, the life-span is the sum of the number of fissions or days from origin to the day it died. The cell used for transfer can be randomly or carefully selected. Different authors define their selection or do not use selection — accounting for some but not all of the variation in life-span reported in the literature.[25]

Senescence is defined as a species-specific decline in the ability of a cell to produce two live cells at the next division. Mean life-span is the average age when a representative sample of the clone dies, and maximal life-span is the longest-lived known representative of that species.

B. THE GENETIC MODEL SYSTEM

The ciliate genetic system has been established as a model for problems of developmental genetics and aging.[26] In general, the ciliates contain two kinds of nuclei, the micronuclei or germ line and the macronuclear somatic line nuclei. The micronuclei and macronuclei have different functions at different times in the life cycle, that is, during the sexual phase of fertilization or the vegetative asexual phase of cell division. During the sexual cycle, the micronuclei undergo meiosis to produce gametes which eventually differentiate the new micro- and macronuclei for the progeny cells. The short period of nuclear development can be synchronized and modulated to discover cytoplasmic and environmental effects on gene determination during this critical interval. During the asexual cycle, the micronucleus undergoes mitosis, exhibits very little transcriptional activity, and functions as the repository of genetic information. The macronucleus dictates the metabolic activity of the cell. The macronucleus is therefore the dominant regulator of the asexual cell cycle phase.

The ciliates offer both an *in vitro* and an *in vivo* model of cellular senescence since these organisms are both single cells and whole organisms at the same time. A single fertilized cell can be considered a newborn cell. All the cells derived from that cell, the clone, show age-specific changes in phenotype and express periods of immaturity (when cells cannot mate), maturity (when cells can mate), senescence (with declined function), and death (when there are no viable representatives of a given clone). The clone is a multicellular unit derived from a single cell with the members dispersed in space rather than joined together as in colonial flagellates or metazoans.

Paramecium are well suited as a model system for cellular aging since cellular decline is a natural process; aging occurs in the absence of experimentally induced changes in the environment or the genotype of cells. The duration of the life-span is short (about 2 months), and the aged cells display many of the senescent traits seen also in mammalian cells. These traits include a decrease in nuclear DNA content, decline in DNA synthesis and template activity, decline in RNA synthesis and endocytic activity, increases in abnormal mitochondria and lysosomal activity, and increased sensitivity to mutagens.[14] The ciliate clonal aging system does not risk possible artifacts that may occur when tissue from a multicellular organism is removed from the whole organism and its normal organ interactions.

C. EVOLUTIONARY CONSIDERATIONS

Since paramecia are eukaryotes, like the cells of higher organisms, the results obtained may have implications for our understanding of the aging process at the cellular level. Although aging is a multistep, multifaceted phenomenon, with senescence occurring at all levels of organization, changes on one level can affect other levels as well. Because nature tends to maintain useful biological mechanisms by survival of those animals which use those strategies, a commonality between mechanisms in such diverse organisms as paramecia and mammalian cells may exist. Those common molecules or mechanisms which are found throughout evolution are at once identified as fundamental by their conservation in surviving species. The evolutionary connections between different protozoans and multicellular organisms are unraveling with the new molecular biology probes.

Recent studies deduce that a branch occurred for plants and animals; the green flagellates and algae became ancestors for vascular plants, while the other branch gave rise to flagellated and ciliated protozoa.[27] Using mutations in conserved molecules as an index of passage of time, eukaryotes are estimated to be 1.5 to 2 billion years old, and *Tetrahymena*, 30 to 40 million years old.[28] Since the ciliates use unique codons for glutamine,[29-30] the ciliates may have preceded all of those organisms which later became fixed for the use of these unique codons as "stop words". As representatives of more primitive species, the ciliates provide a unique insight into an evolutionary perspective of cellular aging. Despite their differences, remarkable similarities exist, and antibodies against ciliate tubulin interact with tubulins of other metazoans and mammals.[31] Because nature tends to maintain useful biological mechanisms by survival of organisms which employ them, it is likely that identification of such mechanisms in both ciliates and human beings constitutes simultaneous identification of fundamental mechanisms.

III. NUCLEOCYTOPLASMIC INTERACTIONS IN THE DETERMINATION OF LIFE-SPAN

Recent studies bear on the role of the nucleus, cytoplasm, and cell membrane in the regulation of cell proliferation capability. A group of mutants isolated for their inability to discharge surface trichocysts (arrow-like organelles) exhibited a subgroup with a short life-span.[32] The short-lived variants shared the inability to divide the macronucleus properly at cell division. Another mutant, with normal trichocyst discharge but unequal macronuclear division, also exhibited the short life-span. The abbreviated life-span was associated with the macronuclear defect rather than the surface discharge trait. Why the inability to discharge trichocysts should sometimes be coupled with the inability to divide the macronucleus equally is still obscure. However, the short-lived subgroup all cannot bring their trichocysts to the "docking site". Defects in assembly of microtubular arrays could affect both positioning of the macronucleus and trichocyst discharge, coupling the defects in certain mutants.[33]

In contrast with the short-lived clones with either trichocyst discharge or macronuclear division defects, Takagi[34] has isolated a short-lived mutant which exhibits a marked reduction in cell division rate. The mutant shows some but not all of the age-related changes in time seen in the wild-type cells. The combined data indicate that the macronuclear division has an important role in cellular aging, but other mutations can affect longevity.

The importance of the macronucleus in the aging process was highlighted in the exquisite nuclear transplantation experiments of Aufderheide.[35] Whereas young macronuclear transplants could significantly extend the life of old recipients, old macronuclear transplants did not. Aufderheide[35] interprets the data as evidence of the paramount role of the macronucleus in aging, especially in view of his negative results using cytoplasmic transfer discussed next.[36]

Cytoplasmic effects on life-span were not detected when 5 to 39% of the cytoplasm was injected from young to old and from old to young cells. In contrast with these negative results, cytoplasmic effects could be detected when cytoplasm was injected from immature to mature cells; maturity was inhibited for about 15 fissions.[37] Therefore, at least some cytoplasmic-injected inhibitors can be expressed. The negative age-related cytoplasmic effects do not exclude the occurrence of membrane-associated inhibitors like those detected in human cells[38] which may not have been transferred, or a cytoplasmic effect dependent on a given macronuclear state, a "responsive" macronucleus. In matings between young and old cells, the young nucleus is usually damaged in the old cell, though if it survives a critical period, rejuvenation can occur.[39] In another species of *Paramecium* (*P. caudatum*), micronuclei transplanted by microinjection from young to old cells were first youthful, but rapidly showed functional loss. Likewise, transplantation of old micronuclei into young cells could rejuvenate the micronuclei only up to a given old-age limit.[40] Thus the nucleocytoplasmic interaction depends on both the cytoplasm and the nuclear states. The micronuclear transplants placed the donor nuclei in the environment of the nucleus, cytoplasm, and cell membrane of the recipient and these could overwhelm the injected micronucleus. Perhaps the injected cytoplasm was also overwhelmed by the recipient cell, and transient effects, even in multiple microinjections, may not have been enough to be detected over the life-span of the recipient. The cumulative data emphasize a paramount role for the nucleus, interacting with the cytoplasm at critical times during the life cycle. There are intervals of time when the nucleus is responsive or nonresponsive to given cytoplasmic signals.

IV. ENVIRONMENTAL EFFECTS ON LONGEVITY

A. RADIATION

Experiments by Smith-Sonneborn[41] showed that when cells were exposed to low doses of ultraviolet irradiation and then photoreactivated with a black light at critical doses and ages, the life-span of the cells could be extended; higher doses killed cells and shortened life-span. Although the molecular basis for the effect has not yet been identified, the beneficial response was postulated to be due to stimulated repair to correct some age-induced damage. One hypothesis would include increase in accessibility of age-induced damage to the repair enzymes.

The question of whether there are age-induced lesions in the genetic material of paramecia was approached by Holmes and Holmes[42,43] using alkaline elution techniques to detect strand breaks in DNA. Double-stranded DNA remains on the filter. Since DNA is denatured by alkali and converts apurinic and apyrmidinic sites into single-strand breaks, the amount of DNA which passes through the filter should be proportional to damaged DNA. Using this technique, Holmes and Holmes found increase of damaged DNA as a function of both the clonal age of the culture and the growth media. These interesting studies must still be considered preliminary since their techniques for obtaining aged cells did not exclude the chance occurrence of autogamy using genetic markers, standard tests proved effective for detecting autogamy, or exclude the role of culture stage or cell density in their results. The occurrence of age-related DNA damage would provide an approach to test whether radiation could lower the amount of damage for cells of a given age. The radiation-induced beneficial effects observed in *Paramecium* were found only at given age-specific low doses of UV. These life-span extension results may belong with the phenomenon of hormesis, a process whereby low doses of an otherwise harmful agent is found to have a beneficial or stimulatory effect.[44] Hormesis is commonly found in nature.[52] The hormesis phenomenon asserts that a small amount of stress stimulates the metabolic processes in a compensatory manner so as to reestablish normal function. If the level of stress remains below the capacity of compensation, beneficial effects may be observed.

A stimulatory effect on cell proliferation by low doses of radiation has been reported.[45] Using cultures of paramecia as a biological tool, investigators grew cells shielded from cosmic radiation or shielded with a γ-radiation source. Cells exposed to cosmic or γ-radiation exhibited an increased capacity for growth. Cultivation at high altitude (with presumed increased cosmic radiation) also stimulated cell growth. Cells grown in *Salyut*,[6] the Soviet spacecraft, and fixed on board showed increased growth rate due to zero gravity or cosmic radiation. In air balloon experiments, without the zero gravity effects, stimulation was still found and was interpreted to be related to the cosmic rays.[45] Telluric radioactivity and cosmic rays both contribute to the stimulatory effects of background radiation. In another experiment, carried out aboard the space shuttle *Challenger*, microgravity effects were the main factor affecting cell division.[46]

The life-span of paramecia was extended when grown in pulsed electromagnetic fields, but was harmful to certain mutant cells.[47,48] The electromagnetic fields affected ion transport by direct or indirect mechanisms and stimulated cell division.[49] The biological processes which underlie the stimulatory effects of radiations and magnetic fields are not understood. Speculations include the production of oxygen-free radicals, inhibition of thymine kinase, and normal DNA replication to favor DNA repair replication.[50]

B. DIET

Dietary changes were found to also influence life-span. Growth in dilute culture medium could shorten the life-span of cells,[25] and supplementing the medium with vitamin E[51] or C[52] could prolong the life-span of the paramecia. The life-span extension could be mediated by adding a nutrient beneficial for survival or as an antioxidant. Age-related loss in endogenous antioxidants was not found in *Paramecium*.[53]

Insight into the role of antioxidants could be examined using the lipid peroxidation bioassay optimized in *Paramecium*.[52] The bioassay uses sonicated cell preparations induced with ascorbic acid and ferrous ion. Lipid peroxidation was detected by a modification of the calorimetric thiobarbituric acid assay of Tappel and Zalkin,[54] which reacts with aldehydes such as malondialdehyde produced during the peroxidation. Antioxidants were tested for their ability to reduce peroxidation in the *in vitro* system, and both lipid-soluble and water-soluble compounds could reduce peroxidation. Dietary supplementation with lipid-soluble antioxidants only could reduce the peroxidation in the *in vitro* assay.

In contrast with programmed aging and attempts to supplement the medium to optimize growth, a deficiency for a required growth factor, vitamin B12, can be used to induce unbalanced growth in a flagellate, *Euglena*. The deficit alters DNA synthesis, RNA synthesis, and causes cells to hypertrophy.[55] Since these responses to a dietary deficiency mimic some of the age-related changes seen in aged cells, the use of this model for aging was proposed. Dobrosielski-Vergona argues that a single signal, restoration of the required substance, can trigger release of the blocked metabolic system and therefore afford a model for regulation of cell division and aging. Cell division studies are most likely relevant to aging but a dietary deficient cell is hardly a model of programmed aging except perhaps in the terminal stages of senescence. Rather the model system may be a model of induced cell death. Likewise, a cell provided with the missing nutrient is not a ''rejuvenated'' cell in the sense the term is used for clonal aging in ciliates but, more accurately, a revived cell. Separation of nutrient-induced cell deterioration should be clearly made from natural model systems which show cell growth decline under constant environmental conditions.

V. INTERACTION OF GENES AND THE ENVIRONMENT

Recent studies further document the rare occurrence of very long-lived paramecia.[25] These cells may reveal the maximal life-span of the species achievable by environmental manipulation. Agents which increase the mean life-span then may be allowing more cells to achieve the possible life-span rather than manipulation of the genetic limit of the species life-span. Animals and cells may well be selected to grow at suboptimum rates and life-spans. The difference between the actual and the possible life-span may be selected to provide a reserve survival component inducible under various environmental stress conditions. If this alternative regulation system is triggered under stress levels insufficient to damage the system (within the repair or restorative capacity of the system), increased survival may be found.

VI. SUMMARY

Aging in ciliates offers a model system of programmed aging in some, not all, species of protozoa. Likewise, in metazoan cells, some but not all cells and organs exhibit a limited life-span potential.[56]

Within the past 2 centuries, an appreciation for the paramount role of cells in the aging process has evolved. Although emphasis continues to be placed on the crucial role of the nucleus in the aging process, compelling evidence dictates that aging is a dynamic process involving interaction of the nucleus with the environment; both intracellular and extracellular influences can alter gene expression.

The nucleus cycles through intervals of responsive and nonresponsive states. Stress may trigger a transition from one metabolic state to another and alter the responsive states. Genetics is a science which must encompass the interaction of the DNA with its environment and the role of each in the determination of the whole. The protozoa offer a valuable reservoir for fundamental studies of cell growth and reproduction and provide an evolutionary perspective to complement studies in higher organisms.

REFERENCES

1. **Maupas, E.,** Recherches experimentales sur la multiplication des infusiores cilies, *Arch. Zool. Exp.,* (2)6, 165, 1888.
2. **Sonneborn, T. M.,** Enormous differences in length of life of closely related ciliates and their significance, in *The Biology of Aging,* Vol. 6, Strehler, B. L., Ed., Waverly Press, Baltimore, 1960, 289.
3. **Carrel, A.,** On the permanent life of tissues outside of the organism, *J. Exp. Med.,* 15, 516, 1912.
4. **Hayflick, L.,** Cell biology of aging, *Bioscience,* 24, 629, 1975.
5. **Hayflick L. and Moorhead, P. S.,** The serial cultivation of human diploid cell strains, *Exp. Cell Res.,* 25, 585, 1961.
6. **Hayflick, L.,** The limited *in vitro* lifetime of human diploid cell strains, *Exp. Cell Res.,* 37, 614, 1965.
7. **Lumpkin, C. K., Jr., McClung, J. K., Pereira-Smith, O. M., and Smith, J.,** Existence of high abundance antiproliferative mRNA in senescent human diploid fibroblasts, *Science,* 232, 393, 1986.
8. **Wang, E.,** A 57,000-mol-wt protein uniquely present in nonproliferating cells and senescent human fibroblasts, *J. Cell Biol.,* 100, 545, 1985.
9. **Krohn, P. L.,** Review lectures on senescence. II. Heterochronic transplantation in the study of aging, *Proc. R. Soc. London Ser. B,* 157, 128, 1962.
10. **Harrison, D. E.,** Normal production of erythrocytes by mouse marrow continuous for 73 months, *Proc. Natl. Acad. Sci. U.S.A.,* 70, 3184, 1973.
11. **Sonneborn, T. M.,** The relation of autogamy to senescence and rejuvenescence in *Paramecium aurelia, J. Protozool.,* 1, 53, 1954.
12. **Nanney, D. L.,** Aging and long term temporal regulation in ciliated protozoa. A critical review, *Mech. Ageing Dev.,* 3, 31, 1974.
13. **Smith-Sonneborn, J.,** Longevity in the protozoa, in *Evolution of Longevity in Animals: A Comprehensive Approach,* Brookhaven Symp. Biology No. 34, Woodhead, A. D. and Thompson, K. H., Eds., Plenum Press, New York, 1987.
14. **Smith-Sonneborn, J.,** Genetics and aging in protozoa, *Int. Rev. Cytol.,* 73, 319, 1981.
15. **Smith-Sonneborn, J.,** Aging in protozoa, in *Review of Biological Research in Aging,* Vol. 1, Rothstein, M., Ed., Alan R. Liss, New York, 1983, 29.
16. **Smith-Sonneborn, J.,** Use of ciliated protozoan as a model system to detect toxic and carcinogenic agents, in *In Vitro Toxicity Testing of Environmental Agents,* Pt. A, Kilber, A. R., Wong, T. K., Grant, L. D., DeWoskin, R. S., and Hughes, T. J., Eds., Plenum Press, New York, 1983, 113.
17. **Smith-Sonneborn, J.,** Protozoan aging, in *Selected Invertebrate Models for Aging Research,* Mitchell, D. and Johnson, T., Eds., CRC Press, Boca Raton, FL, 1984, 2.
18. **Smith-Sonneborn, J.,** Protozoa, in *Non-mammalian Models for Research on Aging,* Lints, F. L., Ed., S. Karger, Basel, 21, 1985, 201.
19. **Smith-Sonneborn, J.,** Aging in protozoa, in *Review of Biological Research in Aging,* Vol. 2, Rothstein, M., Ed., Alan R. Liss, New York, 1985, 13.
20. **Smith-Sonneborn, J.,** Aging in unicellular organisms, in *Handbook of the Biology of Aging,* 2nd ed., Finch, C. E. and Schneider, E. L., Eds., Van Nostrand Reinhold, New York, 1985, 79.
21. **Smith-Sonneborn, J.,** Genome interaction in the pathology of aging, in *CRC Handbook of the Cell Biology of Aging,* Cristofalo, V., Ed., CRC Press, Boca Raton, FL, 1985, 453.
22. **Smith-Sonneborn, J.,** Aging in protozoa, in *Review of Biological Research in Aging,* Rothstein, M., Ed., Alan R. Liss, New York, 1987, 3.
23. **Aufderheide, K. J.,** Cellular aging: an overview, in *Cellular Ageing,* Sauer, H. W., Ed., S. Karger, Basel, 1984, 2.
24. **Takagi, Y.,** *Aging in Paramecium,* Gortz, H. D., Ed., Springer-Verlag, Heidelberg, 1987.
25. **Takagi, Y., Nobuoka, T., and Doi, M.,** Clonal lifespan of *Paramecium tetraurelia:* effect of selection on its extension and use of fissions for its determination, *J. Cell Sci.,* 88, 129, 1987.
26. **Preer, J. R., Jr.,** Nuclear and cytoplasmic differentiation in the protozoa, in *Developmental Cytology,* Pudnick, W., Ed., Ronald Press, New York, 1959, 3.
27. **Kumazaki, T., Hori, H., and Osawa, S.,** Phylogeny of protozoa deduced from 5S rRNA sequences, *J. Mol. Evol.,* 19, 411, 1983.

28. **Van Bell, C. T.,** 5S and 5.8S ribosomal RNA evolution in the suborder Tetrahymena (Coliophora: Hymenostomatida), *J. Mol. Evol.,* 22, 231, 1985.
29. **Preer, J. R., Preer, L. B., Rudman, B. M., and Barnett, A. J.,** Deviation from the universal code shown by the gene for surface protein 51A in *Paramecium, Nature (London),* 314, 188, 1985.
30. **Caron, F. and Meyer, E.,** Does *Paramecium primaurelia* use a different genetic code in its macronucleus?, *Nature (London,)* 314, 185, 1985.
31. **Adoutte, A., Claisse, M., Maunoury, R., and Beisson, J.,** Tubulin evolution: ciliate specific epitopes are conserved in the ciliary tubulin of metazoa, *J. Mol. Evol.,* 22, 220, 1985.
32. **Aufderheide, K. J. and Schneller, M. V.,** Phenotypes associated with early clonal death in *Paramecium tetraurelia, Mech. Ageing Dev.,* 32, 299, 1985.
33. **Ruiz, F., Adoutte, M., Rossignol, M., and Beisson, J.,** Genetic analysis of morphogenetic processes in *Paramecium.* I. A mutation affecting trichocyst and nuclear division, *Genet. Res.,* 27, 109, 1976.
34. **Takagi, Y., Suzuki, T., and Shimada, C.,** Isolation of a *Paramecium tetraurelia* mutant and with novel life-cycle features with short clonal life-span, *Jpn. Zool. Sci. (Toyko),* 4(1), 73, 1987.
35. **Aufderheide, K. J.,** Clonal aging in *Paramecium tetraurelia.* II. Evidence of functional changes in the macronucleus with age, *Mech. Ageing Dev.,* 37, 265, 1986.
36. **Aufderheide, K. J.,** Clonal aging in *Paramecium tetraurelia.* Absence of evidence for a cytoplasmic factor, *Mech. Ageing Dev.,* 286, 57, 1984.
37. **Miwa, I., Haga, N., and Hiwatashi, K.,** Immaturity substances: material basis for immaturity in *Paramecium, J. Cell Sci.,* 19, 369, 1975.
38. **Pereira-Smith, O. M., Fisher, I. F., and Smith, J. R.,** Senescent and quiescent cell inhibitors of DNA synthesis (membrane associated proteins), *Exp. Cell Res.,* 160, 297, 1985.
39. **Sonneborn, T. M. and Schneller, M.,** Age-induced mutations in *Paramecium,* in *The Biology of Aging,* Strehler, B. L., Ed., Waverly Press, Baltimore, 1960, 286.
40. **Karino, S. and Hiwatashi, K.,** Analysis of germinal aging in *Paramecium caudatum* by micronuclear transplantation, *Exp. Cell Res.,* 136, 407, 1981.
41. **Smith-Sonneborn, J.,** DNA repair and longevity assurance in *Paramecium tetraurelia, Science,* 203, 1115, 1979.
42. **Holmes, G. E. and Holmes, N. R.,** Accumulation of DNA fragments in aging *Paramecium tetraurelia* in axenic and nonaxenic media, *Gerontology,* 32(5), 252, 1986.
43. **Holmes, G. E. and Holmes, N. R.,** Accumulation of DNA damage in aging *Paramecium tetraurelia, Mol. Gen. Genet.,* 204(1), 108, 1986.
44. **Luckey, T. D.,** Physiological benefits from low levels of ionizing radiation, *Health Phys.,* 43, 771, 1981.
45. **Planel, H., Soleilhavoup, J. P., Tixador, R., Richoilley, G., Croute, F., Caratero, C., and Gaubin, Y.,** Influence on cell proliferation of background radiation or exposure to very low chronic gamma radiation, *Health Phys.,* 52(5), 571, 1987.
46. **Richoilley, G., Tixador, R., Gasset, G., Templier, J., and Planel, H.,** Preliminary results of the "Paramecium" experiment, *Naturwissenschaften,* 73, 404, 1986.
47. **Smith-Sonneborn, J.,** Programmed increased longevity by weak pulsating current in *Paramecium,* VII, *Bioelectrochem. Bioeng.,* 11, 5, 1984.
48. **Smith-Sonneborn, J. and Darnell, C.,** Differences in longevity of wild type and ion transport mutants in normal and altered environment, 37th Annu. Meet. Gerontological Soc. Am., November 16 to 20, *Gerontology,* 1984, 24.
49. **Dihel, L. E. and Smith-Sonneborn, J.,** Effects of low frequency electromagnetic field on cell division rate and plasma membrane, *Bioelectromagnetics,* 6, 61, 1985.
50. **Feinindegen, L. E., Mühlensiepen, H., Bond, V. P., and Sondhaus, C. A.,** Intracellular stimulation of biochemical control mechanisms by low dose, low LET irradiation, *Health Phys.,* 52, 663, 1987.
51. **Thomas, J. and Nyberg, D.,** Vitamin E supplementation and intense selection increase clonal lifespan in *Paramecium tetraurelia, Exp. Gerontol.,* 23, 501, 1988.
52. **Leibovitz, B.,** Lipid Peroxidation, Antioxidants, and Aging in *Paramecium tetraurelia,* Ph.D. thesis, University of Wyoming, Laramie, 1986.
53. **Croute, F., Vidal, S., Dupouy, D., Soleilhavoup, J. P., and Serre, G.,** Studies on catalase, glutathione peroxidase and superoxidismutase activities in aging cells of *Paramecium tetraurelia, Mech. Ageing Dev.,* 29, 53, 1985.
54. **Tappel, A. L. and Zalkin, H.,** Inhibition of lipid peroxidation in mitochondria by Vitamin E, *Arch. Biochem. Biophys.,* 80, 333, 1959.
55. **Dobrosielski-Vergona, K.,** Vitamin B-12 dependent protozoa, a model system for aging, *Age,* 10(1), 11, 1987.
56. **Smith-Sonneborn, J.,** How we age, in *Promise of Productive Aging,* Butler, R., Ed., Raven Press, New York, 1988.

Chapter 3

LEECHES: ANNELIDS OF MEDICAL AND SCIENTIFIC UTILITY

Charles M. Lent and Stephen L. Adams

TABLE OF CONTENTS

I. INTRODUCTION

When anyone mentions leeches, many people conjure up ugly remembrances of Humphrey Bogart and Katharine Hepburn making disparaging remarks about them and pouring salt on the "filthy" annelids in an effort to detach them in *The African Queen*.[1] In fact, many leeches are beautifully colored and, when swimming, have a sinuous elegance. Leeches make up the class Hirudinea and evolved from earthworm-like ancestors of the class Oligochaeta. Both classes are in the phylum Annelida; that is, they are segmented worms. The bodies of annelids consist of a serial repetition of segments (metameres), and each segment has a complement of most organ systems among which a nervous ganglion is especially prominent. The number of body segments is variable in Polychaetes and Oligochaetes, and these annelids grow throughout their life-span by adding segments in front of their terminal segment: the pygidium. Leech growth does not entail a change in segment number, and all Hirudinea, even juveniles, have exactly 32 metameres.[2] This rigidly fixed number of segments is a likely concomitant to the presence of a sucker at the ends of the leech (Figure 1). The pygidium is part of a functional posterior sucker which seemingly prevents growth by the addition of segments. Oligochaeta and Hirudinea are sometimes treated as a subphylum, the Clitellata. Worms of both classes reproduce hermaphroditically and form cocoons from the mucus secreted by the glandular skin of a few specialized anterior segments: the clitellum.

The leech is useful to experimental biologists and physicians alike. The history of leeching to treat human maladies probably exceeds 2000 years, although the practice fell into disrepute after a period of exuberant abuse in the 18th and 19th centuries. Leeches seem to be witnessing a renaissance in their employment in medical practice.[3] The attachment which *Homo sapiens* has derived from a long association with leeches is reflected in the scientific name which Linnaeus gave to the European leech: *Hirudo medicinalis*.

In this chapter, a history of medicinal leeching is followed by a modern study of leech feeding behavior. The central nervous system, the characteristics of identifiable neurons, and the functions of biochemically specialized neurons are considered next. Excitation-secretion coupling is described in leech salivary glands and is followed by the therapeutic potential of the biochemicals in saliva. The final two sections consider the leech as a model system for investigating the cellular bases of Parkinsonism and the renaissance of the leech in modern medicine.

II. A HISTORY OF MEDICINAL LEECHING

The leech has long been associated with the physician. Etymologists (and perhaps entomologists!) note that the word "leech" may be derived from the Old English "laece", meaning physician.[4] Anglo-Saxon practitioners of medicine were called leeches, and material concerning general medical information could be found in "Leechdoms". It is indeed somewhat amusing to read historical accounts of the "professional or semi-professional leeches who appear to have been literate" in descriptions of the physician![5] The words "doctor" and "leech" have been used as synonyms in Roget's thesaurus.[6]

Bloodletting probably began in the Stone Age, but when leeches were first utilized therapeutically is unclear.[7] Nicander of Colophon (200 to 130 B.C.) and Themison of Laodicea (123 to 43 B.C.), a pupil of Asclepiades, are generally considered to be among the first physicians to have used the leech therapeutically.[8,9] Leeching is mentioned in the works of the Greek physician Galen, and its methods were detailed by the Persian physician Avicenna.[10] Bloodletting was originally believed to aid the body in restoring the balance among its four "humors": the melancholic, choleric, phlegmatic, and sanguine (blood).[7,10] By the Middle Ages, bloodletting developed into a standard treatment for many maladies.

FIGURE 1. A photograph of *Hirudo medicinalis* on the side of a glass aquarium. The large posterior sucker is attached, and the anterior sucker with its sensory structures is detached and flared. The segments of the leech are divided into a number of annuli, which are seen clearly. (Courtesy of John S. Flannery, Utah State University Information Services, Logan, Utah.)

The lithograph in Figure 2 depicts the practice. In addition to the leech, methods of bloodletting included the use of lancets, fleams, scarifiers, and cups.

Leeches were applied to diseased patients when their therapeutic use was indicated. In general, into the 1700s and 1800s, many diseases were believed to be caused by the entity known as "inflammation". Leeches were touted for their ability to ameliorate local inflammation by direct application (e.g., on hemorrhoids or abscesses).[8] They were also thought to have a general therapeutic effect when attached (e.g., gastroenteritis). This general effect was thought by some to be most beneficial in children (with skin of greater vascularity and with less circulating blood volume than adults) and in women.[8]

Examples of diseases amenable to leeching included whooping cough, psoriasis, abscesses, enlargement of the liver, and "congestive headaches". A therapy for whooping cough, for example, was to apply leeches "immediately over the junction of the occiput and atlas vertebra to relieve congestion of the vessels surrounding the origin of the pneumogastric nerve".[11]

Leeches eventually became a popular alternative to the aforementioned mechanical instruments for bloodletting. They were considered especially useful when it was necessary to remove blood from a part of the body that was difficult to bleed with a mechanical device (e.g., inside the nostrils) or a very sensitive area (e.g, the vulva). Concern was expressed that the leecher be wary that the leech did not crawl into proximate orifices![12]

In Europe of the 18th and 19th centuries, leeches became a most popular method of bloodletting. The amount of blood to be removed was prescribed, and the patients were often bled until they fainted from blood loss ("bleeding to syncope").[13,14] In the 1830s in France, the theories of Dr. F. J. V. Broussais became prevalent, and this advocate of bloodletting used leeches to excess.[15] His radical applications of leeches evolved from his belief that every disease was caused by inflammation, and that inflammation was amenable to bloodletting by leeching. Broussais would apply large numbers of leeches to the abdomen of his patients simultaneously.[7] Estimates of the numbers of leeches used in Paris during Broussais's reign are exorbitant. Over the 8-year period from 1829 to 1836, almost 1.5 million pounds of blood were drawn from the ill citizens of France by leeches![17]

Several species of leeches were employed for bloodletting, but the premier worm was unquestionably the European medicinal leech, *Hirudo medicinalis*. *Hirudo* were imported into the U.S. from Europe, a trade which grew to such proportions as to limit their availability. Millions of *Hirudo* were exported from Germany to the U.S. Not surprisingly Germany

FIGURE 2.   A lithograph showing a physician applying leeches onto the neck of a woman who might be afflicted with goiter. The young boy in the foreground is holding a glass in reserve that contains swimming leeches. (Boilly, L., *Les Sangsues,* Chez Aubert, Paris, 1827; Courtesy of the Yale University Medical Library.)

soon depleted its natural reserves of the leech.[17] In an effort to increase the availability of the medicinal leech, American physicians were solicited in journals of the time to breed foreign leeches in the U.S. for medicinal reasons. Other methods of removing blood, such as cupping, were advised.[18,19] Alternative species of leeches employed in Europe at the time included *H. provincialis* and *H. officinalis*.[11,16]

The leech *Macrobdella decora* was sometimes employed as a replacement for *Hirudo* in America, and differences among various species soon became evident.[8] A quarter of a million native leeches were estimated to have been used annually in Philadelphia, but in other regions of the East Coast, imported leeches were preferred.[20] Some species seemed unable to bite; others would cause painful, obstinate wounds. The horse leech, *Haemopis*, reportedly could not penetrate human skin with its jaws; however, its bite was considered to be painful.[13]

Leeches were regularly stocked in many pharmacies of the period. Suggestions for the successful care of leeches were described in publications of the time.[12,13] One included a discussion of the best plant for the leech jar, and the number and type of snails necessary to keep the water clean.[12] Another noted that moss, turf, and fragments of wood should be kept at the bottom of the jar so that the leech could "draw itself through the moss and roots to clear its body from the slimy coat which forms on its skin, and is a principle cause of its disease and death".[13,20]

Similar tests provided the clinician with an array of suggestions on leeching itself. As an example, it was suggested that the area to be leeched should be washed with warm water to promote vascularization and then be shaved if necessary. Sugar water, milk, or raw meat might then be put on the skin to induce the leech to bite. If these were not successful, the area to be leeched could be pricked with a needle and blood smeared about so as to induce the leech to make its incision. The head of the leech would be placed on the skin, or the leech would be put in a cup which was then inverted over the desired region of the body. Placing blotting paper with a hole in it over the area to be leeched was occasionally used to encourage the leech to bite a particular part.[11] The leech-glass, a tube with two open ends, was often used to position the leech within an orifice such as the nostril, throat, or vagina.[20] An eloquent description of the leech at work was provided in an 1884 text:

Having fixed on a suitable spot, the animal applies his oval disc, and firmly fixes it. . . . The three cartilaginous jaws bearing the sharp teeth are now stiffened and protruded through the triradiate mouth against the skin, which they perforate, not at once, but gradually, by a saw-like motion. . . . The wound is not produced instantaneously, for the gnawing pain continues for two or three minutes after the animal has commenced operations. Thus, then, it appears that the leech saws the skin; hence the irritation and inflammation frequently produced around the orifices. The flow of blood is promoted by the suction of the animal, which swallows the fluid as it is evolved. During the whole of the operation the jaws remain lodged in the skin. In proportion, as the anterior cells of the stomach become filled, the blood passes into the posterior ones; and when the whole of the viscous is distended, the animal falls off.[13]

The patient played host to the leech until either the desired amount of blood had been removed or the leech had ingested its fill and detached, falling off the wound. It usually took but 15 to 20 min for a leech to become satiated. It could remove about 5 to 30 cc of blood before becoming engorged.[11] If more blood was prescribed than one leech could remove, several options were available to the clinician. Another leech, of course, could be applied. Some clinicians would cut off the tail of the leech, causing blood to exit through the tail, preventing the leech from becoming engorged, and hence the continued loss of blood. Finally, warm formentations could be applied to promote continued afterbleeding from the wound.[20]

Removing a leech in the course of active feeding could be effected by pouring table salt or vinegar on it, which induced it to vomit and detach. It was recommended not to detach a leech by force since its jaws could become dislodged in the wound.[21,22]

Once filled with blood, the leech could not be used again until it had digested its voluminous meal. Various methods were attempted in order to shorten this period, which often lasted a full year. One was to submerge the leech into a dilute saline or vinegar solution, causing vomiting of the ingested blood and restoration of its appetite.[8] Another method was to empty the leech by squeezing it between the thumb and forefinger from the tail headward and force the blood from its mouth.[8,20] The leeches, whose appetites were so restored, were thought to be susceptible to disease, and it was advised not to store them with fresh leeches.[12]

Complications were sometimes noted to follow leech therapy. The most common problem was that of persistent bleeding after removal of the leech. Bleeding from leech wounds could last for periods of up to 24 h.[9] This was probably due to the combined secretion of an anticoagulant and a vasodilator by the leech.[2] Continued bleeding after removal of a leech was at times worrisome. It was advised not to apply a leech to an infant before bedtime because of the possibility of significant unnoticed blood loss after removal of the leech.[11] Certainly this was a concern in an infant, who has only a fraction of the circulating blood volume that the adult does. Therapy of post-leeching bleeding consisted of firm pressure as well as the various hemostatic agents of the period. These included such remedies as a burned rag and a cobweb.[11]

Waring noted many rules that were to be observed in the practice of leeching and cited "M. Lisfrancs' Rules for the Application of Leeches".[11] The rules included warnings against applying a leech to areas of skin where there are many subcutaneous nerves, "as the pain will be great," and to avoid leeching the scrotum or penis, as "the pain is excessive".[11]

Infection by the repeated use of a leech was recognized as a possibility. Transmission of syphilis and puerperal fever was attributed to the reapplication of leeches previously used in patients with those infectious diseases.[11] One can only but hypothesize that anemia must have been a problem, as this was reported well after leeching was out of vogue.[23]

The tendency of the leech to explore body areas other than those of application was also a cause of some concern. The leech is flexibly elongate and can easily crawl from the tonsils in the throat to the area of the epiglottis and attach itself. As the leech may ingest up to nine times the amount of its body volume, it can obstruct an orifice as it engorges, thus, for example, compromising the airway.[8] This occurs today in those geographic regions where leeches inhabit drinking water.[24] To remove an itinerant leech attached to the pharynx, one was to gargle a strong salt or vinegar solution. The leech would then let loose its hold. Even laryngotomy was reportedly necessary to remove a migrating leech.[8] Similar clinical presentations have occurred even more recently with nontherapeutic exposure to leeches.[25] A leech that was swallowed was destroyed by drinking wine, although the leech probably could not have survived the acidic environment of the stomach. Complications could occur when the leech, which had been placed upon a hemorrhoid, for instance, would migrate into the rectum, and "serious results might ensue".[13]

Scars often developed after leeching.[11,21] The leech left wounds that were apt to become bruised and later developed into a scar[21] that resembles the Mercedes-Benz logo (Figure 3). One advantage of the American leech was that it did not make as large an incision as *Hirudo* and consequently, left a smaller scar.

There were conditions which were exacerbated by leeching. It was known that leech wounds could be serious or even fatal in persons with hemophilia.[13,26] There were also diseases that the rate of blood removal by the leech was considered to be too slow, such as croup, and subsequently venesection was used in its treatment. Leeching was also considered ill advised for erysipelas (streptococcal infections).[12,13]

Leeching, as well as bloodletting in general, declined in popularity as the 19th century came to a close.[27,28] One author noted that at one major hospital in England the number of leeches used dropped off from over 97,000 leeches in 1832 to only 1700 leeches 50 years later.[28] During the first half of the 20th century, leeches were only occasionally employed for medical indications such as thrombosis, post-operative embolism,[29-31] acute coronary thrombosis,[32] and even poliomyelitis.[9]

III. A MODEL OF FEEDING BEHAVIOR

H. medicinalis has the long-known habit of feeding on the blood of mammals, and this behavior is most intriguing when the mammal is human. Despite the fact that the medical history of leech feeding exceeds 2 millennia, its ethology has been described only recently.[33] The behavior alternates between hunger and satiation. Leeches locate and attach to their prey by a series of *appetitive behaviors*. Hungry leeches are usually found resting at the water surface, with both of the suckers attached, and are aroused by several stimuli (surface waves, mechanical shocks, or shadows). Wave ripples from potential prey are an important stimulus. The anterior sucker is released and the head oriented into the ripples. After a few seconds, the leech initiates undulatory swimming by releasing the posterior sucker. Leeches swim in straight lines, and their orientation toward point sources of vibration is accurate; more than half of their orientation trials fall within 25°.

Swimming ceases upon contact with a surface. The leech explores surfaces by crawling in a characteristic, inchworm fashion. If a region of warmth is encountered, the leech bites. If blood does not issue from one bite, the hungry leech explores further and bites again. Leech bites in Parafilm™ constitute the quantitative index of hunger. Each bite is a set of

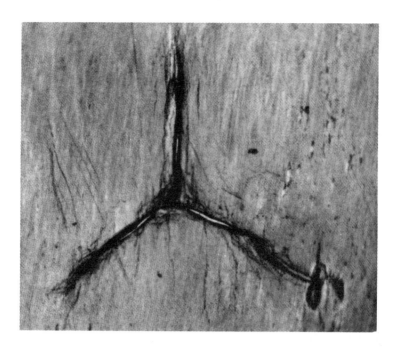

FIGURE 3. A photomicrograph of a single *Hirudo* bite. This triradiate incision was made by the three jaws of a leech which bit into a sheet of paraffin (Parafilm®) resting upon a 35°C surface. Each of the three marks is ~1 mm in length.

three incisions from the leech jaws, and one of these triradiate marks is shown in Figure 3. Biting frequency is highest between 37° and 40°C, the temperature range of their mammalian prey. These appetitive behaviors can be repeated many times until the leech is in a position to obtain its blood meal. Biting is thigmotactic, and leeches usually bite in skin folds; hence their preference for the body orifices noted previously.

When a leech bite draws blood, the consummatory behavior of ingestion begins. Blood is forced into the crop by the peristaltic contractions of the pharynx. The pharyngeal contractions have an initial frequency of 2.3 Hz, which decreases to 1.2 Hz at termination.[34] Ingestion is a highly compelled behavior, and a variety of noxious stimuli fail to interrupt the hungry leech. Pumping lasts 25 to 30 min, after which the leech is extremely distended by its ingested blood: seven to nine times its initial weight. This volume is impressive, even though the leech is an intermittent feeder whose satiation lasts a full year.[35] Satiated behavior differs distinctively from the behavior of hunger. Satiated leeches seek crevices of deep water, tend to crawl, and lack spontaneous bouts of swimming. Most importantly, satiated leeches do not bite, but rapidly lift the head away from warmth.

The stimulus underlying the termination of ingestion is body wall distension.[36] Distending a feeding leech produces an immediate termination of ingestion, and biting returns only when distension is alleviated. If distension is circumvented by cutting through the body wall and allowing the incoming blood to escape, ingestive behavior is prolonged significantly. Distension not only terminates ingestion, but it also produces the behavior of satiation. Removing blood from the crops of fed leeches restores biting. These findings explain the medical practice of prolonging feeding by cutting the tail of the leech off first as suggested by Galen.[10] They also explain how leech practitioners restored hunger. In India, for example, satiated leeches are pierced, the ingested blood removed, and the leeches are recycled quickly for blood letting.[37]

The feeding behavior of the leech has two extremes whose transition is produced by the brief consummatory act of ingestion in one direction and the slow process of digestion in the other. Even though this behavior is rare, it is amenable to laboratory investigation because the stimuli for its initiation and termination are now well defined and can be readily controlled.

IV. IDENTIFIABLE NEURONS AND CELLULAR NEUROBIOLOGY

Without question, the preponderance of research on leeches has been in the field of neurobiology, and many of its findings are important to medicine and biology. The first modern neuroanatomical study of leeches was done in 1891 by Gustav Retzius, who prepared the central nervous systems (CNS) for microscopic examination by staining with methylene blue.[38] Retzius accurately described many of the neurons in *Hirudo* ganglia. The largest of these neurons were the *Kolossale Ganglienzellen* (KZ). These paired colossal neurons (50 to 80 μm) are now usually referred to as Retzius cells (RZ or Rz), and their anatomical properties are shown by his drawing in Figure 4. The Retzius cell bodies (neurosomata) are usually in the anterior of the ganglion, and each cell body issues a single axonal process (i.e., they are monopolar neurons). Each RZ axon courses dorsally into the neuropile, turns laterally, and bifurcates into two major axons (1 to 3 μm). Minor axonal branches also project into the longitudinal connectives.[39] The axons of the Retzius cell leave the ganglion and travel into the body wall via the two pairs of ganglionic lateral roots. The roots branch repeatedly in the periphery, and most branches contain a single RZ axon. This broad innervation of the periphery by RZ contrasts with a limited arborization of terminals in the neuropile, the site of synaptic interactions. RZ are distinguished from the other leech neurons by their large size, but this morphological attribute makes them experimentally accessible and has resulted in a voluminous body of data.[38] The anatomical descriptions of Retzius were above all accurate and foresighted. Nicholls and Baylor used similar staining methods and found neurons which were identical to those described by Retzius 80 years earlier.[40]

The Retzius cell was the first neuron which could be repeatedly identified from animal to animal. The capability to recognize a particular nerve cell before and after experimental manipulation has been a critical factor in the success of invertebrate neurobiology.

The leech's CNS consists of 32 segmental ganglia; 21 are interated in the midbody and are superficially similar.[41] Midbody segmental ganglia intercommunicate via longitudinal connectives and innervate the body wall via paired lateral roots (Figure 5). The fifth and sixth segmental ganglia are larger than the others, and they innervate the complex, hermaphroditic genitalia. At the head end of the leech, a subesophageal ganglion (SubEG; a fusion of four ganglia) communicates with a nonsegmental supraesophageal ganglion (SupraEG) via circumesophageal connectives. These two head ganglia make up the cerebral ganglion, which innervates the anterior sucker, three jaws, the muscular pharynx, and the sensory apparatus of the head. The posterior sucker is larger, used in crawling, and its movements are controlled by the caudal ganglion (a fusion of seven ganglia).[41]

Leech ganglia are composed of a relatively small number of neurons. Each ganglion contains the somata of 350 to 400 neurons,[42] whereas most of the experimentally important mollusks, crustaceans, and insects have 3000 to 7000 neurons per ganglion.[43] Most leech neurons exist in bilaterally symmetrical pairs; however, some of the medially located cells are unpaired. All of the neuron cell bodies are enveloped within six huge glial cells.[44] The morphological organization of the ganglia in the leech CNS has enabled direct physiological investigations of the functional relationships between neurons and glia.[45] In addition, leech ganglia have several characteristics which are especially favorable for fundamental investigations of neurobiology. These include:

1.   Stereotyped morphology and functional properties of identifiable mechanosensory neurons[40] and motor neurons.[46]

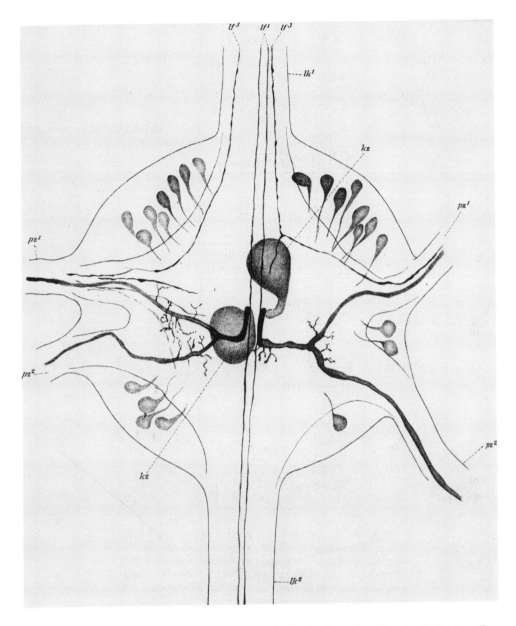

FIGURE 4. An original illustration of the Retzius cells (kz) in a leech ganglion. Note that the Retzius cells as well as the other neurons are monopolar and project a single process into the central neuropile. The Retzius cells are the largest somata in the leech CNS, and each projects axons into the ipsilateral roots. (From Retzius, 1891.)

2. Electronic and chemical synaptic interactions between the identifiable sensory and motor neurons produce some stereotyped reflexive behaviors.[47]

3. Some identifiable neurons are capable of regenerating and reestablishing functional synaptic contacts after interruption.[48]

4. Excised leech neurons develop synaptic connections while maintained in tissue culture, and their interactions can be studied with biophysical and anatomical methods.[49]

5. The cellular mechanisms by which identifiable axons establish electrotonic connections after experimental interruption can be studied morphologically and physiologically.[50]

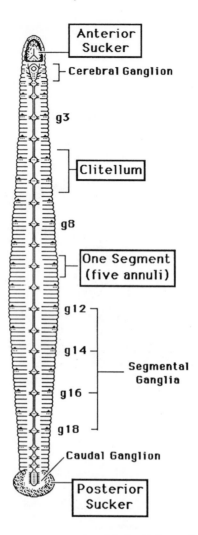

FIGURE 5. The segmentation of the leech. A ventral view of the
somatic seg-ganglion (g1 to g21) located in the central annulus.
The number of annuli per segment varies, but in the midbody there
are five. The cerebral ganglion consists of a nonsegmental supraesophageal ganglion (SupraEG) and a segmental subesophageal ganglion (SubEG), a fusion of four segmental ganglia. The caudal
ganglion is a fusion of seven segments. The clitellum is composed
of two anterior segments, containing the complex hermaphroditic
genitalia and which are innervated by g5 and g6. The skin of the
clitellum is a specialized mucous epithelium which secretes a cocoon.

6.   Synaptic and electrotonic interactions among a small population of interneurons pro-
     duce oscillatory rhythms and are a probable central pattern generator. These synaptic
     neurons activate a pool of motoneurons which generate the sinuous, undulatory swim-
     ming behavior of the leech.[51]
7.   The protostomous developmental patterns of leech embryos facilitate fate-mapping
     experiments which are elucidating the deterministic features of the neuroblasts.[52]
8.   The physiological functions of serotonin neurons integrate the feeding behavior of the
     leech.[53]

V. MONOAMINE NEURONS IN THE LEECH

Individual neurons in the leech CNS utilize most of the major neurotransmitter chemicals. Leech motor neurons, like those of mammals, excite the somatic muscle cells with acetylcholine (ACh). The longitudinal muscles of leech are naturally sensitive to ACh, and the sensitivity is increased by treatment with the anticholinesterase, eserine. Leech body wall is a long-standing bioassay for ACh.[54] Other neurotransmitters in the leech CNS include γ-aminobutyric acid (GABA), several neuropeptides, and three monoamines.[2] The monoamines found in leech neurons include dopamine (DA, a catecholamine); octopamine (OA, an ethanolamine), and serotonin (5-hydroxytryptamine, 5-HT, an indoleamine).

A limited population of leech neurons contain monoamines (MA), and each is repeatedly identifiable. The neurons can be studied individually with various morphological, neurochemical, and electrophysiological techniques.[55] The effects of amine-specific neurotoxins on leech neurons often resemble their effects on mammals. Because leech neurons can be recognized before and after intervention, the neurotoxic effects can be investigated carefully and interpreted unambiguously.[56]

A. DOPAMINE

There are two pairs of lateral anterior roots emerging from each segmental ganglion of the leech. A cluster of 12 neuron cell bodies, located at the first major branch of each anterior root, constitutes a peripheral ganglion. The largest of the somata in this peripheral ganglion is called the anterior root cell (AR), and it fluoresces after monoamine-specific histochemistry (Figure 6). The UV-excited emission of the AR cell is blue-green, and its wavelength (490 nm) is characteristic of catecholamines. The AR cell body projects a single axon into the ganglionic neuropile, where it bifurcates into two axons and ramifies extensive processes throughout the ipsilateral neuropile. These two AR axons diverge into the anterior and posterior longitudinal connectives.

When the AR cell body is dissected from its roots and the amines are extracted, DA is readily detected in the extract by high pressure liquid chromatography (HPLC). There is approximately 1 pmol of dopamine in each 15 to 25 μm AR cell body. Therefore, this catecholamine is stored intracellularly at a minimal concentration of 160 mM. This is the highest concentration of transmitter to have been measured in any neuron.[57] However, any function for dopamine in leech or for the dopamine-containing AR cell has yet to be discerned.

B. OCTOPAMINE

Octopamine in the leech is localized within a pair of electrophysiologically characterized neurons called Leydig cells (Ly). The Ly of the CNS make up a population of extensively intercoupled neurons which receive several synaptic inputs, fire spontaneous impulses at low frequency, and send axons into the periphery only by the ipsilateral roots of adjacent ganglia.[58]

OA has been measured in Leydig cell bodies by means of a sensitive radioenzymatic assay, and it is estimated to have an intracellular concentration of ~8 mM.[59] No function has yet been adduced for Leydig cells.

C. SEROTONIN

Serotonin (5-hydroxytryptamine, 5-HT) is an important transmitter chemical in the leech CNS and has special relevance to this chapter. There are two major classes of serotonin cells (Figure 6): large peripherally projecting effector neurons (50 to 100 μm) and small interneurons (8 to 15 μm).[60] One effector neuron is the biochemically distinctive Retzius cell which synthesizes, stores, accumulates, and releases serotonin.[61-63] In the midbody of the leech, Retzius cell impulse activity or a direct application of 5-HT stimulates the secretion of mucus by dermal glands[64] and decreases the basal tension of the longitudinal body

FIGURE 6.   A collage of two fluorescence-micrographs which shows the glyoxylic-acid-induced fluorescence of leech monoamine neurons. In the midbody ganglion (left), yellow fluorescence is seen in 7 of the 400 neurons, and it is localized to RZ, VL, DL, and M (labels in Figure 7). In the peripheral ganglion on the right, blue-green fluorescence is seen only in the AR cell. The AR soma has a single axon which enters the ipsilateral neuropile, where it bifurcates and enters the anterior and posterior longitudinal connectives. The T-shaped axon of the contralateral AR is also visible in this micrograph.

muscles.[65] The other serotonin effector neuron is the large lateral cell (LL), and a single pair of LL is exclusive to the first segment of the leech in the SubEG.[66] The serotonin cells of the SubEG and the first segmental ganglion are labeled in Figure 7.

There are two pairs of lateral interneurons (VL, DL) found in most ganglia, and they are distinguished by unusually high intracellular concentrations of 100 mM 5-HT.[55] A posteriomedial M interneuron is unpaired in midbody ganglia, is usually paired anteriorly, and is sometimes absent posteriorly. Paired anteriomedial E interneurons are found only in the seven anterior ganglia. These distributional differences produce a rostrocaudal gradient of 5-HT neurons along the leech which decreases from ten to about five cells per ganglion.

The quantity of serotonin in leech ganglia decreases with a similar pattern: each anterior ganglion contains approximately 17 pmol 5-HT; each midbody ganglion, 13 pmol; and each caudal ganglia, 3 pmol.[67] This fivefold gradient in levels of ganglionic 5-HT is a neurochemical correlate for the cellular gradient. Thus, serotonin and its parental neurons are represented disproportionately in the head (encephalon) of the leech.

The serotonin neurons in midbody ganglia are electrically intercoupled and share spontaneous synaptic inputs.[68] Both of these patterns of connectivity result in synchronous impulse activity, and hence leech serotonin cells fire at similar times. Thus, if a recognizable behavior was mediated by 5-HT, it should be expressed with a similar longitudinal gradient as the serotonin. A behavior of *Hirudo* with obvious cephalic dominance is feeding.

VI. NEURONAL SEROTONIN AND FEEDING BEHAVIOR

Recent findings have suggested that serotonin is involved in the expression of one behavior — feeding.[53] In order to discern any involvement of serotonin in the expression of feeding, its effects on the behavior of intact animals were studied. Bathing intact animals in serotonin[69] transiently augments many of the quantifiable components of hunger without injuring the animals. The following behavioral effects of serotonin have been measured.

1.   A reduction by half of the latency for swim initiation into ripples.
2.   A 40% increase in the frequency of biting.
3.   A 25% increase in the frequency of pharyngeal contractions.
4.   A significant increase in the volume of blood ingested.[34,66]

Serotonin then affects feeding behavior of intact leeches and quantitatively shifts it in the direction of hunger. However, it is especially noteworthy that 5-HT changes behavior qualitatively and evokes biting from satiated animals.

The physiological effector organs which generate feeding behavior are accessible in semidissected preparations of the leech head (Figure 8). Irrigating these preparations with micromolar 5-HT stimulates the organs and evokes bite-like movements, a secretion of saliva, and rhythmic, pharyngeal peristalsis. These three physiological responses are also evoked after the CNS has been removed, which suggests that these peripheral organs have receptors for 5-HT. A localized warming of the dorsal lip of the anterior sucker by 3 to 5°C also evokes the three physiological responses; however, thermal stimulation is effective only if the CNS is intact.[66]

The leech pharynx is innervated by yellow fluorescent axons and terminals: histochemical evidence of serotonin nerve fibers. The effector organs which product biting, salivation, and pumping contain more 5-HT than the body wall.[70] Serotonin cell bodies have not been detected in the leech periphery, which makes it likely that the serotonin in these organs derives from axon terminals which project into them from neurosomata in the CNS. Serotonin could be released from the terminals of central neurons into feeding effector organs and directly stimulate physiological components of feeding.

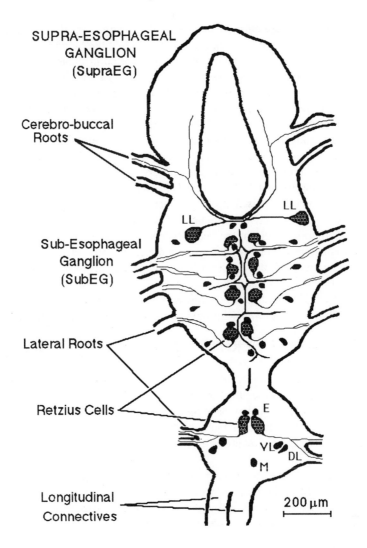

FIGURE 7.   The cerebral and first segmental ganglia of *Hirudo* illustrating the positions of the serotonin-containing neurons. Each segment contains a pair of Retzius cells (RZ), but only the first segment contains a pair of LL cells. There are four classes of interneurons that are labeled in g1. There are two pairs of lateral interneurons found on the ventral (VL) and dorsal (DL) surfaces of the ganglia. DL is found in every segment, but VL is absent from the SubEG. A posteriomedial M cell is paired anteriorly, single in the midbody and sometimes absent from caudal ganglia. E (extra) cells are paired, anteriomedial interneurons restricted to the first seven ganglia (SubEG through g3). Serotonin neurons are not seen in the SupraEG.

In order to demonstrate that neuronal serotonin underlies leech feeding, two criteria must be met.[71] First, impulses by identified 5-HT neurons must evoke the physiological components of feeding (*sufficiency*). Second, the serotonin neurons must be essential for feeding behavior and its physiological components (*necessity*). Anterior LL and RZ somata in these head preparations are readily impaled with microelectrodes, and they can be depolarized into impulse activity as tests of sufficiency. RZ impulse bursts increase salivation and excite the salivary cells to fire calcium-dependent impulses.[72] Impulse bursts of LL cells induce peristaltic movements and oscillatory membrane responses in fibers of the pharynx. Retzius impulses also evoke twitches from the jaw muscles. Hence, the impulse activity of serotonin neurons is *sufficient* for three physiological components of leech feeding.

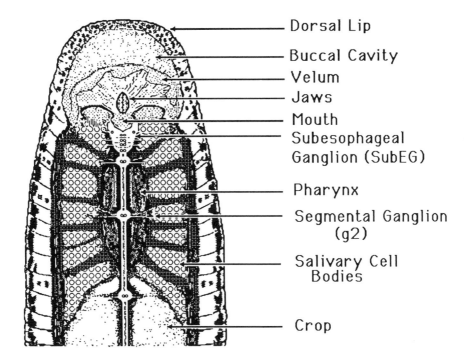

FIGURE 8. A ventrally dissected head preparation of *Hirudo medicinalis*. This preparation allows research into the feeding structures of the leech. The orifice between the three jaws in the buccal cavity (mouth) leads into a muscular pumping pharynx. Intrinsic muscles constrict the pharyngeal lumen and radial, extrinsic muscles dilate the pharynx. These muscles contract antiphasically and are synchronized with the rocking movements of the jaws. The pharynx pumps blood into the crop, a capacious storage organ. Thousands of salivary cell bodies are interspersed among the pharyngeal dilators. Each of the large salivary cells sends a single process, the ductule, to the bases of the jaws. The ductules coalesce and terminate in secretory ducts which alternate with the teeth on the cutting edges of the jaws. This cartoon shows that leech feeding structures are intimately associated with the CNS. The central opening made by the paired connectives between the SupraEG and SubEG (Figure 7) surrounds the mouth.

Membrane responses of the effector cells occur 5 to 10 s after neuronal impulse activity begins. This observation suggests that axonal release sites are distant from the receptors on the target cell membranes and that serotonergic terminals are unlikely to make synaptic contacts with feeding effector cells. Further, the membrane responses of the salivary and pharyngeal cells outlast the neuronal impulse bursts of RZ and LL. This characteristic indicates that serotonin exerts long-term modulatory changes on effector cells of peripheral organs.

Serotonin neurons must be selectively removed in order to demonstrate their requirement to the behavioral circuitry of feeding. Membrane hyperpolarization can sometimes be employed to prevent a neuron from firing, but 16 serotonergic effector neurons in the leech head are implicated in feeding, making an electrophysiological test numerically impossible. However, it is feasible to deplete the transmitter from these serotonergic cells with a lesion which results from a 5-HT-specific neurotoxin.[69] Leeches, which were acting hungry by their biting behavior, were injected with 5,7-dihydroxytryptamine (5,7 DHT).[73] The toxin-treated leeches continued to swim and crawl normally, but when they were tested on a warm surface 2 d later, they acted as if they were satiated and did not bite. Ganglia were examined with fluorescence histochemistry[60] to determine whether 5,7-DHT had changed leech behavior by its specific effects on 5-HT neurons. Both RZ and LL were nonfluorescent in leeches rendered anorexic (nonbiting) by the toxin, and fluorescence is a sensitive indicator of neuronal 5-HT.[69] We also examined the physiological responses of head preparations from toxin-treated animals. Their effector organs did not respond when the lip was warmed; however, when the organs were superfused with 1 μM 5-HT, salivation and pharyngeal peristalsis were activated. Serotonin then is *necessary* for the physiological components of feeding behavior.

Intracellular oxidation of 5,7-DHT produces phenolic compounds, and the nonfluorescent RZ and LL somata in treated leeches were brown and irregularly shaped. The 5-HT interneurons in the SubEG were brown and nonfluorescent; but those in the segmental ganglia appear to be unaffected by these doses of toxin. Serotonin from ganglia and RZ somata was measured by HPLC-EC. Control ganglia contain 21.3 pmol 5-HT, which 5,7-DHT depleted by 49% to 10.9 pmol. Control RZ contain 2 to 5 pmol 5-HT, which the toxin depleted by more than 90% to 0.08 ± 0.03 (5) pmols 5-HT/RZ.[70]

The membrane properties of brown, depleted RZ and LL cells appeared normal, an unexpected result. The functional properties which are retained after transmitter depletion include resting, action, and electronic potentials as well as somatal 5-HT receptors, identified synaptic inputs, and peripheral axons. Thus, 5,7-DHT does not ablate these neurons, it destroys intracellular serotonin. This toxin also fails to ablate serotonin neurons in gastropod mollusks[74] and renders them brown in crustaceans.[75] If serotonin depletion produced anorexic behavior in leech, 5-HT augmentation ought to reverse it. *It does*: serotonin restored biting in depleted leeches, a qualitative alteration in behavior reminiscent of its restoration of biting in satiated animals. It was concluded that neuronal serotonin is *necessary* for both the physiological components of leech feeding, as well as the organismic behavior.

In mammals, serotonin-containing neurons are frequently judged to have been ablated (destroyed) by a neurotoxic lesion if they lack fluorescence. Clearly, the effects of 5-HT neurotoxins on leech neurons should alert mammalian neurobiologists to other possible interpretations of their experimental data.

Fulfilling the requirements of the functional criteria establishes the pivotal role for neuronal 5-HT in the expression of leech feeding. From this, a corollary to the functional criteria is derived logically: stimuli which initiate feeding by the leech ought to excite 5-HT neurons, while the stimulus which terminates feeding ought to inhibit the release of 5-HT. Figure 9 illustrates the preparation with which these sensory issues are addressed experimentally.

Temperature is a sensory determinant for biting and the initiation of ingestion. Warming of the prostomial lip of this preparation by 1 to 3°C excites both RZ and LL, increasing their impulse frequencies substantially (Figure 9). The response is marked by large, unitary EPSPs, and this excitation is blocked reversibly by 20 mM Mg^{++}, which interferes with chemical synaptic transmission in the leech.[41] The prostomium has an array of sensilla, and its thermosensitivity is not seen elsewhere on leech skin. Sensory axons in the cerebrobuccal roots underlie the synaptic activation of the 5-HT neurons. The thermoresponsiveness was the first identified synaptic input to the Retzius cells, which have been long studied.[76] Serotonin interneurons also respond to thermal stimulation, but most of the identified sensory and motor neurons do not. Serotonin neurons usually increase their firing frequency phasically, as lip temperature increases. At certain times, however, the RZ are excited into bursts of high-frequency impulses as the lip is warmed, and they continue to burst for as long as the temperature is elevated. Most of the existing data indicate that bursting is a seasonal phenomenon seen only in the spring and summer. Thus, a stimulus which initiates feeding by the intact leech synaptically excites its serotonergic neurons to fire.

Body wall distension is the sensory determinant in terminating ingestion. When the body wall of this preparation is distended by introducing Ringer into the crop, RZ and LL are tonically hyperpolarized for periods lasting minutes (Figure 7).[35] With sufficient distension, all spontaneous impulses are inhibited, but their activity returns to predistension levels when the fluid is removed. This hyperpolarization abolishes impulses for as long as distension can be maintained, for periods up to 5 min. The inhibitory effect of distension is transmitted

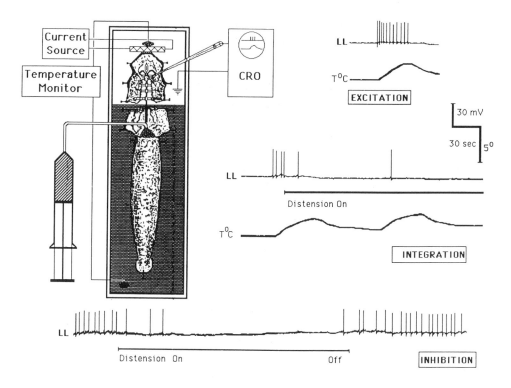

FIGURE 9. The preparation for investigating the sensory responses of leech serotonergic neurons. The anterior sucker can be locally warmed by passing direct current though an insulated resistor. The body can be distended or relaxed by injecting or removing the saline between the syringe and the crop. Warming the lip by 3 to 5°C synaptically excites LL cell (top). Distending the body (bottom) inhibits the LL cell tonically. These synaptic responses are integrated into the final firing frequency of the LL cell (middle). Similar responses have been recorded in the RZ, and the interneurons are also excited by lip warmth.

interganglionically over several segments. Long-duration inhibition is more likely to be mediated by chemical than by electrical synapses.[77] Hence, the stimulus which terminates feeding by the intact leech synaptically inhibits its serotonergic neurons.

Distensive inhibition is integrated with the synaptic excitation evoked by thermal stimulation.[35] Neurons responsible for leech feeding are controlled by behaviorally relevant stimuli and fire in behaviorally appropriate contexts. When the two opposing stimuli are presented simultaneously, the serotonergic effector neurons integrate the excitatory and inhibitory inputs into their final action potential output (Figure 7). Thus, the predictions of the corollary are fulfilled: behaviorally relevant stimuli influence the activity of the serotonin neurons with the appropriate synaptic sign. Integration of disparate synaptic inputs from behaviorally important stimuli appears to be the neuronal locus underlying the expression of leech feeding.

A single neurotransmitter expresses leech feeding behavior by exciting the specific muscles, glands, and nerve cells whose performance represents complex behavior at the organismic level. The expression of feeding behavior by the activity of a limited population of biochemically related neurons is an example of neuronal economy in animals which have only 400 neurons per ganglion and, yet, hunt man and other mammals. Certain of the serotonergic functions may transcend these annelid worms in which they were described. Serotonin evokes salivation and/or pharyngeal movements in animals of three other phyla: nematodes, mollusks, and insects.[66] The involvement of one neurotransmitter in the production of one behavior is a potentially important relationship from which patterns in the evolution of behavioral neurochemicals might be discerned. Should serotonin, other transmitters, or modulators be found to be associated primarily with certain behaviors, unraveling the complex patterns in the regional neurochemistry of mammalian brain might begin in earnest.

VII. STIMULUS-SECRETION COUPLING IN GIANT SALIVARY CELLS

In order to understand the phenomena involved in the biologically important processes of stimulus-secretion coupling, one must describe the membrane properties which lead to changes in membrane potential and elucidate the relationship between the electrical and the secretory events. This goal is technically difficult in mammals because their secretory cells are small and are usually electrically intercoupled into an acinus. However, some leeches have giant salivary gland cells which appear to lack these shortcomings and are well suited for experiments in stimulus-secretion coupling. The leeches which use a proboscis to suck blood belong to the order Rhychobdellida. In rhychobdellids, two pairs of discrete salivary glands are located at the base of the muscular proboscis. In *Haementeria ghilianii* and *H. officinalis* of South and Central America, each anteior gland contains about 200 giant salivary cell bodies which range from 150 to 1000 μm in diameter. These cells are electrically independent of one another, and each salivary cell extends a single long cellular process, or ductule. Many ductules reach a length of 4 to 6 cm. Salivary cells are electrically excitable and fire action potentials of long duration, 100 to 200 ms. The action potentials overshoot zero by 10 to 30 mV and are evoked by either depolarizing currents or by cessation of hyperpolarizing currents (anodal break). The impulse has a plateau, produced by two phases of repolarization (Figure 10A) and is abolished by inorganic calcium channel blockers, cobalt or manganese. Salivary action potential amplitude is unaffected by zero sodium or tetrodotoxin, a sodium channel blocker. Strontium and barium will support salivary action potentials in the absence of calcium. Hence, the upstroke of the action potential of leech giant cells appears to depend exclusively upon calcium as a charge carrier. Because the action potentials overshoot zero, calcium channels in leech salivary cell membranes are likely to have a high density.[72]

Jawed leeches belong to the order of Gnathobdellida, and their salivary cell bodies are located from the base of the three jaws to the beginning of the crop. The somata range from 30 to 200 μm in diameter, and each has a single ductule which runs anteriorly to the base of one of the jaws. These ductules can be 2 to 3 cm long, and they coalesce into fascicles among the muscle fibers which emanate from the jaws. These salivary cells bodies do not form a discrete gland; rather, they are interpersed widely among other tissues of the head. There is no evidence for cell-cell fusion of the secretory ductules nor for the intercellular transport between individual cells. The salivary cells of jawed leeches, like those of proboscis leeches, compose unicellular glands.[72]

Spontaneous action potentials are seen in the salivary cells of the jawed leeches *Hirudo medicinalis* and *Macrobdella decora*. They overshoot zero, and there is a pronounced afterhyperpolarization, or undershoot. These long-duration action potential leeches have a plateau which results from two phases of repolarization. This distinctive shape is similar to that seen in rhynchobdellid salivary cells. Depolarizing currents cause repetitive firing, and action potentials are also elicited by anodal break. Impulses are not seen in the presence of the calcium channel blocker, cobalt, and they appear normal in zero sodium. Hence, these action potentials appear to use calcium as the principal charge carrier, as well.

The salivary glands secrete from the jaws within the buccal cavity. Each jaw has recurvent teeth along the cutting edge. The number and morphology of the teeth reflect the feeding habits of leech species and are important taxonomic characteristics.[79] Within the jaws, ductules terminate in secretory ducts at the bases of the teeth.[80] A minimum of 2000 salivary

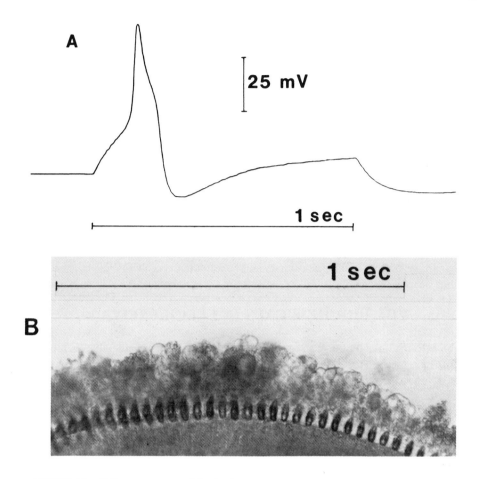

FIGURE 10. Salivary responses of the leech. (A) A calcium-dependent action potential of leech giant salivary cells. These long-duration impulses (200 ms) overshoot zero, have two phases of repolarization and an undershoot (after-hyperpolarization) of 400 to 500 ms. This impulse was recorded in *Haementeria* and is blocked by calcium channel antagonists; (B) Salivary secretion from a jaw of *Hirudo*. This jaw was dissected with many of its salivary cells attached and exposed to micromolar 5-HT for 20 min. Secretory activity is blocked by those agents which block calcium channels and is activated by calcium ionophore.

cells has been estimated for *Hirudo*. Light microscopic examination of living jaw-salivary cell preparations reveals active secretion. Saliva is secreted into the ducts, apparently by exocytosis, and the flow of saliva onto the jaw edge is visible because of numerous salivary granules (Figure 10B).

Serotonin, 10^{-7} to $10^{-6} M$, reliably activates secretion from leech jaws.[72] ACh, GABA, and DA are, however, without effect on salivation. Secretion generally peaks after 20 min and has maximum rate of 230 nl/min per jaw. Cobalt abolishes serotonin-induced secretion completely and reversibly, but zero calcium only slows the rate. Salivary secretion in leech, like the salivary cell action potential, depends upon external calcium.

The effects of serotonin on gland cells were also examined. In quiescent cells, serotonin ($10^{-6} M$) induces membrane potential oscillations. These oscillations increase in amplitude and develop into action potentials at a tonic rate of ~4 Hz for 2 to 3 min. This immediate response is followed by a longer period of impulse bursts lasting for tens of minutes. Bursts occur on the peaks of slow depolarizing waves which occur at intervals of ~20 s. Serotonin appears to alter the membrane conductance of leech salivary gland cells directly and to induce these bursts of action potentials.[72]

Retzius cell (RZ) axons make them the sole neuronal source of peripheral serotonin in somatic segments; hence, the effects of RZ impulses on salivary cell membrane potential were examined. Single impulses did not affect gland cell membrane potential, but high-frequency impulse bursts in RZ elicited depolarizations and action potentials from previously silent gland cells. Once salivary impulses were initiated, they continued for several minutes after the RZ burst ended. Rapid firing by RZ in the CNS appears to release sufficient serotonin in the vicinity of the gland cells to produce repetitive impulses.[72]

Electrical coupling between secretory cells is usually regarded as a mechanism for synchronizing activity and enabling groups of cells to function as a unit. Secretory cells in leeches are not coupled but function in concert, which suggests that electrical coupling is not necessary for coordinated secretory activity. Most salivary glands receive direct inner-vation from the CNS and do not rely on circulating hormones for their activation.[81] Salivary cells of jawed leeches respond to serotonin and impulses of serotonergic cells. Serotonin has been measured in *Hirudo* salivary glands,[70] but any serotonergic innervation has yet to be demonstrated anatomically in gnathobdellids. In rhynchobellids, serotonin-reactive axons and terminals are present in the proboscis sheath.[82] Serotonin is probably released from axon varicosities near salivary cells, at either the level of their somata or their ductule terminals.

VIII. BIOCHEMICALS IN LEECH SALIVA

Leech saliva is an evolutionary armamentarium acting against mammalian hemostasis.[3] Saliva from the medicinal leech contains Hirudin, a powerful natural anticoagulant which was first described over a century ago.[83] Hirudin may have potential as a therapeutic agent in some diseases. Other agents in the saliva of various leech species include hyaluronidase, collagenase, and fibrinase. Fibrinase disrupts clots, an action with additional therapeutic promise. Many of the chemicals in leech saliva are being presently isolated, sequenced, cloned, and patented.

Hirudin is a 65-amino-acid polypeptide of known sequence which has a high affinity for the enzyme thrombin,[84] and it inhibits the thrombin-mediated catalysis of fibrinogen to fibrin. Hirudin may be the most potent natural inhibitor of coagulation known:[84,85] 1 μg inhibits ~10 units of human thrombin, and the secretion of one leech inhibits coagulation of 50 to 100 ml of blood *in vitro*.[9] Hirudin is secreted in the saliva, from the cutting edges of the jaws, and prevents the coagulation of blood as it is ingested. Haycraft first demonstrated the anticoagulant properties of Hirudin, one of a few substances for preventing clotting before heparin was isolated.[84] Experiments have been done with Hirudin as an anticoagulant in blood transfusions.[86]

Hementin is a peptide secreted from the salivary glands of the Amazon leech *Haementeria ghilianii*. Hementin prevents clotting by direct fibrinogenolysis and does not affect throm-bin.[87,88] One *Haementeria* contains sufficient hementin to inhibit the coagulation of 300 ml of blood.[2] Saliva of the related leech, *H. depressa,* may have a plasminogen activator whose mode of action resembles streptokinase.[2]

Leeches also secrete the enzyme hyaluronidase, a potential spreading factors into wounds.[89-92] Other secretions, including proteinase inhibitors such as bdellin (a trypsin-plasmin inhibitor),[93] eglins (inhibitors of chymotrypsin, subtilsin, and the granulocytic neutral proteases), elastase, and cathepsin G, have been isolated from *Hirudo*. Antithrombotic, antifibrinolytic, and antiinflammatory activity has been postulated to exist in these secretions.[94,95]

Fibrinases, apyrases, and collagenase have been isolated in leech saliva.[96,97] Platelet aggregation is apparently prevented by the salivary apyrase and collagenase.[96] It has been suggested that leeches may secrete a vasodilator in the form of an antihistamine.[2,98] Anti-bacterial properties have been attributed to its salivary secretions.[2] In 1913, leech extract was claimed to be efficacious in causing a remission of the growth of carcinoma in the mouse.[99] It has been suggested also that leeches secrete a local anesthetic which deadens the skin to their bites.[2,100] Unfortunately, this suggestion is difficult to substantiate. Not only does there appear to be no experimental evidence for an anesthetic in leech saliva, but there are data that refute its existence.[101]

IX. A MODEL FOR THE CELLULAR BASES OF PARKINSONISM

One can rarely use the simple nervous systems of invertebrates to draw useful lessons for mammals. The neurochemical organization of the leech CNS, however, may provide a useful model for understanding the more complex processes of the mammals. One example concerns the parkinsonogenic neurotoxin MPTP (1-methyl-4-phenyl-1,2,5,6-tetrahydropyridine). MPTP is a neurotoxic byproduct in the illicit synthesis of meperidine (artificial heroin). MPTP was identified after the appearance, in youthful drug addicts, of Parkinsonism. This disorder is normally associated with old age, but the condition develops quickly in addicts after injections of contaminated meperidine. MPTP produces the chemical and behavioral symptoms of Parkinsonism in Rhesus monkeys and provides a primate model of the disease.[102] MPTP selectively destroys dopamine neurons in the substantia nigra. Chemicals which inhibit the enzyme monoamine oxidase-b (MAO-b) interfere with the oxidation of MPTP into MPP+, and the neurotoxicity of MPTP apparently derives from this pyridinium ion.[103] Mice are affected by MPTP in high concentrations,[104] but several other mammalian species are not.[105] Species differences in susceptibility are a major impediment to fundamental investigations into the cellular mechanisms of MPTP toxicity.

The etiology of Parkinsonism in the elderly results primarily from the irreversible loss of dopamine neurons in the substantia nigra. This etiology is morphologically and chemically indistinguishable from the syndrome produced by MPTP.[106] It is necessary to understand the cellular effects of MPTP on dopamine neurons in order to fathom Parkinsonism, and a description of the destruction of dopamine neurons will likely precede prophylaxis.

The nervous system of *Hirudo medicinalis* is surprisingly well suited for investigating the cellular effects of MPTP. Three major effects are seen after leeches are injected with MPTP *in vivo* (4 to 6 $\times$ 10^{-7} mol/g). First, most of the leeches develop body rigidity, very slow movements, and will neither bite nor swim. This rigidity of the body musculature is temporarily reversed by injections of L-DOPA (2 mM, 10 to 20 μl/g).

Second, the serotonin-specific yellow fluorescence of RZ is abolished, and the usually intense fluorescence of the 5-HT interneurons is reduced. The blue-green dopamine fluorescence of AR cells was more intense than controls.[57] The AR cell bodies near the site where the MPTP was injected usually lacked any fluorescence, and their axons in the anterior roots were fragmented.

Third, in segmental ganglia, dopamine is localized to AR terminals and following MPTP, it is reduced by 70%. Dopamine is reduced by 85% in the AR somata. Ganglionic serotonin is found in neurosomata as well as neuropilar processes, and 5-HT is reduced by 69% after MPTP. These results in *Hirudo* are surprisingly similar to the results obtained from primates.

MPTP *in vitro* (10^{-6} to 10^{-5} M) affects leech neurons physiologically and morphologically. Control Retzius cells are transiently depolarized by 5 to 15 mV, which increases their impulse frequency substantially. Vital staining of monoamine neurons in leech is highly selective,[107] and the red color was lost from 5-HT cells reversibly and from DA cells irreversibly. This observation has produced a testable hypothesis that MPTP has the irreversible effect of depleting the membrane-bound dopamine granules from AR cells.[108]

Leech experiments could establish whether parkinsonogenic toxins have a cytological locus of action. One school holds that environmental toxins exacerbate, perhaps cause, Parkinsonism. Identifiable neurons in leech ganglia provide an experimental preparation which can enable an understanding of the cellular and membrane effects of parkinsonogenic neurotoxins.

X. LEECHES IN MODERN MEDICINE

Leeches seem to have made both a clinical and a laboratory resurgence over the past several years. On the clinical front, plastic and microvascular surgeons in both Western Europe and North America have been utilizing the ability of this medicinal worm to inject anticoagulants and remove blood. The use of leeches has been reported as an adjunct for grafted skin flaps[109,110] and breast reconstruction.[111] Reports also indicate their utility in the reattachment of severed digits. Local removal of accumulated blood seems to be a task for which the leech is well suited. The leech has been applied to the suture lines of reimplanted digits in an attempt to reduce venous congestion, as following surgery, venous return may be impaired.[109,112] The anticoagulant in leech saliva may provide additional therapeutic benefits by causing the bites to bleed for many hours after the leech detaches. The leech has also been recently used in the evacuation of periorbital hematomas.[113] Lay persons also have reportedly applied leeches to themselves for conditions such as hemorrhoids and varicose veins.[114]

Systemic utilization of the products of the leech salivary gland appears to be growing as the chemicals produced in their salivary glands are further analyzed. Salivary extracts are available from pharmaceutical firms,[115] and their clinical effects are being studied.[84,116] Hirudin has been reported to prevent some pathological changes of diffuse intravascular coagulation (DIC) such as consumption of clotting factors and development of multiple microthrombi. It also reportedly inhibits localized hemorrhagic necrosis induced by endotoxin injection.[84] Secretions from *Haementeria* salivary cells have been reported to interfere with the metastatic growth of some lung tumors,[116] and the efficacy of a fibrinase on atherosclerotic plaques is reportedly under investigation.[3]

The use of the leech has not been without its shortcomings. In clinical applications, infection has been associated with leech therapy. *Aeromonas hydropha,* normal flora of the gut of *Hirudo medicinalis,* has been suggested as a possible source of infection.[117,118] Recently, a case was reported in which leeches were applied to reduce congestion after breast reconstruction, and the subsequent development of an *Aeromonas* infection occurred.[111] In the experimental laboratory, the leech has shown to be capable as acting as a vector for *Trypanosoma cruzii* and apparently is able to harbor certain viruses for extended periods.[119,120] Clinicians who presently use leeches do not recommend cross application of leeches because of the potential hazard of infection.[121]

In the field of medicine, the popularity curve of the leech has been as undulating as the manner in which it swims. Although even until recently the leech seemed to be but a term of medical antiquity synonymous with sanguinary physicians, *Hirudo* were still available in the pharmacies of many British teaching hospitals throughout this century. The recent resurgence of interest in leeching seems once again to lend credence to the aphorism that the more things change, the more they stay the same!

ACKNOWLEDGMENTS

We are indebted to Mary M. Lent for her extensive editorial assistance. We further thank her for typing this contribution. Much of the research reported here was supported by a PHS grant N.S.24077 and an award from the Willard Eccles Charitable Foundation to Charles M. Lent.

REFERENCES

1. *The African Queen,* John Huston, director, 1951.
2. **Sawyer, R. T.,** in *Leech Biology and Behaviour,* Clarendon Press, Oxford, 1986.
3. **Lent, C.,** New medical and scientific uses of the leech, *Nature (London),* 323, 924, 1986.
4. **Dirckx, J.,** *The Language of Medicine: Its Evolution, Structure, and Dynamics,* 2nd ed., Praeger, New York, 1983, 28.
5. **Grattan, J. H. G. and Singer, C.,** *Anglo-Saxon Magic and Medicine,* Oxford University Press, London, 1952, 17.
6. **Roget, P. M.,** *Roget's International Thesaurus,* Thomas Y. Crowell, New York, 1961, 460.
7. **Glasscheib, H. S.,** *The March of Medicine,* G. P. Putnam's Sons, New York, 1964, 153.
8. **Stille, A. and Maisch, J. M.,** *Hirudo in the National Dispensary,* 3rd ed., Henry C. Lea's Son & Co., Philadelphia, 1884, 766.
9. **Heldt, T. J.,** Allergy to leeches, *Henry Ford Hosp. Med. Bull.,* 9, 498, 1961.
10. **Mettler, C. C.,** Medicine in Rome, in *History of Medicine,* Mettler, F. A., Ed., Blakiston, Philadelphia, 1947, 334.
11. **Waring, E. J.,** Leeches, in *Practical Therapeutics,* 2nd London ed., Lindsay and Blakiston, Philadelphia, 1866, 754.
12. **King, J.,** *Hirudo medicinalis,* in *The American Dispensary,* 8th ed., Wilsatch, Baldwin & Co., Cincinnati, 1870, 424.
13. **Pereira, J.,** Sanguisa, bloodsucking leeches, in *The Elements of Materia Medica and Therapeutics,* Vol. 3, 2nd American ed., Carson, J., Ed., Blanchard and Lea, Philadelphia, 1854, 1106.
14. **Randolph, B. M.,** The bloodletting controversy in the nineteenth century, *Ann. Med. Hist.,* 7, 177, 1935.
15. **Castiglioni, A.,** *A History of Medicine,* 2nd ed., Alfred A. Knopf, New York, 1958, 698.
16. **Singer, C. and Underwood, E. A.,** *A Short History of Medicine,* 2nd ed., Oxford University Press, New York, 1962, 280.
17. **Berghaus, A.,** The medicinal leech, *Pop. Sci. Mon.,* 17, 478, 1880.
18. **Shurtleff, B., Channing, W., and Walker, W.,** Premium for breeding leeches, *Boston Med. Surg. J.,* 12, 322, 1835.
19. **Gillespie, W. A.,** Remarks on the operation of cupping, and the instruments best adapted to country practice, *Boston Med. Surg. J.,* 10, 27, 1834.
20. **Wood, G. B. and Bache, F.,** *Hirudo in the Dispensatory of the United States of America,* 5th ed., Grigg & Elliot, Philadelphia, 1843, 369.
21. **Da Costa, J. C.,** Inflammation, in *Modern Surgery, General and Operative,* 5th ed., W. B. Saunders, Philadelphia, 1907, 73.
22. **Manson-Bahr, P. E. C. and Apted, F. I. C.,** Leeches and leech infestation, in *Manson's Tropical Disease,* 18th ed., Bailliere Tindall, London, 1982, 574.
23. **Glick, S. and Ritz, N. D.,** Hychromic anemia secondary to leeching, *N. Engl. J. Med.,* 256, 409, 1957.
24. **Allmallah, Z.,** Internal hirudiniasis as an unusual cause of haemoptysis, *Br. J. Dis. Chest,* 62, 215, 1968.
25. **Coghlan, C. J.,** Leeches and anaesthesia, *Anaesthesia,* 35, 520, 1980.
26. **Legg, J. W.,** Symptoms of haemophilia, in *A Treatise on Haemophilia: Sometimes Called the Hereditary Haemorrhagic Diathesis,* H. K. Lewis, London, 1872, 43.
27. **Wood, H. C. Jr., and Osol, A.,** *The Dispensatory of the United States of America,* 23rd ed., Lippincott, Philadelphia, 1943, 1424.
28. **Hare, C. J.,** *Good Remedies — Out of Fashion,* J. & A. Churchill, London, 1883, 30.
29. **Mayer, L.,** A proposde l'embolie post operatoire, *Beruxelles Med.,* 10, 312, 1930.
30. **Kretter, K.,** Wartosc lecznicza hirudynizacji przy zakrzepach, *Pol. Gazlek,* 14, 796, 1935.
31. **Dimitriu, V. and Somnea, G. O.,** Action therapeutique de L'Hirunde, *Presse Med.,* 74, 1361, 1931.
32. **Lilienthal, H.,** Coronary thrombosis, proposed treatment by hirudin, *J. Mt. Sinai Hosp. N.Y.,* 10, 135, 1943.
33. **Dickinson, M. H. and Lent, C. M.,** Feeding behavior of the medicinal leech, *Hirudo medicinalis, J. Comp. Physiol. A,* 154, 449, 1984.
34. **Lent, C. M., Fliegner, K. H., Freedman, E., and Dickinson, M. H.,** Ingestive behaviour and physiology of the medicinal leech, *J. Exp. Biol.,* 137, 513, 1988.
35. **Lent, C. M. and Dickinson, M. H.,** On the termination of ingestive behaviour by the medicinal leech, *J. Exp. Biol.,* 131, 1, 1987.
36. **Retzius, G.,** Zur Kenntnis des centralen Nervensystems der Wurmer, *Biol. Unters,* (Neue Folge), 2, 1, 1891.
37. **Bhatia, M. L.,** *Hirudinaria* (the Indian cattle leech), *Indian Zool. Mem. (Lucknow.),* 8, 1, 1941.
38. **Lent, C. M.,** Retzius cells within the central nervous system of leeches, *Progr. Neurobiol.,* 8, 81, 1977.
39. **Leake, L. D.,** Leech Retzius cells and 5-hydroxytryptamine, *Comp. Biochem. Physiol.,* 83C, 229, 1986.

40. **Nicholls, J. G. and Baylor, D. A.,** Specific nodalities and receptive fields of sensory neurons in CNS of the leech, *J. Neurophysiol.,* 31, 740, 1968.
41. **Muller, K. J., Nicholls, J. G., and Stent, G. S., Eds.,** *Neurobiology of the Leech,* Cold Spring Harbor, New York, 1981.
42. **Macagno, E. R.,** Number and distribution of neurons in leech segmental ganglia, *J. Comp. Neurol.,* 190, 283, 1980.
43. **Bullock, T. H. and Horridge, G. A.,** *Structure and Function in the Nervous Systems of Invertebrates,* 2 vols., W. H. Freeman, San Francisco, 1965.
44. **Coggeshall, R. E. and Fawcett, D. W.,** The fine structure of the central nervous system of the leech, *Hirudo medicinalis, J. Neurophysiol.,* 27, 229, 1964.
45. **Kuffler, S. W. and Potter, D. D.,** Glia in the leech central nervous system: physiological properties, and neuron-glia relationship, *J. Neurophysiol.,* 27, 290, 1964.
46. **Stuart, A. E.,** Physiological and morphological properties of motoneurones in the central nervous system of the leech, *J. Physiol. (London),* 209, 627, 1970.
47. **Nicholls, J. G. and Purves, D.,** Monosynaptic chemical and electrical connections between sensory and motor cell in the central nervous system of the leech, *J. Physiol. (London),* 209, 647, 1970.
48. **Baylor, D. A. and Nicholls, J. G.,** Patterns of regeneration between individual nerve cells in the central nervous system of the leech, *Nature (London),* 233, 268, 1971.
49. **Nicholls, J. G.,** *The Search for Connections,* Sinauer Associates, Sunderland, MA, 1986, 100.
50. **Muller, K. J. and Carbonetto, S. T.,** The morphological and physiological properties of a regenerating syapse in the CNS of the leech, *J. Comp. Neurol.,* 185, 485, 1979.
51. **Stent, G. S., Kristan, W. B., Jr., Friesen, W. O., Ort, C. A., Poon, M., and Calabrese, R. L.,** Neuronal generation of the leech swimming movement, *Science,* 200, 1348, 1978.
52. **Stent, G. S., Weisblat, D. A., Blair, S. S., and Zackson, S. L.,** Cell lineage in the development of the leech nervous system, in *Neuronal Development,* Spitzer, N., Ed., Plenum Press, New York, 1982.
53. **Lent, C. M., Dickinson, M. H., and Marshall, C. G.,** Serotonin and leech feeding behavior: obligatory neuromodulation, *Am. Zool.,* in press.
54. **Minz, B.,** Pharmakologische Untersuchungen am Bluegelaparat, zugleich eine Methode zum biologischen Nachweis von Azetlycholin bei answesenheit anderer Pharmakologisch Wirksamer kopereigener Stoffe, *Arch. Exp. Pharamakol.,* 168, 292, 1932.
55. **Lent, C. M., Ono, J., Keyser, K., and Karten, H. J.,** Identification of serotonin in vital-stained neurons from leech ganglia, *J. Neurochem.,* 32, 1559, 1979.
56. **Lent, C. M. and Santamarina, L.,** 6-Hydroxydopamine produces lesions of serotonin-containing Retzius cells in the leech nervous system, *Brain Res.,* 323, 335, 1984.
57. **Lent, C. M., Mueller, R. L., and Haycock, D. A.,** Chromatographic and histochemical identification of dopamine within an identified neuron in the leech nervous system, *J. Neurochem.,* 41, 481, 1983.
58. **Keyser, K. T., Frazer, B. M., and Lent, C. M.,** Physiological and anatomical properties of Leydig cells in the nervous system of the leech, *J. Comp. Physiol.,* 146, 379, 1982.
59. **Belanger, J. H. and Orchard, I.,** Leydig cells: octopaminergic neuons in the leech, *Brain Res.,* 382, 387, 1986.
60. **Lent, C. M.,** Fluorescent properties of monoamine neurons following glyoxyilic acid treatment of intact leech ganglia, *Histochemistry,* 75, 77, 1982.
61. **McAdoo, D. J. and Coggeshall, R. E.,** Gas chromatographic mass-spectrometric analysis of biogenic amines in identified neurons and tissues of *Hirudo medicinalis, J. Neurochem.,* 26, 163, 1976.
62. **Rude, S., Coggeshall, R. E., and VanOrden, L. S., III,** Chemical and ultrastructural identification of 5-hydroxytryptamine in an identified neuron, *J. Cell Biol.,* 41, 831, 1969.
63. **Henderson, L. P.,** The role of 5-hydroxytryptamine as a transmitter between identified leech neurones in culture, *J. Physiol. (London),* 339, 309, 1983.
64. **Lent, C. M.,** Retzius cells: neuroeffectors controlling mucus release by the leech, *Science,* 179, 693, 1973.
65. **Mason, A., Sunderland, A. J., and Leake, L. D.,** Effects of leech Retzius cells on body wall muscles, *Comp. Biochem. Physiol.,* 63C, 359, 1979.
66. **Lent, C. M. and Dickinson, M. H.,** Serotonin integrates the feeding behavior of the medicinal leech, *J. Comp. Physiol.,* 154A, 457, 1984.
67. **Lent, C. M.,** Research in progress, 1988.
68. **Lent, C. M. and Frazer, B. M.,** Connectivity of monoamine-containing neurons in the central nervous system of leeches, *Nature (London),* 266, 844, 1977.
69. **Glover, J. C. and Kramer, A. P.,** Serotonin analogue selectively ablates identified neurons in the leech embryo, *Science,* 216, 317, 1982.
70. **Lent, C. M.,** Quantitative effects of a neurotoxin upon serotonin levels within tissue compartments of the medicinal leech, *J. Neurobiol.,* 15, 309, 1984.
71. **Kupfermann, I. and Weiss, K. R.,** The command neuron concept, *Behav. Brain Sci.,* 1, 3, 1978.
72. **Marshall, C. G. and Lent, C. M.,** Excitability and secretory activity in salivary glands of jawed leeches, *(Hirudinea:Gnathobdellida), J. Exp. Biol.,* 137, 89, 1988.

73. **Lent, C. M. and Dickinson, M. H.,** Retzius cells retain functional membrane properties following "ablation" by the neurotoxin 5,7-DHT, *Brain Res., 300,* 167, 1984.

74. **Gadotti, D., Bauce, L. G., Lukowiak, K., and Bulloch, A. G. M.,** Transient depletion of serotonin in the nervous system of *Helisoma, J. Neurobiol., 17,* 431, 1986.

75. **Glantzman, D. L. and Krasne, F. B.,** 5,7-Dihydroxytryptamine lesions of crayfish serotonin-containing neurons: effect on the lateral giant escape reaction, *J. Neurosci., 6,* 1560, 1985.

76. **Hagiwara, S. and Morita, H.,** Electrotonic transmission between two nerve cells in leech ganglion, *J. Neurophysiol., 25,* 721, 1962.

77. **Spray, D. C. and Bennett, M. V. L.,** Physiology and pharmacology of gap junctions, *Ann. Rev. Physiol., 47,* 281, 1985.

78. **Marshall, C. G. and Lent, C. M.,** Calcium dependent action potentials in leech giant salivary cells, *J. Exp. Biol., 11,* 367, 1984.

79. **Klemm, D. J.,** Freshwater Leeches *(Annelida:Hirudinea)* of North America, U.S. Environmental Protection Agency, Washington D.C., Ident. Manual 8, 1972.

80. **Damas, D.,** Etude histologique et histochemique des glandes salivaire de la sangsue medicinale, *Hirudo medicinalis (Hirudinee, Gnathobdelle), Arch. Zool. Exp. Gen., 115,* 279, 1974.

81. **House, C. R.,** Physiology of invertebrate salivary glands, *Biol. Rev., 55,* 417, 1980.

82. **Marshall, C. G. and Lent, C. M.,** Electrophysiology leech giant salivary gland cells, *Nature (London), 317,* 581, 1985.

83. **Haycraft, J. B.,** On the action of a secretion obtained from the medicinal leech on the coagulation of blood, *Proc. R. Soc. London, 36,* 478, 1884.

84. **Markwardt, F.,** Pharmacology of Hirudin: one hundred years after the first report of the anticoagulant agent in medicinal leeches, *Biomed. Biochim. Acta, 44,* 1007, 1985.

85. **Harvey, R. P., Degryse, E., Stefani, L., Schamber, F., Cazenave, J. P., Courtnery, M., Tolstoshev, and Lecocq, J. P.,** Cloning and expression of a cDNA coding for the anticoagulant hirudin from the bloodsucking leech, *Hirudo medicinalis, Proc. Natl. Acad. Sci. U.S.A., 83,* 1084, 1986.

86. **Satterlee, H. S. and Hooker, R. S.,** The use of Hirudin in the transfusion of blood, *JAMA, 23,* 1781, 1915.

87. **Budzynski, A. Z., Olexa, S. A., and Sawyer, R. T.,** Composition of salivary gland extracts from the leech *Haementeria ghilianii, Proc. Soc. Exp. Biol. Med., 168,* 259, 1981.

88. **Budzynski, A. Z., Olexa, S. A., Brizuela, B. S., Sawyer, R. T., and Stent, G. S.,** Anticoagulant and antifibrinolytic properties of salivary proteins from the leech *Haementeria ghilianii, Proc. Soc. Exp. Biol. Med., 168,* 266, 1981.

89. **Linker, A., Meyer, K., and Hoffman, P.,** The production of hyaluronate oligosaccharides by leech hyaluronidase and alkali, *J. Biol. Chem., 235,* 924, 1960.

90. **Yuki, H. and Fishman, W. H.,** Purification and characterization of leech hyaluronic acid-endo-B-glucoronidase, *J. Biol. Chem., 238,* 1877, 1963.

91. **Claude, A.,** Spreading properties of leech extracts and the formation of lymph, *J. Exp. Med., 66,* 353, 1937.

92. **Favilli, G.,** Mucolytic effect of natural and artificial spreading factors, *Nature (London), 145,* 866, 1940.

93. **Fritz, H., Gebhart, M., Meister, R., and Fink, E.,** Trypsin-plasmin inhibitors from leeches isolation, amino acid composition, inhibitory characteristics, *Int. Res. Conf. Proteinase Inhibitors, 1,* 271, 1971.

94. **Seemuller, U., Meier, M., Ohlsson, K., Muller, H. P., and Fritz, F.,** Isolation and characterization of a low molecular weight inhibitor (of chymotrypsin and human granulocytic elastase and cathepsin G) from leeches, *Hoppe-Seyler's Z. Physiol. Chem., 358,* 1105, 1977.

95. **Baskova, I. P., Nikonov, G. I., and Cherkseva, O. U.,** Antithrombin, antitrypsin and antichymotrypsin activities of the salivary gland secretion and intestinal chyme of medicinal leeches, antichymotrypsin activity of partially purified preparations of hirudin and pseudohirudin, *Folia Haematol. (Leipzig), 111(6),* 831, 1984.

96. **Rigbi, M., Levy, H., Iraqi, F., et al.,** The saliva of the medicinal leech *Hirudo medicinalis* — I. Biochemical characterization of the high molecular weight fraction, *Comp. Biochem. Physiol. B., 87,* 567, 1987.

97. **Baskova, I. P., Khalil, S., and Nikonov, G. I.,** Effect of the salivary gland secretion of *Hirudo medicinalis* on the extrinsic and intrinsic mechanisms of blood clotting, *Bull. Exp. Biol. Med., 98,* 1016, 1985.

98. **Lindeman, B.,** Das Verhalten der Kapillaren in der umbelbung Blutegelbisses, *Arch. Exp. Pathol. Pharmakol., 193,* 490, 1939.

99. **Loeb, L. and Fleisher, M. S.,** Intravenous injections of various substances in animal cancer, *JAMA,* 1857, 1913.

100. **Lenggenhager, K.,** Das Ratsel des Blutegelbisses, *Schweiz. Med. Wochenschr., 9,* 227, 1936.

101. **Rigbi, M., Levy, H., Eldor, A., et al.,** The saliva of the medicinal leech *Hirudo medicinalis* — II. Inhibition of platelet aggregation and of leukocyte activity and examination of reputed anaesthetic effects, *Comp. Biochem. Physiol., 83C,* 95, 1987.

102. **Burns, R. S., Chueh, C. C., Markey, S. P., Ebert, M. H., Jacobowitz, D. M., and Kopin, I. J.,** A primate model of parkinsonism: selective destruction of dopaminergic neurons in the pars compacta of the substantia nigra by *N*-methyl-4-phenyl-1,2,3,6-tetrahydropyridine, *Proc. Natl. Acad. Sci. U.S.A.,* 80, 4546, 1983.

103. **Markey, S. P., Johannessen, J. N., Chiueh, C. C., Burns, R. S., and Herkenham, M. A.,** Interneuronal generation of a pyridinium metabolite may cause drug-induced parkinsonism, *Nature (London),* 311, 464, 1984.

104. **Heikkila, R. E., Hess, A., and Duvoisin, R. C.,** Dopaminergic neurotoxicity of 1-methyl-4-phenyl-1,2,5,6-tetrahydropyridine in mice, *Science,* 224, 1451, 1982.

105. **Chiueh, C. C., Markey, S. P., Burns, R. S., Johannessen, J. N., Jacobowitz, D. M., and Kopin, I. J.,** *N*-methyl-4-phenyl-1,2,3,6-tetra-hydropyridine, a parkinsonian syndrome causing agent in man and monkey, produces different effects in guinea pig and rat, *Pharmacologist,* 25, 131, 1983.

106. **Kopin, I. J. and Markey, S. M.,** MPTP toxicity: implications for research in Parkinson's disease, *Ann. Rev. Neurosci.,* 11, 81, 1988.

107. **Stuart, A. E., Hudspeth, A. E., and Hall, Z. W.,** Vital staining of specific monoamine-containing cells in the leech nervous system, *Cell Tiss. Res.,* 153, 55, 1974.

108. **Lent, C. M.,** MPTP depletes neuronal monoamines and impairs behavior of the medicinal leech, in *MPTP: A Neurotoxin Producing a Parkinsonian Syndrome,* Markey, S., Castagnoli, N., Trevor, A., and Kopin, I. J., Eds., Academic Press, New York, 1986, 105.

109. **Batchelor, A. G. G., Davison, P., and Sully, L.,** The salvage of congested skin flaps by the application of leeches, *Br. J. Plast. Surg.,* 37, 358, 1984.

110. **Henderson, H. P., Matti, V., Laing, A. G., Morelli, S., and Sully, L.,** Avulsion of the scalp treated by microvascular repair: the use of leeches for postoperative decongestion, *Br. J. Plast. Surg.,* 36, 235, 1983.

111. **Dickson, W. A., Boothman, P., and Hare, K.,** An unusual source of hospital wound infection, *Br. Med. J.,* 289, 1727, 1984.

112. Leech in comeback as doctor's helper, *Chicago Sun-Times,* p. 10, November 16, 1987.

113. **Bunker, T. D.,** The contemporary use of the medicinal leech, *Injury: Br. J. Accident Surg.,* 12, 430, 1981.

114. **Gross, A.,** Leeches, anyone?, *Chicago,* 34, 127, January 1985.

115. The leech: rich source of new biochemicals, *Behring Diagn. Biol.,* 12, 1, 1986.

116. **Iwakawa, A., Gasic, T. B., Viner, E. D., and Gasic, G. J.,** Promotion of lung tumor colonization in mice by the synthetic thrombin inhibitor (no. 805) and its reversal by leech salivary gland extracts, *Clin. Exp. Metastases,* 4, 205, 1986.

117. **Bushing, K. H., Doll, W., and Freyteg, K.,** Die Bakterien, Flore der medizinischen Blutegel, *Arch. Mikrobiol.,* 19, 52, 1953.

118. **Whitlock, M. R., O'Hare, P. M., Sanders, R., and Morrow, N. C.,** The medicinal leech and its use in plastic surgery: a possible cause for infection, *Br. J. Plast. Surg.,* 36, 240, 1983.

119. **Marsden, P. D. and Pettitt, L. E.,** The survival of *Trypanosoma cruzi* in the medicinal leech *(Hirudo medicinalis),* *Trans. R. Soc. Trop. Med. Hyg.,* 63, 413, 1969.

120. **Shope, R. E.,** The leech as a potential virus reservoir, *J. Exp. Med.,* 105, 373, 1957.

121. **Green, C. and Gilby, F. A.,** The medicinal leech, *J. R. Soc. Health,* 1, 42, 1983.

Chapter 4

CRUSTACEAN URINARY BLADDER AS A MODEL FOR VERTEBRATE RENAL PROXIMAL TUBULE

David S. Miller

TABLE OF CONTENTS

I. INTRODUCTION: THE NEED FOR SIMPLE RENAL MODELS

During the process of urine formation, renal systems perform two essential functions; (1) they act as one component of an integrated system that regulates body fluid composition and volume, and (2) they remove potentially toxic waste products and xenobiotics from the organism. The nephron is the basic functional unit of the vertebrate kidney, and within each kidney many nephrons work in parallel to produce urine. Most vertebrate nephrons contain two structures: a glomerulus, which ultrafilters plasma, and a tubule, which modifies ultrafiltrate composition to produce final urine. The tubular portion of the nephron is both anatomically and functionally complex. Because of this complexity, formidable technical obstacles are encountered by those seeking a detailed understanding of mechanisms involved in urine formation. Difficulties arise from tissue geometry, tissue heterogeneity, and essential spatial relationships between tubule regions.

First, small size and tubular shape combine to severely limit experimental access to the lumenal compartment. Thus, complex tubule geometry makes analysis of transepithelial solute transport and electrophysiological parameters difficult.

Second, depending on species, each nephron may contain up to 12 morphologically distinct segments, and within a specific segment several cell types may be present.[1] This means that before the mechanisms operating within a particular segment can be characterized, tissue from that segment must be physically or functionally separated from the other components of the nephron and the roles of the various cell types must be eludicated. Along with distinct morphology, tubular segments often exhibit major differences in intracellular enzyme activities and metabolite levels, surface receptors for hormones, and the passive and mediated permeability characteristics of cellular membranes.[2] These biochemical and physiological differences at the cellular level often underlie regional differences in the water and solute transport properties of the tubular epithelium.

Third, the spatial relationships among nephron segments and between segments and the peritubular capillaries also contribute significantly to overall renal function and complexity. This is particularly important for those organisms using a countercurrent multiplier mechanism to produce urine that is considerably more concentrated than their plasma. The countercurrent multiplier mechanism is difficult to probe because techniques that might functionally isolate individual components of the system are bound to disrupt just those relationships that underlie the urine-concentrating ability of the organ.

In view of the complexity of vertebrate kidneys in general and mammalian renal systems in particular, it should not be surprising that the field of renal physiology has a long history of utilizing simple, model systems drawn from lower vertebrates and invertebrates.[3-5] The strategy has been to seek out animals with morphologically and functionally simplified renal systems in the hopes that they will provide easy access to specific processes that are analogous to those found in one or more mammalian tubular segments. In searching for model tissues, several functional and structural considerations are important. The ideal system should have the following characteristics:

1. The model tissue should functionally resemble a single segment of the mammalian nephron.
2. The tissue should be a flat-sheet epithelium. This allows direct access to both surfaces and permits use of flux chamber techniques for the study of solute transport and tissue electrophysiology.
3. The epithelium should be a single cell thick so that mechanisms can be localized unambiguously.
4. It should contain a single-cell type to facilitate electrophysiological studies and the

preparation of isolated cellular components, e.g., membrane vesicles for analysis of transport.

5.   Transport function should be accessible to study *in vivo* as well as *in vitro*.

II. CRUSTACEAN URINARY BLADDER AS A MODEL FOR PROXIMAL TUBULE

The present review is concerned with one model for the proximal segment of the vertebrate renal tubule, the urinary bladder from decapod crustacea. Before discussing the transport properties of crustacean antennal gland and bladder, I shall first describe briefly the structure and transport function of the vertebrate renal proximal tubule. This will provide a set of benchmarks against which the bladder data can be compared. I shall then discuss sources of experimental animals (crustacea), laboratory maintenance, antennal gland (renal system) anatomy, transport methodology, and, finally, describe the transport characteristics of crab bladder and how these relate to the vertebrate proximal segment. In the discussions of transport in both proximal tubule and bladder, emphasis will be on the secretory transport system for organic anions. This is a major route of excretion for negatively charged normal metabolites, xenobiotics, and xenobiotic metabolites. In vertebrate kidney, it is found only in the proximal segment.[6] In crab bladder, it is the transport system for which we currently possess the most complete description.

A. PROXIMAL TUBULE STRUCTURE AND TRANSPORT FUNCTION

The proximal segment is the tubular segment immediately following the glomerulus. Based on morphological criteria, it appears to contain three distinct regions. Major transport functions appear to be distributed over the three regions, and although small regional differences in metabolic requirements and responses to hormones were found in some species,[2] they are outweighed by the structural and functional similarities.

Proximal tubular cell ultrastructure is generally characterized by numerous well-developed microvilli on the apical membrane, basal membrane infoldings, and numerous mitochondria.[1] Overall, the proximal segment performs several important transport functions. These include reabsorption of filtered sugars, amino acids, proteins, bicarbonate, phosphate, NaCl and water, and secretion of protons, organic anions, and organic cations.[2] Electrically, the proximal segment is classified as a "leaky" epithelium; this means that it exhibits low transepithelial resistance and low, spontaneous, electrical potential difference.[7]

In the proximal segment, organic anions are cleared from the plasma into the urine by a process of active tubular secretion.[6] These compounds diffuse from the peritubular capillaries to the interstitial fluid bathing the basal-lateral surface of the tubular epithelial cells. They are then transported through the cells into the tubular lumen. To cross the cells, solutes must pass through two cellular membranes (basal-lateral and apical) and the intervening cytoplasmic compartment. For many charged solutes the secretory process is so efficient that removal from plasma is essentially complete in one pass through the kidney. Clearly, a potent transport mechanism must be present in the membranes of the tubular epithelial cells. Renal slice and membrane vesicles studies show this mechanism to be specific, saturable, powered by cellular metabolism, and dependent on medium Na. Carriers on both the basal-lateral and apical membranes are involved (Figure 1).[6] Recent experiments show that uptake at the basal-lateral carrier is energized indirectly by the cellular Na gradient.[8,9] Efflux at the apical carrier appears to be driven by membrane electrical potential and by anion exchange.[8]

B. DECAPOD CRUSTACEA: SOURCES AND MAINTENANCE

Our studies of urinary bladder function have focused on certain readily available species of marine crabs. We have worked primarily with three species of cancroid crab, *Cancer*

Vertebrate Proximal Tubule

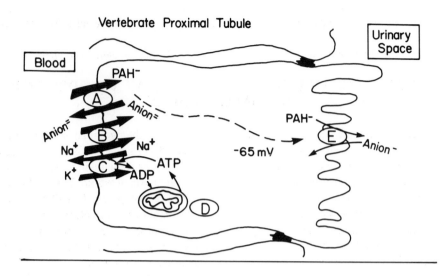

C. Borealis Urinary Bladder

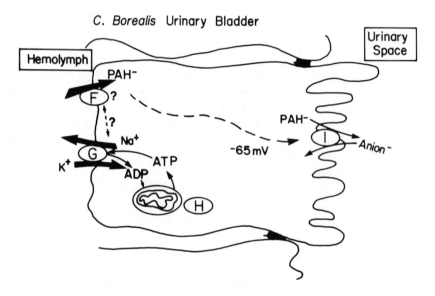

FIGURE 1. Comparison of organic anion (*p*-aminohippurate, PAH) secretion mechanisms vertebrate proximal tubule and *Cancer borealis* urinary bladder. Both tissues utilize a two-step cellular mechanism. In proximal tubule,[6,8,9] entry at the basal-lateral membrane is mediated by a carrier (A) which transports PAH into the cell in exchange for a divalent anion, e.g., glutarate or α-ketoglutarate. PAH entry into the cell is energetically uphill, occurring against an electro-chemical potential gradient. The energy needed to "pump" PAH is supplied by the downhill exit of the divalent anion. The transmembrane divalent anion gradient is maintained by a second carrier (B) that uses the energy in the Na gradient. This transmembrane gradient, in turn, is maintained by the Na,K-ATPase (C), which uses the energy stored in the high energy phosphate bond of ATP to pump NA out of the cell and K in; ATP is produced in the mitochondria (D). The net result is to couple PAH entry indirectly at the basal-lateral membrane to the Na gradient and, ultimately, to cellular metabolism. At the other side of the cell, PAH exit is mediated by another anion exchanger (E), but here the energetics favor PAH efflux. In crab bladder,[16,22] the unidirectional PAH entry step at the basal-lateral membrane (F) is known to be Na dependent and inhibitable by Na,K-ATPase inhibitors, but the intermediate mechanisms which couple anion entry to the Na,K-ATPase (G) have yet to be identified. In bladder, as in proximal tubule, the exit is known to be an exchanger (I).

borealis, C. irroratus, and *C. magister,* and to a lesser extent with the lobster, *Homarus americanus,* the blue crab, *Callinectes sapidus,* and the Florida stone crab, *Menippe mercenaria.* With the exception of *Cancer magister,* all are East Coast species. Our experience has been that the best sources of crabs in quantity are marine laboratories that specialize in supplying live biological specimens, or if geography permits, local fishermen and wholesalers; lobsters are generally available in local markets. For laboratories located some distance from the coast, shipping animals may present special problems. Crabs should be packed in seaweed or moistened absorbent paper, and animals should be kept cool during shipment. The temperature extremes encountered during certain times of the year and long times in transit combine to reduce survival.

In the laboratory, marine crustacea are maintained in large, recirculating aquariums with compressors for cooling, and gravel or charcoal filtration systems. They are fed frozen shrimp or clams occasionally. Intermolt animals can be kept under these conditions for months. During molt, crabs are particularly sensitive to stress, and our experience has been that mortalities increase even when animals in tanks are segregated.

C. CRUSTACEAN ANTENNAL GLAND ANATOMY
1. Coelomosac and Labyrinth
Figure 2 shows a simplified diagram of a typical decapod crustacean antennal gland. The organ consists of three major structures: a coelomosac, a labyrinth, and a urinary bladder. In reality, the first two structures form an integrated unit which cannot be physically separated. That is, the coelomosac is actually a discrete structure embedded within the larger labyrinth. Electron micrographs show that the cells of the coelomosac closely resemble the podocytes of the vertebrate glomerulus.[10] In both coelomosac and glomerulus, the cells and the basement membrane which supports them together act as an ultrafilter for the production of low-protein fluid within the urinary space. In crustacea, ultrafiltrate exits the coelomosac through an opening in the anterior wall and flows into the labyrinth. As its name suggests, this tissue is composed of a maze of interconnecting channels. The channels are lined with a simple epithelium, bathed on one side by hemolymph and on the other by urine. Electron micrographs of labyrinth cells show remarkable structural similarities to proximal tubule cells.[11] In passing through the labyrinth, urine composition is known to be modified by solute reabsorption and secretion. Because of complicated labyrinth geometry, little is known about the details of transport function in that tissue.

2. Urinary Bladder
The final element of the antennal gland and the primary focus of the present review is the urinary bladder. In crustacea, the bladder both stores urine and modifies its composition. In contrast, the vertebrate urinary bladder is primarily a storage organ. Crustacean bladder is actually a fairly complex organ composed of two sets of central and lateral lobes. The central lobes sit over the remainder of the antennal gland and the digestive organs. The lateral lobes extend along the front and sides of the body cavity. Irrespective of location in the animal, bladder tissue is very thin and extremely fragile. Because of this, it is best dissected and handled under saline. The central lobes can be removed intact by careful dissection using watchmaker's forceps and fine scissors. Complete removal yields enough tissue as flat sheets for four or more flux chambers. Additional tissue can be obtained from the lateral lobes of most species; in some large crabs, e.g., *C. magister,* sheets large enough for mounting in chambers are obtainable from the lateral lobes. However, even fragments of tissue from the lateral lobes are useful for biochemical studies, e.g., preparation of membrane vesicles, but care must be taken to remove adhering pieces of hepatopancreas.

a. Bladder Ultrastructure
Electron micrographs show that the crab bladder is a simple epithelium, containing one

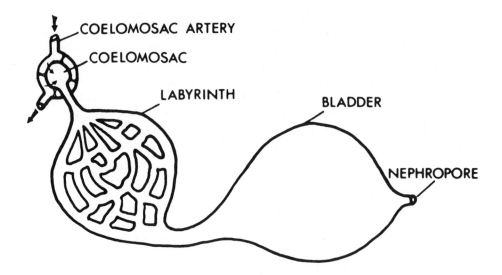

FIGURE 2.   Schematic drawing of a typical decapod crustacean antennal gland. (Reprinted from Holliday, C. W. and Miller, D. S., *Am. Zool.,* 24, 275, 1984. With permission.)

cell type, a columnar cell supported by a thin basement membrane (Figure 3). Several aspects of bladder cell ultrastructure point to a transport function, and these same ultrastructural features are also characteristic of the cells found in the proximal segment of the vertebrate nephron. First, the cells are structurally polarized, suggesting that one cellular function is transepithelial (vectorial) transport; this requires that the basal-lateral and apical plasma membranes possess different selective permeability characteristics. Second, both membrane surfaces are greatly amplified; numerous microvilli increase apical membrane surface area, and deep basal infoldings do the same for the basal-lateral membrane. Third, the cells are rich in mitochondria, and in some regions these appear to be associated with basal membrane infoldings. This arrangement places ATP-producing and -consuming (e.g., ATPases) structures in close proximity, perhaps allowing production to be coupled more tightly to the demands of local consumption.

Other prominent features of bladder cell ultrastructure include an extensive tubular system and numerous large vesicles or vacuoles. The former may actually be continuous with the basal-lateral membrane, representing very deep basal infoldings. The functions of the vesicular structures are unknown. One possibility is that they are involved in fluid transport, providing a pathway for bulk flow from the urinary space to the hemolymph. Irrespective of their functions, the tubular system and vesicular structures together greatly reduce the fraction of the cell volume occupied by cytoplasm.

D. METHODS FOR STUDYING BLADDER TRANSPORT

A wide variety of *in vivo* and *in vitro* techniques were developed for studying solute transport in renal tissue,[7] and many of these were modified for use in experiments with crustacean urinary bladder. This full spectrum of available techniques allows the investigator to focus on mechanisms operating at any single level of organization, from the whole animal to the molecule. In addition, an experimental approach focusing on multiple organizational levels also is possible. Thus, basic findings derived from isolated systems may be related back to function in the intact tissue or animal. This is a distinct advantage.

In this laboratory, we used radioisotopic methods to study bladder transport of organic anions and cations *in vivo,* in tissue slices, tissue sheets mounted in flux chambers, and isolated plasma membrane vesicles. We also used electrophysiological, enzymatic, and polarographic techniques to probe overall cellular function in intact bladders and bladder

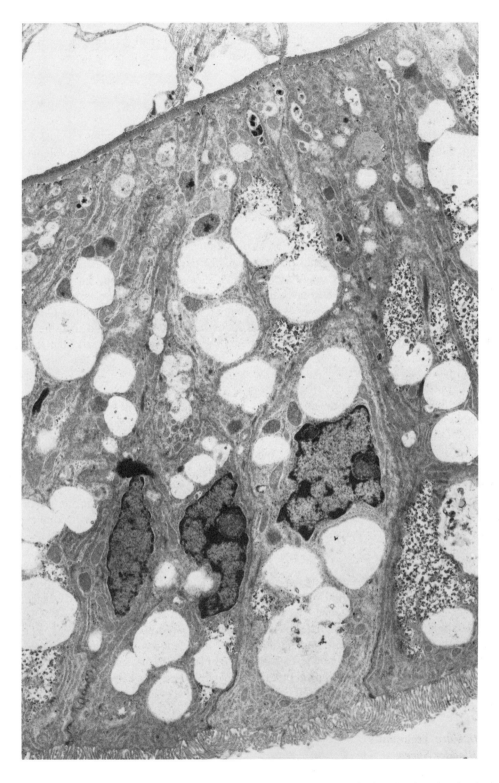

FIGURE 3.   Electron micrograph of urinary bladder tissue from *C. magister*. The tissue is oriented so that the urinary space is at the top right and the hemolymph at the bottom. The epithelial cells are columnar with well-developed brush borders, infolded basal and lateral plasma membranes, and large cytoplasmic vacuoles. (Magnification × 5000). (From Holliday, C., unpublished. With permission.)

homogenates. A discussion of these techniques as they apply to studies of transport function of crustacean bladder follows immediately. In Section II.E.2, I shall give specific examples of the types of information obtained when we used these techniques to probe the mechanism of organic anion secretion in *C. borealis* bladder.

1. *In Vivo* Techniques
a. *Clearance Measurements*

Renal clearance measurements are simple in aquatic crabs. Each animal is injected once with radiolabeled test compound(s), placed in a large beaker containing a known volume of medium, e.g., seawater. The hemolymph and the medium are then sampled periodically.[12] Because of the low rate of urine production (3 to 5 ml/d in 200- to 300-g marine crabs[12]), experiments are usually conducted over a period of days rather than hours as in mammals. We routinely use double-label counting procedures to measure the clearance of a filtration rate marker (conventional vertebrate markers, such as polyethylene glycol or inulin appear to be adequate) along with the test substance. Simultaneous measurements provide two advantages: first, they permit calculation of clearance ratios from paired data, increasing the sensitivity of tests for net solute secretion and reabsorption. Second, since sampling hemolymph and medium measures not only renal clearance but also total clearance, simultaneous measurements provide a way of determining if extrarenal routes are utilized. To do this, urine is also sampled and concentration ratios for the test compound in urine and medium are compared over time. If the ratios are different, an additional extrarenal route, e.g., hepatopancreas, is indicated.[12]

b. *Micropuncture*

One limitation of clearance methods is that they measure overall antennal gland function. Methods also are available to measure the separate contributions of specific gland components *in vivo*. Careful dissection exposes the coelomosac-labyrinth *in situ* for micropuncture sampling of the fluid in the urinary space.[13] When animals are pretreated with labeled solutes before sacrifice, urine concentrations at specific sites, e.g., coelomosac, early and late labyrinth, are then measured by scintillation counting of micropuncture samples. The micropuncture data are used to construct a static picture of solute exchange in the regions between sampling sites. As in vertebrate kidney, micropuncture requires specialized microtechniques for sampling and analysis.

c. *Retrograde Injection*

Urinary bladder transport function can be probed *in vivo* by retrograde injection of labeled solutes (along with fluid volume markers, transport inhibitors, etc.) into the bladder through the nephropore. At some later time, bladder urine is removed for analysis.[14] This procedure is particularly appropriate for solutes that are reabsorbed, e.g., nutrients. Since each of the paired nephropores communicates with a separate lobe of the bladder, an animal may be injected with two different test solutions, and two sets of measurements (control vs. inhibitor) can be made. One technical problem is that because of the extensive lateral lobes, bladder fluid is not usually well mixed, and complete removal of urine may require extensive flushing.

2. *In Vitro* Techniques
a. *Bladder Slices*

Once removed from the animal, bladder tissue is viable at room temperature or lower for many hours. There is no need to take special precautions with regard to oxygenation, since neither solute transport nor electrophysiological parameters change when the tissue is gassed with oxygen rather than air. Slices of bladder tissue are easy to prepare for incubation in a physiological saline with labeled solutes.[14] Slice uptake studies provide a wealth of information on the kinetics, specificities, and ion and metabolism dependence of solute accumulation (for examples, see Table 1). However, with slices, both surfaces of the tissue are exposed to the medium. Thus, there is no simple way to know *a priori* from which side of the tissue the tracer has entered or whether there is any net flux of solute across the tissue. This is clearly a shortcoming of the tissue slice technique, since vectorial transport processes are often of primary interest.

b. Flux Chambers

One major advantage of the crab urinary bladder is that it can be removed from the animal as a flat sheet and then mounted in a flux chamber.[14] This allows the measurement of unidirectional solute fluxes across the bladder. In the chamber, the tissue separates two well-mixed saline-filled compartments. When tracer is added to one, its appearance in the other is followed to establish a unidirectional rate of transepithelial transport. When paired sheets of tissue from the same animal are used (one to measure lumen-to-serosa flux, and the other to measure serosa-to-lumen flux), a value for the net flux can be calculated. Also, by allowing separate access to the two surfaces of the bladder, flux chamber studies permit the investigator to begin to determine the roles of the two surface membranes in the overall transport process.[15,16]

c. Electrophysiology

The simple geometry of the crustacean bladder also greatly facilitates electrophysiological measurements. Transepithelial resistance, potential difference, and short circuit current are measured in a sheet of tissue mounted in a suitably modified flux chamber. These parameters are important aids in understanding the relative roles of cellular and paracellular pathways in transepithelial fluid and solute transport. Moreover, additional information about cellular mechanisms of transport across single membranes can be obtained using intracellular (voltage sensing and ion selective) microelectrodes. Our preliminary studies with these techniques show that crab bladder is a relatively easy tissue in which to impale individual cells and measure transmembrane potentials. Individual impalements can be held for long periods of time, and readings taken in different cells from the same bladder show little cell-to-cell variability.[17]

d. Membrane Vesicles

Over the past 15 years, the field of membrane transport has been revolutionized by the development of techniques for the isolation of specific regions of plasma membranes as vesicles and the study of transport mechanisms in those vesicles.[18] Isolated vesicles appear to retain the transport properties of the membrane from which they were derived. However, with vesicles transport can be studied without the confounding contributions of cellular metabolism, but with the ability to control the composition of the solutions on both sides of the membrane. This level of control is critical to an understanding of transport mechanisms, since in the intact cell, fluxes of nutrients, ions, and metabolites are often energetically coupled. Thus, the transport of one solute may be driven by the transmembrane activity gradient of another. In intact cells, intracellular activities are often difficult to measure. With membrane vesicles, experiments can be performed in which initial chemical and electrical gradients across single membranes are set and then modified in a controlled fashion.

The use of membrane vesicles derived from crustacean tissue has contributed to our understanding of gill[19,20] and hepatopancreas[21] function. The general plan of attack for crustacean tissues has been to use as the starting point standard mammalian procedures for separating epithelial membranes. These are modified as needed to suit the particular tissue used and membrane fraction desired. One recurring problem is the relative paucity of information on membrane-bound enzymes that might be useful as markers for the two sides of the epithelial cells.

TABLE 1
Inhibitors of *p*-Aminohippurate Uptake by Crustacean
Urinary Bladder Slices

Treatment	Cancer irroratus	C. borealis	Menippe mercenaria	Homarus americanus
Anoxia (N$_2$)	—	—	—	+
1 mM NaCN	+ +	—	—	—
1 mM Iodoacetate	+ +	+ +	+ + +	+
1 mM Ouabain	+	+ +	+ +	+
44 mM Na$^+$	+ + +	+ + +	+ + +	+ +
0.1 mM CPR[a]	+ + +	+ +	—	+ +
0.1 mM BCG[b]	+ + +	+ +	—	+ +
1 mM Probenecid	+ + +	+ + +	+ +	+ +

Note: Actual data compiled by Holliday and Miller.[27] Symbols: —, no effect;
+, inhibition <40%; + +, inhibition 41 to 70%; + + +, inhibition
>70%.

[a] CPR: chlorophenol red.
[b] BCG: bromocresol green.

Work with vesicles from crab urinary bladder is at a preliminary stage. Using differential centrifugation, we have been able to isolate two different plasma membrane fractions. One exhibits high Na,K-ATPase activity and little Na-dependent glucose transport; the other, low Na,K-ATPase activity and strong Na-dependent glucose transport. Based on these criteria, the first fractions appear to be enriched in basal-lateral membranes, and the second, in apical membranes.[22]

E. SOLUTE TRANSPORT IN DECAPOD CRUSTACEAN BLADDER
1. General Characteristics

Having shown above that the crustacean urinary bladder is a simple epithelium, with cell architecture resembling that of the vertebrate proximal segment, I shall now present the evidence showing that crab bladder also is functionally analogous to the proximal segment. First, overall bladder transport function will be reviewed and compared with proximal tubule. Then, one specific transport system will be discussed in depth to show that the similarities extend over several levels of organization, from the intact animal to the molecule. This also will furnish an opportunity to show how the bladder techniques surveyed previously have been used to provide detailed information about transport mechanisms. Finally, I shall suggest possible ways in which information obtained from crab bladder might increase our understanding of the regulation of secretory transport mechanisms in vertebrate proximal tubule.

Although transport in crustacean urinary bladder is much less studied, available data show that bladder, like proximal tubule, reabsorbs sugars,[23,24] amino acids,[24] and NaCl[25] and secretes organic cations;[26] organic anions are secreted in bladders from some species and reabsorbed in others[27] (see later discussion). Crab urinary bladders also are electrically leaky epithelia, with initial, spontaneous, transepithelial potential differences (PD) never exceeding a few millivolts and transepithelial resistances averaging approximately 20 Ω-cm^2 in *C. borealis*.[17] In this respect, they resemble the proximal tubule rather than vertebrate urinary bladder, which is an electrically "tight" epithelium.

In marine crustacea, bladder cells are bathed on both surfaces by fluid with a composition that closely resembles seawater. However, the PD across individual bladder cell membranes averages 65 mV, interior negative,[17] which falls well within the range of values obtained for vertebrate proximal tubule cells.[7]

2. Organic Anion Transport

Available data show that the organic anion transport mechanisms found in crustacean antennal gland are remarkably similar to those in vertebrate proximal tubule. For example, in the crab *C. borealis,* renal clearance of the model organic anion, *p*-aminohippurate (PAH), averages approximately ten times the coelomosac filtration rate, indicating strong secretion by the intact gland.[12] Micropuncture experiments show that PAH is freely filterable at the coelomosac and that the PAH concentration in urine increases in the labyrinth and again in the urinary bladder.[13] In both labyrinth and bladder, fluid reabsorption accounts for only a fraction of the increase in PAH concentration. Thus, this model substrate is added to the urine by a process of active tubular secretion in both parts of the antennal gland. Consistent with these findings is the observation that slices of labyrinth and bladder tissue accumulate high intracellular levels of PAH when incubated in a physiological saline containing radio-labeled substrate (with 10 μM PAH in the medium, steady-state tissue-to-medium ratios for *C. borealis* bladder exceed 10).[16]

Experiments with bladder slices were used to study the mechanism by which PAH enters the epithelial cells. They show that PAH uptake is saturable, indicating involvement of a carrier-mediated mechanism.[16] Table 1 shows a summary of uptake data for bladder slices from four species. Although some species differences are apparent, they tend to be minor. The overall picture can be summarized as follows. PAH uptake is inhibitable by competitor organic anions (chlorophenol red, bromocresol green, and probenecid), by reducing medium Na concentration and by compounds which are known to inhibit cell Na,K-ATPase (ouabain) and glycolytic (iodoacetate), but not oxidative (NaCN) metabolism. Except for the source of metabolic energy, similar conclusions have been drawn regarding PAH uptake by tissue from vertebrate proximal segment.[6,8]

When intact bladders from *C. borealis* are mounted in flux chambers and unidirectional fluxes measured, the secretory-to-reabsorptive flux ratio for PAH averages 4,[16] indicating strong net secretion by the tissue *in vitro*. Flux chamber experiments also permit analysis of unidirectional fluxes across the individual basal-lateral and apical membranes of intact tissue. These show that organic anions are transported from the serosal to the luminal bath (secretion) by a two-step mechanism. The first step, across the basal-lateral membrane into the cell, is carrier mediated, energetically uphill (the anion is transported against a concentration gradient into a more electronegative intracellular space), and unidirectional.[16] The basal-lateral step accounts for most of the cellular accumulation in bladder slices. Thus, it is the one that is dependent on medium Na and cellular metabolism. The exit step, across the apical membrane, is also mediated, but is energetically downhill and reversible.

Figure 1 summarizes the mechanisms of organic anion secretion in proximal tubule and *C. borealis* bladder. In all major points they are similar. Both are two-step mechanisms, with an organic anion "pump" on the basal-lateral membrane and an exchanger on the lumenal membrane. In proximal tubule, recent experiments have provided the long-sought-after mechanism for coupling basal-lateral entry of organic anions to the Na gradient. All components of this indirect coupling mechanism have now been identified in *C. borealis* bladder.[28] Experiments are currently under way in this laboratory to determine what contribution this mechanism makes to organic anion secretion by the bladder.

It is interesting to note that *C. borealis* bladder is not fully representative of all crustaceans. We carried out flux chamber experiments with bladders from four species and found that two exhibit net organic anion secretion, and two exhibit net reabsorption (Table 2). With regard to cellular mechanisms, the reabsorptive bladders appear to be mirror images of the secretory bladders.[15,16,27] In the former, the inwardly directed, unidirectional, uphill step is located on the lumenal membrane, and the exchanger, on the basal-lateral membrane. These differences do not extend to other transport systems; that is, bladders from all species studied reabsorb glucose[22] and secrete organic cations,[26] irrespective of the direction of net

TABLE 2
Organic Ion Transport in Crustacean Urinary Bladder

| | Flux ratio (secretory/reabsorptive) | |
Species	Organic anion	Organic cation
Cancer irroratus	0.04	186
C. borealis	4.00	65
Menippe mercenaria	7—14 (F-M)	—
Homarus americanus	0.30	10—15 (M-F)

Note: Flux ratios calculated from unidirectional flux data for the model organic anion, *p*-aminohippurate, and the model organic cation, tetraethylammonium. Original data can be found in references.[14,16,26,27] Flux ratios greater than unity indicate net solute secretion; those less than unity indicate net reabsorption. For both solutes, in all species tested, flux ratios were significantly different from unity ($p < 0.01$). Data are from intermolt males in both *C. borealis* and *C. irroratus*, but from intermolt males (M) and females (F) in the two other species.

organic anion transport (Table 2). In addition, antennal glands in all species tested exhibit overall net PAH secretion.[12,27,29] Apparently, in those species with reabsorbing bladders, strong secretion in the labyrinth is not completely offset by reabsorption in the bladder.

3. Physiological Control of Secretion

One aspect of crustacean urinary bladder function is particularly intriguing: secretory function appears to be a variable, changing with the animal's physiological state. Some evidence for this is shown in Table 2. Most of the year, we were only able to obtain male, intermolt animals for studies of bladder function. However, in two species we did collect enough data from intermolt females to draw some conclusions about male-female differences. In the lobster (*Homarus*) there are no differences in the rates of organic anion transport,[27] but bladders from females exhibit 50% greater flux ratios for organic cations than bladders from males. This is due solely to a higher secretory flux in female bladders. In the crab (*Menippe*) organic anion flux ratios in male bladders are twice those in females; the difference is due to a higher secretory flux in male bladders.

In addition, limited experiments comparing bladder function in intermolt, premolt, and postmolt crabs suggest that rates of organic anion transport are correlated with molt state.[14,30] This indicates, not surprisingly, that bladder transport changes with the molt cycle. It raises the possibility that bladder transport function is under hormonal control. The latter supposition should be of particular interest to students of organic anion transport in vertebrate kidney, since physiological mechanisms that might be involved in regulating organic anion secretion in vertebrate kidney have not yet been identified.

III. CONCLUSIONS

Taken together, available data show that cell structure and function in crustacean urinary bladder closely resemble vertebrate proximal tubule. To date, transport experiments utilizing crab bladder have focused primarily on two general problems: (1) determining the role of the bladder in overall antennal gland function, and (2) validating the bladder as a model for proximal tubule. We are now in a position to begin to take advantage of the groundwork laid in these initial phases and to identify the mechanisms involved in control of transport in bladder with an eye toward unraveling regulatory relationships in proximal tubule.

ACKNOWLEDGMENTS

I thank Drs. J. Pritchard, P. Smith, and C. Holliday for helpful discussions and Dr. Holliday for providing Figure 3.

REFERENCES

1. **Kriz, W. and Kaissling, B.**, Structural organization of the mammalian kidney, in *The Kidney*, Vol. 1, Seldin, D. W. and Giebisch, G., Eds., Raven Press, New York, 1985, 265.
2. **Morel, F. and Chabardes, D.**, Functional segmentation of the nephron, in *The Kidney*, Vol. 1, Seldin, D. W. and Giebisch, G., Eds., Raven Press, New York, 1985, 519.
3. **Forster, R. P.**, Comparative vertebrate physiology and renal concepts, in *Handbook of Physiology*, Sect. 8, *Renal Physiology*, Orloff, J. and Berliner, R. W., Eds., American Physiological Society, Washington, D.C., 1973, 161.
4. **Pritchard, J. B. and Miller, D. S.**, Teleost kidney in evaluation of xenobiotic toxicity and elimination, *Fed. Proc. Fed. Am. Soc. Exp. Biol.*, 39, 3207, 1980.
5. **Miller, D. S.**, Aquatic models for the study of renal transport function and pollutant toxicity, *Environ. Health Perspect.*, 71, 59, 1987.
6. **Wiener, I. M.**, Organic acids and bases and uric acid, in *The Kidney*, Vol. 1, Seldin, D. W. and Giebisch, G., Eds., Raven Press, New York, 1985, 1703.
7. **Ullrich, K. and Greger, R.**, Approaches to the study of tubule transport functions, in *The Kidney*, Vol. 1, Seldin, D. W. and Giebisch, G., Eds., Raven Press, New York, 1985, 427.
8. **Pritchard, J. B.**, Luminal and peritubular steps in renal transport of *p*-aminohippurate, *Biochim. Biophys. Acta*, 906, 295, 1987.
9. **Shimada, H., Moewes, B., and Burckhardt, G.**, Indirect coupling to Na$^+$ of *p*-aminohippuric acid uptake into renal basolateral membrane vesicles, *Am. J. Physiol.*, 253, F795, 1987.
10. **Riegel, J. A.**, *Comparative Physiology of Renal Excretion*, Hafner, New York, 1972, 113.
11. **Schmidt-Nielsen, B. M., Gertz, K. H., and Davis, L. E.**, Excretion and ultrastructure of the antennal gland of the fiddler crab, *Uca mordax*, *J. Morphol.*, 125, 473, 1968.
12. **Holliday, C. W. and Miller, D. S.**, PAH excretion in two species of cancroid crab, *Cancer irroratus* and *Cancer borealis*, *Am. J. Physiol.*, 246, R364, 1984.
13. **Ferraris, J., Holliday, C. W., and Miller, D. S.**, unpublished data.
14. **Holliday, C. W. and Miller, D. S.**, PAH transport in rock crab *(Cancer irroratus)* urinary bladder, *Am. J. Physiol.*, 238, R311, 1980.
15. **Holliday, C. W. and Miller, D. S.**, PAH transport in rock crab urinary bladder. II. Luminal and serosal steps, *Am. J. Physiol.*, 242, R25, 1982.
16. **Miller, D. S. and Holliday, C. W.**, PAH secretion in the urinary bladder of a crab, *Cancer borealis*, *Am. J. Physiol.*, 243, R147, 1982.
17. **Smith, P. M., Pritchard, J. B., and Miller, D. S.**, unpublished observation.
18. **Kinne, R. and Sachs, G.**, Isolation and characterization of biological membranes, in *Physiology of Membrane Disorders*, Andreoli, T. E., Hoffman, J. F., Fanestil, D. D., and Schultz, S. G., Eds., Plenum Press, New York, 1986, 83.
19. **Lee, S. H. and Pritchard, J. B.**, Bicarbonate-chloride exchange in gill plasma membranes of blue crab, *Am. J. Physiol.*, 249, R544, 1985.
20. **Towle, D. W. and Holleland, T.**, Ammonium ion substitutes for K$^+$ in ATP-dependent Na$^+$ transport by basolateral membrane vesicles, *Am. J. Physiol.*, 252, R479, 1987.
21. **Ahearn, G. A.**, Nutrient transport by the crustacean gastrointestinal tract: recent advances with vesicle techniques, *Biol. Rev.*, 62, 45, 1987.
22. **Miller, D. S.**, unpublished data.
23. **Holliday, C. W.**, Glucose absorption by the bladder of the crab, *Cancer magister* (Dana), *Comp. Biochem. Physiol.*, A61, 73, 1978.
24. **Binns, R.**, The physiology of the antennal gland of *Carcinus maenas*, *J. Exp. Biol.*, 51, 1, 1969.
25. **Holliday, C. W.**, Magnesium transport by the urinary bladder of a crab, *Cancer magister*, *J. Exp. Biol.*, 85, 187, 1980.
26. **Miller, D. S. and Holliday, C. W.**, Organic cation secretion by *Cancer borealis* urinary bladder, *Am. J. Physiol.*, 252, R153, 1987.

27. **Holliday, C. W. and Miller, D. S.,** Cellular mechanisms of organic anion transport in crustacean renal tissue, *Am. Zool.,* 24, 275, 1984.

28. **Miller, D. S., Smith, P. M., and Pritchard, J. B.,** Organic anion and cation transport in crab urinary bladder, *Am. J. Physiol.,* in press, 1989.

29. **Burger, J. W.,** The general form of excretion in the lobster, *Homarus, Biol. Bull.,* 113, 207, 1957.

30. **Holliday, C. W. and Miller, D. S.,** unpublished data.

Chapter 5

INSECT MODELS FOR BIOMEDICAL RESEARCH

Richard G. Andre, Robert A. Wirtz, and Yesu T. Das

TABLE OF CONTENTS

"For many problems there is an animal on which it can most conveniently be studied" — the August Krogh Principle.[1]

I. INTRODUCTION

The vast numbers of species of insects and other arthropods offer tremendous possibilities for medical research. Meglitsch included the insects in his statement; "It is no accident that nearly all truly basic zoological discoveries have been based on studies of invertebrates."[2]

The major advantages of using insect models include the ease and low cost of rearing large numbers of specimens. The more commonly used laboratory insects can be reared or purchased at a fraction of the cost of mice, rats, and other laboratory animals. Reduced per diem costs and space requirements also result in significant savings. The rapid reproduction and maturation and the large number of offspring from a single male-female pairing of some species are distinct advantages over vertebrate models. The potential number of descendants from a single pair of insects, such as the housefly *Musca domestica,* is 10^{18} in a matter of months, permitting research designs using multicellular animals which usually are viewed to be restricted to single-cell organisms.[2] The flexibility afforded by use of short-lived species also can be exploited using many insect models. Experiments often can be run in months or a few years using large genetically homogeneous populations. The ability to use large numbers of test specimens can be exploited to arrive at highly significant statistical results and the detection of low-frequency occurrences. Opportunities for studies in embryology are especially promising because of the detailed knowledge of egg and larval development in some insect species. Since it is often easier to isolate physiological or pharmacological systems in insect models, these usually can be studied more simply in insects. Invertebrate tissue cultures, although initially difficult to establish, usually can be handled more easily than vertebrate systems.[3] The use of animals lower on the evolutionary scale also reduces objections by antivivisectionist and animal rights groups, a major concern of scientists today.

II. INSECTS AS A GROUP

A. MORPHOLOGY AND PHYSIOLOGY

Insects differ in their morphology and physiology from mammals in a number of ways.[4] In particular, insects possess an external exoskeleton rather than bone — the only vertebrate tissue they lack.[2] This chitinous structure provides a great deal of protection against a number of environmental stresses, such as desiccating conditions, chemicals, and pressure. Within this exoskeleton is the body cavity (hemocoel) which contains systems for digestion, circulation, respiration, excretion, innervation, and reproduction (see Figure 1). Unlike mammals, there is an open blood system with a dorsal heart and blood (hemolymph) which contains no hemoglobin. The hemolymph is responsible for a variety of transportation and immunological functions. Insect respiration is provided by a branching series of tubes called the tracheal system and by passive diffusion of oxygen to individual cells. Analogous to the vertebrate liver is a tissue known as insect fat body. This group of specialized cells is enclosed in a membranous sheath and is important in insect metabolism. Insects have a well-developed neuromuscular system. The insect organs and muscles are innervated through a series of ganglia that form a ventral nerve cord (see Figure 2). The nervous system is similar to that of mammals in having a blood-brain barrier and cholinergic synapses; however, the neuromuscular junctions are glutaminergic, unlike the vertebrate cholinergic junctions. Reproductive mechanisms in insects are quite species specific, but in general the two sexes mate via a complex chemical, visual, and tactile communication system.[4]

B. HUSBANDRY AND ECONOMICS OF REARING

The development and use of animal models for biomedical research depend upon the

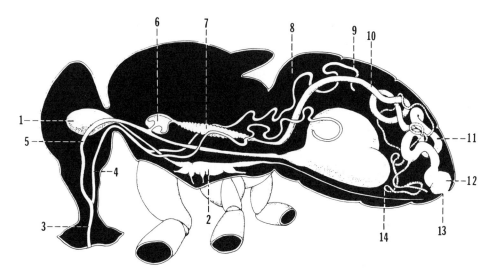

FIGURE 1.   Longitudinal cross-section of an adult housefly (diagrammatic) showing gross internal organization: (1) esophageal ganglion, (2) compound thoracic ganglion, (3) pharynx, (4) salivary duct, (5) esophagus, (6) proventriculus, (7) stomach, (8) hemocoel, (9) salivary gland, (10) proximal intestine, (11) distal intestine, (12) rectum, (13) anus, and (14) Malpighian tubule. (Modified after Patton[66] and West.[67])

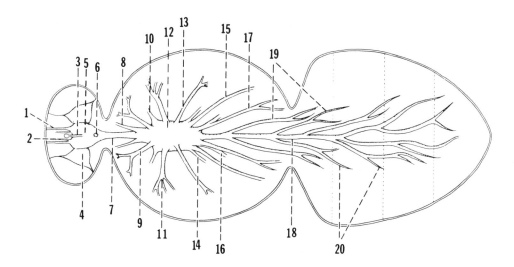

FIGURE 2.   Gross nervous system of an adult housefly (diagrammatic): (1) antennary nerve, (2) pharyngeal nerve, (3) ocellar nerve, (4) optic peduncle, (5) cephalic ganglion, (6) space for esophagus, (7) cephalo-thoracic nerve cord, (8) cervical nerve, (9) prothoracic dorsal nerve, (10) prothoracic crural nerve, (11) mesothoracic dorsal nerve, (12) compound thoracic ganglion, (13) accessory mesothoracic dorsal nerve, (14) mesothoracic crural nerve, (15) metathoracic dorsal nerve, (16) metathoracic crural nerve, (17) accessory metathoracic dorsal nerve, (18) abdominal nerve cord, (19) abdominal nerves of thoracic origin, and (20) abdominal nerves of local origin. (Modified after Hewitt[68] and West.[67])

production of the needed specimens which must meet quality control requirements within specific cost restrictions. The great advantage of insects for use in biomedical studies is the ease with which these biological organisms can be reared.

Successful rearing is dependent upon a detailed knowledge of the biology, behavior, habitat, and nutritional requirements of the insect species selected. This knowledge has been expanded greatly in the past few decades, with numerous descriptions appearing in the literature on rearing methods and diets for selected insects and other arthropods.[5] Some of the most widely used insects are often those most easily reared, such as the flies, Lepidoptera larvae, and other insects of economic importance. The housefly *M. domestica* is reared easily on CSMA (Chemical Specialties Manufacturers Association), a diet that provides year-round rearing on an efficient medium. The commercial availability of artificial and defined diets for some lepidopteran larval species significantly reduces the trouble and expense of feeding. Quality control of diet ingredients is essential to ensure proper insect nutrition at the lowest possible cost. A list of the more important references on insect diets is given by Singh.[5]

The containers and enclosures used for rearing often dictate the success of the operation. Desirable characteristics of rearing containers include economics, barrier to microbial contaminations and pathogens, allowance for gas exchange, moisture regulation, visibility and accessibility, convenience of handling and harvesting, and ease of cleaning and disinfection or disposal.[6]

The rearing procedures usually described are designed for the production of hundreds to thousands of specimens per week. Mechanized mass rearing systems also have been developed where the number of insects reared is measured in millions per week. As part of a sterile-male screwworm eradication program, approximately 500 million flies were produced per week.[7] However, production on this scale requires uniquely designed facilities to meet the needs of controlled environments, mechanized handling methods and control of pathogens, contaminants, and respiratory hazards.[8]

III. DISCOVERIES AND APPLICATIONS IN BIOMEDICAL SCIENCES

A. GENETICS

The study of genetics in multicellular organisms has progressed rapidly during the past 80 years. The fruit fly *Drosophila melanogaster* has become the best-known model for laboratory and field studies of genetics. This insect was used first as the basis for amplifying Mendelian genetics and giving it its present form. In 1910, Morgan[9] at Columbia University reported the crisscross nature of sex linkage in *Drosophila* and, more importantly, set the standards of excellence for experimental work in genetics.[10] Dobzhansky[11] was the first scientist to integrate the results of laboratory and field studies with the predictions arising from mathematical theory such as the Hardy-Weinberg law. Since Morgan's initial report in 1910,[9] it is estimated that over 25,000 articles dealing with *Drosophila* have been published and that the literature would double every 12 years.[12]

The advantages of using the fruit fly *Drosophila* as a model for the study of genetics are many.[12] The adult fly is small, readily handled, and breeds prolifically in the laboratory and in the field. Conditions for rearing the flies are simple, cheap, and readily controlled. The life cycle is short, about 9 d, and thousands can be produced in a small space. There are only four haploid chromosomes, and the polytene chromosomes of the salivary glands of larvae are gigantic and show a characteristic banding pattern. These patterns facilitate the detection of chromosomal rearrangements, the mapping of gene deficiencies, and the subsequent cytological localization of genes. Since the homologous chromosomes do not undergo crossing over in the germ cells of the male,[13] the genetic procedures employed are simplified greatly. This insect can serve as host to a variety of viruses,[14] thereby allowing the study of the genetics of host-parasite interactions. *D. melanogaster* flourishes upon many media; however, a synthetic, minimal medium has been developed upon which flies can be reared aseptically.[15] Schneider[16] has developed a medium for *in vitro* cell cultures of fly-derived cells. Massive collections of hereditary variations in flies have been developed, and stocks of many of the mutants can be obtained from various workers in the field. Finally, an encyclopedic body of information on *Drosophila* genetic studies is readily available through indexed bibliographies, such as those by Herskowitz.[17-21] Without question, these attributes make this insect model one of the more important findings in the field of genetics, as well as in modern science.

B. MUTAGENICITY TESTING

The fruit fly *D. melanogaster* has also proved to be extremely useful in testing materials for mutagenicity, and the literature on this subject is abundant.[22] Studies of mutation induction in *Drosophila* began with Muller's experiments with X-rays in 1927.[23] In the years after World War II, the mutagenic effects of radiation were studied extensively, and Auerbach and co-workers were the first to detect chemical mutagenesis by mustard gas and formaldehyde using *Drosophila*.[23,24]

The wealth of specific test strains, special markers, inversions, and other rearrangements make it possible to test for most of the genotoxic end points relevant to human hazards using *Drosophila*. These range from recessive lethals or visible point mutations and small deletions to translocations, duplications, meiotic or mitotic recombinations, and dominant lethals or chromosome loss as an indication of open, unrepaired breaks, chromosome damage, and aneuploidy.[22,25,26] Testing for the different types of mutations often can be conducted simultaneously if desired. The life cycle of *D. melanogaster* is short enough to permit rapid analysis of many progeny but long enough to distinguish between chronic, acute, and fractionated doses.[22] Since the fruit fly is a multicellular eukaryote, it possesses a cellular and chromosomal organization more akin to mammals than the bacteria sometimes used for the initial screening of mutagens. The overlap between the mutagenic and carcinogenic potential of many classes of chemicals tends to make the distinction between the two an artificial one.[25,26]

Indirect mutagens and carcinogens require activation by the microsomal enzyme systems present in the mammalian liver. Mutagens of this kind register as negative in microbial test systems unless host-mediated assays or plating on microsomal extracts from mammalian tissues are employed. Mammalian-like detoxification pathways have been demonstrated in *Drosophila* and are capable of facilitating similar enzymic reactions to those from mammalian liver. Thus, the use of *Drosophila* is convenient for detecting indirect mutagens and short-lived metabolites. Over 50 compounds, falling into 9 different groups, that all require metabolic activation for the manifestation of their mutagenic and carcinogenic properties have been tested in *Drosophila* and yielded positive responses.[26,27]

Many of the advantages of using the fruit fly listed in the preceding section apply also to mutagenicity testing. Toxicity testing using a housefly model is described in Chapter 6.

C. PATHOGEN PRODUCTION

Many human pathogens, such as bacteria, protozoa, rickettsia, viruses, and helminths, multiply in various insects. These insect hosts may be involved in the natural transmission of certain pathogens to man. Insect-borne diseases, such as malaria, trypanosomiasis, and dengue, account for the loss of millions of people each year, particularly in tropical areas. Scientists, however, have learned to take advantage of this pathogen-insect relationship in disease diagnosis.

A unique application exploiting parasite development in insect vectors is xenodiagnosis. The causative organism of some arthropod-transmitted diseases often occurs only sparsely in human blood, making nonacute forms of the disease difficult to diagnose by recovery of the parasite. Xenodiagnosis involves the feeding of noninfected insects on the patient. After incubation and multiplication in the insect's body, the parasite, if present, may be recovered and examined. Xenodiagnosis is used most commonly in the detection of trypanosomes causing Chagas' disease (American trypanosomiasis) in the gut and feces of conenose bugs fed 1 to 2 weeks earlier on patients.[28] More recently, phlebotomine sand flies and simuliid black flies have been used for the diagnosis of New World leishmaniasis[29] and onchocerciasis,[30] respectively.

Insects are used also in the laboratory confirmation of certain human viral illnesses, such as those caused by the dengue viruses. Dengue is one of the most important arthropod-borne viruses that occurs in man because of the high numbers of individuals infected and because it may cause mortality in children. The four viruses that cause this disease, however, are among the most difficult to detect and propagate in the laboratory. They are not very pathogenic when inoculated into the brain of a newborn mouse and may require many serial passages to produce signs of illness in mice. The application of cell culture techniques for detection led to more sensitive assays, but not all four virus types would produce consistently cytopathic effects which could be detected in the cultured cells. Upon discovery that the dengue virus grew to high titers in certain mosquitoes, workers began inoculating virus into mosquitoes to develop a more sensitive detection system.[31]

The use of mosquitoes to assay dengue viruses offers a considerable advantage in sensitivity whether the viruses are present in mosquitoes, in sera from naturally infected humans, or have been adapted to cell cultures or mice.[32] The discovery that male mosquitoes, such as *Aedes aegypti,* are as sensitive to infection as females, offers a significant advantage in safety, since males cannot transmit the infection should they escape. It was shown also that *Toxorynchites* mosquitoes, a genus that does not feed on blood and is extremely large, could be infected with the virus. This mosquito currently is the genus of choice for the laboratory confirmation of the four dengue viruses.[31]

D. NEUROENDOCRINE CONTROL MECHANISMS

Insect metamorphosis has been a fascinating phenomenon from ancient times. However, it was not until 1922 when an insight into this phenomenon was gained by Kopec.[33] He showed that a chemical factor had to be released from the brain of the gypsy moth larva, *Lymantria dispar,* to cause pupation. This was the first evidence in the animal kingdom that the nervous system was involved in the endocrine control of growth and development. We now know that the vertebrate hypothalamic-hypophyseal complex provides the same coordination of the nervous and endocrine activities as the pars intercerebralis-corpora cardiaca complex of insects. The first evidence on the mode of action of steroid hormones at cellular and molecular level came from the studies of Clever and Karlson in the 1960s on the polytene chromosomes of a fly, *Chironomus* sp.[34] The role of cyclic nucleotides in insect hormone action provides a commonality in the mode of action of insect hormones with those of mammals such as serotonin.[35] The discovery that RNA and protein syntheses were important to the action of insect hormones has yielded basic information of great significance to the mode of action of hormones in general.

E. ANTIMALARIAL DRUGS

Insects have proved very effective in the screening of potential drugs, in particular, with antimalarial compounds. Following World War II and our experience with malaria in vast numbers of military troops serving in tropical areas, malaria research centered on the development of more effective drugs. At this time, there was the need for newer testing methods to seek out compounds with antimalarial activity. The need derived from the fact that testing methods using the vertebrate hosts of avian and simian malarias, ordinarily used for preliminary evaluation of compounds for antimalarial activity, failed to show a consistent relation between the activity in animal models and that in human beings. For example, paludrine had a prophylactic effect against the avian malaria *Plasmodium gallinaceum* but not against the human malaria *P. vivax.* Consequently, preliminary evaluation of this compound required the use of experimentally infected human volunteers. Furthermore, other compounds were not being considered because of their lack of activity against avian or simian malaria and may have been overlooked because they had not been tested against the human malarias.[36]

This need for drug-testing methods which showed drug effects in human malarias was met, in part, by studies in which various antimalarial compounds were administered to the mosquito hosts (*Anopheles quadrimaculatus, Aedes aegypti*) of *Plasmodium falciparum* and other malarias, and in which drug action was evaluated by its morphological and physiological effects on the various stages of the malaria cycle within the mosquito.[37-39] As a result of these studies, a specific relation in drug action between the mosquito and the human liver cycle of malaria was shown. Those compounds that had a prophylactic action in the mosquito, also had a prophylactic effect in the human being. As a consequence of this relation, it became possible to evaluate compounds for prophylactic activity against human malarias by using mosquitoes infected with human malarial strains as the test animal. This reduced the need for tests of drug activity in other animal models.[36]

In addition to the reduced need for animal testing, this insect model made possible a greatly expanded and accelerated malaria drug testing program at a comparatively low cost. With the discovery that drugs tested against avian malaria in the mosquito reliably predicted possible curative activity against *P. vivax*, this insect model was considered even more useful.[36] However, in the 1960s the discovery of several new nonhuman primate models led to decreased utilization of the insect model, although it was and still is a valid and much less expensive model.

F. BIOLUMINESCENCE

Insects have been used also to study the fate of various biochemical components like adenosine triphosphate during bioluminescence. Self-luminescence, not involving bacteria, occurs in insects from the orders Collembola, Homoptera, Diptera, and Coleoptera.[4] Bioluminescence has been characterized best in the common North American firefly, *Photinus pyralis*. Firefly luciferase catalyzes the adenosine triphosphate (ATP)-dependent oxidative decarboxylation of luciferin (LH_2), resulting in the production of light (*hv*) as shown in the reaction where P denotes the product oxyluciferin:

$$LH_2 + ATP + O_2 = P + AMP + CO_2 + hv$$

The reaction catalyzed by this enzyme has a quantum yield of 0.88 with respect to LH_2, making it the most efficient bioluminescent reaction known.[40] Firefly luciferase is useful in a variety of applications. Because of its specificity for ATP, firefly luciferase can be used to measure the amount of ATP present in biological samples without interference from other nucleotide triphosphates.[41] Using luciferase isolated from fireflies, in conjunction with suitably sensitive liquid scintillation counters or biometers, less than 1 fmol (10^{-15} mol) of ATP can be detected.[42] The level of endogenous ATP in a cell may be used as an index of its energy status and is therefore useful in metabolic and physiological studies. Estimates of cell numbers in microbial and tissue cultures may be obtained after determining the ATP per cell under defined conditions and measuring total ATP in a sample of culture.[43] This has served as a basis for rapidly quantitating bacteria in urine, milk, wine, and polluted waters, with sufficient sensitivity to detect the ATP contents of as few as 10 colony-forming units (CFU) per milliliter.[44] Replacing radiolabels (e.g., ^{125}I) with luciferin- or firefly luciferase-conjugated ligands in a bioluminescent immuno- or affinity-assay, can result in increased assay sensitivity, elimination of hazardous radiolabeled compounds, increased speed of the assay, and decreased cost per assay.[45] Commercially available firefly luciferase reagents for use in these assays have been evaluated by Leach and Webster.[46]

IV. AREAS OF POTENTIAL RESEARCH

A. GENERAL CONSIDERATIONS

Insects, by their enormous species diversity and antiquity, present a wide choice of biological parameters. There appears to be no common ancestry between mammals and insects. Interestingly, the basic biological functions are essentially similar in these diverse animal groups. Among the various insect species, cockroaches may be considered as relatively primitive, while bees and flies may be considered more advanced in terms of evolution. Vertebrate evolution, of course, is much more recent than that of insects; however, certain basic mechanisms are conserved throughout the animal kingdom. Therefore, the chances are high of finding a body function or a control mechanism of biomedical interest in insects. Based on the current status of our knowledge on comparative physiology, biochemistry, and molecular biology, the following areas of research appear promising for biomedical purposes.

B. SPECIFIC AREAS

1. Insectan Antibiotics

Because of their long history of survival on this planet, insects may be looked at as the founders of successful defensive mechanisms. They possess a complex, multicomponent, active defensive system that is regulated and coordinated by several distinct cell populations.[47,48] They exhibit cellular and humoral defensive mechanisms as well as the acquisition of a protected ("immune") state to bacterial infections.

The insect immunocytes (hemocytes) are efficient in eliminating bacteria, fungi, nematodes, and other foreign particles by either phagocytosis, nodule formation, or encapsulation. The recognition of foreignness is thought to be mediated by certain hemolymph proteins (agglutinins; lectins) that function as opsonins.[49] Lectins, which may play a role in the receptor-mediated endocytosis, also have been found on the cell surface of insect hemocytes.[49-52]

Insect immunocytes, namely, plasmatocytes and granulocytes, are functionally comparable to vertebrate (mammalian) B- and T-lymphocytes.[48] The plasmatocytes perform the analogous killer function and helper-cell-independent cytotoxic function of the T-lymphocytes. The plasmatocytes also perform the functions of vertebrate macrophages. The granulocytes perform the analogous functions of the B-lymphocytes as well as the suppressor functions of the T-lymphocytes. A detailed hypothesis on the evolution of these immunocytes from a primitive arthropod granulocyte was proposed recently.[48]

Insect hemolymph is rich in a polyphenoloxidase that catalyzes, among others, the oxidation of tyrosine to 3,4-dihydroxyphenylalanine. It has been proposed that the activation of this enzyme may have a role in the recognition of foreign particles.[53]

A broad spectrum of antibiotic proteins and peptides are known to be synthesized by insects in response to bacterial infections. For example, the cecropins (3.5 to 4 kDa) and attacins (20 to 23 kDa) in the hemolymph of silkworm, *Hyalophora cecropia,* are bactericidal.[54-60] The site of synthesis of these proteins or related bactericidal proteins appears to be the fat body. Insect lysozymes exhibit properties (thermostability, pH optima, and ionic strength optima) similar to those of chicken egg white lysozyme.[57,61-63]

In spite of some significant progress made in the past decade in our understanding of invertebrate immunity, our present knowledge of cellular recognition and mediation of immune response is lagging severely behind that of mammalian immunity. Future research, therefore, should concentrate on cell-surface and humoral molecules, their characterization, synthesis, regulation, and possible specificity against human pathogens and toxins. Because of the absence of mammalian-type diversification of cell functions, the insect immunocytes may provide an array of molecules for both basic and applied research in immunology. As a reward, one might be able to identify antibiotic molecules that are very different from

TABLE 1
Homologous/Analogous Aspects Between the Neuroendocrine Systems of Insects and Vertebrates[a]

Insects	Vertebrates
Neuroendocrine System	
Axoplasmic neurosecretion flow	Axoplasmic neurosecretion flow
Paired groups of neurosecretory cells in the protocerebrum	Hypothalamic neurosecretory center
Corpus cardiacum	Posterior lobe of pituitary gland
Corpus allatum	Adenohypophysis
	Anterior lobe of hypophysis
Chemistry of Neuropeptides	
Peptidergic neurosecretions; allatostatin/allatinhibin	Oxytocin, vasopressin, somatostatin
Corpus cardiacum secretions	Substance P, glucagon, insulin, secretin
Proctolin	β-Endorphin
Control of Reproductive Activity	
Synthesis of vitellogenins is extraovarial	Synthesis of vitellogenins is extraovarial
Vitellogenins synthesized in fat body	Vitellogenins synthesized in liver
Reproduction cyclic	Reproduction cyclic
Vivipary/ovovivipary	Pregnancy
Reproductive quiescence terminated by denervation of corpus allatum	Reproductive quiescence terminated by denervation of mammary gland or hypophysectomy
Egg diapause hormone secreted by subesophageal ganglion or other parts	Embryonic diapause (delayed implantation of fertilized egg) controlled by hypothalamic-adenohypophysial system

[a] Modified after Reference 65.

those of mammals and perhaps were never acquired by the mammals through the evolutionary process, either deliberately or accidentally. For example, the inability of the human immunodeficiency virus (HIV) to replicate in insect cells[64] might lead to a novel insectan molecular weapon against this deadly virus now threatening millions of people.

2. Neuroendocrine System

"The episodic events, including the molting cycle and metamorphic transformations that lead to the emergence of adult insects, are programmed with greater precision than the developmental steps leading to maturity in most vertebrates. The cyclicality in the reproductive activity of the females of certain insect species resembles that of mammals."[65] Some of the homologies and analogies are shown in Table 1. The identity and precise biological activity of many neurosecretory materials are currently under investigation in many laboratories. One can conclude at this point that the insects possess a very complex array of neurosecretions that may not be very different from those of mammals. It is hoped that future research efforts will be directed toward a clear understanding of these historical molecules and a better understanding of our own molecular systems.

V. SUMMARY

Insects as models for biomedical research offer attractive alternatives to the use of higher animals, particularly in light of dwindling research dollars and increasing protests by animal rights groups. A major advantage of using insect models is the ease with which large numbers of specimens can be reared; reduced per diem costs and space requirements result in significant savings. The flexibility afforded by use of short-generation species also can be exploited using insect models, and experiments can be run with large genetically homogeneous populations. Although insects differ from mammals in their morphology and physiology in a number of ways, it is often easier to isolate physiological or pharmacological systems in insect models.

Scientists have taken advantage of insect models in the past and have made significant discoveries in the biomedical sciences using them. The fruit fly *Drosophila melanogaster* has become the best-known model for laboratory and field studies of genetics. An encyclopedic body of information on *Drosophila* genetic studies is readily available through indexed bibliographies, proving that this insect model is one of the more important findings in the field of genetics. The fruit fly model has been also valuable in testing materials for their mutagenic and carcinogenic properties. Scientists have learned to use insect-pathogen transmission models to screen antipathogen chemical compounds and to diagnose certain human diseases. In addition, insect models have been used to study such diverse fields as the mode of action of steroid hormones and bioluminescence. For example, the role of cyclic nucleotides in insect hormone action provides a basis for studies on the animal hormone serotonin, and determinations of total ATP using insect luciferase have facilitated the estimation of low bacterial numbers in urine, milk, wine, and water. Current emphasis on utilizing insects as models for biomedical research has been in the fields of immunology and neuroendocrinology.

ACKNOWLEDGMENT

The authors wish to thank Taina Litwak of the Walter Reed Biosystematics Unit for her reinterpretation and illustrations of the internal morphology of the housefly.

REFERENCES

1. **Krebs, A. H.,** The August Krogh principle: "For many problems there is an animal on which it can most conveniently be studied," *J. Exp. Zool.,* 194, 221, 1975.
2. **Kaiser, H. E.,** *Species-Specific Potential of Invertebrates for Toxicological Research,* University Park Press, Baltimore, 1980, 1.
3. **Schneider, I.,** personal communication, 1988.
4. **Chapman, R. F.,** *The Insects: Structure and Function,* Elsevier, New York, 1969.
5. **Singh, P.,** Insect diets, in *Advances and Challenges in Insect Rearing,* King, E. G. and Leppla, N. C., Eds., Agriculture Research Service, New Orleans, 1984, 32.
6. **Burton, R. L. and Perkins, W. D.,** Containerization for rearing insects, in *Advances and Challenges in Insect Rearing,* King, E. G. and Leppla, N. C., Eds., Agriculture Research Service, New Orleans, 1984, 51.
7. **Brown, H. E.,** Mass production of screwworm flies, *Cochliomyia hominivorax,* in *Advances and Challenges in Insect Rearing,* King, E. G. and Leppla, N. C., Eds., Agriculture Research Service, New Orleans, 1984, 193.
8. **Harrell, E. A.,** Engineering for insect rearing, in *Advances and Challenges in Insect Rearing,* King, E. G. and Leppla, N. C., Eds., Agriculture Research Service, New Orleans, 1984, sect. 3.
9. **Morgan, T. H.,** Sex-limited inheritance in *Drosophila, Science,* 32, 120, 1910.
10. **Brown, S. W.,** Genetics — the long story, in *History of Entomology,* Smith, R. F., Mittler, T. E., and Smith, C. N., Eds., Annual Reviews, Palo Alto, CA, 1973, 407.
11. **Dobzhansky, T.,** *Genetics and the Origin of Species,* Columbia University Press, New York, 1937.
12. **King, R. C.,** *Drosophila melanogaster:* an introduction, in *Handbook of Genetics,* King, R. C. Ed., Plenum Press, New York, 1975, 625.

13. **Morgan, T. H.**, Complete linkage in the second chromosome of the male of *Drosophila, Science*, 36, 719, 1912.
14. **L'Heritier, P.**, The *Drosophila* viruses, in *Handbook of Genetics*, King, R. C., Ed., Plenum Press, New York, 1975, 813.
15. **Sang, J. H.**, The quantitative nutritional requirements of *Drosophila melanogaster, J. Exp. Biol.*, 33, 45, 1956.
16. **Schneider, I.**, Cell lines derived from late embryonic stages of *Drosophila melanogaster, J. Embryol. Exp. Morphol.*, 27, 353, 1972.
17. **Herskowitz, I. H.**, *Bibliography on the Genetics of* Drosophila, *Part. 2*, Commonwealth Agricultural Bureau, Farham Royal, Slough, Bucks, England, 1952.
18. **Herskowitz, I. H.**, *Bibliography on the Genetics of* Drosophila, *Part 3*, Indiana University Press, Bloomington, IN, 1958.
19. **Herskowitz, I. H.**, *Bibliography on the Genetics of* Drosophila, *Part 4*, McGraw-Hill, New York, 1963.
20. **Herskowitz, I. H.**, *Bibliography on the Genetics of* Drosophila, *Part 5*, Macmillan, New York, 1969.
21. **Herskowitz, I. H.**, *Bibliography on the Genetics of* Drosophila, *Part 6*, Macmillan, New York, 1974.
22. **de G. Mitchell, I. and Combes, R. D.**, Mutation tests with the fruit fly *Drosophila melanogaster*, in *Mutagenicity Testing: A Practical Approach*, Venitt, S. and Parry, J. M., Eds., IRL Press, Washington, D.C., 1984, chap. 6.
23. **Sankaranarayanan, K. and Sobels, F. H.**, Radiation genetics, in *The Genetics and Biology of* Drosophila, Ashburner, M. and Novitsky, E., Eds., Academic Press, New York, 1976.
24. **Auerbach, C.**, The chemical production of mutations, *Science*, 158, 1141, 1967.
25. **Sobels, F. H. and Vogel, E.**, Assaying potential carcinogens with *Drosophila, Environ. Health Perspect.*, 15, 141, 1976.
26. **Vogel, E. and Sobels, F. H.**, The function of *Drosophila* in genetic toxicology testing, in *Chemical Mutagens, Principles and Methods for Their Detection*, Hollaender, A., Ed., Plenum Press, New York, 1976, chap. 38.
27. **Wurgler, F. E., Sobels, F. H., and Vogel, E.**, *Drosophila* as assay system for detecting genetic changes, in *Handbook of Mutagenicity Test Procedures*, Kilbey, B. J., Ed., Elsevier, Amsterdam, 1977, 335.
28. **Harwood, R. F. and James, M. T.**, *Entomology in Human and Animal Health*, Macmillan, New York, 1979, 123.
29. **Christensen, H. A. and Herrer, A.**, The use of phlebotomine sand flies in xenodiagnosis, in *Ecology of the Leishmaniasis*, Colloq. Int. Cent. Natl. Rech. Sci., Paris, No. 239, 129, 1977.
30. **Perez, J. R.**, Human onchocerciasis foci and vectors in the American tropics and subtropics, *Pan Am. Health Organ. Bull.*, 20, 381, 1986.
31. **DeFoliart, G. R., Grimstad, P. R., and Watts, D. M.**, Advances in mosquito-borne arbovirus/vector research, *Ann. Rev. Entomol.*, 32, 479, 1987.
32. **Rosen, L. and Gubler, D.**, The use of mosquitoes to detect and propagate dengue viruses, *Am. J. Trop. Med. Hyg.*, 23, 1153, 1974.
33. **Kopec, S.**, Studies on the necessity of the brain for the inception of insect metamorphosis, *Biol. Bull.*, 42, 322, 1922.
34. **Clever, U. and Karlson, P.**, Induktion von Puff-veranderungen in den Speicheldrusin Chromosomen von *Chironomus tentans* durch Ecdyson, *Exp. Cell Res.*, 20, 623, 1960.
35. **Smith, W. A. and Combest, W. L.**, Role on cyclic nucleotides in hormone action, in *Comprehensive Insect Physiology, Biochemistry and Pharmacology*, Vol. 8, Kerkut, G. A. and Gilbert, L. I., Eds., Pergamon Press, Oxford, 1985, chap. 8.
36. **Terzian, L. A., Ward, P. A., and Stahler, N.**, A new criterion for the selection of compounds for curative activity in *Plasmodium vivax* malaria, *Am. J. Trop. Med. Hyg.*, 31, 692, 1951.
37. **Terzian, L. A.**, A method for screening antimalarial compounds in the mosquito host, *Science*, 106, 449, 1947.
38. **Terzian, L. A. and Weathersby, A. B.**, The action of antimalarial drugs in mosquitoes infected with *Plasmodium falciparum, Am. J. Trop. Med.*, 29, 19, 1949.
39. **Terzian, L. A., Stahler, N., and Weathersby, A. B.**, The action of antimalarial drugs in mosquitoes infected with *Plasmodium gallinaceum, J. Infect. Dis.*, 84, 47, 1949.
40. **Agosin, M.**, Functional role of proteins, in *Biochemistry of Insects*, Rockstein, M., Ed., Academic Press, New York, 1978, chap. 3.
41. **de Wet, J. R., Wood, K. V., Helinski, D. R., and DeLuca, M.**, Cloning firefly luciferase, in *Methods in Enzymology*, DeLuca, M. A. and McElroy, W. D., Eds., Academic Press, Orlando, FL, 1986, chap. 1.
42. Sigma Chemical Company Catalog, St. Louis, MO, 1988, 924.
43. **Stanley, P. E.**, Extraction of adenosine triphosphate from microbial and somatic cells, in *Methods in Enzymology*, DeLuca, M. A. and McElroy, W. D., Eds., Academic Press, Orlando, FL, 1986, chap. 2.

44. **Hanna, B. A.,** Detection of bacteriurea by bioluminescence, in *Methods in Enzymology,* DeLuca, M. A. and McElroy, W. D., Eds., Academic Press, Orlando, FL, 1986, chap. 3.
45. **Schaeffer, J. M.,** Sensitive bioluminescent assay for alpha-bungarotoxin binding sites, in *Methods in Enzymology,* DeLuca, M. A. and McElroy, W. D., Eds., Academic Press, Orlando, FL, 1986, chap. 5.
46. **Leach, F. R. and Webster, J. J.,** Commercially available firefly luciferase reagents, in *Methods in Enzymology,* DeLuca, M. A. and McElroy, W. D., Eds., Academic Press, Orlando, FL, 1986, chap. 6.
47. **Dunn, P. E.,** Biochemical aspects of insect immunology, *Ann. Rev. Entomol.,* 31, 321, 1986.
48. **Gupta, A. P.,** *Hemocytic and Humoral Immunity in Arthropods,* John Wiley & Sons, New York, 1986.
49. **Amirante, G. A. and Mazzalai, F. G.,** Synthesis and localization of hemagglutinins in hemocytes of the cockroach *Leucophaea maderae* L., *Dev. Comp. Immunol.,* 2, 735, 1978.
50. **Komano, H., Nozawa, R., Mizuno, D., and Natori, S.,** Measurement of *Sarcophaga peregrina* lectin under various physiological conditions by radioimmunoassay, *J. Biol. Chem.,* 258, 2143, 1983.
51. **Amirante, G. A.,** Production of heteroagglutinins in hemocytes of *Leucophaea maderae* L., *Experientia,* 32, 526, 1976.
52. **Rowley, A. F. and Ratcliffe, N. A.,** Insect erythrocyte agglutinins. *In vitro* opsonization experiments with *Clitumnus extradentatus* and *Periplaneta americana* haemocytes, *Immunology,* 40, 483, 1980.
53. **Soderhall, K.,** Prophenoloxidase activating system and melanization — a recognition mechanism of arthropods? A review, *Dev. Comp. Immunol.,* 6, 601, 1982.
54. **Boman, H. G., Faye, I., Pye, A., and Rasmuson, T.,** The inducible immunity system of giant silk moths, in *Comparative Pathobiology,* Vol. 4, *Invertebrate Models of Biomedical Research,* Bulla, L.A. and Cheng, T. C., Eds., Plenum Press, New York, 1978, 145.
55. **Boman, H. G. and Hultmark, D.,** Cell-free immunity in insects, *Trends Biochem. Sci.,* 6, 306, 1981.
56. **Boman, H. G. and Steiner, H.,** Humoral immunity in cecropia pupae, in *Current Topics in Microbiology and Immunology,* Henle, W., Hofschneider, P. H., Koprowski, H., Maaloe, O., and Melchers, F., Eds., Springer-Verlag, Berlin, 1981, 75.
57. **Hultmark, D., Steiner, H., Rasmuson, T., and Boman, H. G.,** Insect immunity. Purification and properties of three inducible bactericidal proteins from hemolymph of immunized pupae of *Hyalophora cecropia, Eur. J. Biochem.,* 106, 7, 1980.
58. **Hultmark, D., Engstrom, A., Bennich, H., Kapur, R., and Boman, H. G.,** Insect immunity: isolation and structure of cecropin D and four minor antibacterial components from cecropia pupae, *Eur. J. Biochem.,* 127, 207, 1982.
59. **Hultmark, D., Engstrom, A., Andersson, K., Steiner, H., Bennich, H., and Boman, H. G.,** Insect immunity. Attacins, a family of antibacterial proteins from *Hyalophora cecropia, EMBO J.,* 2, 571, 1983.
60. **Steiner, H., Hultmark, D., Engstrom, A., Bennich, H., and Boman, H. G.,** Sequence and specificity of two antibacterial proteins involved in insect immunity, *Nature (London),* 292, 246, 1981.
61. **Croizier, G. and Croizier, L.,** Purification et comparison immunologique de 2 lysozymes d'insectes, *C. R. Acad. Sci. (Paris),* 286D, 469, 1978.
62. **Jolles, J., Schoentgen, F., Croizier, G., Croizier, L., and Jolles, P.,** Insect lysozymes from three species of lepidoptera: their structural relatedness to the c (chicken) type lysozyme, *J. Mol. Evol.,* 14, 267, 1979.
63. **Powning, R. F. and Davidson, W. J.,** Studies of insect bacteriolytic enzymes — I. Lysozyme in haemolymph of *Galleria mellonella* and *Bombyx mori, Comp. Biochem. Physiol. B.,* 45, 669, 1973.
64. **Srinivasan, A., York, D., and Bohan, C.,** Lack of HIV replication in arthropod cells, *Lancet,* 8541, 1094, 1987.
65. **Scharrer, B.,** Insects as models in neuroendocrine research, *Ann. Rev. Entomol.,* 32, 1, 1987.
66. **Patton, W. S.,** *Insects, Ticks, Mites and Venomous Animals of Medical and Veterinary Importance. II. Public Health,* Croydon, England, 1930.
67. **West, L. S.,** *The Housefly,* Comstock Publishing (Cornell University Press), Ithaca, NY, 1951, 584.
68. **Hewitt, C. G.,** *The Housefly Musca domestica Linn. Its Structure, Habits, Development, Relation to Disease and Control,* Cambridge University Press, London, 1914.

Chapter 6

HOUSEFLY MODEL FOR ORGANOPHOSPHATE POISONING

Yesu T. Das

TABLE OF CONTENTS

I. INTRODUCTION

Animals cannot sustain even brief periods of nervous dysfunction without losing control of their muscular and various other vital functions. It is understandable, therefore, why there has been a natural selection for the production of lethal chemicals targeted at the nervous system. The venoms of spiders and snakes are common examples of these chemicals, where the median lethal dose (LD_{50}) of these poisons for mammals may be as low as 50 to 100 μg/kg.[1] Although the nerve poisons selectively attack different sites within the nervous system, they all lead to the eventual blockade of the transmission of nerve impulses.

Normal transmission of nerve impulses depends upon several factors. In the cholinergic nervous system of animals, an enzyme, acetylcholinesterase (AChE; EC 3.1.1.7), is responsible for terminating the action of the neurotransmitter acetylcholine (ACh) on the acetylcholine receptor (AChR). The undue presence of ACh at the synapse (or junction) leads to excessive stimulation of the AChR and blockade of nerve function. Once the critical role of AChE was known, scientists began designing chemicals that would selectively attack this enzyme. Through the successful work of Schrader[2,3] in Germany, which began about 1937, a generation of chemical poisons, known as organophosphates, was produced that would have tremendous impact upon public health, agriculture, and chemical warfare in the decades to follow.

The production of these toxic chemicals created the need for research efforts directed toward finding ways of protecting or treating the victims of poisoning — accidental or deliberate. Although some success has been achieved,[4,5] more effective chemicals are still being sought. Progress in large-scale screening of candidate compounds has been limited by the cost and inefficiency of the existing animal (mammalian) models. Smaller animals, raised in large numbers at a nominal cost and requiring very small amounts of expensive test chemicals, offer an attractive alternative to the current models. In this chapter, an insect model is described for testing therapeutic chemicals against organophosphates.

II. SUITABILITY OF A CLASS OF ANIMALS

The possible use of a class of animals in the bioassay of a poison or drug depends upon the presence of physiological, biochemical, and pharmacological attributes that can be equated to those of mammals (or human beings). It is, therefore, essential to satisfy these requirements before any serious testing is launched. Because of the historical use of organophosphates in agriculture and public health, much background information exists in the literature regarding the susceptibility of insects to organophosphate poisoning and the criteria for toxicity evaluation. Based upon the available information on the physiological, biochemical, and pharmacological criteria and the experimental results described in this chapter, the following account establishes the housefly as a suitable insect species to use in bioassays.

A. PHYSIOLOGICAL BASIS

Electrical impulses are conveyed from one nerve cell (neuron) to another through chemical messengers (neurotransmitters) that are released momentarily and promptly destroyed at the junctions (synapses).[6,7] Depending upon the location in the body, several chemicals are known to act as neurotransmitters. An acetyl ester of choline, acetylcholine, is the most important neurotransmitter because of its predominance in the central nervous system.[8] Other neurotransmitters include noradrenaline (norepinephrine), dopamine, 5-hydroxytryptamine (serotonin), γ-aminobutyric acid, glutamic acid, and glycine.[9,10]

When an electrical impulse reaches the end of a neuron, a neurotransmitter is released from its pre-synaptic membrane into the synaptic gap (cleft) in sufficient amount ("quantum") to reach the adjacent neural post-synaptic membrane. Upon acting on the post-synaptic

TABLE 1

Comparative Physiological, Biochemical, and Pharmacological Characteristics in Insects and Mammals Relative to Organophosphate Poisoning

Characteristic	Mammals	Insects	Ref.
Physiology/Toxicology			
Symptoms (or signs) of OP poisoning	Convulsions	Tremors	27, 40
	Coma	Paralysis	27, 40
	Salivation	Salivation	40, 41
Primary cause of death	Respiratory failure	Unclear	11—13, 27
Biochemistry of Acetylcholinesterase			
Molecular form	Polymorphic	Polymorphic	23, 24, 42—44
Subunit molecular weight[a]	83,000	82,000	26, 43, 44
K_m	120 μM	141 μM	26, 27
Substrate inhibition	Millimolar	Millimolar	26—29
Edrophonium	Inhibitory	Inhibitory	40, 41
Relative thermostability	Low	High	45, 46
Relative cryostability	Low	High	45, 46
AE-2 antibody epitope	Present	Absent	27, 47, 48
Therapeutics			
Atropine in therapy	Effective	Ineffective	35, 37, 38, 49
2-PAM in therapy	Effective	Effective	27, 35, 49
Maximum protective ratio	125	162	27, 39

[a] Determined by SDS-PAGE.

membrane, the neurotransmitter is destroyed immediately by an enzyme situated on the membrane. Otherwise, the neurotransmitter continues to stimulate the neuron and eventually blocks the transmission of impulses.

In mammals, the major signs of organophosphate poisoning are convulsions and coma. These are comparable to the tremors and paralysis observed in a poisoned housefly. This syndrome is consistent with the theory that the transmission of nerve impulses is affected. The ultimate cause of death in mammals is always asphyxiation due to failure of the respiratory center in the brain.[11] Insects do not have a mammalian-type respiratory system, but rather breathe by diffusion through a tracheal system. Therefore, the ultimate cause of death in insects remains to be understood.[12,13] Nevertheless, the cholinergic nervous systems in both animal groups appear to be accessible to the poison which may lead to fatal consequences (Table 1).

B. BIOCHEMICAL BASIS

As early as 1914, the occurrence of acetylcholine and the need for its rapid inactivation were discovered by Dale.[14] In 1926, Loewi and Navratil[15] found an enzyme in the aqueous extract of frog tissue that could inactivate the neurotransmitter. In the following 20 years, enzymes that possessed activity against acetylcholine and other choline esters were characterized in horse serum,[16] erythrocyte membrane,[17-20] and nerve and muscle tissue.[21] Compared with the enzyme that was characterized initially in the serum, the enzyme that was found on the erythrocytes and in the neuromuscular tissue has greater specific activity toward acetylcholine and later was called acetylcholinesterase (EC 3.1.1.7).[22]

Acetylcholine, being an ester, is broken down easily into an acid (acetic acid) and an alcohol (choline) as shown here.

$$CH_3-\overset{\overset{\displaystyle O}{\|}}{C}-O-CH_2-CH_2-N^+-(CH_3)_3 \ + \ HOH \rightarrow CH_3-\overset{\overset{\displaystyle O}{\|}}{C}-O^-H^+ \ + \ HO-CH_2-CH_2-N^+-(CH_3)_3$$

Acetylcholine   Water   Acetate   Choline

The enzyme-catalyzed hydrolysis of acetylcholine is shown in the following scheme.

$$EH + AX \underset{k_{-1}}{\overset{k_1}{\rightleftharpoons}} EH \cdot AX \overset{k_2}{\underset{HX}{\searrow}} EA \overset{k_3}{\underset{HOH}{\nearrow}} EH + AOH$$

where E = acetylcholinesterase, H = hydrogen, A = acetyl moiety of acetylcholine, X = choline moiety of acetylcholine, HX = choline, O = oxygen, and k = reaction rate constant. The enzyme and the substrate combine to form a complex (EX.AX) that results in the cleavage of acetylcholine. The acetyl group transfers to AChE, and choline leaves. The acetylated AChE, however, is hydrolyzed quickly, and its activity is restored.

Pentavalent phosphorus esters have the ability to phosphorylate a class of proteins, of which AChE is a member. These proteins are characterized by the presence of a reactive serine residue in the peptide chain that is acetylated or phosphorylated. When the enzyme reacts with an organophosphate, an intermediate similar to that of acetylcholine (EA) is formed. However, the phosphorylated enzyme resists hydrolysis and remains inactive. Consequently, the substrate continues to be active at the synapse.

Acetylcholinesterase is a polymorphic enzyme, i.e., existing in different molecular forms.[23,24] The basic unit (monomer) has a molecular weight of approximately 82 to 83 KDa. Aggregates of two monomers (dimers) or two dimers (tetramers) are usually the abundant forms in the animal tissue extracts. Each monomer has one locus (active center) that is reactive with acetylcholine or some other substrate. The inherent rates at which the active centers from different animal species hydrolyze acetylcholine vary considerably. Also, the relative composition of amino acids that make up the protein chain varies slightly among the animal species.[25,26] Nevertheless, the similarities in certain critical characteristics among the enzymes from various sources are far more striking than the dissimilarities (Table 1). For example, housefly AChE has a K_m value of 141 μM[27] compared with 120 μM of fetal bovine serum AChE.[26] A unique phenomenon of enzyme inhibition by its own substrate at millimolar or higher concentrations is exhibited by all the acetylcholinesterases studied.[26-29] Three amino acids, i.e., serine, histidine, and aspartic acid (or glutamic acid), that are arranged spatially in a strategic fashion, are known to be virtually equivalent in all the serine proteases and esterases studied. Thus, a catalytic mechanism ("charge relay system"[30]) that was highly conserved from microbes to mammals is evident in the hydrolytic mechanism of acetylcholinesterases.

C. PHARMACOLOGICAL BASIS

Several chemical compounds react with the active site of the enzyme. Some covalently react with the serine residue, while others bind to subsites adjacent to it. There were two pharmacological criteria that had to be satisfied to support the theory of cholinergic malfunction in the housefly due to organophosphate poisoning. The first criterion was that the inactivation of AChE by a reversible drug (or ligand) should result in a toxic syndrome similar to organophosphate poisoning. The second criterion was that the introduction of exogenous substrate (neurotransmitter) should have similar consequences. Edrophonium

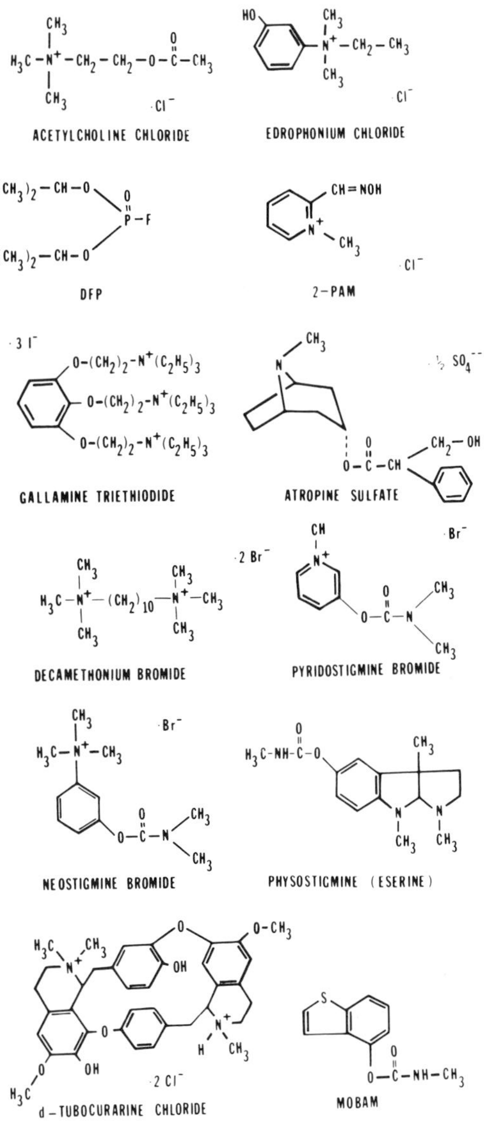

FIGURE 1. Structural formulas of the test compounds.

chloride (Tensilon®) (Figure 1) is a reversible inhibitor of acetylcholinesterase and a drug prescribed as a smooth-muscle relaxant. When this compound was injected into the housefly (88 μg per fly), toxic signs, i.e., leg tremors, paralysis, and death, were identical to those of organophosphate poisoning. A similar situation was encountered when the housefly was injected with acetylcholine chloride ($LD_{50} = 147$ μg per fly), the toxicity being indistinguishable from either that of edrophonium chloride or of an organophosphate. A pharmacological basis thus exists in the housefly for the treatment of organophosphate poisoning.

Other cholinergic ligands, namely, atropine sulfate, decamethonium bromide. d-tubocurarine chloride, and gallamine triethiodide (Figure 1) showed varying levels of paralysis and spasms (Table 2). The folding-in and stretching-out pattern of legs and the prostate or erect position of the wings were also noticeably different among the ligand-induced toxic signs (Figures 2 to 4). Decamethonium bromide appears to be least toxic, presumably due to its inability to penetrate the central nervous system.

TABLE 2
Comparative Effects of Certain Cholinergic Ligands *In Vivo*

Ligand and its known effects in mammals[50-52]	Effects on housefly[27,41]		
	Dose (per fly)	Signs	Recovery
Acetylcholine Chloride			
Cholinergic (AChR agonist) 90—140 mg/min per adult man intravenously causes bradycardia and hypotension	80 μg	Spastic paralysis of proboscis	Spontaneous (15 min)
	160 μg	Spastic paralysis of proboscis and ovipositor; flaccid paralysis of legs and wings	25 min
	240 μg	Spastic paralysis of proboscis, ovipositor, legs, wings, and intestine	None
Atropine Sulfate			
Anticholinergic, smooth-muscle relaxant; mydriatic; suppresses salivary and bronchial secretions	40 μg	Mild spasms; knockdown	15—60 min
Decamethonium Bromide			
Neuromuscular blocking agent (depolarizer of postjunctional membrane)	6.5 μg	Front-leg tremors	10 min
	40 μg	Tremors; no paralysis	15 min
	260 μg	Spastic paralysis of legs, proboscis, ovipositor, and salivation	None
***d*-Tubocurarine Chloride,[a] Gallamine Triethiodide**			
Neuromuscular blocking agent (competitive inhibitor of AChR)	40 μg[b]	Flaccid paralysis; tremors and salivation in case of gallamine	3 h

[a] Complete flaccid paralysis was reported in 16 other insect species.[53]
[b] Less than 40 μg in case of *d*-tubocurarine due to its poor water solubility.

III. CHOICE OF INSECT SPECIES

The use of the housefly for the present purpose was supported by a large number of tests made in the past for insecticidal toxicity evaluation and the mode of action of insecticides. Also, the anatomy of the housefly is known in sufficient detail, and the culture of the flies is an easy and a routine job. The ease with which a large uniform group of test animals can be obtained at a nominal cost and the availability of standard toxicological testing methods make the housefly an immediate choice among more than a million or so insect species that have been described already.

FIGURE 2. Photograph of an adult female housefly receiving an injection of test compound. The syringe needle is shown partly inserted into the thorax.

IV. EXPERIMENTAL STRATEGY

A. CHOICE OF CHEMICALS

The anti-AChE of choice was diisopropylfluorophosphate (DFP) (Figure 1) because of its historical background and its ability to phosphorylate the enzyme stoichiometrically (1:1, based on active site normality).[25,31] Also, the phosphorylated enzyme exhibits a phenomenon known as "aging" in which, with time, reactivation becomes increasingly refractory.[4,32-34] As a reactivating molecule, N-methyl pyridinium-2-aldoxime chloride (2-PAM) has been a prescription drug for several years and, therefore, was an obvious choice. The mechanism of oxime-aided regeneration of the phosphorylated enzyme is shown in Figure 5.

B. DOSE CONSIDERATIONS

By appropriate assays and construction of dose-mortality curves, the LD_{50} values were developed for the test compounds (Figure 6). The LD_{50} of DFP in various studies ranged from 14 ng[27] to 22.5 ng[35,36] per fly (about 0.7 to 1.1 mg/kg based on a body weight of 20 mg per fly). 2-PAM was used at a fixed level of about 30 µg per fly (about 1.5 g/kg), which is approximately $^1/_4$ of the LD_{50}. While the inhibitor could be tested up to levels several-fold the LD_{50} values, restraints were placed on the upper limit of the reactivator levels. It was highly desirable that the maximum dose of the reactivator be well within the LD_{50} and, more importantly, cause no observable signs of toxicity that might interfere with the evaluation.

C. MODE OF ADMINISTRATION

Because of the exoskeleton and its cuticular barrier to the penetration of chemicals,

FIGURE 3.   Photograph of an adult female housefly showing transitory flaccid paralysis of legs and wings after receiving an injection of acetylcholine chloride (160 μg per fly). Note the "folded-in" leg posture.

including several insecticidal compounds, topical applications generally are ruled out. Injection of chemicals into the body cavity (hemocoel) is analogous to the intravenous injection of mammals and is the only sure means of directing the test compound to the target system. Injections, however, can be made into either the abdomen or the thorax or both. In the case of housefly, wherein the abdominal ganglia are fused with those of the thorax, abdominal injections are less effective. Therefore, test compounds were applied effectively via thoracic injections.

The application process is rather simple and easy. The fly was held with a forceps under a binocular microscope using one hand, and the injection is made with the other hand (Figure 4). The microliter syringe needle usually penetrated about 2 mm deep into the thorax via the interscutellar suture. Repeated injections were made through the same puncture without any problem of bleeding (due to low blood pressure).

D. EVALUATION OF TOXICITY AND THERAPEUTIC EFFICACY

When an actively moving housefly contracts the poison from, for example, an insecticide-treated area, it develops severe tremors and goes into total paralysis. This situation is traditionally described by insect toxicologists as a "knockdown" effect. Although it grossly resembles the anesthetic effect of carbon dioxide, unlike it, the knockdown caused by an organophosphate has lethal consequences since the fly does not regain "consciousness". If

FIGURE 4. Photograph of an adult female housefly at death after receiving a fatal dose of 2-PAM (308 μg per fly). Note the "stretched-out" leg posture.

and when it regains consciousness — as in the case of sublethal doses or upon administration of therapeutic chemicals — the fly gradually regains control of the motor function and eventually moves around normally. Depending upon the level of poisoning and/or the efficacy of the drug, recovery may vary from a few minutes to several hours. The paralysis per se does not constitute a lethal effect, unlike the asphyxiation (respiratory failure) in mammals. By this count, insects can withstand the poison much better than mammals and do not undergo the same type of "emergency". Nevertheless, if the poisoning is not alleviated within a reasonable length of time, usually within 24 h, survival chances are not good. For the purpose of scoring the effect of poison or the reactivator, knockdown effect is recorded as positive when the fly does not respond to external stimuli (pricking with a needle or forceps). Flies that do not regain normal activity within 24 h are considered "dead".

E. SCRUTINY OF DATA AND INTERPRETATION OF RESULTS

The efficacy of the therapeutic agent is based on its ability to elevate the normal LD_{50} of the toxicant, i.e., the extent to which the fly would be able to withstand increased levels of the poison under the protection of the drug. The ratio LD_{50} of DFP with 2-PAM over the LD_{50} of DFP without 2-PAM was designated as the "protective ratio" (PR). The larger the value of the PR, the more effective the drug is. The results of tests made with DFP on the housefly by various workers are summarized in Table 3. The maximum PR of 162 is slightly higher than the PR known in mammals; for example, a PR of 125 in mouse (2-PAM plus atropine against paraoxon). However, the level of 2-PAM used in the fly test was twice that used in the mouse test, based on the respective LD_{50} values. (In clinical practice, the 2-PAM dose is usually much lower, 1 to 2 g per adult human being.)

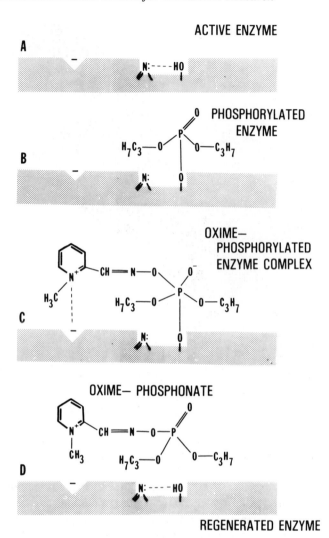

FIGURE 5.   Diagrammatic representation of acetylcholinesterase inhibition and reactivation. (A) The active center of the enzyme is shown as
two subsites: a negative subsite ($-$) serving as the binding or positioning
subsite, and an esteratic subsite (N: $- - -$ HO) serving as the nucleophilic
subsite. (B) DFP covalently attaches itself to the serine oxygen atom, after
losing its fluorine to serine hydrogen. (C) 2-PAM (oxime) covalently
attaches itself to the phosphonate moiety of DFP. The binding of 2-PAM
is aided by the attractive forces between its quaternary nitrogen and the
negative subsite. (D) Oxime-phosphonate complex detaches itself from the
enzyme, leaving the regenerated (active) enzyme. (Modified after Taylor,
P., in *The Pharmacological Basis of Therapeutics*, 6th ed., Gilman,
G. A., Goodman, L. S., and Gilman, A., Eds., Macmillan, New York,
1980, 100.)

Comparisons between animal species must be restricted to studies with comparable methods of treatment, mode of administration, and availability of the baseline LD_{50} values. On
the one hand, in insects, the response to atropine is insignificant.[35,37,38] On the other hand,
in mice, co-administration of atropine could raise the 2-PAM PR against paraoxon from 4.3
to 125.[39] For this reason, 2-PAM (or other reactivator) efficacy in the housefly tests should
be compared only with those mammalian tests wherein atropine was included in the treatment,

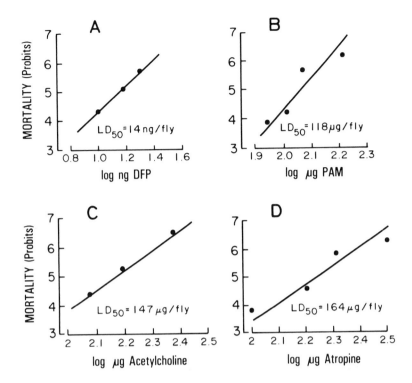

FIGURE 6. Dosage-mortality curves for DFP, 2-PAM, acetylcholine chloride, and atropine sulfate. (A through C are modified after Das, Y. T., Wirtz, R. A., and Andre, R. G., *Comp. Biochem. Physiol.*, 84C, 263, 1986; D is modified after Golenda, C. F., Wirtz, R. A., Andre, R. G., and Roberts, D. R., *Mil. Med.*, 151, 258, 1986; probit analyses were performed as per Finney, D. J., *Probit Analysis — a Statistical Treatment of Sigmoid Response Curve*, Cambridge University Press, London, 1964.)

TABLE 3
Summary of Tests Conducted on Housefly for Diisopropylfluorophosphate Poisoning Therapy

Test conditions	Results

In Vivo Tests

Abdominal injection of 2-PAM	Mild relief from paralysis[54] PR = 1.5[55]
Thoracic injection of 2-PAM	
1 h before DFP	PR = 50[27]
1 h after DFP	PR = 162[27]
Thoracic injection of 2-PAM plus atropine 5 min before DFP	PR = 16[35]
Thoracic injection of physostigmine 15 min before DFP	PR = 4.1[35]
Thoracic injection of pyridostigmine 15 min before DFP	PR = 48.6[36]
Thoracic injection of neostigmine 0—60 min before DFP	No protection[36]
Topical application of Mobam 1 h before DFP	PR = 22.5[36]

In Vitro Tests

2-PAM 1 h after DFP	94% recovery of enzyme activity[55]
2-PAM 3 min before or after DFP	48.5—50.5% recovery of enzyme activity[27]

since the housefly's motor activity entirely correlates with the survival due to 2-PAM and is unaffected by atropine.

V. CONCLUSIONS

The cholinergic nervous system in the housefly appears to play a central role in the muscular activity associated with locomotion (leg and wing movements). When the enzyme that regulates the presence of the neurotransmitter is inhibited, locomotor activity ceases and death ensues. Recovery of the poisoned fly correlates with the regaining of motor activity. Other body functions seem to experience only transitory effects on par with the motor system. An emergency situation, such as respiratory failure in mammals, does not occur in insects, thus obviating the need for a palliative such as atropine.

The cholinergic system in the housefly appears to be comparable with that of mammals based on physiological, biochemical, and pharmacological evidence. The system responds to poisons and therapeutic chemicals in a way comparable to that of mammals. The housefly model is a very useful screening system for organophosphate antidotes.

REFERENCES

1. **Bücherl, W. and Buckley, E. E., Eds.,** *Venomous Animals and Their Venoms,* Vol. 2, Academic Press, New York, 1971.
2. **Schrader, G.,** German Patent 720577, 1942.
3. **Schrader, G.,** *Die Entwicklung ueber insektizider Phosphorsäure-Ester,* Verlag-Chemie, Weinheim, 1963.
4. **Wilson, I. B. and Ginsburg, S.,** A powerful reactivator of alkyl phosphate-inhibited acetylcholinesterase, *Biochim. Biophys. Acta,* 18, 168, 1955.
5. **Ellin, R. I. and Wills, J. H.,** Oximes antagonistic to inhibitors of cholinesterase, *J. Pharm. Sci.,* 53, 1143, 1964.
6. **Eccles, J. C.,** *The Physiology of Synapses,* Springer-Verlag New York, 1964.
7. **Katz, B.,** *Nerve, Muscle and Synapse,* McGraw-Hill, New York, 1966.
8. **Breer, H.,** Neurochemical aspects of cholinergic synapses in the insect brain, in *Arthropod Brain — Its Evolution, Development, Structure and Functions,* Gupta, A. P., Ed., John Wiley & Sons, New York, 1987, 415.
9. **Pitman, R. M.,** Transmitter substances in insects: a review, *Comp. Gen. Pharmacol.,* 2, 347, 1971.
10. **Mercer, A. R.,** Biogenic amines in the insect brain, in *Arthropod Brain: Its Evolution, Development, Structure and Functions,* Gupta, A. P., Ed., John Wiley & Sons, New York, 1987, 399.
11. **De Candole, C. A., Douglas, W. W., Evans, C. L., Holmes, R., Spencer, K. E. V., Torrence, R. W., and Wilson, K. M.,** The failure of respiration in death by anticholinesterase poisoning, *Br. J. Pharmacol.,* 8, 466, 1953.
12. **O'Brien, R. D.,** Esterase inhibition in organophosphorus poisoning of house flies, *J. Econ. Entomol.,* 54, 1161, 1961.
13. **Miller, T. and Kennedy, J. M.,** Flight motor activity of house flies as affected by temperature and insecticides, *Pestic. Biochem. Physiol.,* 2, 206, 1972.
14. **Dale, H. H.,** The action of certain esters of choline and their relation to muscarine, *J. Pharmacol. Exp. Ther.,* 6, 147, 1914.
15. **Loewi, O. and Navratil, E.,** Über humorale Übertragbarkeit der Herznervenwirkung XI über den mechanismus der Vaguswirkung von Physostigmin und Ergotamin, *Pfluegers Arch.,* 214, 689, 1926.
16. **Stedman, E., Stedman, E., and Easson, L. H.,** Choline-esterase. An enzyme present in the blood serum of the horse, *Biochem. J.,* 26, 2056, 1932.
17. **Stedman, E. and Stedman, E.,** The relative choline esterase activities of serum and corpuscles from the blood of certain species, *Biochem. J.,* 29, 2107, 1935.
18. **Alles, G. A. and Hawes, R. C.,** Cholinesterase in the blood of man, *J. Biol. Chem.,* 133, 375, 1940.
19. **Mendel, B., Mundell, D. B., and Rudney, H.,** Studies on cholinesterase. A specific test for true cholinesterase and pseudocholinesterase, *Biochem. J.,* 37, 473, 1943.
20. **Mendel, B. and Rudney, H.,** Studies on cholinesterase. Cholinesterase and pseudocholinesterase, *Biochem. J.,* 37, 59, 1943.

21. **Nachmansohn, D. and Rothenberg, M. A.,** Studies on cholinesterase I on the specificity of the enzyme in nerve tissue, *J. Biol. Chem.,* 158, 653, 1945.

22. **Augustinsson, K.-B. and Nachmansohn, D.,** Distinction between acetylcholinesterase and other choline ester-splitting enzymes, *Science,* 110, 98, 1949.

23. **Massoulié, J. and Bon, S.,** The molecular forms of cholinesterase and acetylcholinesterase in vertebrates, *Ann. Rev. Neurosci.,* 5, 57, 1982.

24. **Brimijoin, S.,** Molecular forms of acetylcholinesterase in brain, nerve and muscle: nature, localization and dynamics, *Prog. Neurobiol.,* 21, 291, 1983.

25. **Rosenberry, T. L.,** Acetylcholinesterase, *Adv. Enzymol.,* 43, 103, 1975.

26. **Ralston, J. S., Rush, R. S., Doctor, B. P., and Wolfe, A. D.,** Acetylcholinesterase from fetal bovine serum — purification and characterization of soluble G_4 enzyme, *J. Biol. Chem.,* 260, 4312, 1985.

27. **Das, Y. T., Wirtz, R. A., and Andre, R. G.,** *In vivo* and *in vitro* assays for organophosphate poisoning therapeutic chemicals using the housefly, *Musca domestica* L., *Comp. Biochem. Physiol.,* 84C, 263, 1986.

28. **Augustinsson, K.-B.,** Substrate concentration and specificity of choline ester-splitting enzymes, *Arch. Biochem. Biophys.,* 23, 111, 1949.

29. **Ellman, G. L., Courtney, K. D., Andres, V., and Featherstone, R. M.,** A new and rapid colorimetric determination of acetylcholinesterase activity, *Biochem. Pharmacol.,* 7, 88, 1961.

30. **Blow, D. M., Birktoft, J. J., and Hartley, B. S.,** The role of buried acid group in the mechanism of action of chymotrypsin, *Nature (London),* 221, 337, 1969.

31. **Oosterban, R. A. and Jansz, H. S.,** Cholinesterases, esterases and lipases, in *Comprehensive Biochemistry,* Vol. 16, Florkin, M. and Stotz, E. M., Eds., Elsevier, New York, 1965, chap. 1.

32. **Davies, D. R. and Green, A. L.,** The kinetics of reactivation, by oximes, of cholinesterase inhibited by organophosphorus compounds, *Biochem. J.,* 63, 529, 1956.

33. **Hobbiger, F.,** Effect of nicotinhydroxamic acid methiodide on human plasma cholinesterase inhibited by organophosphates containing a dialkylphosphato group, *Br. J. Pharmacol.,* 10, 356, 1955.

34. **Hobbiger, F.,** Chemical reactivation of phosphorylated human and bovine true cholinesterases, *Br. J. Pharmacol.,* 11, 295, 1956.

35. **Golenda, C. F., Wirtz, R. A., Andre, R. G., and Roberts, D. R.,** An insect bioassay as a primary screen for nerve agent antidotes, *Mil. Med.,* 151, 258, 1986.

36. **Golenda, C. F., Wirtz, R. A., and Andre, R. G.,** Antagonism of diisopropyl fluorophosphate (DFP) poisoning by carbamate pretreatments in house flies *(Musca domestica* L.), *Comp. Biochem. Physiol.,* 88C, 61, 1987.

37. **Metcalf, R. L. and March, R. M.,** Studies on the mode of action of parathion and its derivatives and their toxicity to insects, *J. Econ. Entomol.,* 42, 721, 1949.

38. **Barker, R. J.,** Cholinesterase reactivators tested as antidotes for use on poisoned honey bees, *J. Econ. Entomol.,* 63, 1831, 1970.

39. **Bošković, B., Tadić, V. and Kušić, R.,** Reactivating and protective effects of pro-2-PAM in mice poisoned with paraoxon, *Toxicol. Appl. Pharmacol.,* 55, 32, 1980.

40. **Taylor, P.,** Anticholinesterase agents, in *The Pharmacological Basis of Therapeutics,* 6th ed., Gilman, G. A., Goodman, L. S., and Gilman, A., Eds., Macmillan, New York, 1980, 100.

41. **Das, Y. T.,** unpublished data, 1988.

42. **Steele, R. W. and Smallman, B. N.,** Acetylcholinesterase from the house fly head — molecular properties of soluble forms, *Biochim. Biophys. Acta,* 445, 131, 1976.

43. **Tripathi, R. K. and O'Brien, R. D.,** Purification of acetylcholinesterase from house fly brain by affinity chromatography, *Biochim. Biophys. Acta,* 480, 382, 1977.

44. **Tripathi, R. K., Telford, J. N., and O'Brien, R. D.,** Molecular and structural characteristics of house fly brain acetylcholinesterase, *Biochim. Biophys. Acta,* 525, 103, 1978.

45. **Das, Y. T., Andre, R. G., and Wirtz, R. A.,** Thermolability and cryostability of acetylcholinesterases, *Fed. Proc. Fed. Am. Soc. Exp. Biol.,* 45, 1653, 1986.

46. **Das, Y. T., Andre, R. G., and Wirtz, R. A.,** Kinetics of thermal inactivation of acetylcholinesterase, in preparation, 1989.

47. **Fambrough, D. M., Engel, A. G., and Rosenberry, T. L.,** Acetylcholinesterase of human erythrocytes and neuromuscular junctions — homologies revealed by monoclonal antibodies, *Proc. Natl. Acad. Sci. U.S.A.,* 79, 1078, 1982.

48. **Chhajlani, V., Olson, C. E., Earles, B., Deschamps, J., Derr, D., and August, J. T.,** Biochemical and immunological studies of the active site of human acetylcholinesterase, 5th Annu. Chemical Defense Biosci. Rev., U.S. Army Med.Res. Dev. Command, Columbia, MD, May 29 — 31, 1985.

49. **Hayes, W. J., Jr.,** *Toxicology of Pesticides,* Williams & Wilkins, Baltimore, 1975.

50. **Taylor, P.,** Cholinergic agonists, in *The Pharmacological Basis of Therapeutics,* 6th ed., Gilman, G. A., Goodman, L. S., and Gilman, A., Eds., Macmillan, New York, 1980, 91.

51. **Taylor, P.,** Ganglionic stimulating and blocking agents, in *The Pharmacological Basis of Therapeutics,* 6th ed., Gilman, G. A., Goodman, L. S., and Gilman, A., Eds., Macmillan, New York, 1980, 211.

52. **Taylor, P.,** Neuromuscular blocking agents, in *The Pharmacological Basis of Therapeutics,* 6th ed., Gilman, G. A., Goodman, L. S., and Gilman, A., Eds., Macmillan, New York, 1980, 220.

53. **Larsen, J. R., Miller, D. M., and Yamamoto, T.,** *d*-Tubocurarine chloride: effect on insects, *Science,* 152, 225, 1966.

54. **Winteringham, F. P. W., Harrison, A., McKay, M. A., and Weatherley, A.,** Biochemistry of diiso-propylphosphorofluoridate poisoning in the adult house fly, *Biochem. J.,* 66, 49P, 1957.

55. **Mengle, D. C. and O'Brien, R. D.,** The spontaeous and induced recovery of fly brain acetylcholinesterase after inhibition by organophosphates, *Biochem. J.,* 75, 201, 1960.

56. **Finney, D. J.,** *Probit Analysis — a Statistical Treatment of Sigmoid Response Curve,* Cambridge University Press, London, 1964.

Chapter 7

FERTILIZATION IN ASCIDIANS

Brian Dale

TABLE OF CONTENTS

I. INTRODUCTION

Over the past 10 years much progress has been made in the treatment of several forms of sterility. *In vivo* culture of human gametes and embryos is a routine procedure practiced in hundreds of laboratories worldwide; nonetheless, the success rates for the variety of protocols employed in gamete and embryo transfer remain low. Basic knowledge in this field is minimal, and the difficulty of obtaining material, together with ethical problems, impede research. Studies on human gametes will be essential for resolving some questions, for example, those related to species-specific surface receptors; however, much information is to be gained on the mechanism of fertilization using an invertebrate model system.

Several areas of research with immediate clinical application spring to mind. Following retrieval, human oocytes are graded and selected for transfer according to their morphological characteristics. Criteria used vary from unit to unit but basically revolve around the meiotic status of the oocyte,[1] and/or the appearance of the oocyte-corona-cumulus complex of accessory cells.[2,3] Inaccurate assessment of the meiotic status of the oocyte may result in failed or polyspermic fertilization;[4,5] of greater importance, however, is the cytoplasmic maturity of the oocyte.[3,6,7] Most *in vitro* fertilization (IVF) clinics use gonadotrophins to stimulate the ovary in order to recruit and develop multiple follicles,[8] which creates asynchrony in follicular growth and both meiotic and cytoplasmic maturation of the oocytes.[2,3] Present knowledge on the mechansim of gonadotrophin-stimulated oocyte maturation is fragmentary, and methods used in clinical situations are not sensitive or sufficiently specific to determine oocyte viability accurately. Cytoplasmic maturity of oocytes is reflected by the electrical properties of the plasma membrane, and noninvasive electrophysiological techniques may be instrumental in helping to design protocols for the classification of human oocytes.[9]

IVF programs, particularly where techniques of gamete intrafallopian transfer (GIFT)[10] are involved, offer a remedy in cases where oligospermy (low sperm count) is a determining factor, since few spermatozoa are required. Where spermatozoa number and motility are extremely poor, fertilization may be achieved by microinjecting a limited number of spermatozoa under the zona pellucida[11] into close contact with the oocyte plasma membrane. It is becoming increasingly apparent that spermatozoa penetrate the egg surface at specific points, and therefore information is required about the distribution, structure, and function of these preferential "hot spots". Furthermore, little is known about the fertilizing capacity of spermatozoa; that is, normal morphology and motility of a spermatozoon do not necessarily assure its fertilizing capacity.

Parthenogenetic activation of eggs from a wide variety of species, leading to pronuclear movements, early cleavage, and then developmental arrest, tells us that early cleavage is not a reliable indication of embryo viability. What seems to be important is that the egg is correctly and sufficiently stimulated by the fertilizing spermatozoon and the signal transduced to set in motion the metabolic processes required to support early embryogenesis until exogenous nutritive sources are established. How a spermatozoon triggers an egg into metabolic activity is unknown, and research in this direction will have profound implications for the study of embryo viability and contraception.

Perhaps the greatest challenge to developmental biology is to understand the mechanisms involved in differential gene expression. Wilson[12] and Morgan[13] were among the first to suggest that cytoplasmic determinants localized in different parts of the zygote promoted specific patterns of gene expression in each cell lineage. Although embryology is still paying the consequences of a classification system drawn from a century of descriptive studies, the advent and application of molecular biological techniques have tended to amalgamate these rigid categories. Terms such as protostome and deuterostome, regulative and mosaic, may remain as taxonomic labels; however, it is becoming clear that embryogenesis in the animal kingdom adheres to a few basic principles and what might differ is simply the time when these mechanisms are expressed.

II. ASCIDIANS

Ascidians, or sea squirts as they are often called, are sessile, barrel-like marine invertebrates commonly found throughout the world. The class Ascidiacea is one of three classes belonging to the subphylum Urochordata and contains about 1600 species. Ascidians lack a backbone, but possess three distinguishing characteristics during their larval stage, a notochord, a dorsal hollow nerve cord, and a pharyngeal cleft, which establish their claims as chordates.

Most sea squirts are found in coastal waters attached to rocks, pilings, or beneath boats by one end of their tubular structure. At the opposite end two siphons extend into the water, the buccal and cloacal siphons, which provide a current of water through the animal. The entire body is covered by a specialized and often hard, colored mantle, hence the name *tunicate*. The principle components of the tunic are a form of cellulose, proteins, and in some cases calcium spicules. When an animal is exposed at low tide or taken out of the water, contractions of the body wall and siphons cause the water in the animal to be forced from the siphons, hence their second common name *sea squirt*.

Most tunicates are hermaphroditic, with a single testis and ovary lying in close contact with the digestive loop (Figure 1). The ovary is located above the stomach and is sac-like with two ventrolateral bands, the germinal areas. Mature eggs are carried into the oviduct that runs parallel to the intestine and opens into the cloaca in front of the anus. The testis lies below the ovary and is composed of a cluster of small sacs opening into a sperm duct. This duct runs parallel to the oviduct and opens into the cloaca.

Perhaps the most common solitary ascidian is *Ciona intestinalis,* which is tubular rather than barrel-shaped with a softish, yellow-to-transparent tunic. Other commonly used species are *Styela partita, Phallusia mammillata, Ascidia nigra, Ascidia malaca, Ascidiella aspersa,* and *Halocynthia pyriformis.* Although the ideal workplace for ascidian embryologists is a marine biology station, these solitary ascidians are quite sturdy and may be kept in artificial aquariums. It must be remembered that tunicates are filter feeders, removing plankton from the current of water that passes through their pharynx. A large volume of water is strained for food; for example, a specimen of *Phallusia* may pass over 150 l of water through its body in 1 d. Thus an adeqate aquarium is required, together with a food source.

A. GAMETES

Spermatozoa and eggs are easily obtained from ascidians kept in the laboratory. Some species spawn at set times, apparently facilitated by light; however, in all animals unfertilized gametes may be obtained by dissection. In *Ciona intestinalis,* ripe gametes in the sperm and oviducts are usually visible through the tunic as white and red/grey lines, respectively. The tunic is removed by hand, and the body wall cut to expose the gonoducts. A Pasteur pipette can then be used to withdraw the eggs and, subsequently, the spermatozoa. *Ciona* is, in the majority of cases, a self-sterile species; therefore, contamination of eggs with its own sperm is not usually a problem. Other species are self-fertile and thus care should be taken during the excision of gametes. Eggs are placed in filtered natural seawater at room temperature (ideally 18 to 22°C), while the spermatozoa are kept in their concentrated form, often called "dry", until required. Spermatozoa at source are packed at a concentration of 10^9 to 10^{10}/ml and have to be diluted at least 1000 times in filtered natural seawater before use. Once diluted, spermatozoa of *C. intestinalis* are viable for up to 2 h; however, it is common practice to use freshly diluted spermatozoa for each insemination.

At the end of its growth period the oocyte leaves the ovary and enters the oviduct, where the germinal vesicle breaks down and the meiotic spindle forms. The egg remains

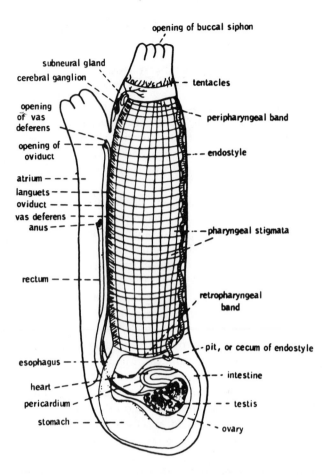

FIGURE 1.   Lateral view of the solitary ascidian *Ciona intestinalis* to
show the position of the gonads and gonoducts. (From Barnes, R. D.,
Invertebrate Zoology, W. B. Saunders, Philadelphia, 1969, 682. With
permission.)

blocked at first metaphase until fertilization. Ascidian eggs are large, sometimes transparent
(see Table 1), and surrounded by several coats or envelopes. Proximal to the egg surface
is a layer of pigmented and vacuolated cells, the *test cells,* which float free in the perivitelline
space. Next is a tough glycoprotein sheet, the *chorion,* which is functionally equivalent to
the *zona pellucida* of mammalian eggs. There is an external layer of long digitate cells which
contain a drop of oil at the distal tip. These *follicle cells* are thought to serve for floatation
of the egg (Figure 2).

Perhaps the most valuable attribute of ascidian eggs is that they permit mechanical
removal of the external surface coats, leaving a ''nude'' egg. Eggs are placed on a glass
slide covered with a thin layer of agar. Two sharpened steel or tungsten needles are used,
one to trap the egg against the surface by the chorion, the second to cut a small lesion in
the chorion. The second needle is then used to apply pressure and thus squeeze the egg
through the damaged chorion. Although deformed, the nude egg rounds up within 2 min
and upon insemination will give rise to a perfectly normal larva. Ascidian spermatozoa are
not untypical of many animals, having an elongated head piece and a long single tail.
Somewhat unusual is the presence of a large mitochondrion which is not situated in the
intermediate section but in the head, practically enveloping the sperm nucleus. Many authors
claim that *Ciona* spermatozoa possess an acrosome-like structure at the tip of the head,
which presumably plays a role in penetration of the egg investments (Figure 3).

TABLE 1
Color and Size of Ascidian Eggs[17]

Egg color and kind	Size (nm)
Clear and colorless	
Ascidiella aspersa	0.17
A. scabra	0.17
Diazona violacea	0.16
Clear with a faint greenish tinge	
Phallusia mammillata	0.16
Ascidia mentula	0.15
Clear with a bright red pigment	
Ascidia conchilega	0.13
Translucent with greenish or reddish pigment	
Ciona intestinalis	0.16
Semiopaque but without pigment	
Boltenia echinata	0.17
Ascidia prunum	0.18
A. obliqua	0.15
Opaque pigmented; yellowish pigment occasionally present	
Polycarpa fibrosa	0.16
Molgula manhattensis	0.11
M. ampulloides	0.11
M. arenata	0.11
M. occulta	0.11
M. bleizi	0.15
Clavelina lepadiformis	0.26
Yellow or reddish pigment	
Tethyum pyriforme americanum	0.26
Molgula citrina	0.20
Stolonica socialis	0.72
Reddish purple pigment	
Molgula robusta	*0.12*
M. oculata	0.11
M. retortiformis	0.18
Polycarpa rustica	0.18
Styelopis grossularia	0.48
Boltenia ovifera	0.16
Distomus variolosus	0.59
Amaroucium nordmanni	0.38

B. DEVELOPMENT

In ascidians, monospermy is the rule; polyspermy only occurs in aged or damaged eggs. The fertilizing spermatozoon normally enters the egg in a limited region of the vegetal hemisphere;[14] however, under experimental manipulation spermatozoa may also enter the animal hemisphere. The first signs of activation are a cortical contraction about 2 min after insemination, followed by expulsion of the first and second polar bodies at 10 and 20 min, respectively, and cleavage at about 50 min. Cleavage is complete, slightly unequal, and leads to a flattened coeloblastula. Gastrulation is accomplished by epiboly and invagination, and the large archenteron completely obliterates the old blastocoel. The blastopore marks the posterior end of the embryo and gradually closes while the embryo elongates along the anterior-posterior axis. Along the mid-dorsal line, the archenteron gives rise to a supporting rod, and notochord. A cord of laterally placed cells along each side of the body is formed from the mesoderm, while the ectoderm along the mid-dorsal line differentiates as a neural tube.

Development then gives rise to a free-swimming larva with a tail containing the notochord and neural tube and an anterior region with sensory organs (Figure 4). The entire larva is

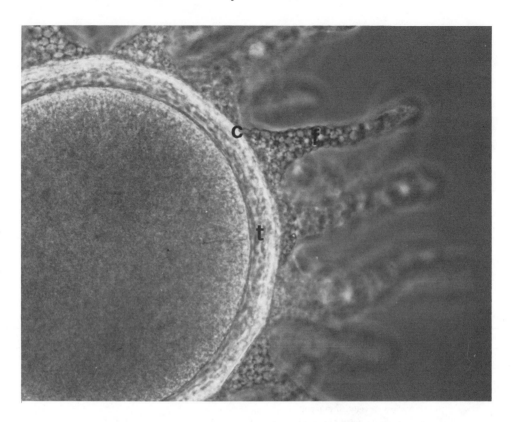

FIGURE 2.   A mature egg of *C. intestinalis* showing the test cells (t), chorion (c), and follicle cells (f). Egg diameter is 140 μm.

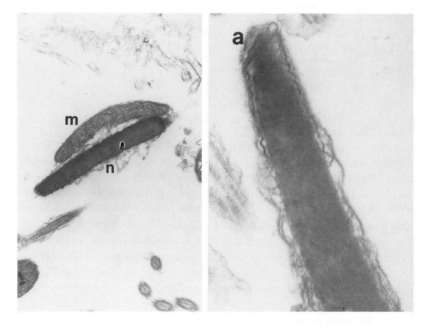

FIGURE 3.   Electron micrographs of *Ciona* spermatozoa showing the nucleus (n) and associated single mitochondrion (m), and at higher magnification to show the acrosome (a). (Courtesy of R. De Santis and R. Pinto.)

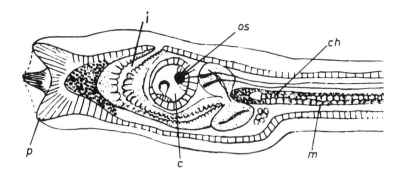

FIGURE 4. Diagrammatic representation of the head of an ascidian tadpole larva; p = palps, c = brain, m = muscle, ch = chorda, os = eye, i = intestine. (From Reverberi, G., *Experimental Embryology of Marine and Freshwater Invertebrates,* North-Holland, Amsterdam, 1971. With permission.)

covered by a tunic secreted by the surface ectoderm. After a short stay in the plankton, the larva settles by its anterior end to a suitable substrate and undergoes metamorphosis. The tail, with neural tube and notochord, are resorbed, and the mouth is carried to the opposite, free end of the animal to become the functional buccal siphon.

III. SPERM-EGG INTERACTION

Spermatozoa are maintained in the testis in a quiescent state. Many factors may be responsible for this metabolic suppression, such as physical restraint, pH, and low oxygen tension of the seminal fluid. In the Japanese tunicates *C. savigny* and *C. intestinalis,* experiments suggest that motility is suppressed by a high K^+ concentration in the seminal fluid.[15] A transmembrane exchange of K^+ and Ca^{2+} at spawning may lead to an increased production of cAMP that directly triggers sperm motility. Once released and motile, the spermatozoa must contact the spawned eggs. For decades it has been assumed that sperm-egg interaction depends upon a random collision of gametes; however, evidence is accumulating for the existence of chemotactic factors. Miller,[16] in a series of very convincing experiments, has assigned a chemotactic role to the follicle cells of *C. intestinalis* eggs, while many researchers from Reverberis Laboratory in Palermo maintain that eggs release factors into the seawater that have the capacity to stimulate spermatozoa into activity (see Reverberi[17]).

The primary event of fertilization is the species-specific recognition and binding of spermatozoa and eggs. It has been shown in ascidians that the chorion is the structure responsible for these processes[18,19] and that specific sites or sperm receptors are to be found on this glycoprotein sheet (Figure 5). Binding has been suggested to be mediated by the interaction of fucosidase on the sperm head and fucosyl sites on the vitelline coat[20,21] Once bound, the ascidian spermatozoon undergoes a number of changes described as sperm activation. First, the single mitochondrion swells and is displaced along the head toward the tail,[22-24] possibly by a mechanism involving actin and myosin[25] (Figure 6). Second, the spermatozoon generates an excess of heat that does not correlate with oxygen consumption and may result from an alteration to its metabolism,[26] linked to the mitochondrial movement. Extracellular Ca^{2+} is required for binding in *Ciona* gametes, and an influx or release of Ca^{2+} from cytoplasmic stores in the spermatozoon is a prerequisite for sperm activation.[24]

The last event of sperm activation is the acrosome reaction. Despite a decade of contention, it is now generally accepted that ascidian spermatozoa, in particular those of *C. intestinalis,* possess an acrosome.[18,26-38] The head of *C. intestinalis* spermatozoa is elongated with a wedge-shaped tip and a single mitochondrion, laterally placed to the nucleus

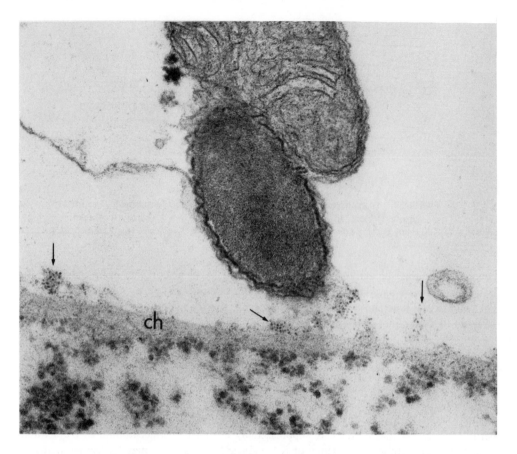

FIGURE 5.   A glycerol treated egg of *C. intestinalis* after treatment with ferritin-conjugated Con A showing a bound spermatozoon. The sperm binding sites, which are positive to Con A, are the tufts of filaments on the outer surface of the chorion (ch—arrows; magnification × 82,000). (From De Santis, R., Jamunno, G., and Rosati, F., *Dev. Biol.* 74, 490, 1980. With permission.)

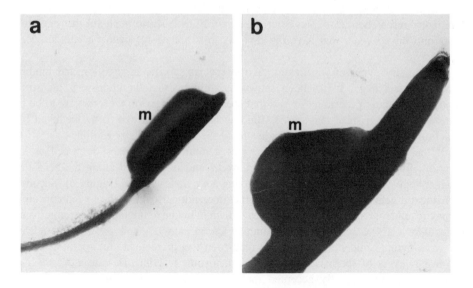

FIGURE 6.   (a) Control spermatozoon (magnification × 13,000); and (b) an activated spermatozoon showing the displaced mitochrondrion (m) (magnification × 28,000). (From Casazza, G., De Santis, R., and Pinto, M., *Exp. Cell Res.*, 148, 508, 1983. With permission.)

(Figure 3). In sagittal and transverse sections through the head, a flattened vesicle 200 nm long, 180 nm wide, and 40 nm high may be observed, which is the acrosome (Figure 3). The outer membrane of the acrosome is in close contact with the plasmalemma, while a gap between the acrosomal inner membrane and the nuclear membrane corresponds to the subacrosomal space of other animals. Electron-dense material fills the acrosomal vesicle, while the subacrosomal space appears to be empty. Although we cannot be exactly certain when the acrosome reaction occurs, spermatozoa of *Ciona* attached to the chorion present an altered acrosomal region.[18,34,36] This was found not to be the case[1] in *Phallusia* spermatozoa.[31] What is of interest is the fact that not all spermatozoa attached to the chorion undergo the acrosome reaction and penetrate this investment,[31] suggesting that a precise molecular match between receptors on the chorion and those on the sperm surface is required for this reaction.[39]

In many animals, lysins are closely associated with the acrosome. These lysins are released at some stage of fertilization to act upon the investments allowing the spermatozoon to penetrate. At present, in *Ciona*, we do not know whether presumptive lysins are bound to the plasmalemma or contained within the acrosome. A second fundamental function of the acrosome reaction is to expose a region of sperm membrane that is capable of fusing with the egg plasma membrane. Once through the chorion into the perivitelline space, the successful spermatozoon makes its way to the egg surface. Conklin was the first to describe sperm entry into ascidian eggs, and according to his classical paper,[14] the fertilizing spermatozoon enters an arc of 30° at the vegetal pole. The process of sperm-egg fusion is by no means understood. Most eggs appear to have preferential entry sites; however, it is also clear that spermatozoa may enter other areas of the egg under certain experimental conditions. For example, the unfertilized egg of *Ciona* may be cut into segments, one originating from the animal hemisphere, the second from the vegetal hemisphere. Both segments may be inseminated and give rise to normal larvae.[40,41] Polyspermy is extremely rare in ascidian eggs, and nude eggs challenged by hundreds of active spermatozoa are invariably monospermic. Dispermy may occur in aged nude eggs, and then the two spermatozoa usually enter the vegetal hemisphere.[14] Much will be gained from the study of sperm entry sites in ascidian eggs, and the simple technique of egg dissection will certainly be instrumental.

IV. ACTIVATION OF THE EGG

A. MORPHOLOGICAL EVENTS

The first indication that an ascidian egg has been successfully fertilized is a rapid modification of its shape. At about 2 min after insemination, a constriction appears at the animal pole which slowly traverses the egg, reaching the vegetal pole after 30 s to 2 min. The egg appears dumbbell-shaped when the constriction reaches the equator, and pear-shaped when at the vegetal hemisphere[42] (Figure 7). After the contraction wave has traversed the egg, a clear cytoplasmic bulge can often be observed at the vegetal pole. Before expulsion of the first polar body, at about 10 min after insemination, the egg assumes a spherical shape.[17,42]

Conklin[14] was one of the first to comment on movements of the egg cortex at fertilization, noting a displacement of the test cells. Ortolani[43-45] took this observation further by placing chalk granules on the surface of unfertilized eggs. The chalk granules attach firmly to the surface and were rapidly displaced at fertilization.[43] Granules at the animal hemisphere spread out, suggesting an expansion of the cortex, while those at the vegetal hemisphere accumulated, indicating a contraction of the surface. Furthermore, these movements were modified by pretreating the eggs with trypsin, suggesting that surface proteins are involved in cortical contraction. Cortical contraction is inhibited in the presence of cytochalasin B,[42,46,47] implicating microfilaments as the motive force, but not by colchicine,[46] which blocks microtubule function.

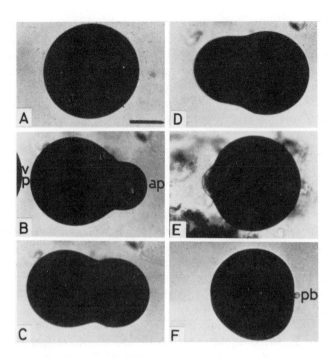

FIGURE 7.   The activation wave of contraction in an egg of *C. intes-tinalis*; vp = vegetal pole, ap = animal pole, pb = 1st polar body; the bar in A = 50 μm. (From Sawada, T. and Osanai, K., *Wilhelm Roux Arch. Entwicklungsmech. Org.*, 190, 208, 1981. With permission.)

The surface of unfertilized nude ascidian eggs is relatively smooth with no[48] or few microvilli.[42] Ascidian eggs do not have cortical granules, nor any comparable organelle, although numerous small vesicles are present in the cortex.[48] Mitochondria are abundant and densely packed in the subcortical region, as are yolk granules, whereas the cytoplasm generally is rich in smooth endoplasmic reticulum, and occasional Golgi complexes are found.[48] In longitudinal sections of unfertilized eggs, the mitochondria appear as a layer in the subcortical region, except for a region at the animal pole.[42] At fertilization, it appears that many of the vesicles present in the cortex fuse with the plasma membrane, since at 5 to 10 min after insemination, the number of vesicles decreases.[48] The surface becomes smoother as the cortical contraction wave passes over the egg (Figure 8), and the subcortical mitochondria and granules are transported to the vegetal pole.[42,49] This movement of organelles is called *cytoplasmic segregation* and will be dealt within the next section.

B. ELECTRICAL EVENTS

Since the egg plasma membrane is distributed between the membranes of the blastomeres, it is not surprising that many ion channels, typical of differentiated tissues, are found in the unfertilized egg. The unfertilized ascidian egg membrane has been the focus of numerous biophysical studies which have characterized endogenous voltage-gated Na, Ca, and K channels.[50-62] Of relevance to early differentiation is the topographical distribution of these channels in the zygote, and this will be dealt with later. As far as fertilization itself is concerned, a new population of ion channels, the ''fertilization channels'', appear in the membrane. Seconds after insemination in *C. intestinalis,* a small step depolarization may be measured, followed some 5 to 7 s later by a much larger overshooting depolarization (Figure 9). The potential remains at a positive value for several minutes and then gradually returns to its original value.[63] Subsequent studies, using the patch clamp technique, showed

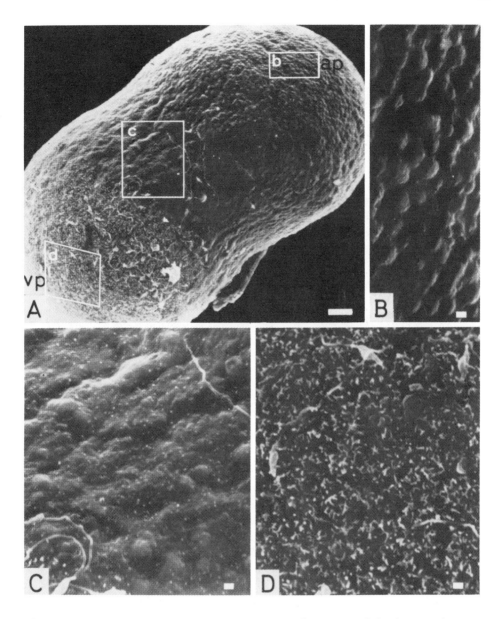

FIGURE 8. (A) Scanning electron micrographs of the surface of a *Ciona* egg during the contraction wave; ap = animal pole, vp = vegetal pole, bar = 10 μm; B, C, and D are micrographs at higher magnification of the areas indicated; bars = 1 μm. (From Sawada, T. and Osanai, K., *Wilhelm Roux Arch. Entwicklungsmech. Org.*, 190, 208, 1981. With permission.)

that these fertilization channels are large, with a single channel conductance of 400 pS[64] (Figure 9). Since their reversal potential was around 0 mV, it was suggested that these channels were not ion specific. Whole-cell currents were shown to peak near −30 mV and approach zero near 0 mV,[65] supporting the preliminary observations on these channels. Knowing the total conductance change at fertilization, the single channel conductance and the probability of a channel being open, it was possible to estimate that the fertilizing spermatozoon opens between 200 and 2000 fertilization channels.

By cutting unfertilized eggs into fragments and inseminating each fragment, it was found that fertilization channel precursors and voltage-gated ion channels are uniformly distributed around the ascidian egg surface.[66-68] Since fertilization currents were similar in whole eggs

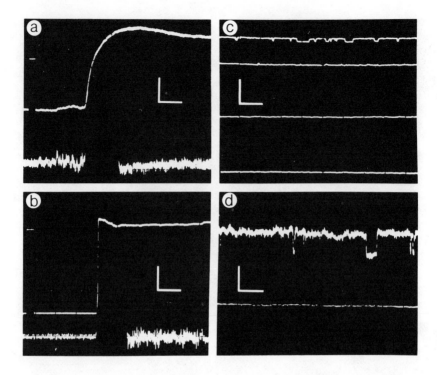

FIGURE 9.   Electrical events during fertilization of *C. intestinalis* eggs. (a) and (b) show
voltage changes from a low resting potential and a high resting potential egg, respectively; (c)
and (d) show the activity of a single fertilization channel measured with the patch clamp
technique. In (a) the vertical bar represents 15 mV, in (b), 30 mV; the horizontal bar represents
10 s for both. In (c) the vertical bar = 100 pA, and 10 pA for (d); the horizontal bar in both
traces = 40 ms. Note that in (a) and (b) the potential depolarizes at fertilization, overshooting
0 mV (the small dashes to the left of each recording); in (a) the small step proceeds the larger
event. (From Dale, B. and De Felice, L., *Dev. Biol.*, 101, 235, 1984. With permission.)

or fragments, irrespective of their size and global origin, it was concluded that the fertilizing
spermatozoon opens a fixed number of fertilization channels limited to an area around its
point of entry.[67] The spermatozoon enters the intact egg at the vegetal pole[14] and creates a
localized ion current[67] that might regulate movements of the cytoskeleton involved in cytoplasmic segregation. Studies on the gating mechanism of fertilization channels have shown
that intracellular Ca^{2+} is not involved,[69] whereas inositol triphosphate (IP_3) is a putative
second messenger (Figure 10). It is possible that IP_3 or other intermediates in phosphoinositide
metabolism are contained in the spermatozoon and delivered to the egg cytoplasm following
gamete fusion.[70]

Within 2 min of insemination there is a large transient increase in intracellular Ca^{2+}
that starts at the animal pole and sweeps across the egg in 10 s to the vegetal pole.[71-73]
Intracellular Ca^{2+} remains high for about 5 to 10 min and then slowly returns to its resting
level before expulsion of the first polar body. Measurements with the fluorescent indicator
Fura-2 show that Ca^{2+} increases from a resting level of 1 nM to 1 μM, a 1000-fold increase.
Although this elevated Ca does not gate fertilization channels, it appears to be involved in
the mechanism of cortical contraction.[42,70,73]

C. METABOLIC EVENTS

To date there is little information on the mechanism of metabolic derepression in ascidian
eggs. In many species the primary signal from the spermatozoon is transduced to the egg
via an increase in intracellular pH. This appears not to be the case in ascidian eggs. Using

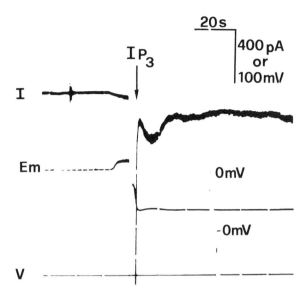

FIGURE 10. Ion current (I) in an egg of *C. intestinalis* whole-cell voltage clamped at $-80\,mV$ (V). A second electrode containing 10 μM inositol triphosphate was inserted into the egg (arrow), at which point the egg contracted and generated a slow inward current of about 100 pA. The second electrode also served to measure the cell potential (E_m). A similar current is recorded during fertilization. (From Dale, B., *Exp. Cell Res.*, 177, 205, 1988. With permission.)

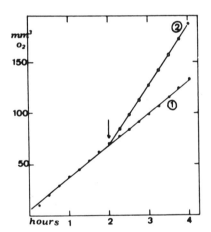

FIGURE 11. Respiration curve of an ascidian egg before (1) and after (2) fertilization. (From Minganti, A., *Acta Embryol. Morphol. Exp.*, 1, 150, 1957. With permission.)

ion-selective intracellular microelectrodes, it has been shown that the pH of unfertilized eggs ranges from 7.2 to 7.5 and that there is no variation in this value at least in the first 10 min of activation.[75] Unfertilized ascidian eggs are metabolically repressed with a relatively low respiratory activity. At fertilization activity increases considerably[76-79] (Figure 11); however the molecular basis for this increase and the trigger mechanisms are unknown. Finally, DNA synthesis in *Ascidia malaca* eggs is initiated after the expulsion of the second polar body, some 30 min after fertilization,[80] and appears to be triggered by a cytoplasmic factor.

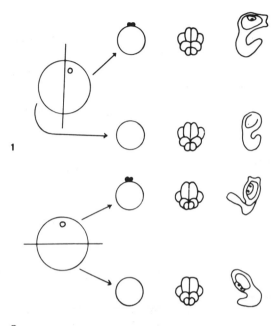

FIGURE 12.   Unfertilized ascidian eggs may be cut into halves (1) meridionally or (2) equatorially, and fertilized. Each segment gives rise to a normal larva. (From Reverberi, G. and Ortolani, G., *Dev. Biol.*, 5, 84, 1962. With permission.)

V. CYTOPLASMIC SEGREGATION

As cleavage progresses, differences arise between blastomeres. Such differences may result from the unequal distribution of cytoplasmic components laid down in the egg or from changes occurring in the blastomeres during development. Each blastomere nucleus will be subjected to a different cytoplasmic environment which, in turn, may differentially influence genome activity. The unfertilized ascidian egg is relatively undetermined. Although an animal-vegetal axis exists, fragmentation experiments show that all segments of the egg have the capacity to give rise to normal larvae[40,41,81,82] (Figure 12). Conklin[14] described three pigmented regions in the unfertilized egg of *Styela partita:* yellow cytoplasm distributed throughout the egg, grey cytoplasm in the vegetal hemisphere, and clear cytoplasm of germinal vesicle extract in the animal hemisphere. Experiments with lectins show that the unfertilized egg plasma membrane is not regionalized,[83,84] and more recently electrophysiological experiments point to a homogeneous distribution of voltage-gated ion channels.[68]

At fertilization the situation changes dramatically. The yellow plasm is segregated in the form of a crescent immediately below the equator of the egg and marks the posterior part of the future embryo. The grey plasm forms a crescent at the opposite side, while the clear cytoplasm segregates to the animal pole. In all, five different regions are recognizable in the fertilized egg. Since these plasms differentiate into musculature, nervous system, notochord, ectoderm, and intestine, they are designated meso-, neuro-, chordo-, ecto-, and endoplasm, respectively (Figure 13). When the whole-cell voltage clamp technique is used on fragments of zygotes, there is evidence to suggest that Na and Ca channels in the plasma membrane also segregate, after first polar body expulsion, accumulating at the animal pole.[68] Con A, a carbohydrate-binding protein, binds strongly to ascidian zygotes after expulsion of the second polar body, indicating a molecular rearrangement of the egg surface,[83] while

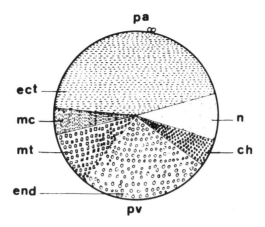

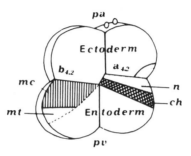

FIGURE 13. The top figure shows an ascidian zygote after ooplasmic segregation with the various plasms; ect = ectoplasm, mc = mesenchyme, mt = myoplasm, end = endoplasm, ch = chordoplasm, n = neuroplasm; pa represents the animal pole; pv, the vegetal pole. (Redrawn from Conklin, E. G., *J. Acad. Natl. Sci. Phil.,* 13, 1, 1905. With permission.) The lower figure shows the distribution of these territories in an eight-cell embryo. (From Ortolani, G., *Experientia,* 11, 445, 1955. With permission.)

Dolichos lectin binds predominantly to the vegetal hemisphere, reflecting a segregation of membrane components.[84]

The mosaic organization of the ascidian zygote is reflected in experiments demonstrating the different developmental potentials of the various plasms. Ortolani and colleagues[40,82] elegantly demonstrated in fragmentation experiments on zygotes that the vegetal cytoplasm contains factors essential for normal development. In addition, blastomers isolated from an eight-cell stage embryo, cultured *in vitro,* give rise to different specific tissue types.[14,17] The muscle cell lineage to ascidian embryos has been the subject of numerous studies,[89-92] and there is evidence that cytoplasmic determinants function by regulating differential gene expression.[93-95] Ascidian myoplasm contains specific proteins, mRNA sequences, and a characteristic cytoskeletal structure,[96-97] which affords some clues to the mechanism of cytoplasmic segregation and differential gene expression in ascidian embryos.

ACKNOWLEDGMENTS

I thank G. Princivalli for typing and G. Falcone for photography.

REFERENCES

1. **McNatty, K., Smith, D., Makris, A., Osathanondth, R., and Ryan, K.,** The microenvironment of the human antral follicle; inter-relationship among steroid levels, in antral fluid, the population of granulosa cells and the status of the oocyte *in vivo* and *in vitro, J. Clin. Endocrinol. Metab.,* 49, 851, 1979.
2. **Laufer, N., Botero-Ruiz, W., De Cherney, A., Halsentine, F., Polan, M., and Behrman, H.,** Gonadotrophin and prolactin levels in follicular fluid surrounding human ova successfully fertilized *in vitro, J. Clin. Endocrinol. Metab.,* 58, 430, 1984.
3. **Testart, J., Frydman, R., De Mouzin, J., Lassalle, B., and Belaisch, J.,** The study of factors affecting the success of human fertilization *in vitro.* I. Influence of ovarian stimulation upon the number and condition of oocytes collected, *Biol. Reprod.,* 28, 415, 1983.
4. **Sandow, B. and Namy, C.,** Light and electron microscopical analysis of human oocytes, in *In Vitro Fertilization,* Jones, H., Jones, G., Hodgin, G., and Rosennaks, Z., Eds., Williams & Wilkins, Baltimore, 1986.
5. **Veeck, L., Worthman, J., Witmeyer, J., Sandow, B., Acosta, A., Garcia, J., Jones, G., and Jones, H.,** Maturation and fertilizability of morphologically immature human oocytes in a programme of *in vitro* fertilization, *Fertil. Steril.,* 39, 594, 1983.
6. **Laufer, N., De Cherney, A., Haseltine, F., Behrman, H.,** Steroid secretion by the human egg-corona-complex in culture, *J. Clin. Endocrinol. Metab.,* 58, 1153, 1984.
7. **Thibault, C.,** Are follicular maturation and oocyte maturation independent processes?, *J. Reprod. Fertil.,* 51, 1, 1977.
8. **Rosennaks, Z. and Muasher, S.,** Recruitment of fertilizable eggs, in *In Vitro Fertilization,* Jones, H., Jones, G., Hodgin, G., and Rosennaks, Z., Eds., Williams & Wilkins, Baltimore, 1986.
9. **Dale, B.,** Mechanism of fertilization, *Nature (London),* 325, 762, 1987.
10. **Asch, R., Balmaceda, J., Ellesworth, L., and Wong, P.,** Gamete intrafallopian transfer (GIFT): a new treatment for infertility, *J. Fertil.,* 30, 41, 1986.
11. **Lasalle, B., Courtot, A., and Testart, J.,** *In vitro* fertilization of hamster and human oocytes by microinjection of human sperm, *Gamete Res.,* 16, 69, 1987.
12. **Wilson, E. B.,** *The Cell in Development and Heredity,* 3rd ed., Macmillan, London, 1925.
13. **Morgan, T. H.,** *Embryology and Genetics,* Columbia University Press, New York, 1934.
14. **Conklin, E. G.,** The organization and cell lineage of the ascidian egg, *J. Acad. Natl. Sci. Phil.,* 13, 1, 1905.
15. **Morisawa, S. and Morisawa, M.,** Initiation of sperm motility in tunicates, *Zool. Sci.,* 2(Abstr.), 91, 1985.
16. **Miller, R.,** Chemotaxis of the spermatozoa of *Ciona intestinalis, Nature (London),* 254, 244, 1975.
17. **Reverberi, G.,** *Experimental Embryology of Marine and Freshwater Invertebrates,* North-Holland, Amsterdam, 1971.
18. **Rosati, F. and De Santis, R.,** Studies on fertilization in the ascidians. I. Self sterility and specific recognition between gametes of *Ciona intestinalis, Exp. Cell Res.,* 112, 111, 1978.
19. **Rosati, F., De Santis, R., and Monroy, A.,** Studies on fertilization in the ascidians. II. Lectin binding to the gametes of *Ciona intestinalis, Exp. Cell Res.,* 116, 419, 1978.
20. **Hoshi, M.,** Role of sperm glycosidases and proteases in the ascidian fertilization, in *Advances in Invertebrate Reproduction,* Engels, W., Ed., Elsevier, Amsterdam, 1984.
21. **Hoshi, M., De Santis, R., Pinto, M., Cotelli, F., and Rosati, F.,** Is sperm α-L-fucosidase responsible for sperm-egg binding in *Ciona intestinalis?,* in *Proc. 4th Int. Symp. Spermatology, Sperm Cell,* Andre, J., Ed., Seillac, France, 1983.
22. **Lambert, C. and Epel, D.,** Calcium mediated mitochondrial movement in ascidian sperm during fertilization, *Dev. Biol.,* 69, 296, 1979.
23. **De Santis, R., Pinto, M., Cotelli, F., Rosati, F., Monroy, A., and D'Alessio, G.,** A fucosyl glycoprotein component with sperm receptor and sperm-activating activities from the vitelline coat of *Ciona intestinalis* eggs, *Exp. Cell Res.,* 148, 508, 1983.
24. **Casazza, G., De Santis, R., and Pinto, M.,** Sperm binding to eggs of *Ciona intestinalis, Exp. Cell Res.,* 148, 508, 1983.
25. **Lambert, C. and Lambert, G.,** The role of acin and myosin in ascidian sperm mitochondrial translocation, *Dev. Biol.,* 106, 307, 1984.
26. **Elia, V., Rosati, F., Barone, G., Monroy, A., and Liguori, A.,** A thermodynamic study of the sperm egg interaction, *EMBO J.,* 2, 2053, 1983.
27. **Franzen, A.,** The fine structure of spermatid differentiation in a tunicate *Corella parallelogramma* (Muller), *Zoologica (N.Y),* 4, 115, 1976.
28. **Villa, L.,** An ultrastructural investigation of normal and irradiated spermatozoa of a tunicate *Ascidia malaca, Bull. Zool.,* 42, 95, 1975.

29. **Kubo, M., Ishikawa, M., and Numakunai, T.,** Differentiation of apical structures during spermiogenesis and fine structures of the spermatozoon in the ascidian *Halocynthia roretzi, Acta Embryol. Exp.*, 3, 283, 1978.

30. **Woollacott, R.,** Spermatozoa of *Ciona intestinalis* and analysis of ascidian fertilization, *J. Morphol.*, 152, 77, 1977.

31. **Honegger, T.,** Fertilization in ascidians: studies on the egg envelope, sperm and gamete interactions in *Phallusia mammillata, Dev. Biol.*, 118, 118, 1986.

32. **Lambert, C. C.,** The ascidian sperm reaction, *Am. Zool.*, 22, 841, 1982.

33. **Villa, L. and Tripepi, S.,** An electron microscope study of spermatogenesis and spermatozoa of *Ascidia malaca, Ascidiella aspersa,* and *Phallusia mammillata* (Ascidiacea, Tunicata), *Acta Embryol. Morphol. Exp.*, 4, 157, 1983.

34. **Fukumoto, M.,** Fertilization in ascidians: acrosome fragmentation in *Ciona intestinalis* spermatozoa, *J. Ultra Struct. Res.*, 87, 252, 1984.

35. **Cloney, R. and Abbott, L.,** The spermatozoa of ascidians; acrosome and nuclear envelope, *Cell Tissue Res.*, 206, 261, 1980.

36. **De Santis, R., Jamunno, G., and Rosati, F.,** A study of the chorion and the follicle cells in relation to the sperm egg interaction in the ascidian *Ciona intestinalis, Dev. Biol.*, 74, 490, 1980.

37. **Cotelli, F., De Santis, R., Rosati, F., and Monroy, A.,** Acrosome differentiation in the spermatogenesis of *Ciona intestinalis, Dev. Growth Differ.*, 22, 561, 1980.

38. **Rosati, F., Pinto, M., and Casazza, G.,** The acrosomal region of the spermatozoon of *Ciona intestinalis*: its relationship with the binding to the vitellin coat of the egg, *Gamete Res.*, 11, 379, 1985.

39. **Dale, B., and Monroy, A.,** How is polyspermy prevented?, *Gamete Res.*, 4, 1981.

40. **Ortolani, G.,** Cleavage and development of egg fragments in ascidians, *Acta Embryol. Morphol. Exp.*, 1, 247, 1958.

41. **Dalcq, A. B.,** Etude des localisations germinales dans l'oeuf vierge d'Ascidie par des experiences de merogonie, *Arch. Anat. Microsc. Morphol. Exp.*, 28, 223, 1932.

42. **Sawada, T. and Osanai, K.,** The cortical contraction related to the ooplasmic segregation in *Ciona intestinalis* eggs, *Wilhelm Roux Arch. Entwicklungsmech. Org.*, 190, 208, 1981.

43. **Ortolani, G.,** I movimenti corticali dell'uovo di Ascidie alla fecondazione, *Riv. Biol.*, 47, 169, 1955.

44. **Ortolani, G.,** Azione della tripsina sul cortex dell'uovo di *Phallusia mamillata, Ric Sci.*, 27, 1175, 1957.

45. **Ortolani, G.,** Modifications of the cortex of ascidian eggs after fertilization obtained by means of different methods, *Boll. Zool.*, 40, 405, 1973.

46. **Zalokar, M.,** Effect of colchicine and cytochalasin B on ooplasmic segregation of ascidian eggs, *Wilhelm Roux Arch. Entwicklungsmech. Org.*, 175, 243, 1974.

47. **Reverberi, G.,** On some effects of cytochalasin B on the eggs and tadpoles of the ascidians, *Acta Embryol. Exp.*, 2, 137, 1975.

48. **Rosati, F., Monroy, A., and De Prisco, P.,** Fine structural study of fertilization in the ascidian, *Ciona intestinalis, J. Ultrastruct. Res.*, 58, 261, 1977.

49. **Mancuso, V.,** The distribution of the ooplasmic components in the unfertilized, fertilized and 16 cell stage of *Ciona intestinalis, Acta Embryol. Morphol. Exp.*, 7, 71, 1964.

50. **Miyazaki, S., Takahashi, K., and Tsuda, K.,** Electrical excitability in the egg cell membrane of the tunicate, *J. Physiol. (London)*, 238, 37, 1974.

51. **Okamoto, H., Takahashi, K., and Yoshii, M.,** Membrane currents of the tunicate egg under the voltage clamp condition, *J. Physiol. (London)*, 254, 607, 1976.

52. **Okamoto, H., Takahashi, K., and Yoshii, M.,** Two components of the calcium current in the egg cell membrane of the tunicate, *J. Physiol. (London)*, 255, 527, 1976.

53. **Ohmori, H. and Yoshii, M.,** Surface potential reflected in both gating and permeation mechanisms of sodium and calcium channels of the tunicate egg cell membrane, *J. Physiol. (London)*, 267, 429, 1977.

54. **Ohmori, H.,** Inactivation kinetics and steady state current noise in the anomalous rectifier of tunicate egg cell membranes, *J. Physiol. (London)*, 281, 77, 1978.

55. **Ohmori, H.,** Duval effects of K ions upon the inactivation of the anomalous rectifier of the tunicate egg membranes, *J. Membr. Biol.*, 53, 143, 1980.

56. **Ohmori, H.,** Unitary current through sodium channel and anomalous rectifier channel estimated from transient current noise in the tunicate egg, *J. Physiol. (London)*, 311, 289, 1981.

57. **Takahashi, K. and Yoshii, M.,** Effects of internal free calcium upon the sodium and calcium channels in the tunicate egg analyzed by the internal perfusion technique, *J. Physiol. (London)*, 279, 519, 1978.

58. **Fukushima, Y.,** Single channel potassium currents of the anomalous rectifier, *Nature (London)*, 294, 368, 1981.

59. **Fukushima, Y.,** Blocking kinetics of the anomalous potassium rectifier of tunicate egg studied by single channel recording, *J. Physiol. (London)*, 331, 311, 1982.

60. **Kozuka, M. and Takahashi, K.,** Changes in holding and ion channel currents during activation of an ascidian egg under voltage clamp, *J. Physiol. (London)*, 323, 267, 1982.

61. **Hirano, T. and Takahashi, K.,** Comparison of properties of calcium channels between the differentiated 1-cell embryo and the egg cells of ascidians, *J. Physiol. (London),* 347, 327, 1984.
62. **Hirano, T., Takahashi, K., and Yamashita, N.,** Determination of excitability types in blastomeres of the cleavage-arrested but differentiated embryos of an ascidian, *J. Physiol. (London),* 347, 301, 1984.
63. **Dale, B., De Santis, A., and Ortolani, G.,** Electrical response to fertilization in ascidian oocytes, *Dev. Biol.,* 99, 188, 1983.
64. **Dale, B. and De Felice, L.,** Sperm activated channels in ascidian oocytes, *Dev. Biol.,* 101, 235, 1984.
65. **De Felice, L. and Kell, M.,** Sperm activated currents in ascidian oocytes, *Dev. Biol.,* 119, 123, 1987.
66. **Talevi, R. and Dale, B.,** Electrical characteristics of ascidian egg fragments, *Exp. Cell Res.,* 162, 539, 1986.
67. **De Felice, L., Dale, B., and Talevi, R.,** Distribution of fertilization channels in ascidian oocyte membranes, *Proc. R. Soc. London Ser. B,* 229, 209, 1986.
68. **Dale, B. and Talevi, R.,** Distribution of ion channels in ascidian eggs and zygotes, *Exp. Cell. Res.,* 181, 238, 1989.
69. **Dale, B.,** Fertilization channels in ascidian eggs are not activated by Ca, *Exp. Cell Res.,* 172, 474, 1987.
70. **Dale, B.,** Primary and secondary messengers in the activation of ascidian eggs, *Exp. Cell Res.,* 177, 205, 1988.
71. **Speksnijder, J., Corson, D., Qiu, T., and Jaffee, L.,** Free calcium pulses during early development of *Ciona* eggs, *Biol Bull.,* 170, 542a, 1986.
72. **Speksnijder, J., Corson, D., Jaffe, L., and Sardet, C.,** Calcium pulses and waves through ascidian eggs, *Biol. Bull.,* 171, 488a, 1987.
73. **Brownlee, C. and Dale, B.,** Intracellular Ca in ascidian zygotes measured with Fura-2 (unpublished data).
74. **Bevan, S., O'Dell, D., and Ortolani, G.,** Experimental activation of ascidian eggs, *Cell Differ.,* 6, 313, 1977.
75. **Russo, P., De Santis, A., and Dale, B.,** pH during fertilization and activation of ascidian eggs (unpublished data).
76. **Tyler, A. and Humason, W.,** On the energetics of differentiation. VI. Comparison of the temperature coefficients of the respiratory rates of unfertilized and of fertilized eggs, *Biol. Bull.,* 73, 261, 1937.
77. **Minganti, A.,** Experiments on the respiration of *Phallusia* eggs and embryos (Ascidians), *Acta Embryol. Morphol. Exp.,* 1, 150, 1957.
78. **Runnström, J.,** Atmungsmechanismus und Entwicklungserregung bei dem Seeigeleie, *Protoplasma,* 10, 106, 1930.
79. **Lentini, R.,** The oxygen uptake of *Ciona intestinalis* eggs during development in normal and experimental conditions, *Acta Embryol. Morphol. Exp.,* 4, 209, 1961.
80. **Ortolani, G., O'Dell, D. S., Mansueto, C., and Monroy, A.,** Surface changes and onset of DNA replication in the Ascidia egg, *Exp. Cell Res.,* 96, 122, 1975.
81. **Abbate, C. and Ortolani, G.,** The development of *Ciona* eggs after partial removal of cortex or ooplasm, *Acta Embryol. Morphol. Exp.,* 4, 56, 1961.
82. **Reverberi, G. and Ortolani, G.,** Twin larvae from halves of the same egg in ascidians, *Dev. Biol.,* 5, 84, 1962.
83. **Monroy, A., Ortolani, G., O'Dell, D., and Millonig, G.,** Binding of Concanavalin A to the surface of unfertilized and fertilized ascidian eggs, *Nature (London),* 242, 409, 1973.
84. **Ortolani, G., O'Dell, D., and Monroy, A.,** Localized binding of Dolichoslectin to the early Ascidia embryo, *Exp. Cell Res.,* 106, 402, 1977.
85. **Ortolani, G.,** The presumptive territory of the mesoderm in the ascidian germ, *Experientia,* 11, 445, 1955.
86. **Nishida, H. and Satoh, N.,** Cell lineage analysis in ascidian embryos by intracellular injection of a tracer enzyme. I. Up to the eight cell stage, *Dev. Biol.,* 99, 382, 1983.
87. **Nishida, H. and Satoh, N.,** Cell lineage analysis in ascidian embryos by intracellular injection of a tracer enzyme. II. The 16- and 32-cell stages, *Dev. Biol.,* 110, 440, 1985.
88. **Zalokar, M. and Sardet, C.,** Tracing of cell lineage in embryonic development of *Phallusia mammillata* (Ascidia) by vital staining of mitochondria, *Dev. Biol.,* 102, 195, 1984.
89. **Whittaker, J.,** Segregation during ascidian embryogenesis of egg cytoplasmic information for tissue specific enzyme development, *Proc. Natl. Acad. Sci. U.S.A.,* 70, 2096, 1973.
90. **Whittaker, J.,** Acetylcholinesterase development in extra cells caused by changing the distribution of myoplasm in ascidian embryos, *J. Embryol. Exp. Morphol.,* 55, 343, 1980.
91. **Whittaker, J.,** Muscle lineage cytoplasm can change the developmental expression in epidermal lineage cells of ascidian embryos, *Dev. Biol.,* 93, 463, 1982.
92. **Denoh, T. and Satoh, N.,** Studies on the cytoplasmic determinants for muscle cell differentiation in ascidian embryos: an attempt at transplantation of the myoplasm, *Dev. Growth Differ.,* 26, 43, 1984.
93. **Whittaker, J.,** Segregation during cleavage of a factor determining endodermal alkaline phosphatase development in ascidian embryos, *J. Exp. Zool.,* 202, 139, 1977.

94. **Crowther, R. and Whittaker, J.,** Differentiation of histospecific ultrastructural features in cells of cleavage-arrested early ascidian embryos, *Wilhelm Roux Arch. Entwicklungsmech. Org.,* 194, 87, 1984.
95. **Meedel, T. and Whittaker, J.,** Development of translationally active mRNA for larval muscle acetyl-cholinesterase during ascidian embryogenesis, *Proc. Natl. Acad. Sci. U.S.A.,* 80, 4761, 1983.
96. **Jefferey, W.,** Identification of proteins and mRNAs in isolated yellow crescents of ascidian eggs, *J. Embryol. Exp. Morphol.,* 89, 275, 1985.
97. **Jefferey, W. and Meier, S.,** A Yellow crescent cytoskeletal domain in ascidian eggs and its role in early development, *Dev. Biol.,* 96, 125, 1983.

Chapter 8

SEA URCHIN FERTILIZATION: A VERSATILE MODEL FOR STUDYING SIGNAL TRANSDUCTION, DEFENSES AGAINST OXIDANT-MEDIATED DAMAGE, AND EXTRACELLULAR MATRIX ASSEMBLY

Jay W. Heinecke and David E. Battaglia

TABLE OF CONTENTS

I. INTRODUCTION AND OVERVIEW OF FERTILIZATION

Many of the events of fertilization involve basic cellular processes such as cell-cell interaction, signal transduction, and assembly of an extracellular protein matrix. A particularly useful model system for studying fertilization is the sea urchin, a marine invertebrate. Sea urchins spawned in the laboratory produce large, homogeneous populations of quiescent gametes. These cells can be activated in synchrony and are easily cultured under physiological conditions. In addition, the sea urchin is one of the most extensively characterized developmental systems in experimental embryology. These factors have facilitated biochemical investigation of the system, making the urchin the best-characterized model of animal fertilization. For example, many observations of general significance in mammalian fertilization were first described using gametes from marine invertebrates. These insights were fundamental to the development of successful *in vitro* fertilization techniques for mammalian eggs and the eventual employment of similar techniques in human clinical situations.

In most animal systems, including the sea urchin, the egg possesses all of the metabolic machinery required for cell division and differentiation, as well as specialized mechanisms for the prevention of polyspermy and protection of the fragile embryo from a hostile environment. Because the events surrounding fertilization of the sea urchin egg occur in the absence of protein synthesis or cell proliferation, this cell is an excellent tool for the study of mechanisms involved in the regulation of protein form and function. In contrast to the totipotent egg, spermatozoa are highly specialized cells, designed solely for the delivery of their haploid genetic material to the egg. This minute cell, which lacks the usual cellular mechanisms for protein and DNA synthesis, represents a relatively simple system for studying cell activation.

The sea urchin releases its gametes into the open ocean, placing the egg and sperm into a demanding environment where fertilization and subsequent development will occur. The release of the sperm into seawater causes rapid alkalinization of its cytoplasm, which stimulates sperm motility through the activation of the flagellar dynein ATPase.[1,2] Vigorous sperm motility, possibly coupled to an egg associated chemoattractant,[3] increases the chance of sperm and egg contact in the often turbulent spawning environment. Unlike the sperm, the newly spawned egg remains metabolically quiescent until it is fertilized. The egg is surrounded by a complex extracellular glycocalyx, the vitelline layer, which possesses species-specific binding sites for the sperm.[4] After the initial attachment of the sperm to the vitelline layer, the sperm penetrates this glycocalyx and fuses with the plasma membrane of the egg.

The fusion of egg with the fertilizing sperm induces a rapid but transient depolarization of the egg plasma membrane.[5] This depolarization marks the onset of several processes that result in the activation of the egg. One effect of depolarization is the prevention of other sperm from fusing with the egg in what has been called the ''fast block'' to polyspermy.[6,7] The change in the egg's membrane potential is followed by a transient rise in intracellular free calcium, which reaches maximum levels approximately 30 s after sperm-egg fusion.[8] One consequence of the rise in free calcium is the exocytosis of thousands of secretory vesicles that reside in the egg cortex. These cortical granules release an assortment of proteins into the small space between the vitelline layer and plasma membrane and initiate one of the most dramatic morphological events of sea urchin fertilization, the assembly of the fertilization envelope.[9] Visible by light microscopy, fertilization envelope assembly involves the elevation of the vitelline layer from the egg surface and subsequent differentiation of this glycocalyx through the activity of numerous cortical granule proteins (Figure 1). The formation of this resilient glycocalyx provides the definitive block to polyspermy and creates a protected environment around the fertilized egg in which early development will occur.

Beginning 3 to 5 min after sperm-egg fusion, other ionic fluxes further influence the

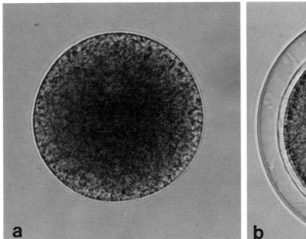
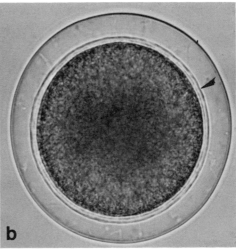

FIGURE 1.   Light micrographs of an unfertilized (a) and a fertilized (b) egg from the sea urchin, *Strongylocentrotus purpuratus*. The eggs of this species are approximately 80 μm in diameter. During fertilization, the egg forms the fertilization envelope (arrow) via the elevation and extensive modification of its tightly apposed extracellular glycocalyx. This process involves the coordinated interaction of the exocytosis of egg secretory vesicles, oxidative metabolism, and covalent cross-linking. The hyaline layer, another product of exocytosis, can also be seen (arrowhead).

activated state of the egg.[10] During this time, an efflux of hydrogen ions from the egg occurs which alkalinizes the cytoplasm by 0.3 to 0.5 pH units.[11] This change in cellular pH eventually results in an increase in protein and DNA synthesis which will be required for subsequent development of the fertilized egg. Other complex biochemical events also begin within minutes after gamete fusion, including a cyanide-resistant respiratory burst.[12] Most of the oxygen consumed during the respiratory burst is used to generate extracellular hydrogen peroxide, which is used as a substrate by a peroxidase involved in the assembly of the fertilization envelope.[12]

Analogous to this system, the mammalian egg is also surrounded by a glycocalyx, the zona pellucida. This matrix, which possesses sperm receptors,[13] becomes modified during fertilization in a process called the zona reaction. Like fertilization envelope assembly in the sea urchin, the zona reaction depends upon the secretion of egg cortical granules.[14] In addition, depolarization of the plasma membrane after sperm-egg fusion has been measured in certain mammalian eggs. While depolarization does not appear to act as a barrier to polyspermy in mammalian eggs,[15] it does precede a series of biochemical events that are analogous to those observed in the activated sea urchin egg. Thus, two widely divergent animal phyla use similar mechanisms to bind sperm, prevent polyspermy, and perhaps to create a special environment for the developing embryo.

In this chapter, we shall focus on the events occurring in the first 10 min after fertilization of the sea urchin egg. Emphasis will be placed on recent findings of potential significance to mammalian fertilization and cell biology. A number of reviews covering other aspects of fertilization have recently appeared.[2,9,10,16]

II. SIGNAL TRANSDUCTION DURING EGG ACTIVATION

A. RISE IN CYTOPLASMIC FREE CALCIUM

Before fusion with the sperm, the sea urchin egg exists in a quiescent, metabolically repressed state. Over 40 years ago, Mazia[17] and Moser[18] suggested that calcium plays an important role in activation of the fertilized egg. This hypothesis has been confirmed using many different experimental approaches, demonstrating that a rise in cytoplasmic free calcium is crucial in triggering such early events in egg activation as cortical granule exocytosis. Initiation of protein and DNA synthesis, which occur later in egg activation, appear to be dependent upon an adequate rise in intracellular pH. These ionic signals have recently been shown to play a role in regulating a wide variety of processes in mammalian cells.

B. IONIC FLUXES VS. EXOCYTOSIS

Exocytosis of the egg's cortical granules begins about 30 s after sperm-egg fusion. This process involves a large population of secretory vesicles, making the sea urchin egg a particularly useful model for studying the influence of ionic events on exocytosis.[19] Numerous lines of evidence indicate that an increase in cytoplasmic free calcium is the immediate signal triggering cortical granule exocytosis. Studies using calcium-sensitive fluorescent probes have shown that the fusion of sperm with the plasma membrane initiates a transient wave of calcium release which precedes cortical granule exocytosis.[20] When intact eggs are treated with the calcium ionophore A23187, cortical granule exocytosis is stimulated, suggesting that a rise in cytoplasmic calcium is the signal that triggers exocytosis.[21] Consistent with this possibility, microinjection of calcium into eggs is adequate to initiate cortical granule exocytosis.[22] Similarly, in isolated cell surface complexes prepared from eggs, fusion of cortical granules with the plasma membrane is stimulated by a rise in the concentration of free calcium in the medium.[23,24] Antibodies to calmodulin block this calcium-dependent effect, suggesting that calmodulin mediates cortical granule fusion in the cell surface complex.[25] When eggs are microinjected with the calcium chelator EGTA, neither sperm nor calcium ionophore can activate the egg, consistent with the hypothesis that both of these agents activate eggs in a calcium-dependent manner.[22,29]

C. SOURCE OF FREE Ca^{2+}

Recent results suggest that the endoplasmic reticulum is the site of both calcium release and sequestration in the egg, analogous to the role of the sarcoplasmic reticulum as the site of calcium uptake and release in muscle cells.[27] In the presence of ATP, egg homogenates lower the concentration of free calcium in the medium, indicating an active uptake process for calcium is present in the homogenate.[24] The ATP-dependent calcium-sequestering activity of the egg homogenate copurifies with the endoplasmic reticulum on density gradient centrifugation.[24] Interestingly, this fraction is also capable of releasing calcium when stimulated with inositol trisphosphate.[24]

D. INOSITOL 1,4,5-*Tris* PHOSPHATE

Inositol 1,4,5-*tris* phosphate (IP₃) appears to be the major regulator of cytoplasmic free calcium in many types of cells. Currently, it is thought that IP₃ is released from phosphoinositol 4,5-*bis* phosphate (PIP₂) of the plasma membrane by the action of a phospholipase C.[28] Several lines of evidence suggest that IP₃ plays a similar role in regulating intracellular free calcium in activated sea urchin eggs. Upon fertilization, the content of trisphosphoinositides radiolabeled with ³²P increased about 30% in sea urchin eggs.[26] This increase in trisphosphoinositides probably occurs before the elevation in cytoplasmic free calcium. These results suggest that IP₃ might be generated from PIP₂, acting as the signal triggering the rise in cytoplasmic calcium. When eggs are microinjected with IP₃, they undergo cortical granule exocytosis, again suggesting IP₃ is a regulator of intracellular free calcium.[29,30] Similarly, IP₃ stimulates calcium release from permeabilized eggs, egg homogenates, and partly purfied egg endoplasmic reticulum.[24] Microinjection of IP₃ into intact eggs previously injected with EGTA fails to stimulate exocytosis,[22] suggesting that IP₃ mediates its effects on egg activation by causing an increase in free calcium. These results strongly support the role of IP₃ as a major regulator of cytoplasmic free calcium in eggs. Application of IP₃ to isolated cell surface complexes, in the absence of exogenous calcium, fails to trigger cortical granule exocytosis.[30] These results are consistent with the proposed role of IP_3 in releasing cellular stores of calcium from the endoplasmic reticulum of the egg.

E. INOSITOL 1,3,4,5-TETRAKISPHOSPHATE

In many cells, the initial IP_3-induced transient rise in calcium is followed by a prolonged period of increased free calcium.[28] This second phase of elevation in intracellular calcium may be dependent upon extracellular calcium rather than release from cytoplasmic stores. Recent studies using sea urchin eggs microinjected with specific inositol phosphates indicate that the signal for the prolonged elevation of intracellular calcium may be inositol 1,3,4,5-tetrakisphosphate[31] (IP_4). It is unclear whether intracellular stores of calcium are adequate to trigger full activation of the egg. For example, microinjection of the IP_3 analog inositol 2,4,5-*tris* phosphate into eggs fails to cause cortical granule exocytosis, despite its ability to stimulate the release of calcium from cellular stores as effectively as IP_3 in other types of cells.[28,31] IP_4 alone is also unable to cause exocytosis. However, coinjection of inositol 2,4,5-*tris* phosphate with IP_4, in the presence of external calcium, results in full activation of the eggs.[31] This activation fails to occur in calcium-free seawater,[31] suggesting that the additional source of calcium was extracellular. Although cytoplasmic calcium levels were not directly measured, these results indicate that IP_4 may also be a second messenger, capable of stimulating the entry of extracellular calcium into cells. To obtain complete egg activation, calcium may have to enter the cytoplasm from both intracellular stores and the extracellular environment.

F. ALKALINIZATION OF THE EGG CYTOPLASM

A second ionic signal crucial to the initiation of protein and DNA synthesis in the fertilized egg is alkalinization of the egg cytoplasm.[32] The elevation in cellular pH can be dissociated from the earlier increase in intracellular calcium, indicating that the rise in cytoplasmic calcium and alkalinization are independent events in egg activation.[10] Alkalinization of the cytoplasm is the result of an efflux of hydrogen ions from the egg, which begins 1 to 3 min after sperm-egg fusion.[2,33] The efflux of hydrogen ions from the egg is accompanied by an equimolar influx of sodium ions, consistent with the hypothesis that hydrogen-sodium exchange is the mechanism for cytoplasmic alkalinization.[2,33] Both the influx of sodium and the efflux of hydrogen ions are amiloride sensitive,[2,33] suggesting that an amiloride-sensitive Na^+/H^+ antiporter located in the plasma membrane regulates intracellular pH.

G. REGULATION OF INTRACELLULAR pH

Recent experiments using phorbol ester, a potent activator of protein kinase C which mimics the action of diacylglycerol, suggest a possible mechanism for the regulation of intracellular pH in sea urchin eggs. Eggs exposed to nanomolar concentrations of phorbol ester underwent cytoplasmic alkalinization and initiated protein synthesis.[33] These effects were dependent upon extracellular sodium and were inhibited by amiloride, suggesting that the rise in intracellular pH may be related to generation of the second messenger diacylglycerol, which in turn might activate protein kinase C. The biochemical targets for phosphorylation by protein kinase C remain unknown.

The regulation of intracellular pH by an amiloride-sensitive Na^+/H^+ antiporter, first described in sea urchin eggs,[34] appears to be important in controlling a wide variety of cellular processes. For example, intracellular pH plays an important role in regulating the volume of red blood cells, initiating the entry from G_o/G_1 into the S phase of the cell cycle, and inducing the differentiation of lymphocytes.[35] Fibroblasts must elevate their intracellular pH to a threshold level in order to begin mitogen-stimulated DNA synthesis, and mitogen-stimulated DNA synthesis is blocked by amiloride.[36] These results indicate a requirement for Na^+/H^+ exchange in growth-factor-induced DNA synthesis in fibroblasts.

III. THE RESPIRATORY BURST

A. HYDROGEN PEROXIDE PRODUCTION

A striking feature of sea urchin egg activation is a dramatic increase in oxygen consumption, first described by Warburg in 1908.[37] This respiratory burst is largely cyanide resistant, indicating that mitochondrial respiration accounts for only a small portion of the oxygen consumed.[38] The metabolic basis for the increase in oxygen use is the generation of large amounts of extracellular hydrogen peroxide. Several lines of evidence indicate that the activated sea urchin egg produces hydrogen peroxide.[38] First, the catalase-dependent oxidation of $[^{14}C]$-formate to $[^{14}C]$-carbon dioxide by activated sea urchin eggs has been demonstrated. This reaction is dependent on hydrogen peroxide and is inhibited by aminotriazole, a reagent which reacts irreversibly with the catalase-hydrogen-peroxide complex, compound I. Second, hydrogen peroxide is detected outside activated eggs using the horseradish-peroxidase-catalyzed oxidation of scopoletin. Finally, upon activation, eggs emit a burst of light. The time courses of light emission and oxygen consumption are similar, suggesting they are related phenomena.[38] Chemiluminescence has been well described in several systems where reactive oxygen species such as hydrogen peroxide are produced. In activated neutrophils, chemiluminescence requires the presence of a peroxidase, myeloperoxidase.[39] Similarly, light emission by eggs requires an egg-associated peroxidase, ovoperoxidase, and is inhibited by peroxidase inhibitors such as aminotriazole.[38]

Because of competing reactions utilizing hydrogen peroxide as substrate (such as ovoperoxidase-mediated cross-linking of the fertilization envelope), exact quantification of the amount of hydrogen peroxide produced during the respiratory burst is difficult. When known amounts of hydrogen peroxide are added to activated eggs, only 20% of the exogenously added hydrogen peroxide is detected.[38] Making the assumption that the added hydrogen peroxide enters the same metabolic pool as that produced by the egg, hydrogen peroxide production was estimated to account for $2/3$ of the oxygen consumed during the first 15 min of the respiratory burst.[38]

B. BIOCHEMICAL BASIS FOR HYDROGEN PEROXIDE PRODUCTION

Recently, an NADPH: O_2 oxidoreductase was isolated from unfertilized sea urchin eggs.[85] Concentrations of free calcium, implicated in regulation of egg activation, modulate the activity of this oxidase. Oxidase activity also requires $ATPMg^{2-}$, suggesting a calcium-dependent protein kinase might regulate the oxidase. Consistent with this suggestion, the kinase inhibitors H-7 and staurosporine are inhibitory in activated eggs and in the isolated oxidase. These results indicate that the respiratory burst oxidase of fertilization is an NADPH-specific oxidase which is regulated by a protein kinase. Interestingly, the initial rise in calcium in activated eggs stimulates an NAD kinase by a calmodulin-dependent process, converting much of the egg's NAD to NADP.[47] Concurrently, an increase in NAD(P)H relative to NAD(P) occurs, providing the reduced NADPH required as an electron donor for reduction of oxygen to hydrogen peroxide.[48]

The increase in oxygen consumption by fertilized eggs bears many similarities to the respiratory burst of activated neutrophils.[39] Upon activation, both cells exhibit a cyanide-insensitive burst of oxygen consumption, chemilumenescence, and the production of extracellular hydrogen peroxide. Both types of cells exocytose a peroxidase, which uses hydrogen peroxide generated by the cell as a substrate for oxidant-mediated reactions. The cellular production of reactive oxygen species such as superoxide and hydrogen peroxide may play a role in the pathogenesis of a wide variety of diseases. For example, oxygen radicals have been implicated in the etiology of cancer, adult respiratory distress syndrome, retrolental fibroplasia, atherosclerosis, and in reperfusion injury.[40] The mechanisms by which cells protect themselves from oxidant damage remain poorly defined. In many cases, the model systems used to study oxygen-radical-dependent toxicity have no clear relevance to normal physiology. In contrast, the egg normally lives in a "radical solution",[41] since seawater contains significant concentrations of partly reduced oxygen species such as superoxide and hydrogen peroxide.[42] Furthermore, at the very beginning of embryogenesis, the egg creates and utilizes high concentrations of hydrogen peroxide. A perplexing question is raised by this sequence of events. Why would a cell, at the beginning of its developmental program, produce large quantities of a potentially deleterious substance like hydrogen peroxide? Because these stresses occur under physiological conditions, it is likely that sea urchin eggs have evolved special mechanisms for protection from oxidant-mediated damage. Hence, the sea urchin system offers many advantages for studying both the beneficial and the deleterious aspects of oxygen physiology.

C. OVOTHIOL AND PROTECTION FROM HYDROGEN PEROXIDE

Recently, an unusual derivative of histidine which contains a free SH group was isolated from sea urchin eggs.[43-45] This amino acid, ovothiol, is present at millimolar concentrations, making it one of the most abundant small-molecular-weight compounds in the egg.[43] Initially, ovothiol was postulated to play a role in the generation of hydrogen peroxide during the respiratory burst. This hypothesis was attractive because ovothiol catalyzed a cyanide-resistant ovoperoxidase and NAD(P)H-dependent oxidase reaction which appeared to be activated by calcium.[44] This scheme suggested that ovothiol might be transported out of the egg, where it would serve as a catalyst for the reduction of oxygen to hydrogen peroxide by ovoperoxidase which had been exocytosed during the cortical reaction. However, several lines of evidence suggest that this proposed mechanism is unlikely. First, no ovothiol is detected outside the egg after fertilization.[43] Second, aminotriazole, a potent inhibitor of the oxidase reaction *in vitro*, has no effect on the rate of oxygen consumption by eggs.[43] Third, the oxidase reaction, although hydrogen peroxide dependent, does not produce detectable hydrogen peroxide.[45] In contrast, fertilized eggs produce large amounts of extracellular hydrogen peroxide (rather than exhibiting a catalase-sensitive source of oxygen consumption). Finally, derivatization of 80% of the egg's ovothiol with iodoacetamide has no effect on the respiratory burst.[43]

Ovothiol may play the primary role in protecting eggs from oxidant damage by directly scavenging reactive oxygen species such as hydrogen peroxide.[43] Sea urchin eggs contain no detectable glutathione peroxidase, the usual cytoplasmic enzyme which detoxifies hydrogen peroxide enzymatically using glutathione as a reducing substrate.[82] In egg lysates, all of the consumption of hydrogen peroxide is heat insensitive and can be accounted for by the amount of ovothiol present. In a cell-free system, ovothiol reacts several times more rapidly than glutathione with hydrogen peroxide.[82] Finally, the presence of ovothiol at millimolar concentrations in eggs is compelling evidence for its possible role as a direct scavenger of reactive oxygen species. We are currently investigating the reactivity of ovothiol with a variety of oxygen radicals, as well as its role in protecting the egg and other cells from oxidant-mediated damage. Potentially, ovothiol represents a nontoxic, naturally occurring amino acid with the ability to protect cells from a variety of free-radical insults.

IV. ASSEMBLY OF THE FERTILIZATION ENVELOPE

A. THE VITELLINE LAYER

During spawning, as gametes are broadcast into the open ocean, the egg finds itself surrounded by thousands of competent spermatozoa and diverse environmental threats. In order to maximize its chances for successful development, the egg assembles the fertilization envelope (FE) to shield itself from environmental insults. Elevation of the FE creates a protected environment, the perivitelline space, in which the earliest stages of development will occur.

Assembly of the FE involves an orchestrated series of biochemical reactions that transform the extracellular matrix of the unfertilized egg, the vitelline layer, into a tough, selectively permeable envelope. The FE provides an excellent model with which to investigate many aspects of extracellular matrix assembly. The vitelline layer contains a 305-kDa protein which is a sperm receptor,[4] as well as binding sites for a cortical granule protein, proteoliasin.[49] However, because the vitelline layer is a complex matrix consisting of multiple proteins that are difficult to dissociate, the majority of its components have not yet been identified.

B. CORTICAL GRANULES

Assembly of the FE depends upon the interaction of the secretory products of the egg's cortical granules with the vitelline layer of the glycocalyx. The unfertilized sea urchin egg cortex contains approximately 18,000 secretory vesicles,[50] the cortical granules. These granules contain several proteins, including trypsin-like proteases,[51,52] ovoperoxidase,[53] an ovoperoxidase binding protein called proteoliasin,[54] B-glucanase,[55,56] hyaline,[57] and the poorly characterized paracrystalline proteins.[58] Evidence indicates that ovoperoxidase, proteoliasin, hyaline, and perhaps B-glucanase are contained within the same vesicles which make up the vast majority of this population of organelles[56,57,83] However, contrary to previous speculation, recent evidence suggests that the cortical granules may be a heterogeneous population, since certain antigens have been detected only in a subpopulation of these vesicles.[59] Because of its many secretory proteins, the sea urchin egg presents a good system for the study of regulation of protein packaging and secretion. Considering the multitude of secretory proteins in the egg, this cell presents an interesting problem regarding the regulation of secretory protein synthesis and packaging.

As described previously, the secretion of the cortical granules depends upon a transient increase in cytosolic free-calcium concentration.[23,27] Secretion begins with the fusion of the cortical granule and the plasma membrane, whereupon the vesicle contents are released into the small perivitelline space between the vitelline layer and plasma membrane. Subsequently, the first proteolytic event occurs, involving the cleavage of the vitelline layer from the plasma membrane by a serine protease released from the vesicles.[52,60] This cleavage permits the vitelline layer to elevate from the egg surface. Elevation of this glycocalyx is assisted by an osmotic gradient created by the influx of the enormous quantity of cortical granule proteins into the confines of the perivitelline space.[61-63] The elevation of the vitelline layer from the egg surface is visually striking and was one of the first events recorded from a fertilized animal egg (Figure 1). Concurrent with vitelline layer elevation is the destruction of the species-specific sperm receptors in this matrix by a cortical granule protease(s), thereby preventing further sperm penetration of this structure.[51,64] During this process, the nascent FE becomes the permanent block to polyspermic penetration.

C. "I TO T TRANSITION"

Several other distinct morphological changes occur during elevation of the FE. The paracrystalline proteins from the cortical granules insert into the inner and outer surfaces of the FE scaffold, creating a paracrystalline array as viewed by the electron microscope.[65,66] In addition, the FE of *Strongylocentrotus purpuratus* retains imprints of the thousands of egg surface microvilli. As assembly of this matrix proceeds, the surface morphology of these microvillar casts changes from a smooth, rounded contour (igloo shaped) to a very angular form (tent shaped), in the "I to T transition".[67] The I to T transition represents an excellent example of a morphological change that is based upon distinct biochemical processes.

D. DITROSYL CROSS-LINKS

The best-characterized reaction in FE assembly involves a secreted cortical granule peroxidase, ovoperoxidase. This enzyme catalyzes the formation of dityrosyl cross-links between adjacent polypeptides on the FE, thus affecting the hardening of this matrix.[53,68] Dityrosyl cross-linking renders the FE resistant to chemical and enzymatic degradation. The activity of ovoperoxidase is controlled in several ways. First, it is inactive in the low pH of the cortical granule environment but becomes converted to an active form approximately 30 s after entering the high pH (8.0) of the perivitelline space.[69] This is an excellent example of pH-controlled hysteresis. As a substrate for the formation of dityrosine, ovoperoxidase utilizes H_2O_2, which is only produced during the respiratory burst by the egg between 2 and 8 min after sperm-egg fusion.[38] Finally, ovoperoxidase insertion into the FE is mediated by a specific binding protein, proteoliasin.[54] Ovoperoxidase, which has no binding receptors on the vitelline layer, binds to proteoliasin in a calcium-dependent manner. Proteoliasin, in turn, binds to specific receptors on this matrix, a reaction which also depends upon calcium and possibly magnesium.[49] Thus, the action of ovoperoxidase is spatially controlled via the binding activity of proteoliasin and temporally regulated by its pH hysteretic behavior and the availability of its oxidizing substrate.

E. TRANSGLUTAMINASE-MEDIATED CROSS-LINKING

On the basis of indirect evidence, a handful of investigators have suggested that transglutaminase-mediated cross-linking of primary amine residues may also play a role in the assembly of the FE.[70-73] Our laboratory recently confirmed this hypothesis[84] and has provided direct evidence that transglutaminase activity is expressed in this process before ovoperoxidase-mediated cross-linking.

Many of the morphological and biochemical properties of this extracellular matrix are integrally tied to the covalent cross-linking of the constituent proteins. For example, we have demonstrated that inhibition of transglutaminase activity during fertilization results in the dramatic alteration of the FE.[84] At the level of light microscopy, the surface area of the FE is increased and its refractile nature is altered in the presence of transglutaminase inhibitors such as putrescine and cadaverine. In addition, inhibition of transglutaminase activity prevents the I to T transition of the microvillar casts that ornament the FE surface. Similarly, the highly ordered nature of the FE is lost, and the deposition of paracrystalline arrays from cortical granule contents is disrupted. Direct evidence of transglutaminase involvement in FE assembly was obtained by the observation that [^{3}H]-putrescine is incorporated into the FE during fertilization.[84] Thus, it appears that the transglutaminase may act directly on the vitelline layer matrix or on cortical granule proteins that are inserted into this scaffold. A similar observation has been made using [^{3}H]-tyramine and [^{14}C]-glycine ethyl ester.[72] Recent data suggest that the enzyme involved is not secreted but exists on the egg surface and exhibits maximal activity 4 min after activation. The permeability of the matrix is also affected by transglutaminase inhibitors as evidenced by the lack of protein aggregates in the perivitelline space and the appearance of increased amounts of ovoperoxidase in supernates of fertilized egg suspensions treated with putrescine. Moreover, in the presence of transglutaminase or ovoperoxidase inhibitors, both the quantity and the specific content of the secreted proteins that pass through the FE are altered. Thus, many key biophysical characteristics of the FE can be altered through the inhibition of a single cross-linking enzyme.

Many questions remain to be answered regarding the effects that the apparent hierarchy of cross-linking activity may have on FE assembly. Ovoperoxidase-induced dityrosyl formation imparts to the FE resistance to chemical and physical disruption. It also results in the covalent attachment of various secretory proteins into the FE. Similarly, the surface-associated transglutaminase may affect the properties of the FE by the direct cross-linking of vitelline layer matrix proteins, by cross-linking cortical secretory products, or by a combination of these activities. Whether these enzymes or other, as yet uncharacterized proteins influence other properties of the envelope and/or perivitelline environment is of great import to the understanding of how this glycocalyx assists in early development.

F. RELEVANCE TO MAMMALIAN FERTILIZATION

Some of the basic concepts provided by studying sea urchin FE assembly are of general significance to mammalian fertilization. For example, mammalian eggs contain a population of cortical granules which are secreted during fertilization and appear to be involved with modifying the zona pellucida. These vesicles contain a trypsin-like protease,[14,74] but it is less clear whether an ovoperoxidase-like protein is present as well.[75,76] It is evident that membrane depolarization and the release of intracellular calcium also coincide with cortical granule exocytosis in mammalian species.[77] However, the difficulty in collecting large quantities of mammalian eggs has hindered the identification of other cortical granule contents and the mechanisms surrounding their secretion. Thus, the sea urchin has been instrumental as a model for identifying the biochemical processes necessary for the modification of egg glycocalyxes.

The zona pellucida surrounding mammalian eggs appears to be less complex than the sea urchin vitelline envelope. The mouse zona pellucida is composed primarily of three large-molecular-weight glycoproteins, ZP1, ZP2, and ZP3.[78] ZP3 possesses sperm receptor activity and the ability to induce the acrosome reaction in sperm.[13] After cortical granule exocytosis the binding activity of these zona proteins is abolished,[79,80] and this matrix acquires an increased resistance to physical and chemical purturbation.[81] It is unknown which components of the cortical granules are responsible for these alterations. Nor is it clear what other properties are imparted to the zona pellucida during fertilization that may have bearing on the earliest stages of embryogenesis. Many of these basic questions are currently being addressed with studies of sea urchin fertilization. Investigations with these invertebrates will continue to provide insight into fundamental biochemical mechanisms that are essential to mammalian fertilization.

V. SUMMARY

Fertilization in the sea urchin lends itself to investigations aimed at a better understanding of many fundamental principles in cell biology. For example, the interplay between ionic fluxes and cell activation has been addressed extensively using this model. In the sea urchin, the control of intracellular calcium concentration via IP_3-mediated events and the role of intracellular alkalinization in cell activation have been explored. Likewise, the influence of these ionic events on the control of exocytosis has been revealed. Sea urchin fertilization has also been a well-suited model for examining the generation and controlled use of potentially lethal oxidants during critical phases of development. Finally, the sea urchin has been a valuable tool in elucidating the biochemistry attendant on the assembly of complex extracellular matrices. The prime example of this is the construction of the fertilization envelope by the sea urchin egg, which involves numerous proteolytic and covalent cross-linking events that are intimately concerned with the survival of the early embryo.

This marine invertebrate has also contributed significantly to our understanding of many other biological phenomena, including (1) the regulation of actin and tubulin polymerization, (2) influences of discrete cytoskeletal elements in cytokinesis and karyokinesis, (3) the control of sperm respiration, flagellar motility, and the acrosome reaction, and (4) the involvement of cGMP regulation in chemotaxis. Thus, sea urchin gametes represent important models for studying questions of wide biological significance. The ease with which the sea urchin can be maintained and manipulated, coupled to the extensive literature generated from their study, makes this system an attractive alternative to mammalian systems in many aspects of biology.

ACKNOWLEDGMENT

This work was supported by grants HD 00798, HD 06967, and GM 23910 from the National Institutes of Health.

REFERENCES

1. **Shapiro, B. M., Schackmann, R. W., Tombes, R. M., and Kazazoglou, T.,** Coupled ionic and enzymatic regulation of sperm behavior, *Curr. Top. Cell. Regul.,* 26, 97, 1985.
2. **Trimmer, J. S. and Vacquier, V. D.,** Activation of sea urchin gametes, *Ann. Rev. Cell Biol.,* 2, 1, 1986.
3. **Ward, G. E., Brokaw, C. J., Garbers, D. L., and Vacquier, V. D.,** Chemotaxis of *Arbacia punctulata* spermatozoa to resact, a peptide from the egg jelly layer, *J. Cell Biol.,* 101, 2324, 1985.
4. **Rossignol, D. P., Earles, B. J., Decker, G. L., and Lennarz, W. J.,** Characteristics of the sperm receptor on the surface of eggs of *Strongylocentrotus purpuratus, Dev. Biol.,* 104, 308, 1984.
5. **Steinhardt, R. A., Lundin, L., and Mazia, D.,** Birelative responses of the echinoderm egg to fertilization, *Proc. Natl. Acad. Sci. U.S.A.,* 68, 2426, 1971.
6. **Jaffe, L. A.,** Fast block to polyspermy in sea urchin eggs is electrically mediated, *Nature (London),* 261, 63, 1976.
7. **Jaffe, L. A., Gould-Somero, M., and Holland, L. Z.,** Studies of the mechanisms of the electrical polyspermy block using voltage clamp during cross-species fertilization, *J. Cell Biol.,* 92, 616, 1982.
8. **Shapiro, B. M. and Eddy, E. M.,** When sperm meets egg: biochemical mechanisms of gamete interaction, *Int. Rev. Cytol.,* 66, 257, 1980.
9. **Shapiro, B. M., Schackmann, R. W., and Gabel, C. A.,** Molecular approaches to the study of fertilization, *Ann. Rev. Biochem.,* 50, 815, 1981.
10. **Whitaker, M. J. and Steinhardt, R. A.,** Ionic regulation of egg activation, *Q. Rev. Biophys.,* 15, 593, 1982.
11. **Shen, S. S. and Steinhardt, R. A.,** Direct measurement of the intracellular pH during metabolic derepression of the sea urchin egg, *Nature (London),* 272, 253, 1978.
12. **Foerder, C. A., Klebanoff, S. J., and Shapiro, B. M.,** Hydrogen peroxide production, chemiluminescence, and the respiratory burst of fertilization: interrelated events in early sea urchin development, *Proc. Natl. Acad. Sci. U.S.A.,* 75, 3183, 1978.
13. **Bleil, J. D. and Wassarman, P. M.,** Sperm-egg interactions in the mouse: sequence of events and induction of the acrosome reaction by a zona pellucida glycoprotein, *Dev. Biol.,* 95, 317, 1983.
14. **Barros, C. and Yanagimachi, R.,** Induction of the zona reaction in golden hamster eggs by cortical granule material, *Nature (London),* 233, 268, 1971.
15. **Jaffe, L. A., Sharpe, A. D., and Wolf, D. P.,** Absence of an electrical polyspermy block in the mouse, *Dev. Biol.,* 96, 317, 1983.
16. **Shapiro, B. M.,** The existential decision of a sperm, *Cell,* 49, 293, 1987.
17. **Mazia, D.,** The release of calcium in *Arbacia* eggs on fertilization, *J. Cell. Comp. Physiol.,* 10, 291, 1937.
18. **Moser, F.,** Studies on a cortical layer response to stimulating agents in the *Arbacia* egg. I. Response to insemination, *J. Exp. Zool.,* 80, 423, 1939.
19. **Vacquier, V. D.,** Dynamic changes of the egg cortex, *Dev. Biol.,* 84, 1, 1981.
20. **Steinhardt, R., Zucker, R., and Schatten, G.,** Intracellular calcium release at fertilization in the sea urchin egg, *Dev. Biol.,* 58, 185, 1977.
21. **Steinhardt, R. A. and Epel, D.,** Activation of sea urchin eggs by a calcium ionophore, *Proc. Natl. Acad. Sci. U.S.A.,* 71, 1915, 1974.
22. **Swann, K. and Whitaker, M.,** The part played by inositol trisphosphate and calcium in the propogation of the fertilization wave in sea urchin eggs, *J. Cell Biol.,* 103, 2333, 1986.
23. **Vacquier, V. D.,** The isolation of intact cortical granules from sea urchin eggs: calcium ions trigger granule discharge, *Dev. Biol.,* 43, 62, 1975.
24. **Clapper, D. L. and Lee, H. C.,** Inositol trisphosphate induces calcium release from nonmitochondrial stores in sea urchin egg homogenates, *J. Biol. Chem.,* 260, 13947, 1985.
25. **Steinhardt, R. A., and Alderton, J. M.,** Calmodulin confers calcium sensitivity on secretory exocytosis, *Nature (London),* 295, 154, 1982.
26. **Turner, P. R., Sheetz, M. P., and Jaffe, L. A.,** Fertilization increases the polyphosphoinositide content of sea urchin eggs, *Nature (London),* 310, 414, 1984.

27. **Eisen, A. and Reynolds, G.T.,** Source and sinks for the calcium released during fertilization of single sea urchin eggs, *J. Cell Biol.,* 100, 1522, 1985.

28. **Berridge, M. J.,** Inositol trisphosphate and diacylglycerol: two intracellular second messengers, *Ann. Rev. Biochem.,* 56, 159, 1987.

29. **Turner, P. R., Jaffe, L. A., and Fein, A.,** Regulation of cortical vesicle exocytosis in sea urchin eggs by inositol 1,4,5,-trisphosphate and GTP binding protein, *J. Cell Biol.,* 102, 70, 1986.

30. **Whitaker, M. and Irvine, R. F.,** Inositol 1,4,5-trisphosphate microinjection activates sea urchin eggs, *Nature (London),* 312, 636, 1984.

31. **Irvine, R. F. and Moor, R. M.,** Microinjection of inositol 1,3,4,5-tetrakisphosphate activates sea urchin eggs by a mechanism dependent on external Ca^{2+}, *Biochem. J.,* 240, 917, 1986.

32. **Dube, F., Schmidt, T., Johnson, C. H., and Epel, D.,** The hierarchy of requirements for an elevated intracellular pH during early development of sea urchin embryos, *Cell,* 40, 657, 1985.

33. **Swann, K. and Whitaker, M.,** Stimulation of the Na/H exchanger of sea urchin eggs by phorbol ester, *Nature (London),* 314, 274, 1985.

34. **Johnson, J., Epel, D., and Paul, M.,** Intracellular pH and activation of sea urchin eggs after fertilization, *Nature (London),* 262, 661, 1976.

35. **Aronson, P. S. and Boron, W. F.,** Eds., *Current Topics in Membranes and Transport,* Vol. 26, *Na-H Exchange, Intracellular pH, and Cell Function,* Academic Press, Orlando, FL, 1986.

36. **L'Allemain, G., Franchi, A., Cragoe, E., and Pouyssegur, J.,** Blockade of the Na^+/H^+ antiport abolishes growth factor-induced DNA synthesis in fibroblasts, *J. Biol. Chem.,* 259, 4313, 1984.

37. **Warburg, O.,** Boebachtungen uber die oxydationsprozesse im seegelei, *Z. Physiol. Chem.,* 57, 1, 1908.

38. **Foerder, C. A., Klebanoff, S. J., and Shapiro, B. M.,** Hydrogen peroxide production, chemiluminescence and the respiratory burst of fertilization: interrelated events in early sea urchin development, *Proc. Natl. Acad. Sci. U.S.A.,* 75, 3183, 1978.

39. **Klebanoff, S. J., Foerder, C. A., Eddy, E. M., and Shapiro, B. M.,** Metabolic similarities between fertilization and phagocytosis. Conservation of a peroxidatic mechanism, *J. Exp. Med.,* 149, 938, 1979.

40. **Halliwell, B. and Gutteridge, J. M. C.,** *Free Radicals in Biology and Medicine,* Oxford University Press, New York, 1985.

41. **Zafiriov, O. C.,** Is sea water a radical solution? *Nature (London),* 325, 481, 1987.

42. **Petasne, R. G. and Zika, R. G.,** Fate of superoxide in coastal sea water, *Nature (London),* 325, 516, 1987.

43. **Turner, E., Klevit, R., Hager, L. J., and Shapiro, B. M.,** Ovothiols, a family of redox-active mercaptohistidine compounds from marine invertebrate eggs, *Biochemistry,* 26, 4028, 1987.

44. **Turner, E., Somers, C. E., and Shapiro, B. M.,** The relationship between a novel NAD(P)H oxidase activity of ovoperoxidase and the CN-resistant respiratory burst that follows fertilization of sea urchin eggs, *J. Biol. Chem.,* 260, 13163, 1985.

45. **Turner, E., Klevit, R., Hopkins, P. B., and Shapiro, B. M.,** Ovothiol: a novel thiohistidine compound from sea urchin eggs that confers $NAD(P)H-0_2$ oxidoreductase activity on ovoperoxidase, *J. Biol. Chem.,* 261, 13056, 1986.

46. **Babior, B. M.,** The respiratory burst oxidase, *Trends Biochem. Sci.,* 12, 241, 1987.

47. **Epel, D., Patton, C., Wallace, R. W., and Cheung, W. Y.,** Calmodulin activates NAD kinase of sea urchin eggs: an early event of fertilization, *Cell,* 23, 543, 1981.

48. **Epel, D.,** Simultaneous measurement of TPNH formation and respiration following fertilization of the sea urchin egg, *Biochem. Biophys. Res. Commun.,* 17, 69, 1964.

49. **Weidman, P. J. and Shapiro, B. M.,** Regulation of extracellular matrix assembly: *in vitro* reconstitution of a partial fertilization envelope from isolated components, *J. Cell Biol.,* 105, 561, 1986.

50. **Schroeder, T. E.,** Surface area change at fertilization: resorption of the mosaic membrane, *Dev. Biol.,* 70, 306, 1979.

51. **Carroll, E. J. and Epel, D.,** Isolation and biological activity of the proteases released by sea urchin eggs following fertilization, *Dev. Biol.,* 44, 22, 1975.

52. **Schuel, H., Wilson, W. L., Chen, K., and Lorand, L.,** A trypsin-like proteinase localized in the cortical granules isolated from unfertilized sea urchin eggs by zonal centrifugation. Role of the enzyme in fertilization, *Dev. Biol.,* 34, 175, 1973.

53. **Foerder, C. A. and Shapiro, B. M.,** Release of ovoperoxidase from sea urchin eggs hardens the fertilization membrane with tyrosine crosslinks, *Proc. Natl. Acad. Sci. U.S.A.,* 74, 4214, 1977.

54. **Weidman, P. J., Kay, E. S., and Shaprio, B. M.,** Assembly of the sea urchin fertilization membrane: isolation of proteoliasin, a calcium-dependent ovoperoxidase binding protein, *J. Cell Biol.,* 100, 938, 1985.

55. **Truschel, M. R., Chambers, S. A., and McClay, D. R.,** Two antigenically distinct forms of B-1,3-glucanase in sea urchin embryonic development, *Dev. Biol.,* 117, 277, 1986.

56. **Wessel, G. M., Truschel, M. R., Chambers, S. A., and McClay, D. R.,** A cortical granule-specific enzyme, B-1,3-glucanase, in sea urchin eggs, *Gamete Res.,* 18, 339, 1987.

57. **Hylander, B. L. and Summers, R. G.,** An ultrastructural immunocytochemical localization of hyaline in the sea urchin egg, *Dev. Biol.,* 93, 368, 1982.

58. **Bryan, J.,** On the reconstitution of the crystalline components of the sea urchin fertilization membrane, *J. Cell Biol.,* 45, 606, 1970.

59. **Anstrom, J. A., Chin, J. E., Leaf, D. S., Parks, A. L., and Raff, R. A.,** Immunocytochemical evidence suggesting heterogeneity in the population of sea urchin egg cortical granules, *Dev. Biol.,* 125, 1, 1988.

60. **Vacquier, V. D., Tegner, M. J., and Epel, D.,** Protease released from sea urchin eggs at fertilization alters the vitelline layer and aids in preventing polyspermy, *Exp. Cell Res.,* 80, 111, 1973.

61. **Loeb, J.,** *Artificial Parthenogenesis and Fertilization,* University of Chicago Press, Chicago, 1913, 218.

62. **Hiramoto, Y.,** Nature of the perivitelline space in sea urchin eggs I, *Jpn. J. Zool.,* 11, 227, 1955.

63. **Chambers, R.,** The intrinsic expansibility of the fertilization membrane of echinoderm ova, *J. Cell. Comp. Physiol.,* 19, 145, 1942.

64. **Carroll, E. J.,** Cortical granule proteases from sea urchin eggs, *Methods Enzymol.,* 45, 343, 1976.

65. **Chandler, D. E. and Kazilek, C. J.,** Extracellular coats on the surface of *Strongylocentrotus purpuratus* eggs: stereo electron microscopy of quick-frozen and deep-etched specimens, *Cell Tissue Res.,* 246, 153, 1986.

66. **Inoue, S. and Hardy, J. P.,** Fine structure of the fertilization membranes of sea urchin embryos, *Exp.Cell Res.,* 68, 259, 1971.

67. **Veron, M., Foerder, C., Eddy, E. M., and Shapiro, B. M.,** Sequential biochemical and morphological events during assembly of the fertilization membrane of the sea urchin, *Cell,* 10, 321, 1977.

68. **Hall, H. B.,** Hardening of the sea urchin fertilization envelope by peroxidase-catalyzed phenolic coupling of tyrosines, *Cell,* 15, 343, 1978.

69. **Deits, T. and Shapiro, B. M.,** pH-induced hysteretic transition of ovoperoxidase, *J. Biol. Chem.,* 260, 7882, 1985.

70. **Lallier, R. A.,** Formation of fertilization membranes in sea urchin eggs, *Exp. Cell Res.,* 63, 460, 1970.

71. **Lallieer, R. A.,** Effects of various inhibitors of protein crosslinking on the formation of fertilization membrane in the sea urchin, *Experientia,* 27, 1323, 1971.

72. **Takeuchi, K.,** Protein cross-linking during the formation of the hardening fertilization envelope in sea urchin eggs; the possible role of transglutaminase, *J. Cell Biol.,* 97, 25a, 1983.

73. **Lorand, L. and Conrad, S. M.,** Transglutaminases, *Mol. Cell. Biochem.,* 58, 9, 1984.

74. **Gwatkin, R. B. L., Williams, D. T., Hartman, J. F., and Kniazuk, M. J.,** The zona-reaction of hamster and mouse eggs production *in vitro* by a trypsin-like protease from cortical granules, *J. Reprod. Fertil.,* 32, 259, 1973.

75. **Gulyas, B. J. and Schmell, E. D.,** Ovoperoxidase activity in ionophore treated mouse eggs: I. Electron microscopic localization, *Gamete Res.,* 3, 267, 1980.

76. **Gulyas, B. J.,** Cortical granules of mammalian eggs, *Int. Rev. Cytol.,* 63, 357, 1980.

77. **Steinhardt, R. A., Epel, D., Carroll, E. J., and Yanagimachi, R.,** Is calcium the universal activator for unfertilized eggs?, *Nature (London),* 252, 41, 1974.

78. **Bleil, J. D. and Wassarman, P. M.,** Structure and function of the zona pellucida: identification and characterization of the proteins of the mouse oocyte's zona pellucida, *Dev. Biol.,* 76, 185, 1980.

79. **Bleil, J. D. and Wassarman, P. M.,** Mammalian sperm-egg interaction: identification of a glycoprotein in the mouse egg zona pellucida possessing receptor activity for sperm, *Cell,* 20, 873, 1980.

80. **Florman, H. M. and Wassarman, P. M.,** O-linked oligosaccharides of mouse egg ZP 3 account for its sperm receptor activity, *Cell,* 41, 313, 1985.

81. **Schmell, E. D., Gulyas, B. J., and Hedrick, J. L.,** Egg surface changes during fertilization and the molecular mechanisms of the block to polyspermy, in *Mechanism and Control of Animal Fertilization,* Hartman, J. F., Ed., Academic Press, New York, 1983, 356.

82. **Turner, E. and Shapiro, B. M.,** unpublished results.

83. **Somers, C. E., Battaglia, D. E., and Shapiro, B. M.,** Localization and developmental fate of ovoperoxidase and proteoliasin, two proteins involved in fertilization envelope assembly, *Dev. Biol.,* 131, 226, 1989.

84. **Battaglia, D. E. and Shapiro, B. M.,** Hierarchies of protein cross-linking in the extracellular matrix: involvement of an egg surface transglutaminase in early stages of fertilization envelope assembly, *J. Cell Biol.,* 107, 2447, 1988.

85. **Heinecke, J. W. and Shapiro, B. M.,** Respiratory burst oxidase of fertilization, *Proc. Natl. Acad. Sci, U.S.A.,* 86, 1259, 1989.

Chapter 9

ELASMOBRANCHS (SHARKS, SKATES, AND RAYS) AS ANIMAL MODELS FOR BIOMEDICAL RESEARCH

Carl A. Luer

TABLE OF CONTENTS

I. INTRODUCTION

A. EVOLUTIONARY SIGNIFICANCE

Typical reactions to the suggestion that sharks, and their close relatives the skates and rays, can be extremely valuable animal models in several areas of biomedical research usually range from disbelief or skepticism to an acknowledgment that, as "primitive" vertebrates, they may be useful in phylogenetic comparisons with mammalian systems. While it is generally accepted that present-day members of class Chondrichthyes (cartilaginous fish) are descendants of animals which branched from the main line of vertebrate evolution during the Devonian period some 400 million years ago,[1,2] it is an unfortunate misconception that sharks, skates, and rays (collectively known as elasmobranchs) represent a "primitive" group of vertebrates. This misconception probably originated from the realization that many anatomical features which have been retained through elasmobranch evolution correspond with features which are present early in mammalian ontogeny but which are not retained in adult mammalian forms. For this reason, the elasmobranch fishes have been represented as being phylogenetically primitive, as well.

In reality, they have evolved to extremely advanced creatures in many respects, having developed certain functional adaptations in common only with the "more highly evolved" birds and mammals, and possessing still other features which are unique among the vertebrate animals.

B. HISTORICAL PERSPECTIVE

The early shark literature, from the late 1800s through the early 1900s, has taken advantage of the anatomical systems which are considered primitive to the mammalian forms. In fact, many of the basic concepts regarding body organization in vertebrates were established from studies of elasmobranch fish. While not intended as an exhaustive review of the early literature, classical elasmobranch studies contributing to the basic understanding of vertebrate systems include investigations into certain muscular arrangements,[3-5] circulatory systems,[6-9] kidney function,[10-13] brain and central nervous system organization,[14-19] vision,[20-24] olfaction (chemoreception),[25-27] hearing (mechanoreception),[28-31] reproduction and reproductive strategies,[32-38] and staging of embryonic development.[39-47]

While these and other early classics have helped to establish a foundation for much of our present knowledge of vertebrate systems, the elasmobranch fishes have remained relatively underutilized as sources of material for biomedical research. Only during the past decade or two have reports begun to increase in the scientific literature, demonstrating that many investigators are gradually appreciating the relevance of body systems and tissue extracts from sharks and their relatives. In some cases, the system or molecule has been employed because of its functional similarity with mammalian counterparts. Still other investigators have taken advantage of systems which cannot be duplicated using traditional mammalian models. The least-appreciated application of elasmobranchs to biomedical research, however, is their value as laboratory animals to be used as experimental models. This chapter, then, will survey some of the more representative recent literature utilizing elasmobranchs as sources of material for biomedically relevant investigations, followed by a more detailed discussion of how sharks and skates can clearly benefit the biomedical researcher as alternative laboratory animal models.

II. ELASMOBRANCH FISHES AS SOURCES OF RESEARCH MATERIAL

A. FEATURES IN COMMON WITH MAMMALIAN MODELS

Perhaps the most accessible biological material from elasmobranch fishes is their blood.

While many aspects of shark blood have been studied through the years, several recent investigations are of biomedical significance since they probe the natural defense mechanisms of this successful group of animals. Recent studies of shark and skate antibody populations have revealed a better understanding of their light and heavy chain structures[48-50] as well as the distribution and characterization of immunoglobulin-forming cells.[51,52] Other serum studies have led to the purification and characterization of a six-component complement system which is functionally indistinguishable from the nine-component system typical of mammals.[53]

Another defense mechanism has been investigated as a result of a better understanding of the peripheral blood cells of elasmobranchs[54,55] and may play a role in the apparent natural resistance to invasive tumors attributed to the elasmobranchs. This mechanism involves the spontaneous cytotoxic activity displayed by shark leukocytes, with the implication that the macrophage-like cell identified in the reaction might closely resemble the activated mammalian tumoricidal macrophage.[56-58] Additional investigations into antitumor substances from shark tissue or tissue extracts have utilized serum, liver, and cartilage to demonstrate varying degrees of inhibitory activity against Rous sarcoma virus,[59,60] Lewis lung carcinoma,[61] PS leukemia,[62] neovascularization,[63,64] and proteolytic enzymes.[64,65]

Recent attention has also been paid to elasmobranch tissue as sources of purified proteins or functional protein activities to be used in comparative studies with their mammalian counterparts. The biomedical relevance of such studies lies in the realization that the elasmobranch proteins, which evolved millions of years before the mammalian proteins, may contain valuable structure-function information. By establishing differences and similarities of such parameters as amino acid composition, primary, secondary, tertiary, and quaternary structures, molecular sizes, conditions for optimum activity, substrate specificities, and tissue distributions, important information can be revealed about fundamental functional mechanisms. Degrees to which individual features are invariant can lead to an understanding of those aspects of a protein which are critical to its function. Also, identification of those properties which have not been retained through evolution to mammalian systems can be recognized as being no longer required under such a different set of environmental conditions.

Representative elasmobranch proteins which have been isolated and purified or whose activities have been characterized during recent years are included here with their tissues of origin. From shark and skate serum, such proteins as retinol-binding protein,[66] transferrin,[67] and ceruloplasmin[68] have been identified, while from shark erythrocytes, hemoglobin[69] and carbonic anhydrase[70] have been purified. Various muscle tissue has lead to the detailed characterization of shark myoglobin,[71-79] while elasmobranch liver has resulted in many protein isolations including metallothioneins for cadmium, zinc, and copper,[80] steroid-binding proteins,[81,82] glutamine synthetase,[83] superoxide dismutase,[84] glutathione S-transferase,[85,86] cytochrome P-450,[87] cytochrome b_5,[87] cytochrome c reductase,[88] and various enzymes in the microsomal mixed function oxidase system.[89-95] Elasmobranch hormones, including insulin,[96] glucagon,[97] somatostatin,[98] and vasoactive intestinal peptide (VIP),[99,100] have been purified from pancreatic and intestinal tissue with interesting degrees of sequence homologies. Another hormone, relaxin, has been purified from the ovaries of live-bearing as well as egg-laying elasmobranchs,[101,102] contrary to the popular belief that this hormone is restricted to the mammals and birds. The successful isolation of enamelin-like proteins from the developing enameloid shark tooth organs has also recently been reported,[103] with implications that this could help elucidate the role of these proteins in the formation of enamel, the hardest tissue in the vertebrate body.

B. FEATURES UNIQUE TO THE ELASMOBRANCH FISHES

Tissues which are unique to elasmobranch systems have also been useful as sources of proteins. The electric organs of the electric rays, for example, have been a valuable source of acetylcholinesterase, the enzyme responsible for the hydrolysis of the neurotransmitter, acetylcholine, at cholinergic synapses. A second type of acetylcholinesterase is also described from extrajunctional areas of nerves and muscles, and on erythrocyte membranes. While catalytically identical, the two types differ in subunit structure and in their substrate and inhibitor specificities. Hydrodynamic analyses have characterized the synaptic form as asymmetrical, with a collagen-like structural subunit disulfide linked to the catalytic subunits, and the outer surface membrane form as being globular. Both forms are readily available from electric ray electric organs, where primary structures have been deduced from complementary DNA clones.[104-106] In addition, monoclonal antibodies have been raised against acetylcholinesterase purified from electric rays[107] which may prove valuable as probes to understand the molecular diversity of this important enzyme.

Another biomedically relevant tissue which is unique to the elasmobranch fishes is the rectal gland. Also known as the digitiform gland because of its finger-like morphology, this gland is not restricted to a particular group of elasmobranchs, as is the case with the electric organ, but rather is a feature common to all sharks, skates, and rays. The rectal gland plays an important osmoregulatory role, being able to remove salts from the blood and secrete them into the intestine.[108] In this regard, the elasmobranch rectal gland has been employed as a useful model for the sodium-coupled electrogenic chloride transport found in a variety of mammalian tissues.[109,110] In particular, studies to date have provided evidence that chloride transport in the rectal gland has many features in common with the electrolyte transport in the mammalian thick ascending limb of Henle's loop. Advantages to experimental usage of this elasmobranch gland include its relatively large size (0.06% of total body weight for the spiny dogfish, *Squalus acanthias*),[111] its high concentration of sodium, potassium-ATPase per weight of gland tissue, and the relative ease with which isolated tubules and membrane vesicles are prepared for experimentation.[112,113] Recent studies have focused their attention on understanding the hormonal[114-120] and neural control of the active chloride transport,[121] with implications of a better understanding of mammalian kidney function.

The elasmobranch kidney is unique among vertebrate excretory systems. Even though its tubules and capillaries exist in a countercurrent multiplier arrangement in much the same fashion that the loops of Henle and the vasa recta are arranged in both birds and mammals,[122] the elasmobranch kidney does not form a hypertonic urine. In an effort to achieve osmotic and ionic homeostasis, a mechanism has evolved in the elasmobranch fishes which allows them to maintain high concentrations of urea and trimethylamine oxide in their plasma and tissues to the extent that their internal osmolality is the same or slightly higher than that of the surrounding seawater.[13,123-127] The countercurrent system in elasmobranchs, then, probably facilitates the net reabsorption of the waste nitrogenous osmolytes.

The detailed physiological understanding of the elasmobranch kidney has been hampered somewhat by the extreme complexity of its anatomical aspects. Only recently have efforts been successful in addressing the enzymatic and chemical differences in the many zones of renal tissue.[128,129] Of special significance are the elegant series of fine structural investigations which, for the first time, allow detailed anatomical knowledge to aid in the understanding of functional mechanisms.[130-133] As a result, it is now apparent that many of the components of shark and skate kidneys share characteristics in common with fetal and newborn mammalian kidneys. Structural features such as the slit membrane,[134] occluding junctions,[135] double basement membrane,[134,136] and endothelial cells with few irregularly distributed pores or fenestrations[137-140] are characteristic of the glomerulus of both the developing mammalian and the elasmobranch kidney.

As a consequence of retaining high tissue concentrations of urea for osmotic homeostasis, the elasmobranchs have placed unique constraints on the environment in which their biomolecules must function. For most mammalian proteins, even very low concentrations of urea can perturb their structures and affect their function. Among the elasmobranchs, whose tissue and plasma concentrations of urea average some 400 mM,[126] there is growing evidence that many of their proteins may have evolved a natural resistance to urea perturbation. For example, high urea concentrations appear to be required to prevent cold precipitation of lens proteins in the dogfish eye.[141] Urea also appears to be necessary in order to observe correct pyruvate binding by lactate dehydrogenase.[142] Other studies have demonstrated that elasmobranch actomyosin ATPase is more urea resistant than those of teleosts,[143] while the activity of this enzyme was shown to increase significantly when 200 to 300 mM urea was added to standard reaction mixtures.[144] Furthermore, reconstitution studies have shown that shark actomyosin and myofibril proteins were about 10 times as resistant to urea as the shark myosin alone.[145] In a very dramatic display of urea resistibility, the oxygen affinity of shark, skate, and ray hemoglobin has been shown essentially to be unaltered at urea concentrations as high as 5 M.[146] In the same studies, it was also demonstrated that the pH dependence of oxygen binding, as well as the degree of heme-heme interaction, remained unaffected at the same high concentrations of urea. While the exact mechanisms responsible for the urea resistance observed for elasmobranch proteins are not yet understood, speculations include an unusual degree of carbamylation of amino groups[146] and counteraction of urea perturbation by trimethylamine oxide.[147,148] A third possible explanation involves increased structural stability through increased disulfide bonds. An example of this proposed mechanism is found in carbonic anhydrase, whose bovine and human forms have no disulfide cross-links. Amino acid compositions of bull and tiger shark carbonic anhydrases, however, show the potential for 12 and 9 disulfide bonds, respectively. This is especially intriguing in light of the observation that the tertiary environment of the metal-binding site is nearly identical in both the mammalian and the elasmobranch enzymes.[70]

As mentioned previously, one of the anatomical systems which has benefited from early work with elasmobranchs is the visual system. While it is recognized that, in most respects, the structural architecture of the elasmobranch retina resembles that of other vertebrates, the skate can offer a unique feature in that its layer of photoreceptor cells is composed entirely of rods.[149] The evidence for such a valuable model system is functional as well as histological, with no indication of the rod-cone discontinuity in the dark adaptation curve characteristic of mixed retinae. In addition, fundus reflectometry,[149] partial bleaching of the isolated retina,[150] and action spectrum comparisons under varying conditions,[151,152] have supported the presence of rhodopsin as the sole visual pigment. The value of an all-rod retinal model has allowed several detailed investigations into mechanisms of scotopic vision which were uncomplicated by the presence of other classes of visual cells.[153-157]

A final uniquely elasmobranch tissue with potential value in biomedical research is the nidamental gland. Also known as the shell gland, this tissue is found along the reproductive tract of female elasmobranchs and serves several functions, including the secretion of an albumin-like substance and a tough collagenous membrane around the fertilized egg. The function of interest to biomedical researchers, however, is its ability to store sperm in a viable state for very long periods of time.[158-160] Following copulation, sperm make their way up the uterus to the nidamental gland, where they can be stored for many months[161] or possibly even years,[162] depending upon the species. The mechanism by which the sperm remain viable for such prolonged periods of time has yet to be investigated fully.

III. SHARKS AND SKATES AS LABORATORY ANIMALS

A. GENERAL COMMENTS

The benefits of maintaining elasmobranch fishes as laboratory animals for biomedically relevant experimentation are certainly underappreciated. While most think of large sharks whenever this group of fish is mentioned, sharks of small size (less than 3 ft total length) are far more common and relatively easy to handle and maintain. These include sharks and skates whose size when fully grown is small, as well as those varieties which remain small enough to handle during their first few years of growth.

While certainly not limited to this number, a list of commonly maintained elasmobranchs for experimental purposes includes the spiny dogfish *(Squalus acanthias)*, smooth dogfish *(Mustelus canis)*, horn shark *(Heterodontus francisci)*, nurse shark *(Ginglymostoma cirratum)*, chain dogfish *(Scyliorhinus retifer)*, leopard shark *(Triakis semifasciata)*, lemon shark *(Negaprion brevirostris)*, Pacific blacktip reef shark *(Carcharhinus melanopterus)* shovelnose guitarfish *(Rhinobatos productus)*, little skate *(Raja erinacea)*, and clearnose skate *(Raja eglanteria)*.

B. SHARKS

Probably the most extensively utilized shark throughout the scientific literature is the spiny dogfish *(Squalus acanthias)*,[163] which migrates northward from the Carolina coast to Maine and Nova Scotia each spring. Nearly every marine laboratory along this range utilizes and maintains this species for scientific investigations. The spiny dogfish is not limited to eastern U.S. shores but is also a regular inhabitant of the northern Pacific[164] and European waters, as well.[165] Long-term maintenance of adults and their embryonic offspring has been well documented.[166-168]

While a ready supply has dictated the utilization of the spiny dogfish, space considerations have popularized the usage of less freely swimming elasmobranchs. Contrary to popular opinion, not all sharks need to swim constantly in order to respire adequately. Those sharks which can exist with a more sedentary lifestyle are of prime consideration for captive maintenance and laboratory usage. For this reason, the horn shark and young specimens of the lemon shark and nurse shark are recommended. In a proper aquarium system,[169,170] these species can easily be maintained for 5 to 10 years, with 25 years having been claimed for the nurse shark.[171]

Much of the early work leading to a better understanding of captive maintenance originated from behavioral studies with the lemon[172-175] and nurse[176-178] sharks. These studies, which demonstrated through classical conditioning studies that sharks were able to learn tasks at quite a respectable rate, progressed from outdoor, tidally fed pens to sophisticated environmentally controlled experimental tanks. State-of-the-art recommendations for keeping lemon sharks in captivity for research purposes have been described previously[169,170] and should be consulted when maintaining that species.

Nurse sharks are commonly maintained at marine laboratories and display aquaria alike. They have the distinct advantage of having a more docile disposition than the lemon sharks, an important consideration when selecting a laboratory animal. Nurse sharks are found in tropical waters and can be captured as newborn pups or as juveniles. Collection can be made by hand while tank diving, by hook and line, or by setting traps which provide protective cover. In the laboratory, nurse sharks can be maintained in environmentally controlled recirculating seawater tanks (Figure 1). Compartmentalization of animals for purposes of separating experimental treatments is easily achieved. Rectangular tanks or sections of tanks approximately 3 × 4 ft in area with about 14 in. of water depth (total volume of about 105 gal, or 400 l) are sufficient for maintaining up to four nurse sharks in the 1- to 1.5-kg range, or two to three animals weighing 2 to 2.5 kg each.

Adequate filtration is critical in the captive maintenance of any shark. For nurse sharks, it is recommended that each animal-holding compartment be constructed with its own undergravel biological filter in addition to an external biological filter, which acts as a central filter for several individual compartments. With this configuration, water leaving each tank is exposed to two biological filter beds before returning to the animal contact area. No more than four to six experimental compartments are recommended per external filter, assuming the external filter volume is approximately that of a single holding tank. Water quality should be monitored regularly, keeping temperature around 25°C, salinity between 31 and 35 parts ‰, nitrite (NO_2-N) concentrations below 0.1 mg/l, and nitrate (NO_3-N) levels not to exceed

FIGURE 1. Nurse sharks *(Ginglymostoma cirratum)* up to about 2.5 kg total body weight can be easily maintained in environmentally controlled, recirculating seawater holding tanks similar to the one pictured. Because of their sedentary behavior, nurse sharks can be compartmentalized for experimental treatments at densities which are described in the text.

approximately 50 mg/l. Occasional additions of deionized water can be made to help control salinity increase due to evaporation, while periodic seawater exchanges can be employed to minimize nitrate accumulation.

Nurse sharks grow slowly, but even up to the age of about 3 years, can still be easily handled for routine laboratory manipulations, such as weighing, measuring, blood sampling, chemical injections, oral dosing, cannulation, and anesthetic treatment. Nurse sharks respond well to repeated anesthesia using 3-aminobenzoic acid ethyl ester methanesulfonate (MS-222, Sandoz).[179] In fact, no adverse effects, as judged by growth and behavior, were apparent following weekly anesthesia of nurse sharks for a period of 3 years.[180]

Experimentally, nurse sharks have been utilized as laboratory animals in aquarium conditions similar to those described previously. In an effort to examine the paucity of tumors in elasmobranchs through biochemical mechanisms, nurse sharks have been exposed to a chemical carcinogen (aflatoxin B_1) under controlled environmental conditions. Acute and chronic exposures have utilized oral dosing by way of capsules concealed in a piece of cut fish as part of a regular feeding. Intramuscular injections of carcinogens have also been employed with success. In either case, metabolic byproducts, mutagenic activity, detoxifying adduct formation, and DNA binding have been monitored.[95,181-183] Monooxygenase activity in this species appears to be very low compared with mammalian systems. The principle metabolite is aflatoxicol, produced by a cytosol reductase rather than the mixed function oxidases. Consistent with the low monooxygenase activity is the low ability of nurse shark liver to activate aflatoxin B_1 to a mutagen as monitored by the Ames mutagenicity assay. Also, no direct evidence for binding of aflatoxin B_1 or its metabolites to hepatocyte DNA has been observed.

C. SKATES

A potential disadvantage to both the lemon and nurse shark is the fact that, as useful laboratory animals, they are in their early stages of juvenile or immature development. They eventually grow too large for convenient handling and maintenance. Also, some questions which are asked experimentally may require the use of sexually mature specimens. For these reasons, an alternative elasmobranch model for consideration is the skate. A close relative of the shark, skates are found in every ocean of the world and are therefore widely available. Skates have been extensively studied in both Japan and Europe but do not have the same visibility in the U.S. Since their biochemistry and physiology closely resemble those of the sharks, virtually any research application considered for the sharks could also be suited to the skates.

Except for those species which are exceptionally large, skates are easily maintained in captivity for experimental needs. The same experimental tank setup described previously for nurse sharks is just as applicable for skates. The only change of any significance is the holding density of animals. Because of their body morphology and since they occupy the bottom surface of the tank, it is recommended that the tank area might be slightly larger.

The two most common species of skate from U.S. waters utilized for experimental purposes are the little skate, *Raja erinacea,* and the clearnose skate, *Raja eglanteria* (Figure 2). The little skate ranges from the Gulf of St. Lawrence to Virginia[158] and is regularly studied by marine laboratories in that area. The clearnose skate ranges from Texas to central Florida in the Gulf of Mexico and in the Atlantic Ocean from central Florida to Cape Cod. Their appearance inshore in the areas of distribution overlap is temperature dependent, the little skate preferring the colder temperature ranges.[184] Temperatures in the range of 20 to 22°C are recommended for the captive maintenance of the clearnose skate.[161]

The primary advantage to using skates as laboratory animals lies in the realization that they will reproduce in captivity,[161] something not yet achieved with any species of shark routinely maintained for research purposes. As experimental animals, then, skates are available as adults to be used for tissue and blood sampling, chemical exposure studies, or reproduction studies; as juveniles, to be used as bioassay organisms or in studies utilizing larger numbers of individuals in a given tank space (Figure 3); or, as embryos, to follow the normal development of the various organ, sensory, or tissue systems, or to assess altered development resulting from direct or indirect (maternal) exposures to toxins, carcinogens, or teratogens.

While the entire range from viviparity to oviparity is represented among the elasmobranch fishes, all skates are egg layers. In this regard, they have many reproductive aspects in common with the chick. However, there are many features unique to skate reproduction and development which contribute to their value as experimental animal models. The elasmobranch reproductive system has the distinction of being more like that of human beings than is found in any other group of nonmammalian vertebrates.[185] Anatomically, female skates possess paired ovaries, a common ostium, paired oviducts, paired nidamental glands, and a Y-shaped uterus having a single external opening (Figure 4). Using a mechanism similar to that found in mammals, individual ova are released into the body cavity. With the aid of cilia strategically placed along the peritoneal wall, outer surface of the oviducts,

FIGURE 2. Clearnose skates *(Raja eglanteria)* are excellent laboratory animals and are available for biomedical research as adults, juveniles, and embryos. The holding tank pictured (4 × 4 ft, with about 14 in. water depth) is sufficient space for a mature female (left) to breed with a mature male (right) and lay fertile eggs.

and portions of the liver, bile duct and hepatic portal vein,[38] ova are directed to the ostium, where they enter and travel down the oviducts. In reality, two eggs are released, presumably one from each ovary, so that fertilization and egg laying occurs in pairs. Fertilization takes place in the nidamental gland, where sperm have been stored from a recent or previous copulation, followed by the encapsulation of the fertile egg in a leathery egg case.

For the clearnose skate,[161]egg pairs are laid every 4 to 5 d during the egg-laying season (January through May) to the extent that each female will lay between 30 and 40 pairs of eggs. Each egg case, bearing prominent tendrils, or horns, at each corner, contains a single egg, or yolk (Figure 5). On the day of laying, each yolk already has a visible blastodisc on which the embryo will develop. Figures 6 through 8 show the changes which are observed between 4 and 6 d after egg deposition. Migration of cells at the edge of the blastodisc to form the primitive streak followed by the gradual closure of the neural tube and head region all take place during this time period. By 2 weeks after laying, only the middle section of the embryo is attached to the yolk mass, and blood can be seen circulating (Figure 9). Developing eyes, brain, gill arches, heart, alimentary tract, and myomeres are clearly visible at this stage.

While the embryo continues to grow over the next few weeks, the spreading germ ring completes its encompassing of the yolk mass to form the yolk sac membrane. Until this membrane is completed, the yolk mass is not structurally self-supportive but is stabilized in the egg case by a clear gel-like material. This material fills the space in the egg case not occupied by the yolk mass and serves as a plug for the respiratory canals formed along the horns of the egg case. As the yolk sac membrane is developing, the gel becomes less viscous until it either dissolves or attains the same viscosity as the surrounding seawater. It takes from 25 to 28 d for the canals to become unplugged and for seawater to enter the egg case. After this time, the embryo can be removed from its protective environment and will continue to develop, provided proper seawater conditions are maintained. Clearnose skate embryos

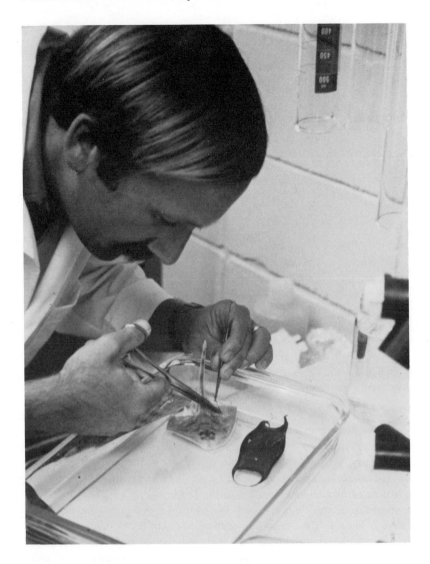

FIGURE 3.   Following an incubation period of about 12 weeks at 20 to 22°C, clearnose skates will hatch from their protective egg case (right). Newly hatched specimens can be tagged, as shown here, and used as laboratory raised research animals.

4 weeks into development, then, can be removed from their egg cases for experimental manipulation. By this stage of development (Figure 10), tiny gill filaments can be seen protruding from the gill slit area. These will continue to proliferate over the next few weeks and become a distinctive feature of embryos during the second trimester of their incubation. Also during the fourth week, pigment appears in the eyes, and the pectoral fin, or disc, so characteristic of the skates and rays, begins its lateral growth. By the fifth week (Figure 11), the disc is about as long as it is wide, and the pelvic fins of male embryos show their first signs of differentiation into claspers.

At 7 weeks, embryos (Figures 12 and 13) begin to assume features more typical of newborn skates. The rostral cartilage has expanded anteriorly, gills are more prominent at the expense of resorbing gill filaments, the mouth and nares are easily distinguishable, and the earliest indications of body pigment are visible in isolated areas of the dorsal surface. By 8 to 9 weeks of development (Figure 14), embryos possess their newborn coloration and have totally resorbed their external gill filaments. From this time until hatching, little happens

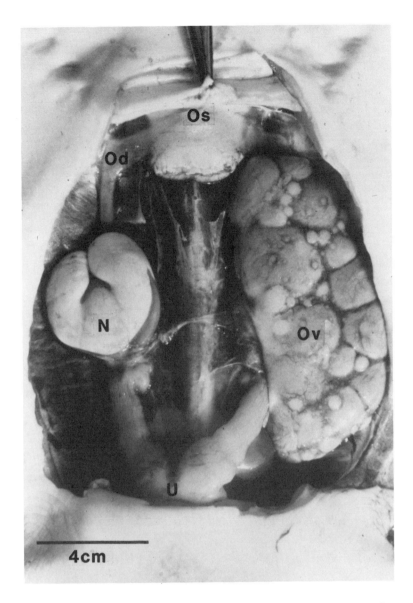

FIGURE 4.   Ventral view of the reproductive tract of a female clearnose skate. One
ovary has been removed so that the oviduct and nidamental gland on that side can be
viewed. In the remaining ovary, ova at varying stages of maturation are present. (Ov,
ovary; Os, ostium; Od, oviduct; N, nidamental gland; U, uterus).

externally except an increase in body size at the expense of a gradually diminishing yolk
sac (Figure 15). Considerable internal yolk is also present and will help to nourish the
animals for a short time after hatching.

Egg cases allowed to incubate at a constant temperature (20 to 22°C) and photoperiod
(12 h light, 12 h dark) develop and hatch after an average of 12 weeks (85 ± 6 d; range,
74 to 94 d). Body dimensions for newly hatched clearnose skates average 13.7 ± 0.5 cm
in total length (range, 12.6 to 14.9 cm), and 9.4 ± 0.5 cm in disc width (range, 7.5 to
10.6 cm). No statistical difference in body dimensions is found between male and female
newborns.

The novel reproductive physiology and behavior typical of the skate underscore its

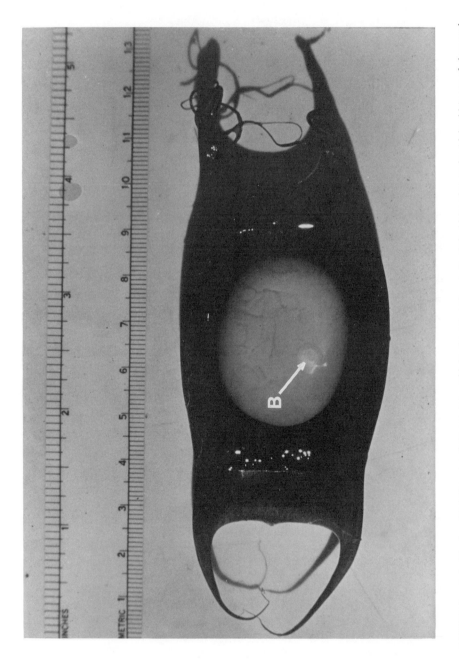

FIGURE 5.   A freshly deposited clearnose skate egg case, with its characteristic anterior (left side) and posterior (right side) sets of elongated horns. A window has been cut into one side of the case to reveal the yolk mass, which already possesses an easily visible blastodisc (B).

133

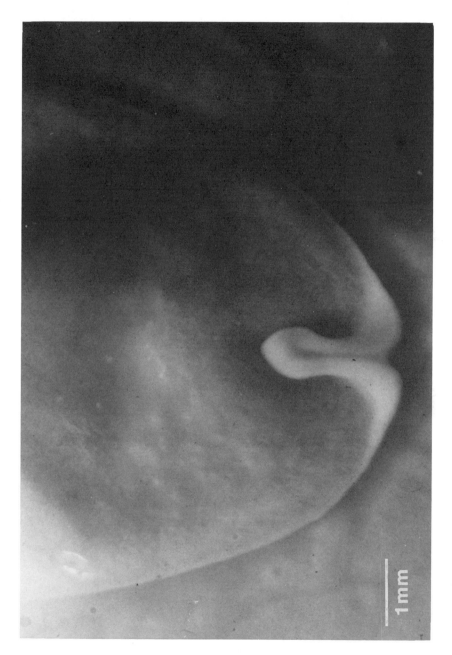

FIGURE 6. Blastodisc 4 d (96 h) after egg deposition, showing the migration of cells to form the primitive streak.

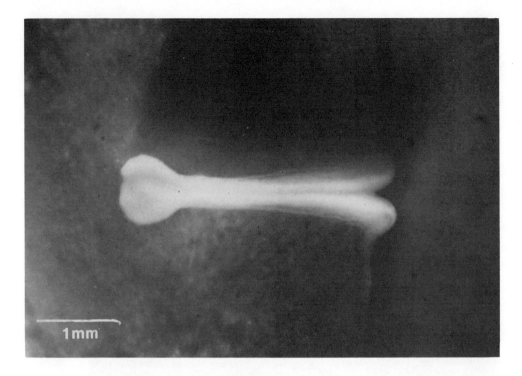

FIGURE 7.   Embryonic tissue 5 d (120 h) after egg deposition, showing the neural groove as well as the broadening and upward folding of tissue in the head region (left).

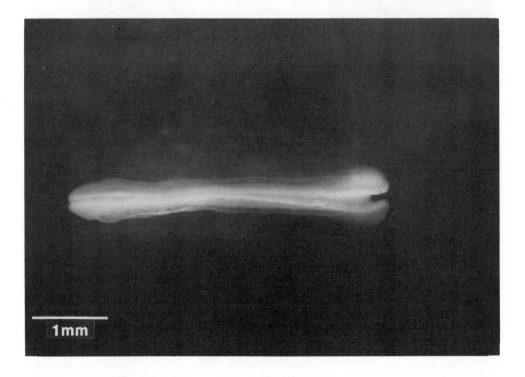

FIGURE 8.   Embryonic tissue 6 d (144 h) after egg deposition. Closure of the neural tube in the head region (left) is complete, with the anal region (right) being the last to close.

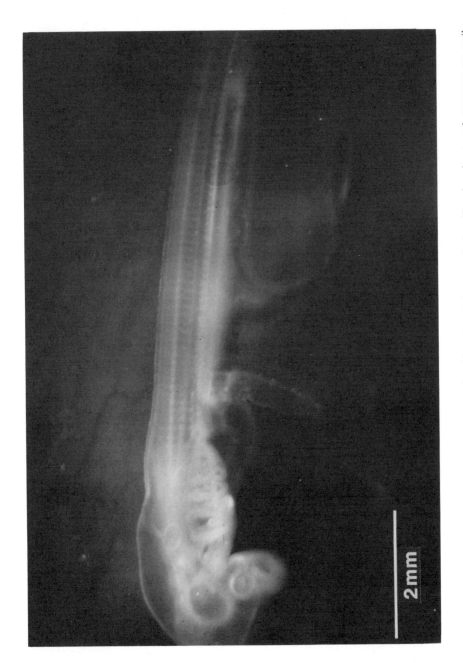

FIGURE 9. Clearnose skate embryo 2 weeks after egg deposition. Developing eyes, brain, gill arches, heart, and myomeres are easily distinguishable. Also, blood is freely circulating between the embryo and its yolk sac. (Total length of embryo is 1.8 cm.)

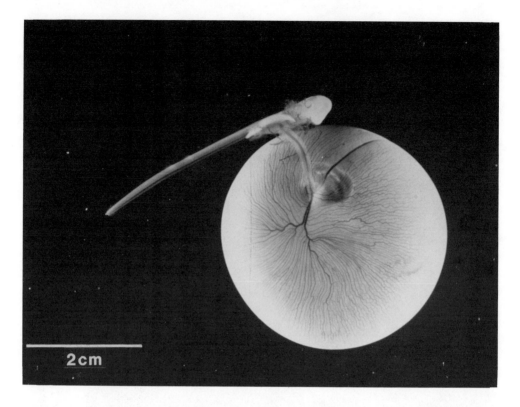

FIGURE 10.   Clearnose skate embryo 4 weeks after egg deposition. Pectoral and pelvic fins are beginning to develop, and external gill filaments are beginning to proliferate from the developing gill region. After approximately 4 weeks of incubation, the respiratory pores in the egg case horns become unblocked, allowing seawater to circulate through the egg case. At this stage of development, embryos can survive outside of their egg case and be utilized as experimental embryonic models. (Total length of embryo is 4.3 cm.)

significance as an experimental animal. In particular, this animal would be an excellent model for characterizing the mechanisms of toxicological and teratological effects of environmental chemicals on reproduction and embryonic development. Studies with the clearnose skate have identified several features of skate reproduction especially applicable to experimental designs which cannot be achieved with traditional animal models. Examples are discussed next.

Toxicological effects at different stages of embryonic development — During the egg-laying season, fertilized females lay eggs in pairs, one from each side of the paired reproductive tract described earlier. Pairs of eggs are laid every 4 to 5 d to the extent that each female can lay up to 30 to 40 pairs of eggs. Since all eggs laid could be the result of a single controlled laboratory breeding, sibling embryos covering a broad range of development are available at any one time. Studies designed to characterize the effects of an environmental agent on various stages of embryonic development are well suited to this model.

Mechanisms involved in both normal and abnormal development — With the unusual nature of fertilizing and laying eggs in pairs, several unique experimental designs are obvious. As each pair is laid, one egg can be maintained as a control, while the other can be manipulated experimentally, resulting in an many as 30 to 40 control and 30 to 40 experimental embryos from each adult mated pair. Being the same age and from the same parents, relative effects during embryogenesis of each pair can be assessed with confidence.

Effects of environmental agents on gametogenesis — Since eggs are ovulated and fertilized over an approximately 5-month period, it is not unexpected to find that the ovaries

FIGURE 11. Clearnose skate embryo 5 weeks after egg deposition, showing continued growth of the pectoral disk and early development of the rostral area. (Total length of embryo is 5.5 cm; disc width is 1.2 cm.)

of a clearnose skate contain ova of vastly different stages of maturation. The ova range in size from those which are ready to be ovulated to many others of decreasing sizes which will grow to mature size by the time they are ready to be released from the ovary. The smallest eggs found in the skate ovaries are eggs which will not mature until the following year's breeding season. Experimentally, cumulative uptake and long-term storage of chemical substances as well as relative sensitivities of eggs at different stages of maturation can be determined.

Sperm survivability — One copulatory act is all that is required for females to produce fertilized eggs throughout the entire egg-laying season (January through May). This means that sperm are stored in the female and remain viable for at least 5 months. The implication is that the effects of environmental agents on the storage of viable sperm can be explored with this animal model.

Embryonic development effects through maternal exposure — For studies involving maternal dosing, the female can be allowed to lay several pairs of eggs before exposure. These would be considered control eggs and embryos to be compared with those eggs laid after dosing commences. Control and experimental offspring would then be siblings produced as the result of a single breeding act. An additional control would be to allow the same male skate to fertilize several female skates so that the male gamete would be common to the entire set of offspring encompassing the range of experimental treatments. Studies using this experimental design have already been undertaken.[186]

Readily accessible whole-animal embryonic model — Research has determined that normal embryogenesis of the clearnose skate spans a period of approximately 12 weeks when eggs are maintained at 20 to 22°C.[161] As with the chick, development occurs external to the mother in a protective shell, known as an egg case. Windows can be cut through the leathery case for observation or chemical exposure. During the first 4 weeks of development,

FIGURE 12.   Dorsal view of clearnose skate embryo 7 weeks after egg deposition. Early formation of spiracles, visible just posterior to the eyes, accompanies the development of internal gills. (Total length of embryo is 8.8 cm; disc width is 3.4 cm.)

FIGURE 13. Ventral view of clearnose skate embryo shown in Figure 12. External gill filaments are still visible but will be resorbed over the next few weeks as the internal gills complete their functional development. The presence of developing claspers in the pelvic region indicates that this embryo is a male.

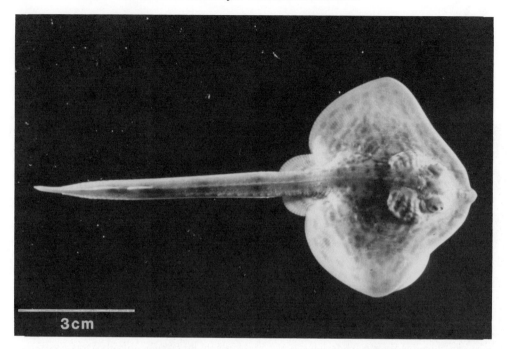

FIGURE 14.   Dorsal view of clearnose skate embryo 9 weeks after egg deposition. Many features characteristic of the skate species, including external pigmentation, are beginning to take shape. (Total length of embryo is 11.5 cm; disc width is 6.0 cm.)

the yolk sac is not self-supportive and manipulations must be achieved through the window. After 4 weeks, the developing embryos with their yolk sac can be removed from the egg case. If left in circulating seawater, the embryo will continue to grow and develop normally until its yolk sac is completely resorbed. This provides researchers with approximately 8 weeks for direct observation of embryonic effects as well as the opportunity to design studies which would employ use of embryos for direct exposure or injection. An important advantage of skate embryos, therefore, is that many experiments and manipulations can be performed without sterile environments or sophisticated experimental chambers. Investigations have demonstrated that embryos can be handled, anesthetized for microscopic examination or photographic documentation, and then revived without altering their developmental process.[64]

Effects on sexual differentiation — A final feature takes advantage of the sexual dimorphism which is characteristic of sharks, skates, and rays. Among this group of fish, all males possess a pair of copulatory organs known as claspers. Claspers are actually modified portions of the pelvic fin which begin to differentiate sometime during the fourth week of development, providing a method by which the sex of these "hands-on" embryos can be determined.[187]

These features, plus many features in common with typical vertebrate species, strongly support the value of the clearnose skate and, potentially, many other species of skate as useful models for biomedical research. It is also anticipated that as marine holding facilities expand their interests in the biomedical fields, other shark species to complement the traditional usage of lemon and nurse sharks might be identified. Of prime interest would be a species of shark which is not too large at sexual maturity, is easily maintained in captivity, is not dangerous to the researcher, and which will breed in captivity and produce offspring. At the present time, such a species is not available. Until this shark is found, the various species of skates which are common inhabitants of nearshore waters throughout the world should be recognized for their potential value and should be considered as alternative animal models for biomedical research.

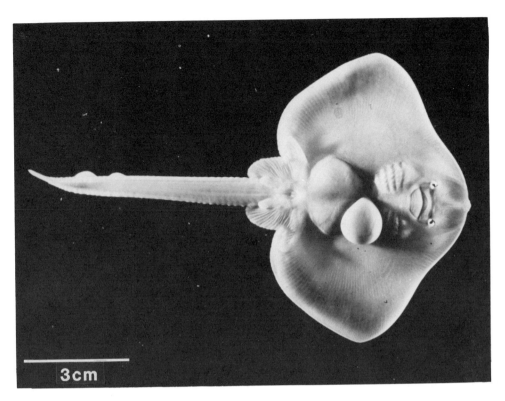

FIGURE 15. Ventral view of clearnose skate embryo 10 weeks after egg deposition. External gill filaments are now totally resorbed, as is most of the yolk sac. The swelling in the abdominal area indicates the presence of internal yolk. (Total length of embryo is 12.7 cm; disc width is 7.7 cm.)

ACKNOWLEDGMENT

The author wishes to extend his appreciation to Patricia Blum, Jose Castro, Allen Horton, Robert Pelham, and Richard Pierce for their assistance in the photographic contributions to this chapter.

REFERENCES

1. **Schaeffer, B.,** Comments on elasmobranch evolution, in *Sharks, Skates, and Rays,* Gilbert, P. W., Mathewson, R. F., and Rall, D. P., Eds., Johns Hopkins Press, Baltimore, 1967, 3.
2. **Schaeffer, B. and Williams, M.,** Relationships of fossil and living elasmobranchs, *Am. Zool.,* 17, 293, 1977.
3. **Marion, G. E.,** Mandibular and pharyngeal muscles of *Acanthias* and *Raia, Am. Nat.,* 39, 891, 1905.
4. **Müller, E.,** Untersuchungen über die Muskeln und Nerven der Brustflosse und der Korperwand bei *Acanthias vulgaris, Anat. Anz.,* Hefte 43, 1911.
5. **Gregory, W. K. and Camp, C. L.,** Studies in comparative myology and osteology, no. III, *Bull. Am. Mus. Nat. Hist.,* 38, 1918.
6. **Daniel, J. F.,** The lateral blood supply of primitive elasmobranch fish, *Univ. Calif. Publ. Zool.,* 29, 1, 1926.
7. **Coles, E.,** The segmental arteries in *Squalus sucklii, Univ. Calif. Pub. Zool.,* 31, 93, 1928.

8. **O'Donoghue, C. H. and Abbot, E. B.,** The blood vascular system of the spiny dogfish, *Squalus acanthias* and *Squalus sucklii, Trans. R. Soc. Edinburgh,* 55 (part 3), 823, 1928.

9. **Daniel, J. F. and Bennett, L. H.,** Veins of the roof of the buccopharyngeal cavity of *Squalus sucklii, Univ. Calif. Pub. Zool.,* 37, 35, 1931.

10. **Leydig, F.,** *Beiträge zur mikroskopischen Anatomie und Entwicklungsgeschichte der Rochen und Haie,* Engelmann Verlag, Leipzig, 1852.

11. **Bates, G. A.,** The pronephric duct in elasmobranchs, *J. Morphol.,* 25, 345, 1914.

12. **Marshall, E. K.,** The comparative physiology of the kidney in relation to theories of renal secretion, *Physiol. Rev.,* 14, 133, 1934.

13. **Smith, H. W.,** The retention and physiological role of urea in elasmobranchii, *Biol. Rev.,* 11, 49, 1936.

14. **Shaper, A.,** The finer structure of the selachian cerebellum *(Mustelus vulgaris), J. Comp. Neurol.,* 8, 1, 1898.

15. **Houser, G. L.,** The neurons and supporting elements of the brain of a selachian, *J. Comp. Neurol.,* 11, 65, 1901.

16. **Burkhardt, R.,** Das Zentral-Nervensystem der Selachier als Grundlage für eine Phylogenie des Vertebratenhirns. I. Teil: Einleitung und *Scymnus lichia, Nova Acta Acad. Caesureae Leopold.-Carolinae Nat. Curios.,* 73, 241, 1907.

17. **Ariëns Kappers, C. V.,** The structure of the teleostean and selachian brain, *J. Comp. Neurol.,* 16, 1, 1909.

18. **Johnston, J. B.,** The telencephalon of selachians, *J. Comp. Neurol.,* 21, 1, 1911.

19. **Ariëns Kappers, C. V., Huber, G. D., and Crosby, E. C.,** *The Comparative Anatomy of the Nervous System of Vertebrates Including Man,* Macmillan, New York, 1936.

20. **Boll, F.,** Zur Anatomie und Physiologie der Retina, *Mber. Berl. Akad. Wiss.,* 12, 783, 1876.

21. **Schiefferdecker, P.,** Studien zur vergleichenden Histologie der Retina, *Arch. Mikrosk. Anat.,* 28, 305, 1886.

22. **Ramon y Cajal, S. R.,** La retina de vertebres, *Cellule,* 9, 17, 1893.

23. **Neumayer, L.,** Der feinere Bau der Selachier Retina, *Arch. Mikrosk. Anat.,* 48, 83, 1897.

24. **Young, J. Z.,** Comparative studies on the physiology of the iris, I. Selachians, *Proc. R. Soc. London, Ser. B,* 112, 228, 1933.

25. **Sheldon, R. E.,** The reactions of dogfish to chemical stimuli, *J. Comp. Neurol.,* 19, 273, 1909.

26. **Sheldon, R. E.,** The sense of smell in selachians, *J. Exp. Zool.,* 10, 51, 1911.

27. **Parker, G. H.,** The relation of smell, taste, and the common chemical sense in vertebrates, *Proc. Acad. Nat. Sci. Philadelphia,* 15, 219, 1912.

28. **Retzius, G.,** *Das Gehororgan der Wirbeltiere. I. Das Gehororgan der Fische und Amphibien,* Samson and Wallin, Stockholm, 1881.

29. **Johnson, S. E.,** Structure and development of the sense organs of the lateral canal system of selachians *(Mustelus canis* and *Squalus acanthias), J. Comp. Neurol.,* 28, 1, 1917.

30. **Lowenstein, O. and Sand, A.,** The mechanism of the semicircular canal. A study of the responses of single-fibre preparations to angular accelerations and to rotation at constant speed, *Proc. R. Soc. London Ser. B,* 129, 256, 1940.

31. **Lowenstein, O. and Roberts, T. D. M.,** The localization and analysis of the responses to vibration from the isolated elasmobranch labyrinth. A contribution to the problem of the evolution of hearing in vertebrates, *J. Physiol. (London),* 114, 471, 1951.

32. **Beard, J.,** The yolk sac, yolk and merocytes in *Scyllium* and *Lepidosteus, Anat. Anz.,* 12, 334, 1896.

33. **Woods, F. A.,** Origin and migration of the germ cell in *Acanthias, Am. J. Anat.,* 1, 307, 1902.

34. **Gudger, E. W.,** Natural history notes on some Beaufort, N. C. fishes, 1910-11., No. 1., Elasmobranchs — with special reference to uterogestation, *Proc. Soc. Wash.,* 25, 141, 1912.

35. **Leigh-Sharpe, W. H.,** The comparative morphology of the secondary sex characters of elasmobranch fishes, Memoirs 1, *J. Morphol.,* 34, 245, 1920.

36. **Leigh-Sharpe, W. H.,** The comparative morphology of secondary sex characters of elasmobranch fishes, Memoirs 3, 4, and 5, *J. Morphol.,* 36, 191, 1922.

37. **Hisaw, F. L. and Abramowitz. A. A.,** Physiology of reproduction in the dogfishes, *Mustelus canis* and *Squalus acanthias, Rep. Woods Hole Oceanogr. Inst.,* 1938, 22, 1939.

38. **Metten, H.,** Studies on the reproduction of the dogfish, *Philos. Trans. R. Soc. London,* 230, 217, 1939.

39. **Balfour, F. M.,** *A Monograph on the Development of Elasmobranch Fishes,* Macmillan, London, 1878.

40. **Marshall, A. M.,** On the head cavities and associated nerves of elasmobranchs, *Microsc. J.,* 21, 72, 1881.

41. **Platt, J. B.,** A contribution to the morphology of the vertebrate head, based on a study of *Acanthias vulgaris, J. Morphol.,* 5, 79, 1891.

42. **Hoffmann, C. K.,** Beiträge zur Entwicklungsgeschichte der Selachii, *Morphol. Jahrb.,* 24, 209, 1896.

43. **Lamb, A. B.,** The development of the eye muscles in *Acanthias, Am. J. Anat.,* 1, 185, 1902.

44. **Dean, B.,** *Chimaeroid Fishes and Their Development,* Carnegie Institute, Publ. 32, Washington, D.C., 1906.

45. **Scammon, R. E.,** *Normal Plates of the Development of* Squalus acanthias, Gustav Fischer, Jena, 1911.
46. **Goodrich, E. S.,** On the development of the segments of the head in *Scyllium, Q. J. Microsc. Sci.,* 63, 1, 1918.
47. **van Wijhe, J. W.,** Fruhe Entwicklungsstadien des Kopf- und Rumpf-Skeletts von *Acanthias vulgaris, Bijdr. Dierkd.,* 22, 271, 1922.
48. **Hagiwara, K., Kobayashi, K., Kajii, T., and Tomonaga, S.,** J-chain-like component is 18-S immunoglobulin of the skate *Raja kenojei,* a cartilagenous fish, *Mol. Immunol.,* 22, 775, 1985.
49. **Rosenshein, I. L. and Marchalonis, J. J.,** The immunoglobulins of carcharhine sharks: a comparison of serological and biochemical properties, *Comp. Biochem. Physiol. B.* 86, 737, 1987.
50. **Marchalonis, J. J., Schluter, S. F., Rosenshein, I. L., and Wang, A. C.,** Partial characterization of immunoglobulin light chains of carcharhine sharks: evidence for phylogenetic conservation of variable region and divergence of constant region structure, *Dev. Comp. Immunol.,* 12, 65, 1988.
51. **Tomonaga, S., Kobayashi, K., Kajii, T., and Awaya, K.,** Two populations of immunoglobulin-forming cells in the skate, *Raja kenojei:* their distribution and characterization, *Dev. Comp. Immunol.,* 8, 803, 1984.
52. **Kobayashi, K., Tomonaga, S., Teshima, K., and Kajii, T.,** Ontogenic studies on the appearance of two classes of immunoglobulin-forming cells in the spleen of the Aleutian skate, *Bathyraja aleutica,* a cartilagenous fish, *Eur. J. Immunol.,* 15, 952, 1985.
53. **Jensen, J. A., Festa, E., Smith, D. S., and Cayer, M.,** The complement system of the nurse shark: hemolytic and comparative characteristics, *Science,* 214, 566, 1981.
54. **Morrow, W. J. W. and Pulsford, A.,** Identification of peripheral blood leucocytes of the dogfish *(Scyliorhinus canicula* L.) by electron microscopy, *J. Fish Biol.,* 17, 461, 1980.
55. **Hyder, S. L., Cayer, M. L., and Pettey, C. L.,** Cell types in peripheral blood of the nurse shark: an approach to structure and function, *Tissue Cell,* 15, 437, 1983.
56. **Pettey, C. L. and McKinney, E. C.,** Temperature and cellular regulation of spontaneous cytotoxicity in the shark, *Eur. J. Immunol.,* 13, 133, 1983.
57. **Obenauf, S. D. and Hyder Smith, S.,** Chemotaxis of nurse shark leukocytes, *Dev. Comp. Immunol.,* 9, 221, 1985.
58. **McKinney, E. C., Haynes, L., and Droese, A. L.,** Macrophage-like effector of spontaneous cytotoxicity from the shark, *Dev. Comp. Immunol.,* 10, 497, 1986.
59. **Bliznakov, E. G.,** Effect of reticuloendothelial system stimulation on resistance to Rous sarcoma virus infection in chicks, *Int. J. Cancer,* 3, 336, 1968.
60. **Sigel, M. M., Voss, E. W., Rudikoff, S., Lichter, W., and Jensen, J. A.,** Natural antibodies in primitive vertebrates: the sharks, in *Miami Winter Symp.,* 2, North-Holland, Amsterdam, 1970, 409.
61. **Snodgrass, M. J., Burke, J. D., and Meetz, G. D.,** Inhibitory effect of shark serum on the Lewis lung carcinoma, *J. Natl. Cancer Inst.,* 56, 981, 1976.
62. **Pettit, G. R. and Ode, R. H.,** Antineoplastic agents L: isolation and characterization of sphyrnastatins 1 and 2 from the hammerhead shark *Sphyrna lewini, J. Pharm. Sci.,* 66, 757, 1977.
63. **Lee, A. and Langer, R.,** Shark cartilage contains inhibitors of tumor angiogenesis, *Science,* 221, 1185, 1983.
64. **Luer, C. A.,** Inhibitors of angiogenesis from shark cartilage, *Fed. Proc. Fed. Am. Soc. Exp. Biol.,* 45, 949, 1986.
65. **Lee, A. K., van Beuzekom, M., Glowacki, J., and Langer, R.,** Inhibitors, enzymes, and growth factors from shark cartilage, *Comp. Biochem. Physiol. B,* 78, 609, 1984.
66. **Shidoji, Y. and Muto, Y.,** Vitamin A transport in plasma of the nonmammalian vertebrates: isolation and partial characterization of piscine retinol-binding protein, *J. Lipid Res.,* 18, 679, 1977.
67. **Burch, S. J., Lawson, R., and Davies, D. H.,** The relationships of cartilagenous fishes: an immunological study of serum transferrins of holocephalans and elasmobranchs, *J. Zool. (London),* 203, 303, 1984.
68. **Bodine, A. B., Luer, C. A., and Gangjee, S.,** Determination of ceruloplasmin and other copper transport ligands in the blood sera of the nurse shark and clearnose skate, *Comp. Biochem. Physiol. B,* 77, 779, 1984.
69. **Aschauer, H., Weber, R. E., and Braunitzer, G.,** The primary structure of the hemoglobin of the dogfish shark *(Squalus acanthias), Biol. Chem. Hoppe-Seyler,* 336, 589, 1985.
70. **Maynard, J. R. and Coleman, J. E.,** Elasmobranch carbonic anhydrase. Purification and properties of the enzyme from two species of shark, *J. Biol. Chem.,* 246, 4455, 1971.
71. **Fisher, W. K. and Thompson, E. O. P.,** Myglobin of the shark *Heterodontus portusjacksoni,* isolation and amino acid sequence, *Aust. J. Biol. Sci.,* 32, 277, 1979.
72. **Fisher, W. K., Koureas, D. D., and Thompson, E. O. P.,** Myoglobins of cartilagenous fishes. II. Isolation and amino acid sequence of myoglobin of the shark *Mustelus antarcticus, Aust. J. Biol. Sci.,* 33, 153, 1980.
73. **Fisher, W. K., Koureas, D. D., and Thompson, E. O. P.,** Myglobins of cartilagenous fishes. III. Amino acid sequence of myoglobin of the shark *Galeorhinus australis, Aust. J. Biol. Sci.,* 34, 5, 1981.

74. **Suzuki, T., Suzuki, T., and Yata, T.,** Shark myoglobin. II. Isolation, characterization, and amino acid sequence of myoglobin from *Galeorhinus japonicus*, *Aust. J. Biol. Sci.,* 38, 347, 1985.

75. **Wilson, M. T., Lalla-Maharajh, W., Darley-Usman, V., Bonaventura, J., Bonaventura, C., and Brunori, M.,** Structural and functional properties of cytochrome c oxidases isolated from sharks, *J. Biol. Chem.,* 255, 2722, 1980.

76. **Georgevich, G., Darley-Usmar, V. M., Malatesta, F., and Capaldi, R. A.,** Electron transfer in monomeric forms of beef and shark heart cytochrome c oxidase, *Biochemistry,* 22, 1317, 1983.

77. **Bickar, D., Lehninger, A., Brunori, M., Bonaventura, J., and Bonaventura, C.,** Functional equivalence of monomeric (shark) and dimeric (bovine) cytochrome c oxidase, *J. Inorg. Chem.,* 23, 365, 1985.

78. **Kanoh, S., Watabe, S., and Hashimoto, K.,** Isolation and some physicochemical properties of requiem shark myosin, *Bull. Jpn. Soc. Sci. Fish.,* 49, 757, 1983.

79. **Kanoh, S., Watabe, S., and Hashimoto, K.,** ATPase activity of requiem shark myosin, *Bull. Jpn. Soc. Sci. Fish.,* 51, 973, 1985.

80. **Hidalgo, J., Tort, L., and Flos, R.,** Cd-, Zn-, Cu-binding protein in the elasmobranch *Scyliorhinus canicula, Comp. Biochem. Physiol.,* 81C, 159, 1985.

81. **Idler, D. R. and Kane, K. M.,** Cytosol receptor glycoprotein for 1 α-hydroxycorticosterone in tissues of an elasmobranch fish *(Raja ocellata), Gen. Comp. Endocrinol.,* 42, 259, 1980.

82. **Burton, M. and Idler, D. R.,** The cellular location of 1 α-hydroxycorticosterone binding protein in skate, *Gen. Comp. Endocrinol.,* 64, 260, 1986.

83. **Smith, D. D., Ritter, N. M., and Campbell, J. W.,** Glutamine synthetase isozymes in elasmobranch brain and liver tissues, *J. Biol. Chem.,* 262, 198, 1987.

84. **Galtieri, A., Natoli, G., Lania, A., and Calabrese, L.,** Isolation and characterization of Cu, Zn superoxide dismutase of the shark *Prionace glauca, Comp. Biochem. Physiol.,* 83B, 555, 1986.

85. **Foureman, G. L. and Bend, J. R.,** The hepatic glutathione transferases of the male little skate, *Raja erinacea, Chem. Biol. Interact.,* 49, 89, 1984.

86. **Foureman, G. L., Hernandez, O., Bhatia, A., and Bend, J. R.,** The stereospecificity of four hepatic glutathione S-transferases purified from a marine elasmobranch *(Raja erinacea)* with several K-region polycyclic arene oxide substrates, *Biochim. Biophys. Acta,* 914, 127, 1987.

87. **Philpot, R. M. and Arinc, E.,** Solubilization, separation, and partial purification of cytochrome P-450 and cytochrome b_5 from hepatic microsomes of the little skate, *Raja erinacea, Bull. Mt. Desert Isl. Biol. Lab.,* 15, 62, 1975.

88. **Serabjit-Singh, C. J., Bend, J. R., and Philpot, R. M.,** Purification of hepatic microsomal NADPH-cytochrome P-450 reductase from little skate, *Raja erinacea, Bull. Mt. Desert Isl. Biol. Lab.,* 21, 57, 1981.

89. **Bend, J. R., Pohl, R. J., and Fouts, J. R.,** Some properties of the microsomal drug-metabolizing enzyme system in the little skate, *Raja erinacea, Bull. Mt. Desert Isl. Biol. Lab.,* 12, 12, 1972.

90. **Bend, J. R., Pohl, R. J., and Fouts, J. R.,** Further studies of the microsomal mixed-function oxidase system of the little skate, *Raja erinacea,* including its response to some xenobiotics, *Bull. Mt. Desert Isl. Biol. Lab.,* 13, 9, 1973.

91. **Boyer, J. L., Bend, J. R., Pohl, R., Swarz, J., and Smith, N.,** Effects of phenobarbital on bile secretion, glutathione-S-aryltransferase activity and hepatic mixed function oxidase (MFO) pathways in the small skate *Raja erinacea, Bull. Mt. Desert Isl. Biol. Lab.,* 13, 14, 1973.

92. **Bend, J. R., Pohl, R. J., Davidson, N. P., and Fouts, J. R.,** Response of hepatic and renal microsomal mixed-function oxidases in the little skate, *Raja erinacea,* to pretreatment with 3-methylcholanthrene or TCDD (2,3,7,8-tetrachlorodibenzo-p-dioxin), *Bull. Mt. Desert Isl. Biol. Lab.,* 14, 7, 1974.

93. **Pohl, R. J., Fouts, J. R., and Bend, J. R.,** Response of hepatic microsomal mixed-function oxidases in the little skate, *Raja erinacea,* and the winter flounder, *Pseudopleuronectes americanus,* to pretreatment with TDCC (2,3,7,8-tetrachlorodibenzo-p-dioxin) or DBA (1,2,3,4-dibenzanthracene), *Bull. Mt. Desert Isl. Biol. Lab.,* 15, 64, 1975.

94. **Bend, J. R., Hall, P., and Foureman, G. L.,** Comparison of benzo[a]pyrene hydroxylase (aryl hydrocarbon hydroxylase, AHH) activities in hepatic microsomes from untreated and 1,2,3,4-dibenzanthracene (DBA)-induced male little skates *(Raja erinacea), Bull. Mt. Desert Isl. Biol. Lab.,* 16, 3, 1976.

95. **Bodine, A. B., Luer, C. A., and Gangjee, S.,** A comparative study of monooxygenase activity in elasmobranchs and mammals: activation of the model pro-carcinogen aflatoxin B_1 by liver preparations of calf, nurse shark and clearnose skate, *Comp. Biochem. Physiol.,* 82C, 255, 1985.

96. **Bajaj, M., Blundell, T. C., Pitts, J. E., Wood, S. P., Tatnell, M. A., Falkmer, S., Endin, S. O., Gowan, L. K., Crow, H., Schwabe, C., Wollmer, A., and Strassburger, W.,** Dogfish insulin. Primary structure, conformation and biological properties of an elasmobranchial insulin, *Eur. J. Biochem.,* 135, 535, 1983.

97. **Conlon, J. M. and Thim, L.,** Primary structure of glucagon from an elasmobranch fish, *Torpedo marmorata, Gen. Comp. Endocrinol.,* 60, 398, 1985.

98. **Conlon, J. M., Agoston, D. V., and Thim, L.,** An elasmobranchian somatostatin: primary structure and tissue distribution in *Torpedo marmorata, Gen. Comp. Endocrinol.,* 60, 406, 1985.

99. **Dimaline, R., Thorndyke, M. C., and Young, J.,** Isolation and partial sequence of elasmobranch VIP, *Regul. Peptides,* 14, 1, 1986.

100. **Dimaline, R., Young, J., Thwaites, D. T., Lee, C. M., Shuttleworth, T. J., and Thorndyke, M. C.,** A novel vasoactive intestinal peptide (VIP) from elasmobranch intestine has full affinity for mammalian pancreatic VIP receptors, *Biochim. Biophys. Acta,* 930, 97, 1987.

101. **Reinig, J. W., Daniel, L. N., Schwabe, C., Gowan, L. K., Steinetz, B. G., and O'Byrne, E. M.,** Isolation and characterization of relaxin from the sand tiger shark *(Odontaspis taurus), Endocrinology,* 109, 537, 1981.

102. **Bullesbach, E. E., Schwabe, C., and Callard, I. P.,** Relaxin from an oviparous species, the skate *(Raja erinacea), Biochem. Biophys. Res. Commun.,* 143, 273, 1987.

103. **Graham, E. E.,** Isolation of enamelinlike proteins from blue shark *(Prionace glauca)* enameloid, *J. Exp. Zool.,* 234, 185, 1985.

104. **Schumacher, M., Camp, S., Maulet, Y., Newton, M., MacPhee-Quigley, K., Taylor, S. S., Friedman, T., and Taylor, P.,** Primary structure of *Torpedo californica* acetylcholinesterase deduced from its cDNA sequence, *Nature (London),* 319, 407, 1986.

105. **Bon, S., Chang, J.-Y., and Strosberg, A. D.,** Identical N-terminal peptide sequences of asymmetric forms and of low-salt-soluble and detergent-soluble amphiphilic dimers of *Torpedo* acetylcholinesterase, *FEBS Lett.,* 209, 206, 1986.

106. **Sikorav, J.-L., Krejci, E., and Massoulie, J.,** cDNA sequences of *Torpedo marmorata* acetylcholinesterase: primary structure of the precursor of a catalytic subunit; existence of multiple 5'-untranslated regions, *EMBO J.,* 6, 1865, 1987.

107. **Musset, F., Frobert, Y., Grassi, J., Vigny, M., Boulla, G., Bon, S., and Massoulie, J.,** Monoclonal antibodies against acetylcholinesterase from electric organs of *Electrophorus* and *Torpedo, Biochimie,* 69, 147, 1987.

108. **Burger, J. W. and Hess, W. N.,** Function of the rectal gland in the spiny dogfish, *Science,* 131, 670, 1960.

109. **Silva, P., Stoff, J., Field, M., Fine, L., Forrest, J. N., and Epstein, F. H.,** Mechanism of active chloride secretion by shark rectal gland: role of Na-K-ATPase in chloride transport, *Am. J. Physiol.,* 233, F298, 1977.

110. **Frizzell, R. A., Field, M., and Schultz, S. G.,** Sodium-coupled chloride transport by epithelial tissues, *Am. J. Physiol.,* 236, F1, 1979.

111. **Burger, J. W.,** Further studies on the function of the rectal gland in the spiny dogfish, *Physiol. Zool.,* 35, 205, 1962.

112. **Eveloff, J., Kinne, R., Kinne-Saffran, E., Murer, H., Silva, P., Epstein, F. H., Stoff, J., and Kinter, W. B.,** Coupled sodium and chloride transport into plasma membrane vesicled prepared from dogfish rectal gland, *Pfluegers Arch.,* 378, 87, 1978.

113. **Dubinsky, W. P. and Monti, L. B.,** Resolution of apical from basolateral membrane of shark rectal gland, *Am. J. Physiol.,* 251, C721, 1986.

114. **Stoff, J. S., Rosa, R., Hallae, R., Silva, P., and Epstein, F. H.,** Hormonal regulation of active chloride transport in the dogfish rectal gland, *Am. J. Physiol.,* 237, F138, 1979.

115. **Forrest, J. N., Wang, F., and Beyenbach, K. W.,** Perfusion of isolated tubules of the shark rectal gland. Electrical characteristics and response to hormones, *J. Clin. Invest.,* 72, 1163, 1983.

116. **Shuttleworth, T. J.,** Haemodynamic effects of secretory agents on the isolated elasmobranch rectal gland, *J. Exp. Biol.,* 103, 193, 1983.

117. **Shuttleworth, T. J. and Thorndyke, M. C.,** An endogenous peptide stimulates secretory activity in the elasmobranch rectal gland, *Science,* 225, 319, 1984.

118. **Silva, P., Stoff, J. S., Leone, D. R., and Epstein, F. H.,** Mode of action of somatostatin to inhibit secretion by shark rectal gland, *Am. J. Physiol.,* 249, R329, 1985.

119. **Marver, D., Lear, S., Marver, L. T., Silva, P., and Epstein, F. H.,** Cyclic AMP-dependent stimulation of Na, K-ATPase in shark rectal gland, *J. Membr. Biol.,* 94, 205, 1986.

120. **Silva, P., Stoff, J. S., Solomon, R. J., Lear, S., Kniaz, D., and Greger, R.,** Atrial natriuretic peptide stimulates salt secretion by shark rectal gland by releasing VIP, *Am. J. Physiol.,* 252, F99, 1987.

121. **Erlij, D., Lodenquai, S., and Rubio, R.,** Nervous control of secretion of rectal gland of the dogfish: pharmacological evidence, *Bull. Mt. Desert Isl. Biol. Lab.,* 21, 74, 1981.

122. **Lacy, E. R., Reale, E., Schlusselberg, D. S., Smith, W. K., and Woodward, D. J.,** A renal countercurrent system in marine elasmobranch fish: a computer-assisted reconstruction, *Science,* 227, 1351, 1985.

123. **Marshall, E. K.,** The comparative physiology of the kidney in relation to theories of renal secretion, *Physiol. Rev.,* 14, 133, 1934.

124. **Kempton, R. T.,** Studies on the elasmobranch kidney. II. Reabsorption of urea by the smooth dogfish *Mustelus canis, Biol. Bull.,* 104, 45, 1953.

125. **Schmidt-Nielsen, B.,** Renal transport of urea in elasmobranchs, in *Transport Mechanisms in Epithelia,* Ussing, H. H. and Thorn, N. A., Eds., Munksgaard, Copenhagen, 1973, 608.

126. **Yancey, P. H., Clark, M. E., Hand, S. C., Bowlus, R. D., and Somero, G. N.,** Living with water stress: evolution of osmolyte systems, *Science,* 217, 1214, 1982.

127. **Groninger, H. S.,** The occurrence and significance of trimethylamine oxide in marine animals, *U.S. Fish Wildl. Serv. Fish. Bull.,* 333, 1, 1959.

128. **Endo, M.,** Histological and enzymatic studies on the renal tubules of some marine elasmobranchs, *J. Morphol.,* 182, 63, 1984.

129. **Hentschel, H., Elger, M., and Schmidt-Nielsen, B.,** Chemical and morphological differences in the kidney zones of the elasmobranch, *Raja erinacea* Mitch, *Comp. Biochem. Physiol.,* 84A, 553, 1986.

130. **Lacy, E. R. and Reale, E.,** The elasmobranch kidney. I. Gross anatomy and general distribution of the nephrons, *Anat. Embryol.,* 173, 23, 1985.

131. **Lacy, E. R. and Reale, E.,** The elasmobranch kidney. II. Sequence and structure of the nephrons, *Anat. Embryol.,* 173, 163, 1985.

132. **Lacy, E. R. and Reale, E.,** The elasmobranch kidney. III. Fine structure of the peritubular sheath, *Anat. Embryol.,* 173, 299,1986.

133. **Lacy, E. R., Castellucci, M., and Reale, E.,** The elasmobranch renal corpuscle: fine structure of Bowman's capsule and the glomerular capillary wall, *Anat. Rec.,* 218, 294, 1987.

134. **Reeves, W., Caulfield, J. P., and Farguhar, M. G.,** Differentiation of epithelial foot processes and filtration slits, *Lab. Invest.,* 39, 90, 1978.

135. **Humbert, F., Montesano, R., Perrelet, A., and Orci, L.,** Junctions in developing human and rat kidney: a freeze-fracture study, *J. Ultrastruct. Res.,* 5b, 202, 1976.

136. **Schaeverbeke, J. and Chiegnon, M.,** Differentiation of glomerular filter and tubular reabsorption apparatus during foetal development of the rat kidney, *J. Embryol. Exp. Morphol.,* 58, 157, 1980.

137. **Hay, D. A. and Evan, A. P.,** Maturation of the glomerular visceral epithelium and capillary endothelium in the puppy kidney, *Anat. Rec.,* 193, 1, 1979.

138. **Ichikawa, I., Maddox, D. A., and Brenner, B. M.,** Maturational development of glomerular ultrafiltration in the rat, *Am. J. Physiol.,* 236, F465, 1979.

139. **Larsson, L. and Maunsbach, A. B.,** The ultrastructural development of the glomerular filtration barrier in the rat kidney: a morphometric analysis, *J. Ultrastruct. Res.,* 72, 392, 1980.

140. **Larsson, L. and Maunsbach, A. B.,** Quantitative ultrastructural changes of the glomerular capillary wall in the developing rat kidney, in *The Kidney During Development — Morphology and Function,* Spitzer, A., Ed., Masson, New York, 1982, 115.

141. **Zigman, S., Munro, J., and Lerman, S.,** Effect of urea on the cold precipitation of protein in the lens of dogfish, *Nature (London),* 207, 414, 1965.

142. **Yancey, P. H. and Somero, G. N.,** Urea requiring lactate dehydrogenases of marine elasmobranch fishes, *J. Comp. Physiol.,* 125, 135, 1978.

143. **Arai, K., Hasnain, A., and Takano, Y.,** Species specificity of muscle proteins of fishes against thermal and urea denaturation, *Bull. Jpn. Soc. Sci. Fish.,* 42, 687, 1976.

144. **Nishimoto, J.,** Effect of urea on the Mg^{2+}-ATPase activity of myofibrils prepared from marine elasmobranch muscle, *Bull. Jpn. Soc. Sci. Fish.,* 47, 1391, 1981.

145. **Kanoh, S., Watabe, S., Takewa, T., and Hashimoto, K.,** Urea-resistibility of myofibrillar proteins from the requiem shark, *Triakis scyllia, Bull. Jpn. Soc. Sci. Fish.,* 51, 1713, 1985.

146. **Bonaventura, J., Bonaventura, C., and Sullivan, B.,** Urea tolerance as a molecular adaptation of elasmobranch hemoglobins, *Science,* 186, 57, 1974.

147. **Yancey, P. H. and Somero, G. N.,** Counteraction of urea destabilization of protein structure by methylamine osmoregulatory compounds of elasmobranch fishes, *Biochem. J.,* 183, 317, 1979.

148. **Yancey, P. H. and Somero, G. N.,** Methylamine osmoregulatory solutes of elasmobranch fishes counteract urea inhibition of enzymes, *J. Exp. Zool.,* 212, 205, 1980.

149. **Dowling, J. E. and Ripps, H.,** Visual adaptation in the retina of the skate, *J. Gen. Physiol.,* 56, 491, 1970.

150. **Brin, K. P. and Ripps, H.,** Rhodopsin photoproducts and rod sensitivity in the skate retina, *J. Gen. Physiol.,* 69, 97, 1977.

151. **Dowling, J. E. and Ripps, H.,** S-Potentials in the skate retina. Intracellular recordings during light and dark adaptation, *J. Gen. Physiol.,* 58, 163, 1971.

152. **Dowling, J. E. and Ripps, H.,** Adaptation in skate retinas, *J. Gen. Physiol.,* 60, 698, 1972.

153. **Dowling, J. E. and Ripps, H.,** Effect of magnesium on horizontal cell activity in the skate retina, *Nature (London),* 242, 101, 1973.

154. **Green, D. G. and Siegel, I. M.,** Double branched flicker fusion curves from the all-rod skate retina, *Science,* 188, 1120, 1975.

155. **Green, D. G., Dowling, J. E., Siegel, I. M., and Ripps, H.,** Retinal mechanisms of visual adaptation in the skate, *J. Gen. Physiol.,* 65, 483, 1975.

156. **Ripps, H., Shakib, M., and MacDonald, E. D.,** Peroxidase uptake by photoreceptor terminals of the skate retina, *J. Cell Biol.,* 70, 86, 1976.

157. **Szamier, R. B. and Ripps, H.,** The visual cells of the skate retina: structure, histochemistry, and disc-shedding properties, *J. Comp. Neurol.,* 215, 51, 1983.

158. **Richards, S. W., Merriman, D., and Calhoun, L. H.,** Studies on the marine resources of southern New England, IX. The biology of the little skate, *Raja erinacea* Mitchill, *Bull. Bingham Oceanogr. Collect.,* 18, 5, 1963.

159. **Prasad, R. R.,** The structure, phylogenetic significance and function of the nidamental glands of some elasmobranchs of the Madras coast, *Proc. Natl. Inst. Sci. India,* 11, 282, 1944.

160. **Prasad, R. R.,** Further observations on the structure and function of the nidamental glands of a few elasmobranchs of the Madras coast, *Proc. Indian Acad. Sci.,* 22, 368, 1945.

161. **Luer, C. A. and Gilbert, P. W.,** Mating behavior, egg deposition, incubation period, and hatching in the clearnose skate, *Raja eglanteria, Environ. Biol. Fish.,* 13, 161, 1985.

162. **Pratt, H. L.,** Reproduction in the blue shark, *Prionace glauca, Fish. Bull.,* 77, 445, 1979.

163. **Jensen, A. C.,** Life history of the spiny dogfish, *U.S. Fish Wildl. Serv. Fish. Bull.,* 65, 527, 1966.

164. **Jones, B. C. and Geen, G. H.,** Reproduction and embryonic development of spiny dogfish *(Squalus acanthias)* in the strait of Georgia, British Columbia, *J. Fish. Res. Board Can.,* 34, 1286, 1977.

165. **Compagno, L. J. V.,** FAO species catalogue. Vol. 4. Sharks of the world. An annotated and illustrated catalogue of shark species known to date. Part 1. Hexanchiformes to Lamniformes, *FAO Fish. Synop.,* 125, 111, 1984.

166. **Jones, R. T. and Price, K. S.,** A method for maintaining spiny dogfish, *(Squalus acanthias),* pups artificially, *Copeia,* 1967 (2), 471, 1967.

167. **Jones, R. T. and Price, K. S.,** Osmotic responses of spiny dogfish *(Squalus acanthias,*L.) embryos to temperature and salinity stress, *Comp. Biochem. Physiol.,* 47A, 971, 1974.

168. **Jones, R. T., Hudson, E. A., and Andrews, J. C.,** Methods for transport and long-term maintenance of spiny dogfish sharks, *Laboratory Animal Science,* 33, 388, 1983.

169. **Gruber, S. H.,** Keeping sharks in captivity, *J. Aquaric.,* 1, 6, 1980.

170. **Gruber, S. H. and Keyes, R. S.,** Keeping sharks for research, in *Aquarium Systems,* Hawkins, A. D., Ed., Academic Press, London, 1981, chap. 14.

171. **Clark, E.,** Maintenance of sharks in captivity, *Inst. Oceanogr. Monaco Bull.,* 1A, 7, 1963.

172. **Clark, E.,** Instrumental conditioning of the lemon shark, *Science,* 130, 217, 1959.

173. **Clark, E.,** Maintenance of sharks in captivity with a report on their instrumental conditioning, in *Sharks and Survival,* Gilbert, P. W., Ed., D. C. Heath, Lexington, MA, 1963, 115.

174. **Gruber, S. H. and Schneiderman, N.,** Classical conditioning of the nictitating membrane response of the lemon shark *(Negaprion brevirostris), Behav. Res. Methods Instrum.,* 7, 430, 1975.

175. **Gruber, S. H. and Myrberg, A. A.,** Approaches to the study of the behavior of sharks, *Am. Zool.,* 17, 471, 1977.

176. **Aronson, L. R., Aronson, F. R., and Clark, E.,** Instrumental conditioning and light-dark discrimination in young nurse sharks, *Bull. Mar. Sci.,* 17, 249, 1967.

177. **Graeber, R. C. and Ebbesson, S. O.,** Visual discrimination learning in normal and tectal-ablated nurse sharks *(Ginglymostoma cirratum), Comp. Biochem. Physiol.,* 42A, 131, 1972.

178. **McManus, M. E., Johnson, C. S., and Jeffries, M. M.,** Using operant conditioning to train nurse sharks *(Ginglymostoma cirratum),* paper presented at Sharks: Recent Advances at Captive Biology, Baltimore, March 31 to April 3, 1985, 15.

179. **Gilbert, P. W. and Wood, F. G.,** Methods of anesthetizing large sharks and rays safely and rapidly, *Science,* 126, 212, 1957.

180. **Luer, C. A. and Gilbert, P. W.,** unpublished data, 1988.

181. **Luer, C. A. and Luer, W. H.,** Acute and chronic exposure of nurse sharks to aflatoxin B_1, *Fed. Proc. Fed. Am. Soc. Exp. Biol.,* 41, 925, 1982.

182. **Luer, C. A., Bodine, A. B., and Gangjee, S.,** *In vitro* metabolism of the proximate carcinogen aflatoxin B_1 by liver preparations of the calf, nurse shark, and clearnose skate, *Fed. Proc. Fed. Am. Soc. Exp. Biol.,* 46, 2290, 1987.

183. **Bodine, A. B. and Luer, C. A.,** unpublished data.

184. **Fitz, E. S. and Daiber, F. C.,** An introduction to the biology of *Raja eglanteria* Bosc 1802 and *Raja erinacea* Mitchill 1825 as they occur in Delaware Bay, *Bull. Bingham Oceanogr. Collect.,* 18, 69, 1963.

185. **Moss, S. A.,** *Sharks. An Introduction for the Amateur Naturalist,* Prentice-Hall, Englewood Cliffs, NJ, 1984, preface and chap. 8.

186. **Luer, C. A. and Luer, W. H.,** Toxic effects of aflatoxin B_1 on reproduction and embryonic development in the clearnose skate, *Raja eglanteria, Fed. Proc. Fed. Am. Soc. Exp. Biol.,* 43, 1953, 1984.

187. **Blum, P. and Luer, C. A.,** Sexual maturation in the clearnose skate, *Raja eglanteria, Fl. Sci.,* 47, 19, 1984.

Chapter 10

BIOMEDICAL APPLICATIONS OF MARINE ANIMALS

Albert C. Smith

TABLE OF CONTENTS

I. INTRODUCTION

The use of animals in research has become a major issue of our times. Many articles concerning this topic have appeared with increasing frequency in a variety of sources, among them, "Animal Experimentation — When Do the Ends Justify the Means?",[1] "Lab Animal Welfare Issue Gathers Momentum",[2] " 'Animal Slavery' Leading to Civil War?",[3] *The Case for Animal Rights,*[4] "A Pivotal Year for Lab Animal Welfare",[5] and "Public Relations Tactics in the Debate Over Animal Experimentation".[6]

The concern for considerate treatment of research animals has reached a new height, and such concern is commendable. However, it has created certain new problems for the biomedical researcher in the form of a proliferation of bureaucratic and socioeconomic impediments. The problems are particularly great for the researcher who has become dependent on one class of laboratory animal, the mammal.

In this climate, it would clearly benefit biomedical researchers to be able to free themselves from dependency on the laboratory mammal by including alternative, nonmammalian species. Laboratory mammals have, of course, served biomedical research well and will almost certainly continue to do so; but it appears that the time has come for the frontiers of laboratory animal resources to be expanded by including nontraditional animal species. I propose that marine animals can help significantly in this effort.

II. THE LIFE OF THE SEA

Marine animals represent a vast, nearly untapped resource for biomedical researchers.[7-9] However, over the years, a number of researchers have pointed out the potential benefits to biomedicine of using marine animals, viz., fish, protochordates, and invertebrates. For example, Elie Metchnikoff, perhaps the most venerable name in marine biomedicine, working in the late 1800s with starfish larvae, discovered a fundamental process found in all metazoan animals: phagocytosis.[10] The importance of this discovery cannot be overemphasized because it laid the cornerstone of modern knowledge about the inflammatory response. More recent contributions have been made by a few, generally isolated pioneers, but today it is still rare to find marine animals used in the biomedical research laboratory.

With the present and increasing need to identify alternative species to the traditional laboratory mammal, it is now more important than ever to realize the vast potential of marine animals as experimental animals. The present report (1) examines possible reasons why marine animals have not gained greater acceptance and (2) presents a phylogenetic approach to introducing these animals to the biomedical research community.

III. EXISTENT OBSTACLES

The acceptance of nonmammalian species by biomedical researchers will almost certainly come slowly, as mammals have a long tradition of use and researchers are simply more comfortable with them. Marine animals, in contrast, are so greatly different that they are likely to seem particularly alien, and the hostile environment (to the researchers, not to the animals) of the sea may also lower their appeal.

Other problems with acceptance seem to be (1) the belief by some investigators that the more distantly related species have less to offer in the way of information relevant to humans and (2) the rarity to date of actual clinical applications from research with lower animals.

A common denominator to understanding why the foregoing situation exists is the educational system, which, for the most part, produces biomedical scientists trained in mammalian biology. Ironically, scientists who are educated in nonmammalian biology are rarely trained for biomedical research. By eliminating this narrow educational channeling and providing training with a more comparative approach to the animal kingdom, the biomedical scientist will become more comfortable with nonmammalian species and gain a more realistic understanding of system relationships between mammals and lower animals. A natural consequence of this knowledge would, predictably, be greater use of lower animals and increased benefits to humankind.

IV. BIOMEDICAL ASPECTS OF MARINE BIOLOGY

The sea is a challenging frontier containing the greatest and most diverse biomass on this planet. The animals range from the simplest to the most complex and specialized forms known. Nevertheless, conservative evolution has produced many systems that are basic and shared with mammals. Examples of such systems are DNA, RNA, oncogenes, mitochondria, ribosomes, endoplasmic reticulum, prostaglandins, metabolic enzymes, eye lens proteins, the major histocompatibility complex, and nerve impulse transmission. Basic processes may, in fact, be more clearly visualized in simpler animals due to the absence of evolutionary overlay found in mammals. Seen in the unmasked state, such processes may be more accessible to experimentation. Additionally, new processes may be more readily discovered and their existence sought and identified for the first time in mammals (e.g., phagocytosis).

Marine animal diversity is so great that at first this group may appear as a nearly hopelessly complex collection of widely varying species. Since this impression will not encourage use of these animals by biomedical researchers, it is important that the researchers gain familiarity with the relationships of these animals to each other and to mammals. A phylogenetic approach can meet this need.

V. HUMAN EVOLUTIONARY ROOTS IN THE SEA

According to evolutionary theory, fish have a distant place in the ancestry of humans.[11] Their presence on earth antedates our ape-like ancestors by some 500 million years and all other vertebrates by more than 100 million years. Without piscine ancestry, humans might never have evolved. Many human features were originated or were already present eons ago in fishy ancestors. These features include the ground plans and basic functions of the major organ systems, including those for vision, internal fertilization, intrauterine (placental) nourishment, live birth, learning, and memory.

In fish, it is possible to see vertebrate diversity at its greatest.[11] The approximately 20,000 living species show incomparable variety with respect to shape, color, behavior, personality, and type of environment occupied. Most of these fish are of the boney variety, exemplified by such familiar species as salmon, flounder, perch, and barracuda. A much smaller group of fish, showing a more conservative evolution into only about 800 species, is the sharks, skates, and rays. All have a cartilaginous skeleton and an interesting combination of primitive and specialized features (discussed in Chapter 9).

Preceding fishes on the evolutionary pathway are a transition, or protochordate, group of animals phylogenetically placed between vertebrates (fishes) and invertebrates. Animals of this group, represented, for example, by the sac-like tunicate or sea squirt, show the vertebrate body plan in its most elemental form. Organs so represented include the brain and spinal cord, pituitary gland, vertebral column, thyroid gland, and pharyngeal gill slits.

The pathway below protochordates leads to the echinoderm group of invertebrates (starfish, sea urchins, sea cucumbers). This group does not exhibit obvious vertebrate affinities, but these are given away primarily by developmental and larval similarities.

The echinoderm-protochordate axis is thought to have evolved from the flatworms (flukes, tapeworms), which also generated most of the remaining major invertebrate groups: annelids (worms, leeches), mollusks (shellfish, octopi, squid), and arthropods (crustaceans,

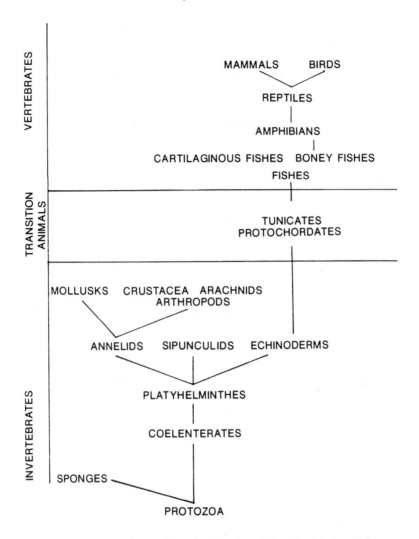

FIGURE 1. Phylogenetic tree of the animal kingdom. (Adapted mainly from Reference 12.)

insects, arachnids). Additionally, the flatworms may also have given rise to several so-called minor groups, including sipunculids (peanut worms), bryozoa (moss animals), and nematodes (roundworms).

The evolutionarily prolific and pivotal flatworms themselves apparently evolved from coelenterates (jellyfish, sea anemones, coral), which, in turn, were the products of certain loosely aggregated protozoa. These protozoa probably also gave rise to the sponges, which are considered a dead-end in evolution as they did not give rise to any other group.

Figure 1 summarizes the preceding phylogenetic information.

VI. SPECIAL FEATURES FOR RESEARCH

The value of marine animals in biomedical research lies in appreciating and utilizing their unique features. These include systems which are specialized, especially accessible, or missing by comparison with mammals. Selected examples from the recent literature are given in Table 1.

The table reviews some of the biomedically useful features of marine animals. The effort

TABLE 1
Biomedical Research Applications of Marine Animals

Animal group	Special features	Areas of biomedical application	Ref.
I. Boney fishes			
Angler fish	Fusion of male to female	Transplantation and rejection	13
Toadfish	High binding affinity for vitamin B_{12}	Vitamin B_{12} radioassay	14
Tuna	Can manipulate extracted lens protein to induce specific reactivity with selected molecules and cells	Possible new diagnostic reagents	15—17
Pinfish, king whiting, sea catfish	Sera may identify sex-specific structures on human erythrocytes	Identification of previously unknown cell surface structures	18, 15
Swordtail, goldfish, damselfish	Natural development of pigmented tumors	Models of human pigmented tumors, e.g., melanoma and neurofibroma	19
Damselfish	Natural development of pigmented lesions	Model of human von Recklinghausen neurofibromatosis	20
Swordtail	Propensity to develop neoplasms	Carcinogenesis	8
	Ease of hybridization	Oncogene expression	
Milkfish	Catalase absent	Enzyme deficiency	21
Pacific salmonid	Natural regression of severe coronary artery disease (arteriosclerosis)	Control of arteriosclerosis	22
General	Rapid degenerative changes	Aging, arteriosclerosis	
	Produce only IgM	Immunodeficiency	8
	Temperature dependence of immune response	Control of immune response	
	Accessibility of simple organ (gill) involved in gas diffusion	Membrane physiology	
	Melanin concentration fluctuates with temperature	Melanin function and production	
	All developmental stages available	Carcinogenicity testing	
	Large numbers of individuals produced; can be of one sex	Studies requiring many genetically homogeneous individuals	
	Live over wide temperature range	Temperature effects on metabolic processes	
	Rarity of metastases from malignant tumors	Metastatic process	
II. Cartilaginous fishes			
Dogfish	Large leukocyte lysosomes in which urate crystal uptake and processing can be easily followed	Mechanism of gout	7
General	High blood urea	Hepatic disease	22
	High liver fat	Fatty liver	
	Abundant cartilage	Inhibitors of tumor angiogenesis	23
	Low incidence of malignant tumors	Carcinogenesis	24—26
III. Jawless fishes			
Lamprey, Hagfish	Endocrine pancreas is separate from exocrine pancreas and appears to possess insulin-secreting beta cells but no alpha cells or glucagon	Diabetes	22
	Elemental thyroid and thyroid function	Thyroid control	22

TABLE 1 (continued)
Biomedical Research Applications of Marine Animals

Animal group	Special features	Areas of biomedical application	Ref.
IV. Protochordates			
Tunicates	Elemental thyroid and thyroid function	Thyroid control	22
V. Echinoderms			
Sea cucumber	High level of carcinoembryonic antigen (CEA)	Biological origin and elemental function of CEA	27
	Coelomic fluid may preferentially react with non- or weak antigenic proteins	Reagent for non- or weak antigens	28
	Possible prototype immunologic organ (Polian vesicles)	Biological origin of lymphoreticular system, elemental immunologic and immunopathologic responses	29
	Natural resistance to infection (peritonitis)	Natural resistance, peritonitis	30
Sea urchin	Ease of collecting ova and sperm	Fertilization and development	7
Starfish (seastar)	Transparent projections from limbs through which one can directly visualize circulating amoebocytes	Phagocytosis and inflammation	31
VI. Mollusks			
Clam	High eosinophil count	Eosinophilia, allergy	32
	Kidney concretions	Nephrolithiasis	33, 34
Sea hare	Large, accessible ganglion	Neuronal organization and interaction	22
Snail	Externally released secretion reacts with human proteins	Possible new diagnostic reagent	35, 36
VII. Crustacea			
Shrimp	Externally released secretion reacts with human proteins	Possible new diagnostic reagent	37
Crab	Sera may identify sex-specific structures on human erythrocytes	Identification of previously unknown cell surface structures	18, 15
	Natural lectins which differentiate pathogenic strains of *Escherichia coli*	Laboratory diagnosis of *E. coli* strains	38
VIII. Arachnids			
Horseshoe crab	Hemocyte susceptibility to endotoxin	Endotoxin test (Limulus Amebocyte Lysate test)	39, 40, 41
	Natural agglutinins for human cells	Analysis of blood for circulating tumor cells	42
IX. Sipunculids			
Sipunculus	Mucus-secreting cell complex in coelomic fluid	Mucus secretion	43
X. Flatworms	Neoplasm production under laboratory conditions	Initiators and promoters in carcinogenesis	44
XI. Coral			
General	Porous microstructure	Prosthetics	45
XII. Variety of invertebrates	Rich in lectins	Mitogenesis, endocytosis, and tubulin	7
	Natural lectins against human cells and bacteria	Laboratory diagnosis and prognosis	46

to date has been relatively small, leaving much opportunity for the innovative biomedical researcher.

Those individuals who have worked with marine animals often develop an ever-increasing list of potential biomedical studies based on increasing familiarity with the unique characteristics of various species. Some selected personal examples follow:

1. The absence of hemoglobin in the antarctic icefish could serve as a model of human anemia.
2. The absence of certain antigens on fish tissue may allow its transplantation to mammals.
3. The high mucus production and accessibility of fish skin make it ideal for studies of mucus secretion.
4. The superficial location of the "brain" — really a gland/ganglion complex — in tunicates suggests its use in studies of elemental pituitary/hypothalamic function.
5. The superficial location of a large nerve (radial nerve) on the underside of the starfish arm could facilitate studies of the influence of nerves on regeneration.
6. The always-exposed, large fibromuscular mass (foot) in certain mollusks (abalones, limpets) seems ideal for studies of wound healing.

VII. PATHOLOGY RECAPITULATES PHYLOGENY

Another value of marine animals to the biomedical scientist is that by providing the phylogenetic dimension, these animals can give insight into otherwise perplexing phenomena in mammalian pathology. Three examples follow.

A. C-REACTIVE PROTEIN (CRP)

In humans it is well known that the concentration of a specific serum protein, CRP, may increase 1000-fold in concentration a few hours after the onset of inflammation and tissue destruction.[47] The biological function of CRP is still uncertain in that it binds nonspecifically to a large variety of cell products (e.g., phosphorylcholine, DNA, lecithin, and carbohydrate) and does not appear to be basically protective. However, working with North Atlantic flatfish, Baldo and Fletcher[48] demonstrated that CRP-like precipitins are highly reactive against extracts of a variety of pathogenic organisms, viz., several fungal species, *Streptococcus pneumoniae,* and an ascarid nematode.

It thus appears that CRP in fish plays a role in chemical defense, and this is likely to be vital since humoral antibody production ceases when temperatures are low as in the North Atlantic. It is clearly advantageous in this instance to have serum proteins (CRP) with broad specificities for determinants on surface structures of invading organisms and on cell debris resulting from this invasion.

In mammals, with the evolution of homeothermy and a sophisticated immunologic system, CRP was no longer needed to protect the animal. However, CRP continues to respond to inflammation and tissue destruction as if it is still protective, and so is simply recapitulating its original, ancestral role.

B. BLOOD CLOTTING

Another example of how studies of marine animals can provide insight into a human pathological situation is the rare but life-threatening one of clot development in a fully anticoagulated patient. In this seemingly paradoxical circumstance, there is evidence from the work of Laki and colleagues[49] at the National Institutes of Health for activation of a primitive clotting system found normally in fish and lobster. This system does not respond to the standard anticoagulants of coumarin and heparin, and is normally nonoperational in humans. However, when blood vessels are damaged as by the development of atheromatous plaques or sclerosis, the enzyme transglutaminase in the walls of the vessels may reach critically high levels which directly produce clots of fibrinogen. In this process, the complex clot-forming (cascade) system, which does respond to the standard anticoagulants, is bypassed.

Hence, a normally nonexpressed clotting system remains as a hallmark of our aquatic ancestry and, under certain pathological conditions, reveals its presence. As pointed out by Laki, it would behoove us to search in fish and invertebrates for the needed new anticoagulant which, based on discoveries of other anticoagulants, is likely to exist there.

C. SCHIZOPHRENIA

A study of sharks may provide information relevant to our understanding about mental illnesses like schizophrenia. Schizophrenics, particularly with the endogenous type, are often concerned with violent matters, e.g., killing, and life and death; and their hallucinations often involve primitive senses, viz., smell and vision. Further, there is characteristically a singleness and fixation of purpose and thought, with inflexible, stereotyped behavior. This is often maladaptive, but when the stimulus is removed, the behavior continues and can be injurious.

These characteristics of schizophrenics reflect the distorted expression of primitive brain centers, especially the limbic system. Such behavior is also characteristic of instinctive control.

A good example of an animal driven by instincts is the shark, which in a feeding frenzy, exhibits (1) preoccupation with killing and eating, (2) dominance of smell and vision senses, (3) fixed, purposeful behavior, and (4) inflexible, often harmful, behavior. In the shark, the entire cerebral cortex is a paleopalium (primitive form of gray matter) from which the human limbic system evolved. In the human, this system is reduced and tucked under the brain where it is concerned with smell and repetitive behavior (jaw movements).

The behavior of schizophrenics suggests a reassertion of the type of control seen normally in sharks and the concept of phylogenetic regression. Although ontogenetic, or developmental, regression is a familiar concept in psychiatry, the phylogenetic aspect is not.

As the preceding examples (CRP, clotting, and schizophrenia) demonstrate, marine biology and medicine can interact in a way that is mutually enriching and insightful.

VIII. MARINE NATURAL PRODUCTS

Marine animals are a virtual pharmacopoeia of drugs and other bioactive substances which have hardly been touched. Antineoplastic, antiviral, antibacterial, and other substances are being obtained from them with increasing frequency,[50,51] and their rich content of immunoreactive substances offers promise of new reagents for medical diagnosis.

IX. HUMAN DISEASE COUNTERPARTS IN MARINE ANIMALS

Since human pathologists have not commonly studied marine animals, little is known of counterparts to human disease in these animals. However, there are reports of gastritis,[52] allergy or allergy-like conditions,[53,54] intravascular clotting,[55] coronary artery disease,[22] and skeletal deformities[56] in fishes, and cancer in both fishes[57] and invertebrates.[58]

X. MAINTENANCE OF MARINE ANIMALS

Many marine animals are readily collected in the wild and available for experimentation. For example, animals such as sponges, worms, barnacles, and tunicates are often scraped from boat hulls and docks, and discarded as undesirable species. Still other species, both fish and invertebrate, can be purchased from suppliers such as pet stores, or from their suppliers. Aquaculture has become very sophisticated, and several species (e.g., catfish and oysters) are commercially available from aquaculture farms. Maintenance of species selected for research can be relatively easy and inexpensive. Aquarium systems today are highly refined and, when properly equipped, require little attention.

XI. CONCLUSION

Marine animals clearly have much to offer the biomedical researcher. Ranging from specific studies with direct clinical goals to intellectually provocative projects with future applications, addition of the phylogenetic dimension is clearly worthwhile. The effort to control disease — the ultimate goal of medical research — will never be an all-out one until the animal life of the sea is included. Considering the benefits to humankind, the time to start is now.

ACKNOWLEDGMENT

I wish to thank Dr. Herman Chmel, Division of Infectious and Tropical Diseases, University of South Florida, Tampa, for critical reading of the manuscript.

REFERENCES

1. **Garfield, E.,** Animal experimentation — When do the ends justify the means?, *Curr. Contents,* 5, 3, 1984.
2. **Fox, J. L.,** Lab animal welfare issue gathers momentum, *Science,* 223, 468, 1984.
3. **Dickinson, J.,** ''Animal slavery'' leading to civil war?, *Gov. Lab.,* 1, 12, 1985.
4. **Regan, T.,** *The Case for Animal Rights,* University of California Press, Berkeley, 1985.
5. **Holden, C.,** A pivotal year for lab animal welfare, *Science,* 232, 147, 1986.
6. **Silverman, J. and Barber, L. G.,** Public relations tactics in the debate over animal experimentation, *Lab. Anim.,* 16, 21, 1987.
7. **Cheng, T. C.,** The (re-)emerging area of marine biomedicine (editorial), *J. Invertebr. Pathol.,* 40, 153, 1982.
8. **Wolke, R.,** The use of fish in biomedical research, *Comp. Pathol. Bull.,* 16, 1, 1984.
9. **Smith, A. C.,** Marine animals in medical research, *JAMA,* 242, 2847, 1979; reprinted in *Environs* (Marine Biomedical Center, Duke University Marine Laboratory, Beaufort, NC), 3, 2, 1980; *Gov. Lab.,* 1, 21, 1985.
10. **Karnovsky, M. L.,** Metchinkoff in Messina, *N. Engl. J. Med.,* 304, 1178, 1981.
11. **Lagler, K. F., Bardach, J., Miller, R. R., and Passino, D. R. M.,** *Ichthyology,* John Wiley & Sons, New York, 1977, 1.
12. **Kukalová-Peck, J.,** *A Phylogenetic Tree of the Animal Kingdom (Including Orders and Higher Categories),* Publications in Zoology, No. 8, National Museums of Canada, Ottawa, 1973, 18.
13. **Beer, A. E. and Billingham, R. E.,** Transplantation in nature, *Perspect. Biol. Med.,* 22, 155, 1979.
14. **Kubiatowicz, D. O., Ithakissios, D. S., and Windorski, D. C.,** Vitamin B_{12} radioassay with oyster toadfish *(Opsanus tau)* serum, *Clin. Chem.,* 23, 1037, 1977.
15. **Smith, A. C.,** Alternatives to mammalian antisera in the evaluation/identification of human proteins, *Am. Zool.,* 25, 380, 1985.
16. **Smith, A. C.,** Fish eye-lens reagents: a possible new class of reagents for molecular and cellular identification, *Comp. Biochem. Physiol.,* 80B, 719, 1985.
17. **Smith, A. C.,** Fish eye-lens reagents: sex-specific agglutination of human erythrocytes, *J. Nat. Prod.,* 49, 163, 1986.
18. **Smith, A. C.,** Sex-specific agglutination of human erythrocytes by body fluids from marine invertebrates and fishes, in *Recognition Proteins, Receptors, and Probes,* Cohen, E., Ed., Alan R. Liss, New York, 1984, 97.
19. **Riccardi, V. M. and Eicher, J. E.,** Pathogenesis, in *Neurofibromatosis: Phenotype, Natural History, and Pathogenesis,* Johns Hopkins Press, Baltimore, 1986, 184.

20. **Schmale, M. C., Hensley, G. T., and Udey, L. R.,** Neurofibromatosis in the bicolor damselfish *(Pomacentrus partitus)* as a model of von Recklinghausen neurofibromatosis, *Ann. N.Y. Acad. Sci.,* 486, 386, 1986.

21. **Smith, A. C.,** Catalase in fish red blood cells, *Comp. Biochem. Physiol.,* 54B, 331, 1976.

22. **Wolf, S. G.,** Contributions of the sea to medicine, in *Drugs and Food from the Sea: Myth or Reality?,* Kaul, P. N. and Sindermann, C. J., Eds., University of Oklahoma, Norman, 1978, 7.

23. **Lee, A. and Langer, R.,** Shark cartilage contains inhibitors of tumor angiogenesis, *Science,* 221, 1185, 1983.

24. **Mawdesley-Thomas, L. E.,** Neoplasia in fish, in *The Pathology of Fish,* Ribelin, W. E. and Migaki, G., Eds., University of Wisconsin Press, Madison, 1975, 805.

25. **Harshbarger, J. C., Charles, A. M., and Spero, P. M.,** Collection and analysis of neoplasms in sub-homeothermic animals from a phyletic point of view, in *Phyletic Approaches to Cancer,* Dawe, C. J., Harshbarger, J. C., Kondo, S., Sugimura, T., and Takayama, S., Eds., Japan Scientific Societies Press, Tokyo, 1981, 357.

26. **Kaiser, H. E.,** Phylogeny and paleopathology of animal and human neoplasms, in *Neoplasms — Comparative Pathology of Growth in Animals, Plants, and Man,* Kaiser, H. E., Ed., Williams & Wilkins, Baltimore, 1981, 725.

27. **Smith, A. C.,** Oncofetal proteins in marine animals, *Comp. Biochem. Physiol.,* 61B, 499, 1978.

28. **Smith, A. C.,** Reactions of fish nuclear lens proteins with sea cucumber coelomic fluid, *Dev. Comp. Immunol.,* 3, 557, 1979.

29. **Smith, A. C.,** Immunopathology in an invertebrate, the sea cucumber, *Holothuria cinerascens, Dev. Comp. Immunol.,* 4, 417, 1980.

30. **Dybas, L. and Fankboner, P. V.,** Holothurian survival strategies: mechanisms for the maintenance of a bacteriostatic environment in the coelomic cavity of the sea cucumber, *Parastichopus californicus, Dev. Comp. Immunol.,* 10, 311, 1986.

31. **Bang, F. B.,** Disease processes in seastars: a Metchnikovian challenge, in *Phagocytosis-Past and Future,* Karnovsky, M. L. and Bolia, L., Eds., Academic Press, New York, 1982, 567.

32. **Smith, A. C.,** Search among Invertebrates for Eosinophils Useful in Medical Research, presented at 8th Ann. Meet., Society for Invertebrate Pathology, Corvallis, OR, August 18 to 22, 1975.

33. **Tiffany, W. J., III,** Analysis of renal calculi from a marine mollusc *(Macrocallista nimbosa), Invest. Urol.,* 17, 164, 1979.

34. **Reid, R. G. B., Fankboner, P. V., and Brand, D. G.,** Studies on the physiology of the giant clam *Tridacna gigas* Linné. II. Kidney function, *Comp. Biochem. Physiol.,* 78A, 103, 1984.

35. **Smith, A. C.,** A precipitin for human serum proteins as released under stress by a marine invertebrate, the marsh snail, *Littorina irrorata, Dev. Comp. Immunol.,* 8, 273, 1984.

36. **Smith, A. C.,** Snailine: a possible diagnostic reagent from a common marine snail, the southern periwinkle, *Littorina angulifera, Dev. Comp. Immunol.,* 10, 489, 1986.

37. **Smith, A. C.,** A precipitin for human serum proteins as released into the environment by stressed bait shrimp, *Penaeus duorarum, Comp. Biochem. Physiol.,* 87B, 659, 1987.

38. **Ravindranith, M. H. and Cooper, E. L.,** Crab lectins: receptor specificity and biomedical applications, in *Recognition Proteins, Receptors, and Probes: Invertebrates,* Cohen, E., Ed., Alan R. Liss, New York, 1984, 83.

39. **Cohen, E.,** *Biomedical Applications of the Horseshoe Crab (Limulidae),* Alan R. Liss, New York, 1979.

40. **Watson, S. W., Levin, J., and Novitsky, T. J., Eds.,** Endotoxins and their detection with the limulus amebocyte lysate test, *Progress in Clinical and Biological Research,* Vol. 93, Alan R. Liss, New York, 1982.

41. **Friberger, P.,** A new method of endotoxin determination, *Am. Clin. Prod. Rev.,* 6, 12, 1987.

42. **Cohen, E.,** Detection of membrane receptors by arthropod agglutinins, in *Hemocytic and Humoral Immunity in Arthropods,* Gupta, A. P., Ed., John Wiley & Sons, New York, 1986, 493.

43. **Bang, F. B. and Bang, B. G.,** Pathologic principles revealed by study of natural disease of invertebrates, in *Physiology and Biology of Horseshoe Crabs: Studies on Normal and Environmentally Stressed Animals,* Bonaventura, J., Bonaventura, C., and Tesh, S., Eds., Alan R. Liss, New York, 1982, 289.

44. **Hall, F., Morita, M., and Best, J. B.,** Neoplastic transformation in the planarian. I. Cocarcinogenesis and histopathology. II. Ultrastructure of malignant reticuloma, *J. Exp. Zool.,* 240, 211-227, 1986.

45. **Dunn, E., Brooks, T., Gordon, B., Resnert, S., White, E., and Tarhay, L.** (no title), *Trans. 24th Ann. Meet. Orthopaed. Res. Soc.,* 3, 301, 1978.

46. **Cohen, E.,** Summation; biomedical significance of invertebrate lectins, in *Recognition Proteins, Receptors, and Probes: Invertebrates,* Cohen, E., Ed., Alan R. Liss, New York, 1984, 193.

47. **Putnam, F. W.,** Alpha, beta, gamma, omega — the structure of the plasma proteins, in *The Plasma Proteins: Structure, Function, and Genetic Control,* Putnam, F. W., Ed., Academic Press, New York, 1984, 45.

48. **Baldo, B. A. and Fletcher, T. C.,** C-reactive protein-like precipitins in plaice, *Nature (London)*, 246, 145, 1973.
49. **Stern, M.,** New theory of blood clotting sparks search for new inhibitor. Adult chinook salmon used in blood research, *News Features NIH*, 7, 11, 1974.
50. **Colwell, R. R.,** Biotechnology in the marine sciences, *Science*, 222, 19, 1983.
51. **Colon-Urban, R., Reyes, L., and Winston, J. E.,** Antibiotic substances from several antarctic bryozoans, *Am. Zool.*, 25, 52a, 1985.
52. **Smith, A. C.,** Pathology and biochemical genetic variation in the milkfish, *Chanos chanos, J. Fish Biol.*, 13, 173, 1978.
53. **Smith, A. C.,** Reactions of fish red blood cells with mucus and sera from other fish(es), *Calif. Fish Game*, 63, 52, 1977.
54. **Henderson-Arzapalo, A. and Stickney, R. R.,** Immune hypersensitivity in intensively cultured tilapia species, *Trans. Am. Fish. Soc.*, 109, 244, 1980.
55. **Smith, A. C.,** Formation of fatal blood clots in fishes, *J. Fish Biol.*, 16, 1, 1980.
56. **Tucker, B. W. and Halver, J. E.,** Vitamin C metabolism in rainbow trout, *Comp. Pathol. Bull.*, 28, 1, 1986.
57. **Budd, J. and Roberts, R. J.,** Neoplasia of teleosts, in *Fish Pathology*, Roberts, R. J., Ed., Bailliere Tindall, London, 1978, 105.
58. **Sparks, A. K.,** *Synopsis of Invertebrate Pathology Exclusive of Insects*, Elsevier, New York, 1983, 91.

Chapter 11

NERVOUS CONTROL OF VENTILATION AND HEART RATE IN ELASMOBRANCH FISH, A MODEL FOR THE STUDY OF THE CENTRAL NEURAL MECHANISMS MEDIATING CARDIORESPIRATORY INTERACTIONS IN MAMMALS

Edwin W. Taylor

TABLE OF CONTENTS

I. INTRODUCTION

As vertebrates, fish and mammals share common morphological and physiological features in all their major organ systems, but because most fish are aquatic gill breathers while mammals breathe air using lungs, there are major differences in the construction of their respiratory and cardiovascular systems. In mammals lung ventilation is tidal and accomplished by coordinated contractions of diaphragmatic, intercostal, and abdominal muscles[1] which contract during the active, inspiratory phase of ventilation. These respiratory muscles are innervated from the spinal cord, and only some accessory respiratory muscles (e.g., for control of the glottis) are innervated by cranial nerves and have their motor control circuits located in the brainstem. By contrast, in fish, water is propelled continuously over the gills in one direction by the ventilatory muscles which operate around the jaws and skeletal elements in the gill arches lining the pharynx.[2-6] As far as the buccal pump is concerned, the active phase of ventilation is expiration, when the jaw closes and water is forced over the gills. In fish the respiratory muscles are all cranial muscles, innervated by motor neurons with their cell bodies in the brainstem located close to the site of the central pattern generator (CPG), which generates the respiratory rhythm.

The cardiovascular systems of fish and mammals have in common a chambered heart, with a myogenic pacemaker, delivering blood into a closed system of arteries supplying the major organs, which are in turn drained by a venous system returning blood to the heart. Major differences are again related to their respective modes of respiration. The mammals have a completely divided blood system, with separate pulmonary and systemic circuits, while fish typically have a single system supplying the branchial vasculature directly from the heart, with an arterio-arterial respiratory route conducting blood directly from the gills to the systemic circuit. A subsidiary arteriovenous route drains the branchial circulation in fishes,[7] and both the fish and the mammalian systems are complicated by the formation and drainage of lymph, but these need not be considered further in this account.

Despite these major differences in the construction and mode of operation of the respiratory and cardiovascular systems of gill and lung breathers, there are important similarities in the nervous control of the respiratory rhythm and heart rate and more particularly in the central nervous generation of cardiorespiratory interactions. These have prompted the author to propose that the study of nervous regulation of the cardiorespiratory system in fishes may provide important insights into the control of the more complicated mammalian system. These studies are potentially rewarding because fish have certain advantages over mammals as experimental animals which will be discussed here.

The following account begins with a comparison of the central nervous generation of the respiratory rhythm and its control in fish and mammals. This is followed by a description of the nervous control of heart rate in elasmobranch fish, and finally there is a consideration of the control of cardiorespiratory interactions. The emphasis is quite clearly on a discussion of the basis of apparent similarities and important differences between the control systems in these two vertebrate groups, which are present-day representatives of the two ends of the vertebrate evolutionary scale. The chapter concludes with a summary of the possible advantages of the fish model for investigating the fundamentals of central nervous control of respiration and cardiorespiratory interactions in vertebrates and discusses its possible use in interpretation of some elements of control of the more complex mammalian system.

II. CONTROL OF RESPIRATION

Valuable comparisons of the neural control of respiration in fish and mammals have been provided by Ballintijn.[8,9] This account is firmly based on his excellent reviews, and I strongly recommend the reader to consult them. The first[8] concentrates on a comparison of the better-known fish and mammalian systems, while the second[9] also considers the relatively neglected subject of respiratory control in amphibians, reptiles, and birds.

A. THE CENTRAL PATTERN GENERATOR

Rhythmic ventilatory movements continue in fish following brain transection to isolate the medulla oblongata; though changes in pattern indicate that there are influences from higher centers.[10] Central recording and marking techniques have identified a longitudinal strip of neurones with spontaneous respiration — related bursting activity, extending dorsolaterally throughout the whole extent of the medulla.[11-14] These neurones make up elements of the trigeminal Vth, facial VIIth, glossopharyngeal IXth, and vagal Xth motor nuclei, which drive the respiratory muscles, together with the descending trigeminal nucleus and the reticular formation. All of the motor nuclei are interconnected, and each receives an afferent projection from the descending trigeminal nucleus and has efferent and afferent projections to and from the reticular formation (Figure 1A). The intermediate facial nucleus, which receives vagal afferents from the gill arches that innervate a range of tonically and phasically active mechanoreceptors,[15] projects to the motor nuclei.[16] Finally, areas in the midbrain such as the mesencephalic tegmentum have efferent and afferent connections with the reticular formation.[17,18] The respiratory rhythm apparently originates in a diffuse CPG in the reticular formation, and this remains functional under anesthesia.[9]

Central recordings from the medulla oblongata of the carp by Ballintijn and Alink[19] suggested that adjacent neurones have different firing patterns; they identified the target muscle for individual motoneurones by simultaneous recordings of neuronal activity and electro-myograms (emgs) from the respiratory muscles. However, retrograde intraaxonal transport of horseradish peroxidase (HRP) along nerves that innervate the respiratory muscles revealed that in the brainstem of elasmobranchs the neurones in the various motor nuclei are distributed in a sequential series.[20] Further, in accordance with these findings, recordings of efferent activity from the central cut ends of the nerves innervating the respiratory muscles of the dogfish *Scyliorhinus canicula* and the ray *Raia clavata*[79] have revealed that the branches of the Vth, VIIth, IXth, Xth and spino-occipital branches constituting the hypobranchial nerve, fire sequentially in the order of the sequential rostrocaudal distribution of their motonuclei in the brainstem and rostral spinal cord. The resultant coordinated contractions of the appropriate respiratory or feeding muscles may relate to their original segmental arrangement before cephalization — an arrangement which is retained in the hindbrain of the fish in the sequential topographical arrangement of the motor nuclei, including the subdivisions of the vagal motonucleus (Figure 2).

The central elements determining respiratory rhythm generation in mammals have been reviewed by Hugelin,[21,22] Cohen,[23] and von Euler.[24] Two populations of respiration-related neurones have been recognized: a dorsal, bulbar respiratory nucleus in the solitary tract (NTS) and a ventral bulbar respiratory nucleus containing the nucleus para-ambigualis and merging with the nucleus ambiguus (NA). The accessory respiratory motor nuclei (trigeminal Vth, facial VIIth, vagal Xth, and hypoglossal) all contain respiration-related neurones. A CPG in the reticular formation is active in the waking state.[22] In addition mammals possess a "pneumotaxic center" located in the pons and more correctly designated the pontine respiratory complex (Figure 1B), which also contains neurones showing respiration-related activity.[22] This area of the brain is absent in fish. Bertrand et al.,[25] have suggested that the mammalian brainstem contains a large number of local respiratory oscillators whose dominance varies.

In mammals nearly all neurones in the dorsal and ventral bulbar nuclei send axons, mostly contralaterally, down the spinal chord to innervate spinal motoneurones that supply the thoracic respiratory musculature (Figure 1). In addition, the NA sends axons to the IXth, Xth and XIth motor nuclei.[26] The dorsal bulbar nucleus receives pulmonary mechanoreceptor

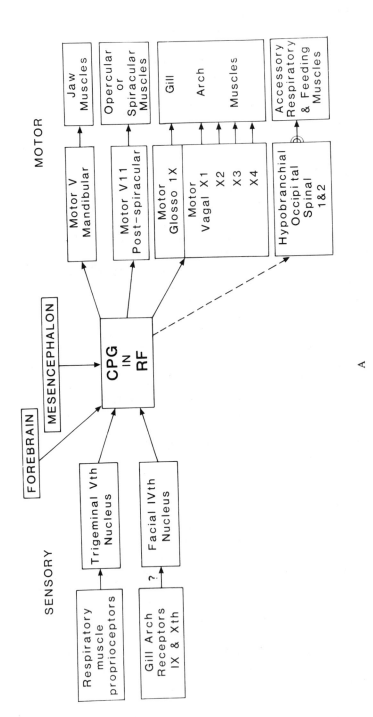

FIGURE 1. Summary diagrams of the functional connections involved in the central nervous control of ventilation in (A) fish and (B) mammals. The arrows merely indicate functional connections for which there are various levels of direct and indirect evidence and say nothing about synaptic pathways. The arrows with a spiral in the shaft indicate contralateral connections. More details of CPG and their connections are given in the text. CPG, central pattern generator; DBRN, dorsal bulbar respiratory nucleus; KF, Kollike Fuse nucleus; NA, nucleus ambiguus; NPA, nucleus para-ambigualis; NPBM, nucleus parabrachialis medialis; NTS, nucleus tractus solitarius; PRC, pontine respiratory complex; RF, reticular formation; VBRN, ventral bulbar respiratory nucleus.

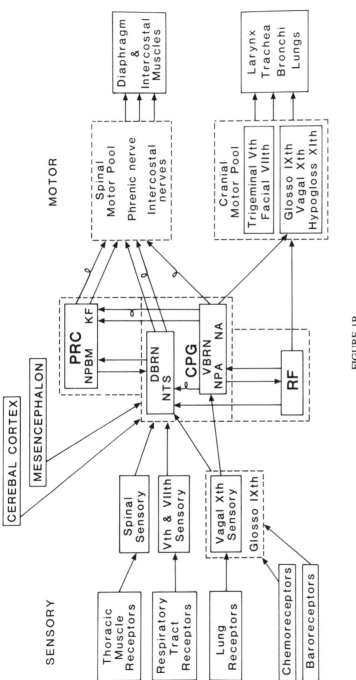

FIGURE 1B.

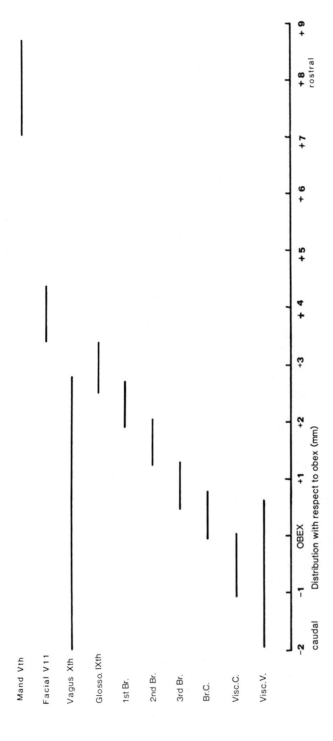

FIGURE 2. The topographical organization of the vagal motor column and respiratory motor nuclei of the Vth and VIIth cranial nerves in the dogfish. The rostrocaudal distribution either side of obex of the pools of motoneurones that supply the mandibular branch of the Vth cranial nerve, the facial branch of the VIIth cranial nerve, the glossopharyngeal (IXth) cranial nerve, and the vagus (Xth) cranial nerve together with its constituent branches (the first three branchial branches, 1st Br. to 3rd Br.; the branchial cardiac, Br. C., and visceral cardiac, Visc. C., branches, and the visceral branch, Visc. V.) are indicated by solid lines.

information via the vagus and afferent input from the cerebral cortex. It sends efferent information to the limbic system in the forebrain and from dorsal, inspiratory neurones to the phrenic motor pool. Finally, the pontine area has afferent and efferent connections with the bulbar respiratory nuclei and directly to the spinal cord.[26]

B. GENERATION OF THE RESPIRATORY RHYTHM

Experiments in which ventilatory movements have been prevented with curare have demonstrated that the CPG in fish is self-contained and generates rhythmic bursts of activity which are slower than the respiratory rhythm before paralysis in teleosts,[27] but faster than it is in elasmobranchs.[28] These changes in rate following paralysis indicate that activity in the CPG is normally modulated by peripheral receptors. This is consistent with the fact that the CPG received afferent information from a range of mechanoreceptors, proprioceptors, and skin stretch receptors around the gill arches and jaws.[8] Electrical stimulation of a vagal ganglion, containing the cell bodies of vagal afferent fibers supplying the gill arches, can terminate inspiration while bursts of stimuli can drive the CPG either faster or slower than its intrinsic rhythm.[16] This suggests that the vagal afferents are coupled to the CPG in the reticular formation, possibly via the NA to which they project.[16] Furthermore, the CPG in unanesthetized fish is known to be influenced by higher, mesencephalic neurones, some of which have direct connections with the reticular formation and may drive the CPG.[29] These mesencephalic neurones are responsible for initiating bursts of ventilation during the periodic or ''bout'' respiration shown by some inactive or hyperoxic fish.[9]

In mammals, the respiratory rhythm continues in animals which are anesthetized, vagotomized, and artificially ventilated (Figure 3), indicating that the intrinsic rhythmicity of the CPG is independent of lung receptor feedback. Bilateral lesioning of the pontine respiratory complex stops this respiratory rhythm, indicating that higher centers are implicated in the CPG. However, anesthetized animals with intact vagi still breath rhythmically following lesions of the pons, indicating that lung receptor feedback may substitute for input from higher centers in the CNS.[8] Nevertheless, vagotomized mammals that are unanesthetized still breathe rhythmically even following lesioning of the pons, implying the existence of a further CPG in the reticular formation, which is suppressed by anesthesia. Consequently, the normal breathing rhythm arises from three or possibly more mutually coupled CPG,[22] which are influenced by peripheral feedback.

In the mammalian CPG, inspiration is the dominant phase. The duration of inspiration is determined by an inspiratory off-switch in the ventrolateral NTS which is activated by input from the pons[30] phasic vagal input from lung receptors,[31] and input from the reticular formation.[22] Inspiration, as indicated by motor activity recorded from the phrenic nerve, always has the same time course, whether occurring spontaneously or triggered by vagal or pontine stimulation, indicating that the inspiratory neurones are tightly coupled. However, the duration of expiration is more variable and is matched with inspiration via the pontine centers.[31] The input from pulmonary stretch receptors with vagal afferents can extend the period of expiration, but rapid lung inflation or stimulation of lung irritant receptors reduces expiration time. In addition, thoracic receptors in the diaphragm and intercostal muscles provide spinal afferent inputs which converge with the vagal input from the lungs to affect the inspiratory off-switch.

The respiratory nuclei in the mammalian medulla contain large numbers of interneurones and motor neurones with a respiration-related activity, firing in the inspiration, postinspiration, or expiration phase (Figure 4), and resembling the motor pattern in intercostal muscles and the diaphragm. Many of these are preganglionic vagal respiratory neurones that innervate the larynx, trachea, bronchi, and lungs. Others are interneurones which project to spinal motor nuclei that innervate the thoracic respiratory structures.[32] Together the activity in these various preganglionic motor neurones constitutes the detailed pattern of motor discharge

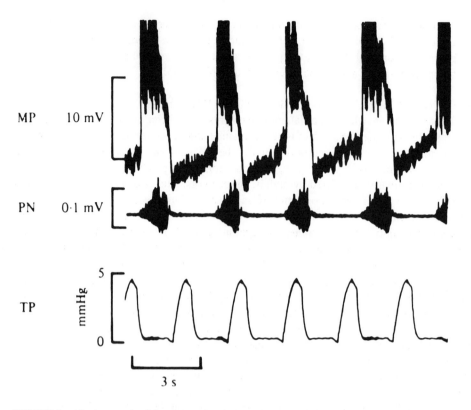

FIGURE 3.   Membrane potential pattern (MP) of an early inspiratory interneurone. PN, phrenic nerve activity; TP, tracheal pressure of the pentobarbital anesthetized, thoracotomized, and artificially ventilated cat. The central respiratory rhythm continues independently of the imposed artificial ventilation of the lungs. (From Richter, D. W., *J. Exp. Biol.*, 100, 93, 1982. With permission.)

during ventilation via reticular neurones which send fibers to all levels of the spinal cord and influence activity in respiratory motoneurones.[33] Part of the population of motoneurones that innervates the respiratory muscles is silent in the paralyzed animal in both fish[8] and mammals[34] and may be stimulated to fire by artificially induced proprioceptive information.

C. CHEMORECEPTOR RESPONSES

The aquatic environment contains less oxygen per unit volume than air (approximately 2% at 20°C). Consequently, the ventilation requirement (respiratory medium passed over the respiratory surfaces in proportion to oxygen consumed) is high in active fish compared with mammals.[35] As a result, fish are effectively hyperventilating with respect to carbon dioxide excretion, and this, coupled with the high solubility coefficient for CO_2 in water, results in Pa,CO_2 levels in fish never rising above 5 mmHg.[36] Although fish do show ventilatory responses to carbon dioxide,[12] this is not normally an important reflex for them. Rather, ventilation is controlled with respect to oxygen levels, which is appropriate as hypoxia due to reduced supply or increased demand (e.g., exercise) is a common problem for water breathers. This, of course, is in sharp contrast to the situation in mammals, in which circulating Pa,CO_2 levels are high relative to those occurring in fish, and changes in Pa,CO_2 are one of the strongest influences on respiration, this influence being exerted by changes in the pH of the extracellular fluid or in intracellular/extracellular pH gradient over chemosensory neurones which are presumed to lie in the superficial layers of the ventrolateral medulla.[37]

The oxygen receptors in mammals are the carotid and aortic chemoreceptors, which are

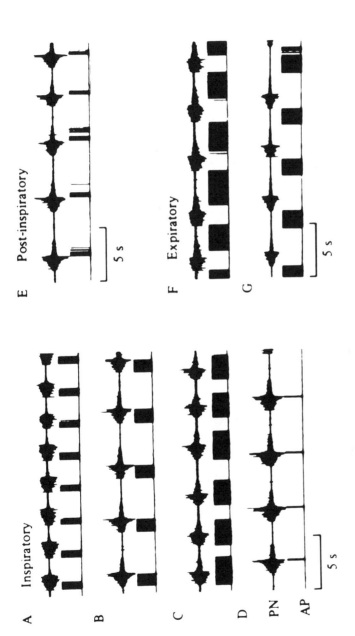

FIGURE 4. Firing patterns of different types of respiratory neurones located in the medulla of pentobarbital anesthetized cats. Neurones are classified as being inspiratory when they fire action potentials (AP) during the phrenic nerve discharge (PN), as postinspiratory when they fire only during stage 1 of expiration, and as expiratory when they discharge predominantly during stage 2 of expiration. (From Richter, D. W., *J. Exp. Biol.*, 100, 93, 1982. With permission.)

innervated by the IXth and Xth cranial nerves and are sensitive to deviations in Pa,O_2 in the blood supply to the brain and systemic circulation. However, their influence on ventilation rate is secondary to that imposed by the central effects of CO_2 level.[38] Moreover, stimulation of the carotid bodies is only effective in increasing inspiratory effort if it is timed to coincide with the phase of inspiration (the same stimulus has little effect when delivered in expiration), which implies that there is some central gating of input to central inspiratory neurones.[39,40]

Studies of the behavioral and physiological (largely cardiac) responses to hypoxia in fish have provided indirect evidence that they also have oxygen receptors at various sites in the ventilatory stream or blood system.[12,41-43] The gill arches in fishes are innervated by cranial nerves IX and X, and it is these nerves which innervate the carotid and aortic bodies of mammals. Bilateral section of IX and X abolished the reflex decrease in heart rate (bradycardia) evoked by hypoxia in the trout[44] but did not affect the hypoxia-induced bradycardia in elasmobranchs.[45] In fact, Butler et al.[46] had to bilaterally section cranial nerves V, VII, IX, and X to abolish the hypoxia-induced bradycardia in the dogfish and concluded that O_2 receptors are distributed diffusely in the orobranchial and parabranchial cavities. In the tench, Laurent[47] recorded oxygen-sensitive chemoreceptor activity from branches of cranial nerve IX that innervate the pseudobranch. This organ is derived from the spiracle, which is open in elasmobranchs, and as it receives arterialized blood flowing from the gills to the brain, it is ideally sited to monitor Pa,O_2 levels and may be the evolutionary homolog of the mammalian carotid body.

However, bilateral denervation of the pseudobranch in the trout had no effect on the increase in ventilation evoked by exposure to hypoxia or hyperoxia;[48] so this field is still somewhat confused. More recent evidence has suggested the existence of venous as well as arterial receptors in fish. Thus, exposure of the dogfish to progressive hyperoxia caused an initial tachycardia, toward the level of heart rate seen in the atropinized animal, indicating a reduction in vagal tone. This was followed by a secondary reflex bradycardia at high ambient P,O_2 levels, which corresponded with an increase in P_v,O_2. This latter response was mimicked by injection of hyperoxic blood into the venous system.[49] Receptors that monitor P,O_2 levels may therefore lie on both the afferent and efferent sides of the blood supply to the gills in dogfish.[43]

III. CONTROL OF THE HEART AND CIRCULATION

A. MECHANISMS OF CONTROL

The fish heart like the mammalian heart is composed of typical vertebrate cardiac muscle fibers, with contraction initiated by a propagated muscle action potential from a myogenic pacemaker and generating a characteristic EKG wave form.[50,51] Further, the elasmobranch heart is influenced by intrinsic mechanisms, such as the relationship between the force of contraction and stretch applied to the muscle fibers, which is identical to the Frank-Starling relationship that exists in mammals. Thus, the increase in diastolic filling time, which accompanies cardiac slowing in the dogfish, can result in an increase in cardiac stroke volume. Indeed Short et al.[52] concluded that the compensatory increases in cardiac stroke volume, which maintained cardiac output in the dogfish during a hypoxia-induced bradycardia, were wholly attributable to the Frank-Starling relationship.[41]

In mammals, it is established that the cardiac pacemaker in the sinoatrial node and therefore the heart rate, are under the influence of vagal and sympathetic fibers, the former innervating muscarinic cholinoceptors, and the latter β-adrenergic receptors. In addition the contractility of the heart and consequently stroke volume operate under the influence of sympathetic fibers. In elasmobranchs the heart is supplied with inhibitory parasympathetic innervation via the vagus nerve, which terminates on the heart as a plexus of fibers limited to the sinus venosus and innervating muscarinic cholinoreceptors on the pacemaker and myocardium.[53,54] There is little morphological or physiological evidence for sympathetic innervation of the elasmobranch heart, though it remains possible that some degree of cardioregulation may be exercised by catecholamines, released by activity in sympathetic preganglionic fibers supplying the chromaffin tissue located in the posterior cardinal sinus, which drains directly into the heart.[41] The elasmobranch heart operates under a degree of inhibitory vagal tone which varies with physiological state and environmental conditions. In the dogfish, cholinergic vagal tone, assessed as the proportional change in heart rate following injection of atropine, which blocks muscarinic cholinoreceptors, increased with an increasing temperature of acclimation.[55] As indicated previously, in fish, the typical response to systemic hypoxia is an increase in ventilation accompanied by a reflex brady-cardia. If temperature is constant at 15 to 17°C, then heart rate in dogfish restrained in a standard set of experimental conditions varies directly with ambient P,O_2, so that hypoxia induces a reflex bradycardia, and a normoxic vagal tone was released by exposure to moderate hyperoxia.[49,55] These data imply that the level of vagal tone on the heart in dogfish is determined peripherally by graded stimulation of P,O_2 receptors and that variations in this inhibitory tone may result in a reflex bradycardia or tachycardia.

In mammals a reflex bradycardia, predominantly mediated by vagal efferent fibers, plus peripheral vasoconstriction under sympathetic control, typifies the so-called primary response to stimulation of carotid chemoreceptors, and this is seen in anesthetized animals with lung ventilation artificially controlled by a respiratory pump.[56] However, during spontaneous ventilation, the increase in lung ventilation evoked by hypoxia leads to a secondary tach-ycardia and vasodilatation, elicited predominantly by stimulation of lung receptors, and this typically overcomes the primary response.[57]

In addition, stimulation of carotid chemoreceptors can activate higher centers such as the defense areas in mammals, to evoke the defense or visceral alerting response. This includes a tachycardia, vasoconstriction in splanchnic regions, and vasodilatation in skeletal muscle, which is not secondary to hyperventilation and is the response that occurs generally in mammals in response to novel noxious stimuli and is accompanied by behavioral arousal which may culminate in fight or flight.[58] Not only are the secondary effects of changes in ventilation upon the CNS not evident in fish, but there is also no evidence that chemoreceptor stimulation produces behavioral arousal. In fact, the unrestrained dogfish responds to en-vironmental hypoxia with a reduction in activity, which remains suppressed throughout the hypoxic period, despite an increase in circulating catecholamines.[59] This would seem to be the opposite of a defense or alerting response and is reflected in the apparent absence or suppression of a flight syndrome in dogfish, which often remain motionless on the seabed when approached or even handled by a scuba diver[80] and do not struggle even when held by lateral tail clamps in an experimental chamber, after recovery from a general anesthetic.[60]

In mammals their relatively high blood pressures are monitored by baroreceptors located in the carotid sinus and innervated by the sinus nerve. Stimulation of the receptor or the sinus nerve causes a reflex bradycardia, primarily affected by the vagal innervation of the heart, though responses to baroreceptor stimulation are modulated by the respiratory cycle, being silenced during inspiration.[40] Historical evidence indicating that there are baroreceptor-type reflexes in elasmobranchs, responsible for regulating blood pressure by reflex alteration of heart rate and peripheral vascular resistance, were reviewed by Johansen.[41] However, recent studies on the dogfish, which involved repeated sampling and reinjection of blood, revealed that the resultant changes in blood pressure were largely uncompensated.[49] These observations indicated that baroreceptor reflex responses are weak or nonexistent in elas-mobranch fishes. The absence of clear visceral alerting or baroreceptor responses in dogfish precludes their interference in chemoreceptor-induced changes in ventilation or heart rate.

B. ANATOMY OF THE VAGAL SUPPLY TO THE HEART

The heart in the dogfish is supplied with two pairs of vagi — a branchial cardiac branch leaves the fourth branchial nerve and a visceral cardiac branch is supplied from the visceral branch of the vagus — on either side of the animal.[55] Experiments involving transection and electrical stimulation of these nerves toward the heart revealed that the branchial cardiac branches are more effective in producing cardioinhibition than the visceral cardiac branches[52] and account for the majority of normoxic vagal tone and for the reflex bradycardia during hypoxia.[55] The topography of the vagal motor column in the dogfish has been described in detail by Withington-Wray et al.[61] The majority of vagal motoneurones are located medially, close to the wall of the fourth ventricle, and constitute the dorsal vagal motor nucleus (DVN), as described in mammals. The DVN includes respiratory vagal preganglionic motoneurones (RVM) contributing axons to four branchial branches of the vagus, which innervate respiratory muscles in the second, third, fourth, and fifth gill arches (the first gill arch is innervated by the glossopharyngeal, IXth cranial nerve). The RVM that supply axons to each of these gill arches have a sequential, overlapping distribution either side of obex in the vagal motor column (Figure 2). In a continuous series with the RVM are the cardiac vagal motoneurones (CVM) which supply the branchial cardiac branch of the vagus, followed by the visceral branch, which supplies the gut and other viscera, and the separate visceral cardiac branch (Figure 2). In addition, a clearly distinguishable lateral group of cells with a rostrocaudal extent of about 1 mm rostral of obex was identified. These cells are CVM which contribute axons solely to the branchial cardiac branch of the vagus[62] (Figure 5). They compose 8% of the total population of vagal motoneurones but supply 60% of the efferent axons running in the branchial cardiac branch, with the other 40% being supplied by cells in the DVN. When the medial cells that contribute axons to the visceral cardiac branch are taken into account, then the lateral cells are found to supply 45% of vagal efferent output to the heart. We have argued that the lateral cells constitute a primordial nucleus ambiguus (NA) as identified in the mammalian brainstem (see later).[61,63]

In the mammal the vagus innervates the larynx, trachea, bronchi, lungs, heart, and viscera. The vagal preganglionic motoneurones have most of their cell bodies in the DVN and the NA, but others are found in adjacent areas, such as the nucleus of the solitary tract (NTS), the nucleus retroambigualis (nRA), and the spinal nucleus of the accessory nerve (nspA).[64,65] Neurones in both the DVN and the NA supply fibers to all of the vagally innervated structures, but there is no discernible tographic representation of the thoracic and abdominal organs within either nucleus.[66] The proportion of vagal motoneurones with their cell bodies in the DVN varies between species. In the rat, 70% are located medially, whereas in the pig the majority are located in the NA. In the cat, both respiratory (RVM) and cardiac (CVM) vagal motoneurones are found predominantly in the NA (70% of RVM and 80% of CVM).[67,68] This compares with 45% of CVM in the NA of dogfish and indicates that a progressive ventrolateral migration of vagal motoneurones has occurred in the evolution of mammals from lower vertebrates. This is borne out by studies on amphibians and reptiles where a progressively larger proportion of motoneurones are located outside the DVN.[63] This migration seems to be associated particularly with the evolution of lung breathing when the lateral CVM were apparently joined by RVM. Presumably this progressive relocation of vagal motoneurones has a functional relevance, and this may be revealed by comparative studies of central connections and interactions.

IV. CARDIORESPIRATORY INTERACTIONS

Recordings from a branchial cardiac branch of the vagus in the dogfish revealed high levels of spontaneous efferent activity, which could be attributed to two types of unit. Some units fired sporadically and increased their firing rate during hypoxia (Figure 6A). These

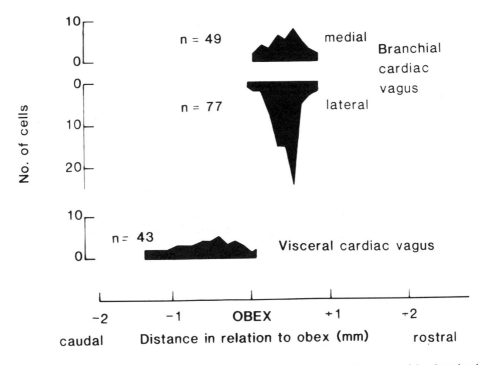

FIGURE 5. A diagrammatic representation of the rostrocaudal distribution either side of obex of the visceral and branchial cardiac vagal motoneurones. The extent of the vagal motor column is indicated by the top line. Labeled cell bodies were counted (at 120-μm intervals) from the best backfills of each branch in single preparations. Labeled branchial cardiac motoneurones were found in two locations (medial and lateral); n is the number of cells identified in each location. (From Barrett, D. J. and Taylor, E. W., *J. Exp. Biol.*, 117, 449, 1985. With permission.)

nonbursting units seem to play the major role in initiating the reflex hypoxic bradycardia and may also determine the overall level of vagal tone on the heart, which seems to vary according to oxygen supply.[49] Other, larger units fired in rhythmical bursts which were synchronous with ventilatory movements.[69] These bursts were synchronous with the efferent activity in branchial branches of the vagus that innervate respiratory muscles and continued in decerebrate dogfish even after treatment with curare, which stopped ventilatory movements (Figure 6B). This evidence suggests that the bursting activity recorded from the cardiac vagus originates in the CNS through some interaction, either direct or indirect, with the CPG.

Extracellular recordings from CVM identified in the hindbrain of decerebrate, paralyzed dogfish by antidromic stimulation of a branchial cardiac branch revealed that neurones located in the DVN were spontaneously active, firing in rhythmical bursts which contributed to the bursts recorded from the intact nerve (Figure 7). Neurones located in the NA were either spontaneously active, firing regularly or sporadically but never rhythmically, or were silent.[70] Thus it seems that the two types of efferent activity described in the branchial cardiac branches of the vagus have separate origins in the CNS, which may indicate a separation of function. All of the spontaneously active CVM from both divisions and some of the silent CVM fired in response to mechanical stimulation of a gill arch, which implies that they could be entrained to ventilatory movements in the spontaneously breathing fish and that

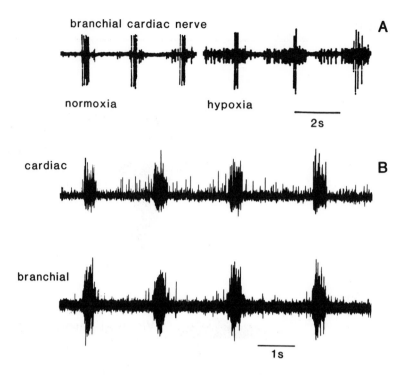

FIGURE 6.   (A) Efferent activity recorded from the left branchial cardiac branch of the vagus of a decerebrate dogfish in normoxia (P_IO_2 150 mmHg) and hypoxia (P_IO_2 45 mmHg). The smaller, sporadically active units markedly increased firing rate during hypoxia, while the larger, rhythmically bursting units were less active. (From Taylor, E. W. and Butler, P. J., *J. Appl. Physiol.*, 53, 1330, 1982. With permission.) (B) Simultaneous recordings of efferent activity in the 3rd branchial branch (branchial) and the branchial cardiac branch (cardiac) of the left vagus from a decerebrate, paralyzed dogfish. The persistence of rhythmic bursting activity in both nerves indicates that it is generated in the CNS and not by stimulation of mechanoreceptors on the gill arches. Sporadically active units which are present in recordings from the cardiac nerve are less evident in the branchial nerve. (From Barrett, D. J. and Taylor, E. W., *J. Exp. Biol.*, 117, 433, 1985. With permission.)

both may be responsible for the reflex bradycardia recorded in response to any form of mechanical stimulation in the dogfish.[69]

The respiration-related bursting activity recorded from the medial group of CVM is of particular interest because respiratory modulation of CVM has been observed in mammals.[39] This modulation, which is the central origin of the variation in heart rate with breathing movements known as "sinus arrhythmia" in mammals, is thought to arise from direct, inhibitory synaptic contact between collaterals from respiratory vagal motoneurones (RVM) and CVM; this relationship is described by Jordan and Spyer.[40] Direct connections between bursting CVM and RVM are possible in the dogfish hindbrain, as both are located in the DVN with an overlapping rostrocaudal distribution.[61] As the bursts are synchronous (Figure 6B), the innervation of CVM is likely to be excitatory rather than inhibitory, and it is equally possible that a direct drive from the CPG operates both on the RVM and the CVM.

The functional role of the bursting CVM located in the DVN is not yet clear, though it seems probable that they serve to correlate heart rate with ventilation and may be responsible for the generation of cardiorespiratory synchrony in fish. A link between heart rate and ventilation in fish was first noted in 1895 by Schoenlein,[71] who described a 1:1 synchrony in the electric ray *Torpedo marmorata*. This observation has recently been repeated.[43] Lutz[72]

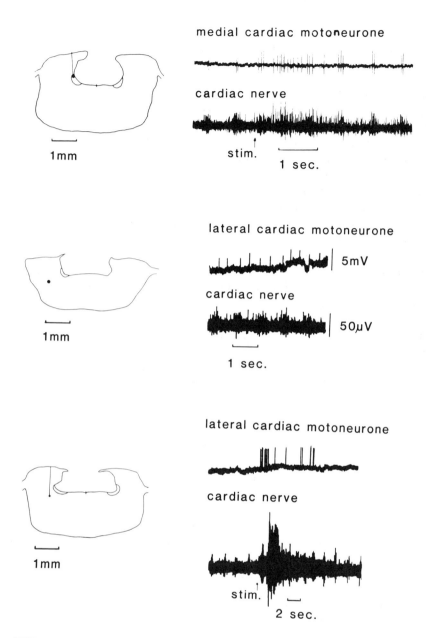

medial cardiac motoneurone

cardiac nerve

1mm

stim.

1 sec.

lateral cardiac motoneurone

5mV

cardiac nerve

50µV

1mm

1 sec.

lateral cardiac motoneurone

cardiac nerve

1mm

stim.

2 sec.

FIGURE 7.   The location and properties of antidromically identified, spontaneously active branchial cardiac motoneurones in the hindbrain of the dogfish *Scyliorhinus canicula*. (A) Transverse section (T.S.) through the medulla just rostral of obex, the recording position of the electrode tip is signified by the filled circle (●) which was located within the dorsal motor nucleus of the vagus. The recordings from this medial neurone displayed rhythmic activity which contributed to the bursts of activity recorded from the cardiac nerve. It also responded to mechanical stimulation of a gill arch (stim.). (B) T.S. 240 µm rostral of obex to show the location of a laterally placed neurone which was continuously active, firing within and between bursts of activity in the cardiac nerve. (C) T.S. of medulla just rostral of obex to show the location of a laterally placed silent neurone which fired in response to mechanical stimulation of a gill arch (stim.). (These recordings are from Barrett, D. J. and Taylor, E. W., *J. Exp. Biol.,* 117, 459, 1985. With permission.)

demonstrated that the cardiac vagus was involved in the maintenance of this link, and a full account of the reflex coordination of heart beat with ventilation in the dogfish was published by Satchell,[71] who concluded that coordination was reflexly controlled, with mechanoreceptors on the gill arches constituting the afferent limb, and the cardiac vagus the efferent limb of a reflex arc. It is of interest that sinus arrhythmia in mammals was previously attributed to stimulation of lung receptors and baroreceptors but is now known at least partly to be generated centrally[39] by a direct inhibitory action of respiratory inputs onto the vagal preganglionic cardiomotor neurones (CVM). It is therefore not unreasonable to propose that cardiorespiratory synchrony in the dogfish could originate centrally, with the respiration-related bursting units in the cardiac branches of the vagus causing the heart to beat in a particular phase of the ventilatory cycle.

In fish the supposed functional significance of cardiorespiratory synchrony relates to the importance of matching the relative flow rates of water and blood over the countercurrent created at the gill lamellae. Although virtually continuous, both water and blood flows over the lamellae are markedly pulsatile. Recordings of differential blood pressure and gill opacity in the dogfish revealed a brief period of rapid blood flow through the lamellae early in each cardiac cycle.[71] As each cycle of EKG activity, indicating the genesis of cardiac contraction, tended to occur at or near the mouth-opening phase of the ventilatory cycle, this could result in coincidence of the periods of maximum flow rate of blood and water during each cardiac cycle[12] and a consequent increase in the relative efficiency of respiratory gas exchange (i.e., maximum exchange for minimum work).

One problem with this proposal is that ventilation rate is often faster than heart rate in the restrained dogfish.[73] The absence of synchrony, or even consistent close coupling, as opposed to a drifting phase relationship, was most often attributable to the fact that heart rate was more variable than ventilation rate.[73,74] Changes in heart rate in the dogfish may be reliably attributed to variations in cardiac vagal tone, possibly exerted by changes in the rate of firing of the nonbursting units recorded from the branchial cardiac nerves; certainly activity in these units is high in the restrained dogfish when cardiorespiratory synchrony is absent.[67,69] Moreover, a decrease in vagal tone on the heart, such as that recorded during exposure to moderately hyperoxic water,[49] causes heart rate to rise toward ventilation rate, suggesting that in the undisturbed fish 1:1 synchrony may occur. Furthermore, unrestrained dogfish which were allowed to settle in large tanks of running, aerated seawater at 23°C showed 1:1 synchrony between heart beat and ventilation for long periods,[43] and this relationship was abolished by atropine (Figure 8A), confirming the role of the vagus in the maintenance of synchrony and providing a hypothetical role for the bursting units recorded from the cardiac vagi. When the dogfish was spontaneously active or disturbed, the relationship broke down due to a reflex bradycardia and acceleration of ventilation (Figure 8B). Synchrony may therefore serve to maximize the efficiency of respiratory gas exchange in the "resting" animal. In the dogfish at least this probably represents a large proportion of their life-span because they are observably rather inactive animals and spend a lot of time resting on the substratum, both in aquaria and on the seabed.

What emerges from current work is that a potent mechanism for the generation of cardiorespiratory synchrony in fish may exist in the form of the bursting units present in recordings of efferent activity in the cardiac vagi. Interestingly, in the dog, stimulation of the vagus nerves toward the heart with brief bursts of stimuli, similar to those recorded from efferent cardiac vagal fibers, caused heart rate to synchronize with the stimulus, beating once for each vagal stimulus burst over a wide frequency range.[75] Similar entrainment with the bursts of activity recorded from the cardiac vagi could explain the 1:1 synchrony observed in "settled" dogfish. The apparent loose coupling observed in active or restrained animals may arise from interactions between the effects of the bursting and nonbursting units when vagal tone is relatively high and the nonbursting units are active.

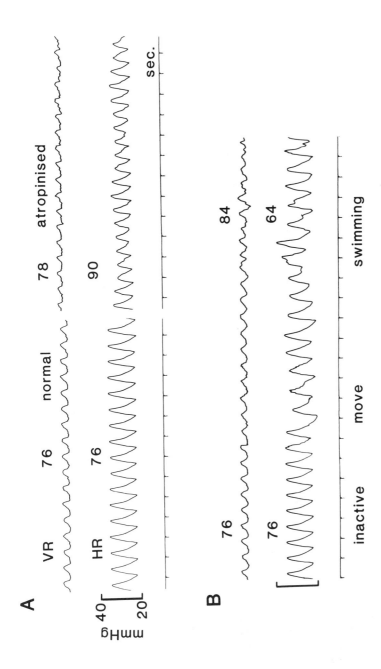

FIGURE 8. (A). Ventilation rate (VR), measured as orobranchial water pressure, and heart rate (HR), measured as ventral aortic blood pressure (beats min⁻¹), recorded from an unrestrained dogfish enclosed in a large tank of running seawater at 23°C. When the animal was stationary, resting on the bottom of the tank (normal), the two rates were identical, and there were clear signs of maintained synchrony. Atropinization (atropinised) caused an increase in heart rate and loss of synchrony. (B) When the normal, inactive animal moved (move) and then spontaneously commenced swimming, it showed a bradycardia and then an increase in ventilation rate so that ventilation became considerably faster than heart beat, a condition previously observed in disturbed or restrained animals. (Taken from Taylor, E. W., *Symp. Soc. Exp. Biol.*, 39, 123, 1985. With permission.)

Consequently, it can be argued that the dogfish provides a valuable model for the study of the central nervous interactions resulting in cardiorespiratory coupling. We have described bursting CVM in the DVN, which have a sequential overlapping distribution with RVM and have located their cell bodies with extracellular microelectrodes for central recordings. Subsequent intracellular recordings will reveal the postsynaptic events associated with afferent inputs from elsewhere in the CNS, and the use of multibarreled micropipettes will allow a study of the neuropharmacology of the observed interactions. These studies are complicated in the mammal by the complex topography of the vagal motonucleus, by central respiratory drive originating from multiple CPG in the brainstem, by the interaction of central respiratory modulation of activity in CVM with lung and baroreceptor inputs, and by possible modulation from the brainstem defense areas. Presently, we have no clear indication of the possible physiological role of sinus arrhythmia in adult mammals. However, study of the phenomenon in the unborn fetus may have clinical implications, for variations in fetal heart rate with fetal breathing movements *in utero* may prove to be a useful indicator of a normal fetus.[76] A clearer idea of the possible functions and underlying mechanisms of sinus arrhythmia could emerge from the study of cardiorespiratory synchrony in dogfish as the central interactions may represent a simplified model of the mammalian system (Figure 9), and this may prove a virtue from an experimental point of view!

V. CONCLUSION

The mechanisms of control of respiration and heart rate in fish, particularly elasmobranchs, differ from those of mammals in several, possibly fundamental respects which may truly reflect their primitive status. Some interesting differences are listed in the following:

1. The heart in elasmobranch fishes receives only a vagal (inhibitory) efferent supply. Consequently, reflex control of heart rate and cardiorespiratory interactions is affected solely by variations in vagal tone, which may thus be studied in isolation from sympathetic innervation.
2. Respiratory motoneurones (RVM) have an overlapping sequential topography in the brainstem of fish which reflects their segmental origins. Their sequential patterns of firing, to cause coordinated contractions of the respiratory muscles, appear to reflect their topography in the CNS. In mammals the inspiratory, postinspiratory, and expiratory neurones innervate thoracic muscles via interneurones and are not sequentially located.
3. All respiratory neurones, as well as the cardiomotor neurones, supply axons to cranial nerves in fish and are located together in the brainstem. This may simplify the study of central cardiorespiratory interactions, as there is no involvement of spinally projecting interneurons as seen in mammals. Also their sequential topography, in a regular rostrocaudal series in the medulla, renders them more accessible to location with microelectrodes.[77]
4. Vagal cardiomotor neurones (CVM) are found in two locations in the brainstem: in the DVN, where they overlap with RVM, and in a ventrolateral location which may represent a primordial nucleus ambiguus (NA). Parallel studies on amphibians and reptiles suggest a progressive ventrolateral migration of vagal preganglionic neurones from fish to mammals, possibly associated with the evolution of lung breathing. The functional significance of this migration is likely to be revealed by comparative studies.
5. Fish show the primary responses to peripheral chemoreceptor stimulation by hypoxia, i.e., an increase in ventilation accompanied by a reflex bradycardia. As they are gill breathers, there is little or no interference from variations in P,CO_2 levels. This enables the central nervous mechanisms involved in the response to be studied in unanesthetized, normally breathing animals.

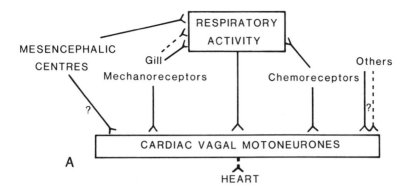

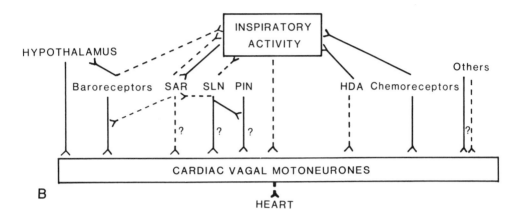

FIGURE 9. A summary diagram of the possible mechanisms that interact at the level of the cardiac vagal motoneurones in (A) dogfish and (B) mammal. Excitatory mechanisms are shown as solid lines, inhibitory mechanisms as dashed lines. These lines indicate possible pathways, not individual neurones, and say nothing of the number of synapses involved. HDA, hypothalamic defense area; PIN, postinspiratory neurones; SAR, slow-adapting lung stretch receptors; SLN, superior laryngeal nerve afferents; ?, postulated pathways. (Part B adapted from Jordan, D. and Spyer, K. M., in *The Neurobiology of the Cardiorespiratory System*, Taylor, E. W., Ed., Manchester University Press, Manchester, 1988, chap. 17. With permission.)

6.   This chemoreceptor-induced response seems not to be interacting with baroreceptor responses or a visceral alerting response.
7.   A component of the functional chemoreceptor, which responds to changes in Pa,O_2 is innervated by the IXth and Xth cranial nerves and may be morphologically and functionally homologous with the carotid and aortic bodies in mammals.

There are also important similarities between the fish and the mammalian systems which may justify the study of fish as a model system for mammalian mechanisms of control:

1.   The functional properties of dogfish RVM and CVM resemble mammalian RVM and CVM, including the fact that there is a population of normally silent cells.
2.   The interactive effects of the CPG and inputs from peripheral mechanoreceptors onto respiratory motoneurones have functional similarities but are less complex in fish (Figure 1).
3.   Cardiorespiratory synchrony in fish is likely to be generated centrally by interactions between RVM and CVM. This parallels the central generation of sinus arrhythmia in mammals.
4.   Some of the additional inputs to CVM in fish resemble those in mammals, but the

various influences are likely to be simpler than the extremely complex influences on mammalian CVM (Figure 9).

At a practical level the study of elasmobranch fish has certain clear advantages over mammalian studies:

1. Fish are relatively cheap and easy to obtain and maintain.
2. Dogfish in particular are hardy animals and survive well in experimental regimes.
3. Elasmobranch fish have a cartilaginous skeleton, rendering the exposure of portions of the CNS relatively easy.
4. Although holding the animals in flowing seawater during experiments may seem troublesome, it has the great advantage that it may be electrically grounded, thus minimizing problems with electrical interference during neurophysiological recordings.
5. Dogfish have a relatively large circulating blood volume which makes them resistant to blood loss during experiments.
6. Dogfish have a relatively low blood pressure (40/20 mmHg) even in the ventral aorta leading directly from the heart which falls to 30/20 mmHg in the systemic circulation. Thus the problem of the brainstem "bouncing" during central recordings, as it does in mammals due to their high blood pressures, is avoided when working on dogfish.

This account has not addressed the complex area of the alternative roles for respiratory muscles in mammals for functions such as posture and expulsive motor acts such as coughing, micturition, or defecation and the complex use of the breathing apparatus for vocalization.[24] All of these imply complex changes in the patterns of central control, and the general principle that fish are likely to show less complex modifications of their control mechanisms to accommodate activities such as coughing, ram ventilation, and feeding seems likely to apply and is the subject of current research in our laboratory.

There are clearly many practical advantages to working on elasmobranch fish, and it seems possible that these studies could yield important new insights into some of the fundamental mechanisms determining cardiorespiratory control in vertebrates which may be of help in interpreting the more complex mammalian system.

ACKNOWLEDGMENTS

I wish to record my gratitude to Dr. Janice Marshall of the Physiology Department, University of Birmingham, for her helpful criticism of this manuscript and to Mrs. Jean Hill for processing and patiently reprocessing the text.

REFERENCES

1. **Campbell, E. J. M., Agostoni, E., and Newson, D. J.,** *The Respiratory Muscles,* Lloyd-Luke, London, 1970.
2. **Ballintijn, C. M.,** Functional anatomy and movement coordination of the respiratory pump of the carp *(Cyprinus carpio* L.), *J. Exp. Biol.,* 50, 547, 1969.
3. **Ballintijn, C. M.,** Muscle coordination of the respiratory pump of the carp *(Cyprinus carpio* L.), *J. Exp. Biol.,* 50, 569, 1969.
4. **Ballintijn, C. M.,** Movement pattern and efficiency of the respiratory pump of the carp *(Cyprinus carpio* L.), *J. Exp. Biol.,* 50, 593, 1969.
5. **Ballintijn, C. M. and Hughes, G. M.,** The muscular basis of the respiratory pumps in the trout, *J. Exp. Biol.,* 43, 349, 1965.

6. **Hughes, G. M. and Ballintijn, C. M.,** The muscular basis of the respiratory pumps in the dogfish, *J. Exp. Biol.,* 43, 363, 1965.
7. **Cooke, I. C. R.,** Functional aspects of the morphology and vascular anatomy of the gills of the Endeavour dogfish *Centrophorus scalpratus, Zoomorphologie,* 94, 167, 1980.
8. **Ballintijn, C. M.,** Neural control of respiration in fishes and mammals, in *Exogenous and Endogenous Influences on Metabolic and Neural Control,* Addink, A. D. F. and Spronk, N., Eds., Pergamon Press, Oxford, 1982.
9. **Ballintijn, C. M.,** Evolution of central nervous control of ventilation in vertebrates, in *The Neurobiology of the Cardiorespiratory System,* Taylor, E. W., Ed., Manchester University Press, Manchester, 1988, chap. 1.
10. **Shelton, G.,** The respiratory centre in the tench *(Tinca tinca* L.). I. The effects of brain transection on respiration, *J. Exp. Biol.,* 36, 191, 1959.
11. **Shelton, G.,** The respiratory centre in the tench *(Tinca tinca* L.). II. Respiratory neuronal activity in the medulla oblongata, *J. Exp. Biol.,* 38, 79, 1961.
12. **Shelton, G.,** The regulation of breathing, in *Fish Physiology,* Vol. 4, Hoar, W. S. and Randall, D. J., Eds., Academic Press, New York, 1970.
13. **Waldron, I.,** Spatial organisation of respiratory neurones in the medulla of tench and goldfish, *J. Exp. Biol.,* 57, 449, 1972.
14. **Bamford, O. S.,** Respiratory neurons in the rainbow trout, *Comp. Biochem. Physiol.,* 48, 77, 1974.
15. **de Graaf, P. J. F. and Ballintijn, C. M.,** Mechanoreceptor activity in the gills of the carp. II. Gill arch proprioceptors, *Respir. Physiol.,* 69, 183, 1987.
16. **Ballintijn, C. M., Roberts, B. L., and Luiten, P. G. M.,** Respiratory responses to stimulation of branchial vagus nerve ganglia of a teleost fish, *Respir. Physiol.,* 51, 241, 1983.
17. **Ballintijn, C. M., Luiten, P. G. M., and Juch, P. J. W.,** Respiratory neuron activity in mesencephalon, diencephalon and cerebellum of the carp, *J. Comp. Physiol.,* 133, 131, 1979.
18. **Juch, P. J. W. and Luiten, P. G. M.,** Anatomy of respiratory rhythmic systems in brainstem and cerebellum of the carp, *Brain Res.,* 230, 51, 1981.
19. **Ballintijn, C. M. and Alink, G. M.,** Identification of respiratory motoneurones in the carp and determination of their firing characteristics and interconnections, *Brain Res.,* 136, 261, 1977.
20. **Levings, J. J. and Taylor, E. W.,** The central organisation of the hindbrain visceromotor column in the elasmobranch fishes *Raja clavata* and *Raja microocellata, Neurosci. Lett.,* 32, S80, 1988.
21. **Hugelin, A.,** Anatomical organisation of bulbopontine respiratory oscillators, *Fed. Proc. Fed. Am. Soc. Exp. Biol.,* 36, 2390, 1977.
22. **Hugelin, A.,** Does the respiratory rhythm originate from a reticular oscillator in the waking state?, in *The Reticular Formation Revisited,* Hobson, J. A. and Brazier, A. B., Eds., Raven Press, New York, 1980.
23. **Cohen, M. I.,** Central determinants of respiratory rhythm, *Ann. Rev. Physiol.,* 43, 91, 1981.
24. **von Euler, C.,** On the central pattern generator for the basic breathing rhythmicity, *J. Appl. Physiol.,* 55, 1647, 1983.
25. **Bertrand, F., Hugelin, A., and Vibert, J. F.,** Quantitative study of anatomical distribution of respiration related neurons in the pons, *Exp. Brain Res.,* 16, 383, 1973.
26. **Merrill, E. G.,** The lateral respiratory neurones of the medulla: their associations with nucleus ambiguus, nucleus retroambigualis, the spinal accessory nucleus and the spinal cord, *Brain Res.,* 24, 11, 1970.
27. **Ballintijn, C. M.,** Efficiency, mechanics and motor control of fish respiration, *Respir. Physiol.,* 14, 125, 1972.
28. **Barrett, D. J. and Taylor, E. W.,** Spontaneous efferent activity in branches of the vagus nerve controlling heart rate and ventilation in the dogfish, *J. Exp. Biol.,* 117, 433, 1985.
29. **Juch, P. J. W. and Ballintijn, C. M.,** Tegmental neurons controlling medullary respiratory centre neuron activity in the carp, *Respir. Physiol.,* 51, 95, 1983.
30. **Bertrand, F., Hugelin, A., and Vibert, J. F.,** A stereologic model of pneumotaxic oscillator based on spatial and temporal distributions of neuronal bursts, *J. Neurophysiol.,* 37, 91, 1974.
31. **Feldman, J. L. and Gautier, H.,** Interaction of pulmonary afferents and pneumotaxic centre in the control of respiratory pattern in cats, *J. Neurophysiol.,* 39, 31, 1976.
32. **Cohen, M. I., Piercey, M. F., Gootman, P. M., and Wolotsky, P.,** Synaptic connections between medullary inspiratory neurons and phrenic motoneurons as revealed by cross-correlation, *Brain Res.,* 81, 319, 1974.
33. **Anderson, P. and Sears, T. A.,** Medullary activation of intercostal fusimotor and alpha motoneurones, *J. Physiol. (London),* 209, 739, 1970.
34. **Bennett, J. A., Ford, T. W., Kidd, C., and McWilliam, P. N.,** Characteristics of cat dorsal motor vagal motoneurones with axons in the cardiac and pulmonary branches, *J. Physiol. (London),* 351, 27P, 1984.
35. **Dejours, P.,** *Principles of Comparative Respiratory Physiology,* 2nd ed., Elsevier, Amsterdam, 1981.
36. **Rahn, H.,** Aquatic gas exchange: theory, *Respir. Physiol.,* 1, 1, 1966.

37. **Kiwull-Schone, H. and Kiwull, P.,** Hypoxic modulation of central chemosensitivity, in *Central Neurone Environment,* Schlafke, M. E., Koepchen, H. P., and See, W. R., Eds., Springer-Verlag, Berlin, 1983.

38. **Heymans, C. and Neil, E.,** *The Reflexogenic Areas of the Cardiovascular System,* Churchill Livingstone, London, 1958.

39. **Spyer, K. M.,** Central nervous integration of cardiovascular control, *J. Exp. Biol.,* 100, 109, 1982.

40. **Jordan, D. and Spyer, K. M.,** Central neural mechanisms mediating respiratory-cardiovascular interactions, in *The Neurobiology of the Cardiorespiratory System,* Taylor, E. W., Ed., Manchester University Press, Manchester, 1988, chap. 17.

41. **Johansen, K.,** Comparative physiology: gas exchange and circulation in fishes, *Ann. Rev. Physiol.,* 33, 569, 1971.

42. **Taylor, E. W.,** Control and co-ordination of ventilation and circulation in crustaceans: responses to hypoxia and exercise, *J. Exp. Biol.,* 100, 289, 1982.

43. **Taylor, E. W.,** Control and coordination of gill ventilation and perfusion, *Symp. Soc. Exp. Biol.,* 39, 123, 1985.

44. **Smith, F. M. and Jones, D. R.,** Localisation of receptors causing hypoxic bradycardia in trout, *(Salmo gairdneri), Can. J. Zool.,* 56, 1260, 1978.

45. **Satchell, G. H.,** Respiratory reflexes in the dogfish, *J. Exp. Biol.,* 36, 62, 1959.

46. **Butler, P. J., Taylor, E. W., and Short, S.,** The effect of sectioning cranial nerves V, VII, IX and X on the cardiac response of the dogfish, *Scyliorhinus canicula* to environmental hypoxia, *J. Exp. Biol.,* 69, 233, 1977.

47. **Laurent, P.,** La pseudobranchie des teleosteens: preuves electrophysiologique de ses fonctions chemoreceptrice et baroreceptrice, *C. R. Acad. Sci. Ser. D,* 264, 1879, 1967.

48. **Randall, D. J. and Jones, D. R.,** The effect of deafferentation of the pseudobranch on the respiratory response to hypoxia of the trout *(Salmo gairdneri), Respir. Physiol.,* 17, 291, 1973.

49. **Barrett, D. J. and Taylor, E. W.,** Changes in heart rate during progressive hyperoxia in the dogfish *Scyliorhinus canicula* L.: evidence for a venous oxygen receptor, *Comp. Biochem. Physiol.,* 78A, 697, 1984.

50. **Randall, D. J.,** Functional morphology of the heart in fishes, *Am. Zool.,* 8, 179, 1968.

51. **Satchell, G. H.,** *Circulation in Fishes,* Cambridge University Press, Cambridge, 1971.

52. **Short, S., Butler, P. J., and Taylor, E. W.,** The relative importance of nervous, humoral and intrinsic mechanisms in the regulation of heart rate and stroke volume in the dogfish *Scyliorhinus canicula, J. Exp. Biol.,* 70, 77, 1977.

53. **Young, J. Z.,** The autonomic nervous system of selachians, *Q. J. Microsc. Sci.,* 75, 571, 1933.

54. **Holmgren, S.,** Regulation of the heart of a teleost, *Gadus morhua,* by autonomic nerves and circulating catecholamines, *Acta Physiol. Scand.,* 99, 62, 1977.

55. **Taylor, E. W., Short, S., and Butler, P. J.,** The role of the cardiac vagus in the response of the dogfish *Scyliorhinus canicula* to hypoxia, *J. Exp. Biol.,* 70, 57, 1977.

56. **Daly, M. de B. and Scott, M. J.,** An analysis of the primary cardiovascular reflex effects of stimulation of the carotid body chemoreceptors in the dog, *J. Physiol. (London),* 162, 555, 1962.

57. **Daly, M. de B. and Scott, M. J.,** The cardiovascular responses to stimulation of the carotid body chemoreceptors in the dog, *J. Physiol. (London),* 165, 179, 1963.

58. **Marshall, J. M.,** Contribution to overall cardiovascular control made by the chemoreceptor-induced alerting/ defense response, in *The Neurobiology of the Cardiorespiratory System,* Taylor, E. W., Ed., Manchester University Press, Manchester, 1988, chap. 12.

59. **Metcalfe, J. D. and Butler, P. J.,** Changes in activity and ventilation in response to hypoxia in unrestrained, unoperated dogfish *(Scyliorhinus canicula), J. Exp. Biol.,* 108, 411, 1984.

60. **Butler, P. J. and Taylor, E. W.,** Response of the dogfish *(Scyliorhinus canicula* L.) to slowly induced and rapidly induced hypoxia, *Comp. Biochem. Physiol.,* 39A, 307, 1971.

61. **Withington-Wray, D. J., Roberts, B. L., and Taylor, E. W.,** The topographical organisation of the vagal motor column in the elasmobranch fish, *Scyliorhinus canicula* L., *J. Comp. Neurol.,* 248, 95, 1986.

62. **Barrett, D. J. and Taylor, E. W.,** The location of cardiac vagal preganglionic neurones in the brainstem of the dogfish, *Scyliorhinus canicula, J. Exp. Biol.,* 117, 449, 1985.

63. **Withington-Wray, D. J., Taylor, E. W., and Metcalfe, J. D.,** The location and distribution of vagal preganglionic neurones in the hindbrain of lower vertebrates, in *The Neurobiology of the Cardiorespiratory System,* Taylor, E. W., Ed., Manchester University Press, Manchester, 1988, chap. 16.

64. **Kalia, M.,** Brain stem localization of vagal preganglionic neurons, *J. Auton. Nerv. Syst.,* 3, 451, 1981.

65. **Kalia, M. and Sullivan, J. H.,** Brainstem projections of sensory and motor components of the vagus nerve in the rat, *J. Comp. Neurol.,* 211, 248, 1982.

66. **Kalia, M. and Mesulam, M. M.,** Brain stem projections of afferent and efferent fibres of the vagus nerve in the cat. II. Laryngeal tracheobronchial, pulmonary, cardiac and gastrointestinal branches, *J. Comp. Neurol.,* 193, 467, 1980.

67. **Bennett, J. A., Kidd, C., Latif, A. B., and McWilliam, P. N.,** A horseradish peroxidase study of vagal motoneurones with axons in cardiac and pulmonary branches of the cat and dog, *Q. J. Exp. Physiol.,* 66, 145, 1981.

68. **Jordan, D., Gilbey, M. P., Richter, D. W., Spyer, K. M., and Wood, L. M.,** Respiratory-vagal interactions in the nucleus ambiguus of the cat, in *Neurogenesis of Central Respiratory Rhythm,* Bianchi, A. L. and Denavit-Saubié, M., Eds., MTP Press, Lancaster, England, 370, 1985.

69. **Taylor, E. W. and Butler, P. J.,** Nervous control of heart rate: activity in the cardiac vagus of the dogfish, *J. Appl. Physiol.,* 53, 1330, 1982.

70. **Barrett, D. J. and Taylor, E. W.,** The characteristics of cardiac vagal preganglionic motoneurones in the dogfish, *J. Exp. Biol.,* 117, 459, 1985.

71. **Satchell, G. H.,** The reflex coordination of the heart beat with respiration in the dogfish, *J. Exp. Biol.,* 37, 719, 1960.

72. **Lutz, B. R.,** Reflex cardiac and respiratory inhibition in the elasmobranch *Scyllium canicula, Biol. Bull. Woods Hole, Mass.,* 59, 170, 1930.

73. **Taylor, E. W. and Butler, P. J.,** Some observations on the relationship between heart beat and respiratory movements in the dogfish *Scyliorhinus canicula* L., *Comp. Biochem. Physiol.,* 39A, 297, 1971.

74. **Hughes, G. M.,** The relationship between cardiac and respiratory rhythms in the dogfish, *Scyliorhinus canicula, J. Exp. Biol.,* 57, 415, 1972.

75. **Levy, M. N. and Martin, P.,** Parasympathetic control of the heart, in *Nervous Control of Cardiovascular Function,* Randall, W. C., Ed., Oxford University Press, Oxford, 1984.

76. **Gee, H. and Taylor, E. W.,** Changes in human fetal heart rate with fetal breathing movements as an indicator of fetal condition, *J. Physiol. (London),* 373, 59P, 1986.

77. **Taylor, E. W. and Elliott, C. J. H.,** Neurophysiological techniques, in *Techniques in Comparative Physiology,* Bridges, C. R. and Butler, P. J., Eds., Cambridge University Press, Cambridge, 1988.

78. **Richter, D. W.,** Generation and maintenance of the respiratory rhythm, *J. Exp. Biol.,* 100, 93, 1982.

79. **Taylor, E. W. and Levings, J. J.,** unpublished data.

80. **Taylor, E. W.,** personal observations.

Chapter 12

USE OF SMALL FISH IN BIOMEDICAL RESEARCH, WITH SPECIAL REFERENCE TO INBRED STRAINS OF MEDAKA

Yasuko Hyodo-Taguchi and Nobuo Egami

TABLE OF CONTENTS

I. INTRODUCTION

In the history of biomedical research, many important findings were made with fish as experimental animals. For example, as early as 1910 it was found that goiter in the brook trout *(Salvelinus alpinus)* was due to a lack of iodine in the environment, a finding that was directly applicable to human disease.[1] The biological effects of some hormones (e.g., androgen) and the physiological mechanisms of color vision have been analyzed with teleost fishes. For quantitative studies of the toxic, carcinogenic, mutagenic, and teratogenic effects of various chemicals, fish have proved very valuable as animal models. One important advantage in the use of fish is that certain species exist naturally as clones, reproducing gynogenetically, and also inbred strains or clones can be established in the laboratory. There is no doubt that recent progress in biomedical research, such as cancer research, molecular immunology, radiation biology, and geriatrics, has resulted, in part, from the development of laboratory animal science, particularly the establishment of inbred strains of mice. Because the genetic background of inbred animals is uniform, genetical analyses of biological phenomena have become possible. Small teleost fishes, such as *Oryzias latipes* (the medaka) and *Poeciliidae* (guppies, mollies, and platyfishes), are excellent experimental materials because of their small size, relatively short life cycle, reproductive capacity, and ease of rearing in the laboratory. Each of these species has been used in experimental laboratories around the world, and there is a long history of breeding. The medaka has also been used extensively for teaching biology in elementary and middle schools in Japan.[2] The orange-red variety of the medaka, the so-called "himedaka", is readily available in pet shops and can be obtained from experimental animal dealers in Japan. The himedaka is not a pure orange-red variety but is a heterozygous stock from orange-red, white, and variegated races. To establish pure strains for experiments, the medaka must be bred in the laboratory. In the U.S. and in European laboratories,[3,4] the counterpart of the medaka is species of *Poeciliidae,* small live-bearing fishes from the Atlantic slope of Mexico and adjacent parts of Central America. Inbred strains of these species have been established and used in cancer research and toxicological studies. One of the most important problems in using the medaka, the *Poeciliidae,* and such species for biomedical research is to set up a supply system that will provide the large number of inbred individuals required for testing carcinogens and toxic chemicals.

Besides these small fishes, some moderate-sized fish, such as goldfish, rainbow trout, and mudminnow, have been useful in experimental biology. They each offer quite specific advantages as experimental animals. Generally, it is difficult to breed these fishes in the laboratory, but cultivated stocks or wild fish are readily available, and the ease with which they can be kept in the laboratory makes them highly suitable for biomedical research. For instance, goldfish have been used extensively to analyze the mechanisms of various radiation effects in our laboratory.[5-7] Radiation effects on cell renewal systems, such as the intestine, hematopoietic tissue, and skin, were examined by histopathological and autoradiographical techniques. Moreover, goldfish are useful for surgical experiments[8] and biochemical research.[9,10] The value of rainbow trout embryos in carcinogenesis research is well established.[11] A cytogenetic system for short-term screening with the mudminnow *(Umbra limi)* has been developed; this fish possesses a small number of large chromosomes (2n = 22), enabling easy and rapid quantification of cytogenetic damage.[12]

Cultured cells from several species, including the bluegill sunfish, also have been used in studies of organic pollutants.[13] Moreover, many important findings on carcinogenesis and oncogene expression were carried out using a model of hereditary melanoma in hybrids from crosses between swordtails and platyfish.[14,15] The usefulness of clones of *Poecilia formosa* was pointed out by Setlow et al.[16] However, since advantages of these particular fish as animal models have been reviewed elsewhere, we shall not elaborate on them this chapter. We shall review in more detail the usefulness of the medaka and goldfish for fundamental biomedical research.

II. THE MEDAKA AS A TOOL IN BIOMEDICAL RESEARCH

A. GENERAL BIOLOGY

The medaka *Oryzia latipes,* a small freshwater oviparous fish, is common in Japan and in some parts of Asia. Adult fish are 3 to 4 cm in length and 0.5 to 0.7 g in weight. Besides the wild type (brown or black), colored varieties and orange-red varieties (himedaka) are available from goldfish breeders.

The sex of fully grown fish can be easily determined by the outline of the anal and dorsal fins. The female is characterized by having less well developed anal and dorsal fins, but has a prominent urinogenital papilla. The anal fin rays of the male on the posterior region of the fin have numerous small papillar processes, and the dorsal fin has a deep cleft between the last ray and the one preceding it.

This fish is very prolific. Under natural daylight and temperature conditions, the breeding season extends from April to September in Tokyo; a pair of fish lay a cluster of eggs almost every morning and fertilization takes place immediately after laying. The number of eggs in a single cluster varies from 1 to 67, being 12 to 19 in the majority of cases. The total number of eggs laid by a single female during a season usually is 1000 to 2000.[17] Under favorable laboratory conditions, the fish lay eggs all year round.

The egg of the medaka is about 1 mm in diameter and transparent so that embryonic development is easily observable. For embryological studies, techniques have been established for artificial fertilization, surgical operation of embryos by removing the chorion from the eggs, and microinjection into fertilized eggs.[2] Artificial induction of spawning is readily accomplished by the regulation of temperature and light.

Fundamental knowledge of the morphology, embryology, physiology, and ecology of this fish already has been accumulated. The normal developmental stages of this fish were illustrated by Matui[18] and Gamo and Terajima.[19] Recently, Parenti[20] published the specific position of medaka from a phylogenetic viewpoint: he described the structure and development of the oral and pharyngeal jaws of the medaka and assessed the medaka as a model higher teleost.

These fish have been used in many fields of fundamental and applied biology and are becoming established as a "standard animal" for the biological assay of pharmacological effects of various drugs and of water pollution.

FIGURE 1.   The aquatic laboratory and ponds for culture of the medaka (National Institute of Radiological Sciences, Chiba, Japan).

B. HUSBANDRY
1. Reproductive Habits

Since the medaka is native to Japan and it can withstand winter without supplementary heating, the stocks are able to be kept outdoors throughout the whole year. Medaka can survive at a wide range of temperatures, from 0 to 37°C; however, sudden changes of temperature are harmful. Temperatures of 25 to 28°C are suitable for breeding. Fry hatched in spring reach maturity at the end of the summer; fry hatched in late summer reach maturity next spring. This species lives about 3 years, and its maximum lifespan is 5 years in outdoor conditions.[21] If the temperature is kept at 25 to 28°C, fry can reach sexual maturity in 3 months, so that 4 generations may be reared during 1 year.

2. Rearing and Breeding

The medaka is very small; therefore, many individuals can be held in a limited space. For the past 20 years we have been rearing and breeding the medaka on a large scale in outdoor ponds and in the indoor facility of the National Institute of Radiological Sciences (Figures 1 and 2).

a. Outdoor Mass Breeding

In outdoor mass breeding, about 200 fish (100 males and 100 females) are kept in a concrete pond measuring 1 × 2 × 0.20 m. The water temperature falls to 0°C during winter days, while it goes up to almost 30°C on hot summer days. To prevent excessive rises in temperature and to refresh the water, well water runs in and out of the ponds slowly. During the natural breeding season, progeny are obtained by natural mating. Fertilized eggs are collected in a mass of cut-off nylon nets on the water surface. It is recommended that water hyacinths *(Pontederia crassips)* be grown in the ponds. When eggs are deposited on the nets or on the plant roots, the nets and plants can be transferred into other ponds to prevent the newly hatched fry from being preyed upon by their parents.

FIGURE 2.   The laboratory facility for culture of medaka for breeding and rearing (National Institute of Radiological Sciences, Chiba, Japan).

b. Indoor Breeding

In indoor breeding and rearing, the size of the container and the quality of the water, temperature, and lighting must be controlled. Generally, we use a small plastic aquarium (18 × 29 × 12 cm) containing 3 l of water for breeding and for experiments: aeration is unnecessary. Usually 35 aquariums are placed on a single rack, so that about 350 experimental fish can be kept in a floor space of 0.5 × 2 m. The temperature of the room is controlled at 26°C ± 1°C. We use about 10 adult fish in the aquarium in experiments such as survival tests and observation of tumorigenesis. Water in the aquariums is renewed every 2 or 3 d; well water is aged for at least 1 d to equilibrate with room temperature before use.

Because the medaka likes plenty of light, a daylight fluorescent lamp is set about 15 cm above the aquariums and switched on and off to give a 16-h light/8-h dark cycle. The more light the better. A daily light-dark cycle is necessary for keeping the rhythm of reproduction, although the fish will breed under continuous light.

For breeding, one male and one female are put into an aquarium, and an adequate amount of unicellular green algae (such as *Chlorella* spp.) is added. After mating for several days, the parents are transferred into a new aquarium. Newly hatched fry swim up in the original aquarium within a week. For collecting a large number of synchronously developing embryos, males and females are separated for 24 h, and the following morning they are allowed to mate for 30 min and the eggs that are laid then are collected.

The medaka takes both animal and vegetable food. Aquatic worms (living freshwater oligochaetes) and algae are favored by the fish. Live foods may be carriers of various kinds of infection; therefore, in indoor breeding, adult fish are fed Tetramin® (Tetra Werke Co., Melle, FRG). Roughly measured amounts of the food (10 to 13 mg per fish per day) are given daily, and the food is consumed within 30 min of feeding. Newly hatched fry and young are given powdered Tetramin® or *Daphnia*. To get a high growth rate, the addition of unicellular green algae and sufficient powdered Tetramin® are recommended. Residual food is removed daily from the water surface by pipetting or with a fine net.

C. GENETICS

The diploid chromosome number of medaka collected in the wild in Japan is 48.[22] Studies on sex-linked inheritance demonstrate that the sex-chromosome constitution is XX for females and XY for males.[23] However, a heteromorphic chromosome pair has not been detected in karyotypic analyses.[24] Recent advances in the culture of fish cells and differential staining techniques have made it possible to obtain a number of ideal metaphase figures and to demonstrate characteristic banding patterns in fish chromosomes. Fin cells from the medaka were cultured, and detailed karyotype analyses undertaken by differential C-banding and silver-staining techniques. The orange-red varieties and wild medaka showed the same karyotype, characterized by 48 chromosomes consisting of 2 pairs of metacentrics, 8 pairs of submetacentrics, 1 pair of subtelocentrics, 13 pairs of acrocentrics (FN = 68). One pair of the smaller submetacentrics had stalks on their short arms, with the appearance of the satellite chromosomes. These chromosomes also had C-bands. The centromeric regions of almost all chromosomes were C-band-positive. No sexual differences in the chromosomes were detected.[25]

Some 65 to 70 mutant characters (genes) are established in medaka, among which the *b* and *r* alleles concerned with body colors are particularly well known.[23] The chromatophores responsible for body color of the fish are melanophores, xanthophores, and leucophores. According to studies by Toyama,[26] Ishiwara,[27] and Aida,[23] the pigmentation of melanophores is controlled by the *b* alleles ($B > B' > b$). The *B* produces black melanin granules; the *b* induces colorless melanophores; and the *B'* causes black variegation. The *r* alleles ($R > r$) govern the deposition of the orange-red pigment in xanthophores. The *r* gene and the *R* gene produce colorless and orange-red xanthophores, respectively. For experimental embryology, the gene *B* of the wild type is an excellent visible marker. This gene manifests itself within 48 h after fertilization when eggs are kept at 25 to 28°C; at that time, embryonic melanophores appear on the yolk sac of embryos. For instance, *b* eggs were inseminated by *B* sperm that had been treated with UV and inactivated: embryos without black melanophores developed. The results show the success of gynogenetic development.[28] Such *b* eggs inseminated with *B* sperm also can be used for quantitative estimation of the X-ray-induced mutation rate of visible genes in fish.[29] This system is valuable for assessing environmental mutagens in the aquatic environment. Many mutants of body colors are now bred in mass-cultured colonies. For example, the *bR* type is homozygous for the *b* and *R* genes, but the state of other genes in this stock are unknown. Some mutants, for instance, *pl* (lack of pectoral fins), are useful as animal models of human diseases, particularly teratological studies.

As experimental models, isogenic strains of fish are useful tools for research. Therefore, efforts to develop a standard inbred strain of the medaka were started in 1974 by Hyodo-Taguchi, and several inbred strains of the medaka have been successfully established.[30,31] In the following section, we review the process of the establishment of the inbred strains of the medaka and the characteristics of some of the strains. Following this, we discuss the use of the inbred strains to study the effects of radiation and of chemical carcinogens.

D. INBRED STRAINS IN STUDIES OF RADIATION AND CHEMICAL CARCINOGENESIS

1. Establishment of the Inbred Strains

a. Procedures

To establish inbred strains of the medaka, we carried out sister-brother mating of pairs of fish for successive generations since 1974. In this system, one male and one female are selected from among the fish born from one pair of parents and are mated in each generation. The process of inbreeding is summarized in Figure 3. During inbreeding of many pedigrees, the reproductive potential was reduced, or there was high mortality. However, two pedigrees,

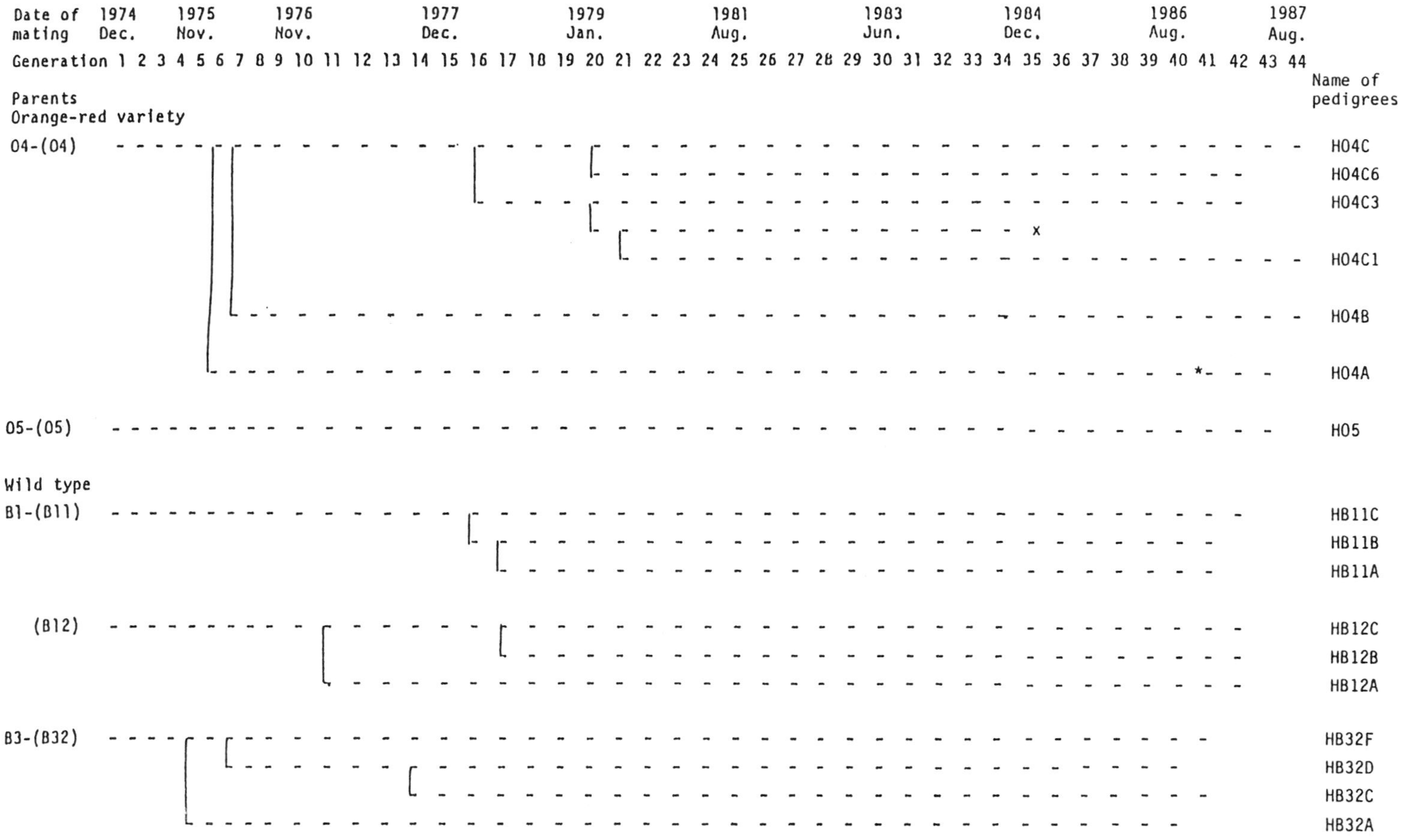

Date of mating — 1974 Dec. — 1975 Nov. — 1976 Nov. — 1977 Dec. — 1979 Jan. — 1981 Aug. — 1983 Jun. — 1984 Dec. — 1986 Aug. — 1987 Aug.

Generation 1 2 3 4 5 6 7 8 9 10 11 12 13 14 15 16 17 18 19 20 21 22 23 24 25 26 27 28 29 30 31 32 33 34 35 36 37 38 39 40 41 42 43 44

Name of pedigrees

Parents
Orange-red variety

O4-(O4)

HO4C
HO4C6
HO4C3
HO4C1
HO4B
HO4A

O5-(O5)

HO5

Wild type

B1-(B11)

HB11C
HB11B
HB11A

(B12)

HB12C
HB12B
HB12A

B3-(B32)

HB32F
HB32D
HB32C
HB32A

FIGURE 3. Diagram showing the pedigrees of inbred fish by full-sibling mating.

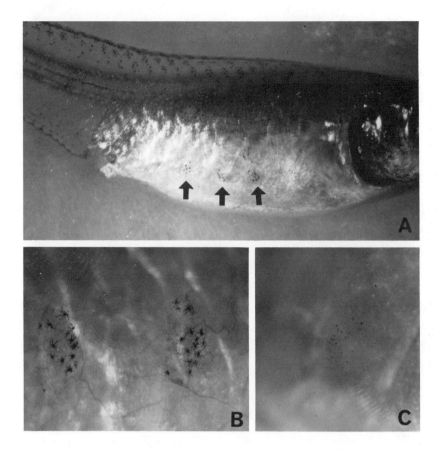

FIGURE 4. Photographs of the wild-type fish showing the successful and unsuccessful transplants of the scale. (A) Three scale autografts on the 30th day after transplantation; (B) Two scale isografts on the day 30 after transplantation; (C) Homograft on the 5th day after transplantation. This scale shows the breakdown of the pigment cells.

HO4 and HO5, of the orange-red variety of *Oryzias latipes,* and three pedigrees, HB11, HB12, and HB32, of the wild-type (brown) fish, were successfully inbred by full sister-brother mating for 22 successive generations during 5 years from 1974 to 1979. Three substrains HO4A, HO4B, and HO4C were separated from HO4 after 5, 6, and 14 generations and then inbred for a further 17, 16, and 8 generations, respectively. These inbred strains now have been further inbred by brother-sister matings for more than 38 generations in the laboratory. Moreover, to establish different kinds of isogenic strains, two families, one from Niigata in northern Japan and the other from a Chinese population in Shanghai, have been inbred by brother-sister mating since 1980.

b. Tissue Transplantation

Demonstration of genetical homogeneity — If a part of a fin or a scale from one fish is transplanted to a different individual in an outbred colony of *O. latipes,* so-called "allograft rejection" takes place.[33] The fundamental mechanism of allograft rejection is the same in fish and mammals.[34,45] Therefore, transplantation experiments with scales were made between siblings during the course of successive generations of sister-brother mating. Scales with chromatophores, taken from the dorsal side of the donor, were transplanted to the recipient's ventral, nonpigmented region (Figure 4A). A quantitative estimate of graft survival was made by counting the number of intact pigment cells on the scale grafts. Siblings from the 5th to the 14th generation of the HB32 pedigree from the B-3 family were used

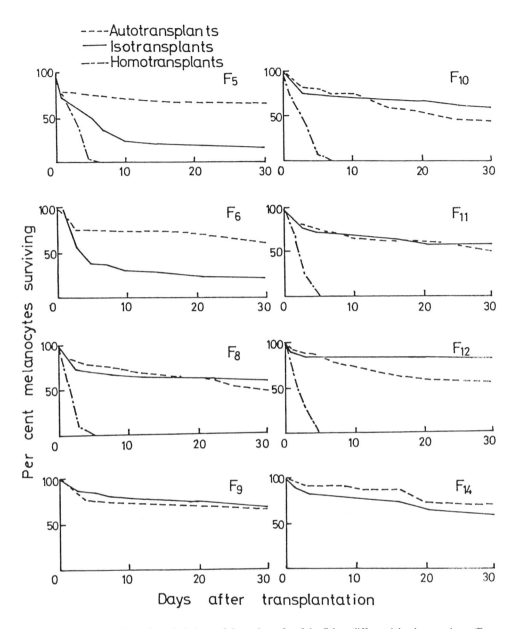

FIGURE 5. A comparison of survival times of the scale grafts of the fish at different inbred generations. (From Hyodo-Taguchi, Y. and Egami, N., *Zool. Sci.*, 2, 305, 1985. With permission.)

for this survey (Figure 5). In autografts, 73 to 48% of the melanophores in the scales at the time of transplantation survived in a healthy state for over 30 d (Figure 4B). Homogeneic grafts (allografts), however, were rejected completely within 6 d at 25°C in homotransplants. Melanophores in homogeneic grafts were destroyed, and the melanin granules were dispersed within a week (Figure 4C). This finding showed that fish strains highly homogeneous in most histocompatibility loci could be obtained after 8 sibling matings.

Genetics of scale transplantation — Tissue transplantation between fish of different highly inbred strains provides the necessary material for genetic analysis.[36] Four inbred strains of the medaka, HB11C, HB12C, HB32C, and HO4C, were used in studies of scale transplants. The F_1 was obtained from the crosses: HO4C × HB11C, HB12C × HB11C, and HB11C × HB32C. The F_2 and backcrosses (BC) generations were obtained in each

TABLE 1
Graft Survival and Estimated Minimum Number of
Histocompatibility Loci Involved in Graft Rejection Between
Different Inbred Strains of Medaka

Crossing	Donor	Graft survival, no.[a] (%) F$_2$ hosts	Graft survival, no.[a] (%) BC hosts	No. of loci[b]
(HO4C × HB11)^2	HB11	1/44 (2)		
(HO4C × HB11)^2	HO4C	4/16 (25)		7—8
(HO4C × HB11) × HO4C	HB11		0/44 (0)	
(HO4C × HB11) × HB11	HO4C		1/16 (6)	
(HB11 × HB32)^2	HB32	8/38 (21)		
(HB11 × HB32)^2	HB11	2/14 (14)		5—6
(HB11 × HB32) × HB11	HB32		0/39 (0)	
(HB11 × HB32) × HB32	HB11		0/19 (0)	
(HB11 × HB12)^2	HB12	1/11 (9)		
(HB11 × HB12)^2	HB11	7/23 (30)		5—6
(HB11 × HB12) × HB11	HB12		2/24 (8)	
(HB11 × HB12) × HB12	HB11		4/14 (28)	

[a]   Number of grafts surviving on the 30th day per total number of grafts.
[b]   Results obtained from F$_2$ generation.

cross. All intrastrain grafts (strain HO4C to HO4C, HB11C to HB11C, HB12C to HB12C, and HB32C to HB32C), all grafts from either inbred strain to F$_1$ hosts (e.g., strain HO4C or HB11C to F$_1$ of the cross HO4C × HB11C), and all grafts among F$_1$ hybrids were successful. All interstrain grafts and all grafts from F$_1$ donors to parental (P) hosts, however, were clearly rejected. Their rate of the rejection of P grafts in the F$_2$ and BC hosts varied, depending the degree of genetic relationship between the two inbred strains of the P generation (Table 1). The number of histocompatibility genes involved in determining the fates of transplants between two inbred strains was estimated from the percentages of successful P grafts in the hosts of the F$_2$ and BC generations, depending upon their Mendelian segregation. Based upon the F$_2$ results, the HO4C strain appears to differ from the HB11C strain at approximately nine histocompatibility loci, and from the HB32C and HB12C strains at five loci. These experimental results are reasonable because of the degree of genetic relationship between these inbred strains. The HB11C, HB12C, and HB32C strains originated from same wild-type population, while the HO4C was derived from a quite different population.

Suggestion of H-Y antigenes — If the scales of males from these inbred strains were transplanted to females, often the transplants failed to take. However, in reciprocal transplantation from female to male, the transplant was accepted. This finding suggests that histocompatibility genes located on the Y chromosome of the male medaka may play a role in tissue rejection, as has been reported in mice.

c. Allozymic Variation in the Inbred Strains

Allozymic variation was observed in at least 9 out of 21 loci in the wild populations collected at about 70 localities in Japan and in the outbreeding colonies of this species.[38,39] No variations, however, were detected in any individual within the same inbred strains. The fact strongly indicated genetic homogeneity in the inbred strain.

Three types of LDH isozymes (type I, II, and III) were found in the muscle of adult medaka from a commercial stock. Electrophoretic analyses showed that there were two kinds of subunits (A and A') constituting the most anodally migrating band and that they were present in individuals separately or together. Subunit B, which constitutes the most cathodally migrating band, was common to all individuals. Thus the patterns of type I, II, and III are considered to be composed of subunits A and B; A, A′, and B; and A′ and B, respectively. The type II was the most predominant pattern in the commercial stock. The LDH patterns of type I and type III were found to have been fixed in HO4C and HB12C, respectively.[40] In HO4C and HB12C, moreover, different types of esterase isozymes were fixed. The electrophoretic patterns of five enzymes, alcohol dehydrogenase (Adh), sorbitol dehydrogenase (Sdh), phosphoglucomutrase (Pgm), tetrazolium oxidase (To), and esterase 1 (Es-1), of the HO4C inbred strains were the same as that of fish in southern populations.[39]

Numerous biological studies were made with the himedaka, orange-red variety, but its origin is not yet known. The himedaka were painted by Ukiyoe artists of the Yedo era (100 to 300 years earlier), so the variety must have arisen by mutation from the wild type several hundred years ago or more. Studies of geographic variation in the number of anal fin rays of the medaka[41] and biochemical studies of allozyme suggest that the orange-red variety of medaka (himedaka) was derived from the southern population, especially, the eastern sub-population from the eastern part of the Pacific coast of Japan.

d. Laboratory Stocks of Inbred Strains

Several different inbred strains of the medaka are maintained in our laboratory at the National Institute of Radiological Sciences. The following is a list of the stocks available.

Wild Type (Brown or Black) Fish

HB32C (*BBRR*) — Derived from offspring of fish obtained from a stock at Chiba University, which had been collected near Chiba City around 1970; inbred for 39 generations. Substrains HB32A and HB32F were separated from HB32C after 4 and 6 generations and have been inbred for 34 and 32 generations, respectively.

HB12C (*BBRR*) — Derived from a different pair in the same population of HB32C and inbred for 39 generations.

HB11C (*BBRR, of*) — Derived from a pair of HB12C, separated from the HB12C after the 1st generation, and inbred for 39 generations. The fish are selected by the characteristics of the eggs. The eggs laid by the female mutant are characterized by an unusual pattern of oil-globule fusion after fertilization. The genetic symbol *of* (oil globule fusion delay) has been proposed, and the gene is recessive and autosomal.[42]

Orange-Red Fish

HO4C (*bbRR*) — Derived from a stock in our institute which was obtained originally from a dealer in Chiba Prefecture; inbred for 43 generations.

HO4B (*bbRR*) — Derived from a pair of HO4C, separated from the HO4C after 6 generations, and inbred for 37 generations.

HO4A (*bbRR, lf-2*) — Derived from a pair of HO4C, separated from the HO4C after 5 generations and selected by their pale body color. The fish have reduced numbers of xanthophores and leucophores; corresponding locus to *lf-2* found by Tomita.[43]

HO5 (*bbRR*) — Derived from a different pair in the same population of HO4C and inbred for 40 generations.

In general, the inbred strains of the medaka have a normal level of hatchability and a normal growth rate. However, the life-span of the inbred strain HB12 is shorter than that of the medaka of the outbred colony. There are differences in the enzyme activity in some organs, such as superoxide dismutase in the liver between the inbred strains, suggesting that longevity may be under genetic control. Thus, the inbred strains of the medaka may be valuable for gerontological studies.

In recent years, we began to inbreed the d-rR strain in which the females are white (*bb*, $X^\gamma X^\gamma$) and males orange-red (*bb*, $X^\gamma Y^R$). They have now been inbred for 19 generations. A strain designated d-rR has been used in experiments on sex differentiation.[2] In addition, inbreeding of newly collected medaka from Niigata (NI) and Shanghai (SH) was started in 1984, and now there are 14 and 11 inbred generations. Since they have different allozymic variations from those of HO4, HB32, and HB12, the establishment of the new inbred strains will be valuable for genetic studies of allozyme strain differences.

2. Responses to Radiation of the Inbred Strains

The medaka was extensively used in laboratory analyses of radiation effects.[44-46] The fish were obtained from cultivated stocks available from dealers or from a large closed colony. After the establishment of our inbred strains of medaka, we began to examine the differences in the effects of radiation on the adults and embryos of different strains; the findings have been compared with those obtained with the outbred orange-red variety fish from mass matings.

a. Effects on Lethality on Adults

Adult fish of the HO4C and HB32C strain of the medaka are more sensitive to X-rays than outbred populations of the orange-red variety of the medaka (himedaka) (Figure 6). The lethal effects of radiation were compared in terms of the $LD_{50/30}$ at 25°C: the value of the HO4C fish was 15 Gy, that of the HB32C was 17.5 Gy, and that of the himedaka was 20 Gy.[47] The $LD_{50/30}$ of F_1 hybrid from a cross between HO4C and HB32C fish was also 20 Gy. Moreover, the median survival time of inbred fish irradiated with a sublethal dose was shorter than that of the outbred or hybrid medaka.

b. Effects on Embryos

Effects of UV-irradiation — The experiment was designed to compare the effects of UV irradiation on embryos of different inbred strains.[48] Embryos at the morula stage were irradiated with UV, and their mortality and hatchability examined as criteria for biological effects. The photoreactivation (PR) of UV damage in embryos of these strains was also studied. Because the eggs of the fish are transparent, visible light can penetrate into the embryonic cells. The results, in Figure 7, show that: (1) the strain most sensitive to UV light was the HO4C, the 50% survival dose at the optic-bud formation stage (Stage 19) being 50 J/m^2, whereas in the HB32C (resistant strain), the 50% survival dose was as high as 150 J/m^2; (2) the survival rates in embryos of all strains increased upon postirradiation illumination with visible light, the photoreactivable fraction varying from 0.4 to 0.7. The survival of irradiated embryos, when examined at hatching, was lower than that examined at Stage 19 in all strains; in this case also the highest sensitivity to UV of the HO4C embryos was clearly demonstrated. However, no close correlations between UV sensitivity and photoreactivability were demonstrated among the embryos of different inbred strains.

Effects of X-irradiation — Embryos of 2 different inbred strains (HO4C and HB32C) and outbred fish (himedaka) were irradiated with X-rays at the early morula stage (at 4.5 h after fertilization at 25°C). Dose-dependent decreases in the survival rates at Stage 19 and in the hatchability after X-irradiation were observed (Figure 8). The strain most sensitive to X-rays was the HO4C, the 50% hatching dose being 230 rad, whereas in the HB32C and the orange-red variety, the 50% hatching doses were 340 and 670 rad, respectively.

c. Effects on Mutation Induction

Genetic effects of radiation have been analyzed mainly with microorganisms, with insects such as *Drosophila,* and with mice; however, in recent years there have been several quantitative studies on radiation-induced mutations in teleosts.[49-52] In most experiments on the medaka, germ cells at various stages of spermatogenesis were irradiated, and the frequency of the induction of mutations was examined in subsequent generations. The dominant lethal mutation rate was ascertained in terms of the hatchability of fertilized eggs, and the

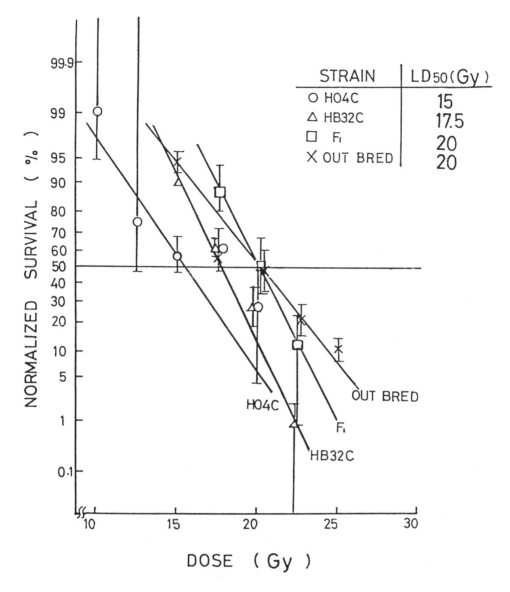

FIGURE 6. Dose-effect relationship for 30-d survival in X-irradiated different strains of medaka.

recessive visible mutation was demonstrated by the so-called specific loci method, as in experiments with mice.

In the dominant lethal test with γ-irradiated male fish, there was no apparent difference in radiosensitivity in the induction of dominant lethals or in the production of sterility between the orange-red variety and HB12C strain.[53]

As to visible mutations, heritable changes in body color, *b* and *r* alleles are useful because the *B* gene of the wild type manifests itself within 48 h after fertilization when eggs are kept at 25°C. HO eggs inseminated with HB sperm or vice versa should be used. In recent years, strains of medaka with several visible mutants were established by Shimada et al.[54] A multiple recessive tester stock, homozygous for five loci, was generated that was mated with treated fish normally carrying wild-type alleles. These recessive loci were colorless melanophores *(b)*, double anal fins *(Da)*, guanineless *(gu)*, pectoral finless *(pl)*, and colorless xanthophores *(r)*. If the phenotypic expression of each locus was observed during

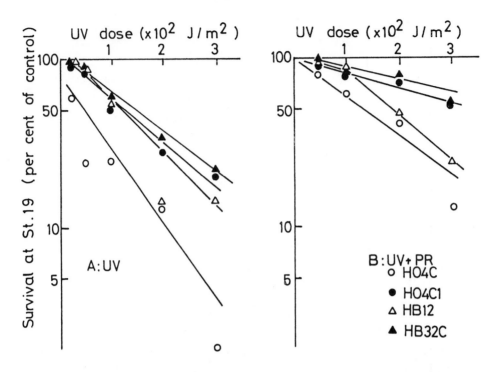

FIGURE 7. Ultraviolet dose-survival relationship at Stage 19 in various inbred strain embryos. (A) UV alone; (B) UV and visible light. These results show photoreactivation. (From Hyodo-Taguchi, Y. and Egami, N., *Zool. Sci.*, 2, 305, 1985. With permission.)

the development, the *b/b* and *gu/gu* embryos could be distinguished from wild-type ones at an early developmental stage and the *pl/pl* and *Da/Da* fry could be distinguished just after hatching. Mutation rates thus are detectable in inbred fish more easily and earlier than in mice.

3. Response of the Inbred Strains to Chemical Carcinogens

The medaka is very sensitive to chemicals even though it is more resistant to ionizing radition than mammalian species. Males of the inbred strain (HB12) were treated with various chemicals, including mitomycin C and MMS and then mated with normal females. The dominant lethal effects of these chemicals were estimated by their effects on hatchability of the eggs. The mutagenic effects of chemicals are similar to those in mammals, and the medaka therefore is useful for the detection of environmental mutagens.[55] One of our interests lay in the difference in response to chemical carcinogens between inbred and outbred strains.

a. Induction of Transplantable Melanoma in Adults

There are differences among the inbred strains in sensitivity to the chemical carcinogen *N*-methyl-*N'*-nitro-*N*-nitrosoguanidine (MNNG) in terms of acute toxicity and carcinogenicity.[56] Adult fish of the two inbred strains (HO4C and HB32C) were exposed for 2 h to an aqueous solution of MNNG at concentrations from 20 to 100 mg/l. No tumor was produced in the HO4C strain fish, although these fish were sensitive to the acute toxicity of MNNG, the median lethal dose being 28 mg/l 48 h after the treatment. In contrast, a dose-related tumorigenic response was observed in the HB32C strain, with the median lethal dose being 38 mg/l 48 h after treatment. Most MNNG-induced tumors developed 3 to 6 months after treatment; these tumors were considered to be amelanotic melanomas on the basis of histologic findings and a positive dopa reaction. Serial transplantation of tumor tissue into the

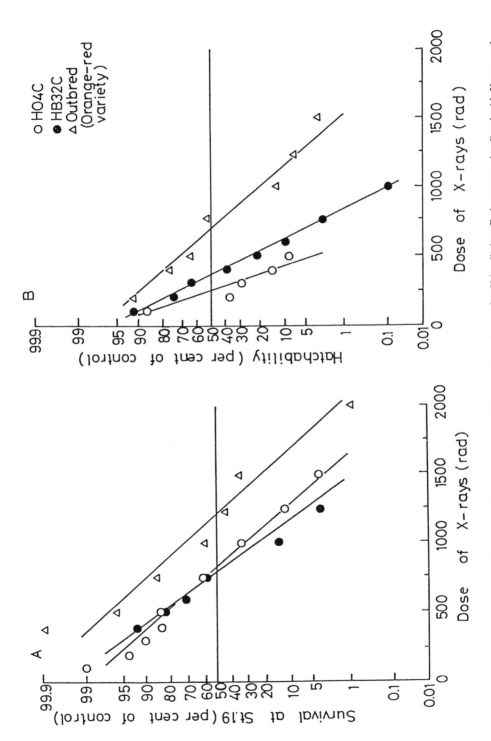

FIGURE 8. Dose-survival relationship at Stage 19 and hatching in different strain embryos after X-irradiation. Embryos were irradiated with X-rays at the early morula stage. (A) Survival at Stage 19; (B) Hatchability. (From Hyodo-Taguchi, Y. and Egami, N., *Zool. Sci.*, 2, 305, 1985. With permission.)

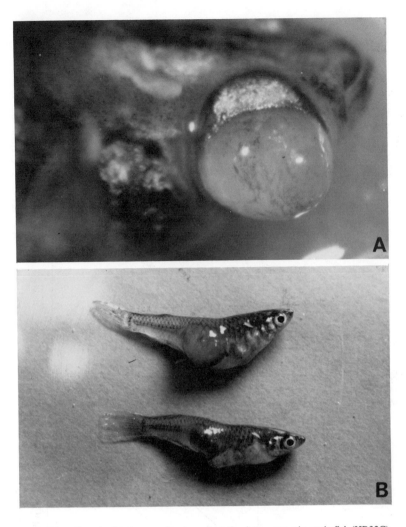

FIGURE 9.   Growth of melanoma after transplantation into syngeneic-strain fish (HB32C). (A) Transplant at 30 d after transplantation into eye chamber; (B) Transplants at 35 d after transplantation into intraperitoneal cavity.

eye chamber and the intraperitoneal cavity of fish of the syngeneic strain (melanoma tissue induced in HB32C fish into the eye of an HB32C host) was successfully accomplished (Figure 9). Table 2 shows the results of the serial transplantability of this fish melanoma into the anterior chamber of the eye of syngeneic, allogeneic, and xenogeneic strains. Tumor transplantation into one of the allogeneic strains (HO5) also was successful.

b. Transplantable Hepatic Tumors Induced by Treatment with MAM Acetate

Liver tumors were induced in himedaka by treatment with chemical carcinogens such as DENA and MAM acetate;[57,58,69] and the similarities and differences in tumorigenesis between fish and mammals were analyzed. There was a high incidence of liver tumors within 1 to 3 months after 1- to 24-h exposure to high levels (0.5 to 10 mg/l) of MAM acetate.[60] The neoplastic changes in the liver caused by MAM acetate were similar to those resulting from exposure to DENA, indicating that the MAM-acetate-medaka system may be a convenient model in which to analyze basic problems of chemical carcinogenesis. However, there is insufficient evidence to show that the lesions in fish liver are malignant. To gain a better understanding of neoplastic changes in the fish, we designed an experiment to examine

TABLE 2
Serial Transplantability of Fish Tumors into the Anterior Chamber of the Eye of Syngeneic, Allogeneic, and Xenogeneic Strains[61]

Transplant designation	Passage no.	Recipients	No. of fish transplanted	No. of takes Positive	No. of takes Negative
AA	2—9	HB32C	108	108	0
	4	HO5	4	4	0
	4	HO4C	6	2	4
AB	9	Orange-red variety	10	4	6
	2—5	HB32C	40	39	1
	2—5	HO5	31	31	0
	2,7	HO4C	9	1	8
	4	Goldfish	9	0	9

From Hyodo-Taguchi, Y. and Matsudaira, H., *J. Natl. Cancer Inst.*, 55, 909, 1975. With permission.

TABLE 3
Transplantability of HO4C Fish-Derived Hepatoma into Anterior Eye Chamber of Syngeneic Strain

Individual no.	Days after MAM-acetate treatment	No. of fish transplanted	No. of positive takes
1	55	14	0
2	62	19	0
3	120	15	14[a]
4	135	5	0
5	135	5	0
6	142	7	0
7	142	9	0
8	162	17	12[a]

Tumor donors

a   Tumor tissue developed well within 1.5 to 2 months after transplantation.

transplantability of liver tumors induced by treatment with MAM acetate in an inbred strain of medaka, HO4C.

Five-month-old medaka were treated with 0.5 mg/l MAM acetate for 3 d and then maintained in well water at 25°C for 6 months. Adult fish of HO4C strain were more sensitive than fish of the orange-red variety. Almost half of treated fish died within 2 months after treatment. Within the 2 to 5 months after treatment, there were seven fish bearing liver tumors that were used for the transplantation experiments. The MAM-acetate-induced hepatic tumors survived and grew in the anterior eye chamber and peritoneal cavity of syngeneic hosts (Table 3), and the tumors were malignant (Figure 10).

c. Higher Incidence of MNNG-Induced Melanoma in F_1 Fish

Inbred fish strains are most desirable for analysis of hereditary factors determining the sensitivity to carcinogens. Extensive studies by Anders et al.[61] showed the usefulness of interspecific hybrids in demonstrating the presence of tumor genes, regulatory genes, and their interactions, as reflected in differences in the incidence of spontaneous and induced melanoma.

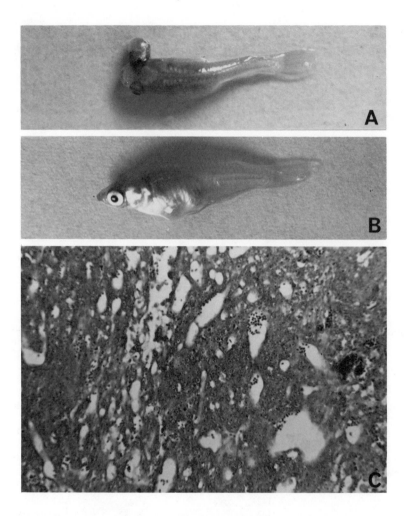

FIGURE 10.   Growth of hepatoma after transplantation into syngeneic-strain fish (HO4C).
(A) Transplant at 35 d after transplantation into eye chamber; (B) Transplants at 25 d
after transplantation into intraperitoneal cavity; (C) Histologic section of hepatoma that
developed in the eye chamber at 35 d after transplantation.

We looked at the value of F_1 hybrids of inbred strains of medaka for this purpose. Adult
hybrids (F_1) from crosses between HO4C and HB32C were exposed for 2 h to an aqueous
solution of the carcinogens MNNG at concentrations of 15, 20, 25, 30, and 35 mg/l: their
survival and neoplastic changes were examined for 6 months. A large variety of neoplasms
were induced, including melanomas, papillomas, ovarian tumors, olfactory epitheliomas,
branchioblastomas, and fibromas in various organs (Figure 11). More than 60% of the tumors
were classified as melanomas on the basis of histological examinations. A markedly higher
cumulative incidence of melanomas and dose-related response was found in the F_1 hybrid
fish compared with the parental strains (Figure 12). The results indicate the usefulness of
the highly sensitive F_1 hybrids.[62]

d. Effects of MNNG on Embryos
In general, the embryonic stages of the lifespan are the most sensitive to chemicals or
mutagens in fish as well as in mammals. MNNG, a direct-acting carcinogen, was used in
this experiment because there seems to be no metabolic activation system in embryos at
their early developmental stages. Embryos at 2 h, 3 d, and 7 d after fertilization from

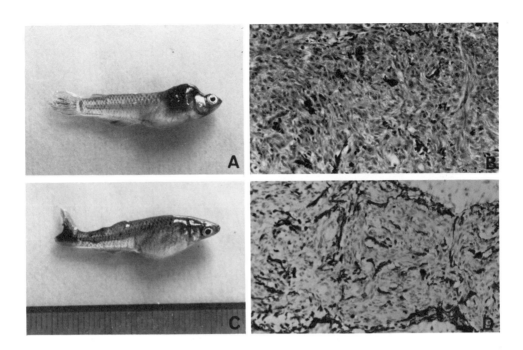

FIGURE 11. Typical melanomas induced by MNNG treatment in the F_1 hybrid fish. (A) The fish was exposed to 25 ppm MNNG for 2 h at 3 months old and then survived for 5 months. (B) Section of the tumor shown in A. (C) The fish (3 months old) was exposed to 25 ppm MNNG for 2 h. A melanotic lesion in tail was observed around 3 months after exposure, and the heavily pigmented area grew until 5 months. (D) Section of melanoma shown in C.

different inbred strains were exposed to MNNG at concentrations from 1 to 100 mg/l and for 2 or 20 h at 25°C.

Mortality and hatchability at early developmental stages — The dose-survival and dose-hatchability relationships summarized in Figure 13 show differences among the embryos of the different strains. A high mortality and a marked decrease in hatchability were found in the HO4C strain after treatment with a low concentration of MNNG (5 to 10 mg/l). In the HB32C strain, 90% of the embryos treated with 10 mg/ml of MNNG survived through Stage 19 and hatched, even when they had been exposed to 40 and 50 mg/l. The 50% survival doses of MNNG at Stage 19 of the HO4C, HB32C, and outbred fish embryos were 12, 25, and 32 mg/l, respectively. The 50% hatchability doses were around 5 mg/l in the HO4C, 16 mg/l in the HB32C, and 21 ppm in the outbred embryos. If we compare the slope of the dose-effect relationship curve in the inbred-strain embryos, particularly HO4C, with that of the outbred embryos, it is clear that there is a substantial increase in slope in the inbred strains (Figure 13). The use of the inbred strains reduces the variation in individual response to carcinogens.

Sensitivity to MNNG at different developmental stages — Table 4 summarizes the hatchability, mortality, and numbers of tumor-bearing fish in the HO4C and HB32C strains after a single exposure to MNNG at different embryonic stages. It is evident from the hatchability data, that the cleavage stage is the most sensitive to MNNG, and at all stages the HO4C embryos are more sensitive than are the HB32C embryos. The mortalities of fry were very high in all groups at 1 month after hatching. From 1 to 3 months, the mortality rates were low; thereafter, the mortalities again became high. After a 2-h treatment with MNNG during the cleavage stage, no neoplastic changes were produced in the HO4C and HB32C during the 6 months of observation. In contrast, tumors were observed in a number of the HB32C fish exposed to MNNG for 20 h at 3 or 7 d after fertilization. The 20-h

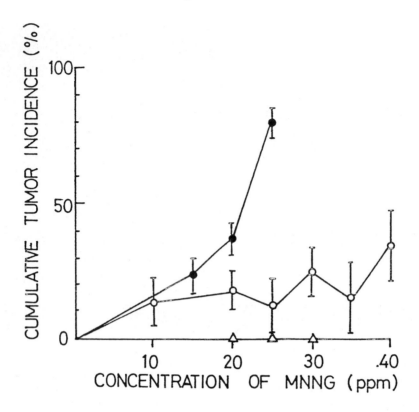

FIGURE 12.   The cumulative incidence of melanoma in the inbred strains of medaka, HO4C and HB32C, and in their F_1 hybrids. $\triangle$: HO4C; $\circ$: HB32C; $\bullet$: F_1 between HO4C $\times$ HB32C. (From Hyodo-Taguchi, Y. and Matsudaira, H., *J. Cancer Res. (Gann.)*, 78, 487, 1987.

treatment at 7 d induced a few tumors in the HO4C fish. These findings show that (1) the acute toxicity and carcinogenicity of MNNG differ between two inbred strains at the embryonic stages, and (2) tumorigenic sensitivity to MNNG increases with increasing age of the embryo.

E. STOCKS OF MEDAKA AND MODELS OF TRANSGENIC ANIMALS

Besides inbred strains, about 100 strains of wild medaka collected at different localities in Japan, Korea, and China are maintained in the Zoological Institute, Faculty of Science, Tokyo University.[32] The laboratory of Freshwater Fish Stocks, Nagoya University, keeps more than 50 strains having mutant genes.[43]

Cultured cell lines from the medaka derived from normal fins, heptomas, and melanomas are maintained at the Zoological Institute, Tokyo University, and at the National Institute of Radiological Sciences. The establishment of cell lines from various strains of the medaka is valuable in analyzing at the molecular level the genetic background of the differences in response to chemicals.

Ozato et al.[63] produced a model of transgenic fish: recombinant plasmids containing chicken δ-crystallin gene were microinjected into the oocyte nucleus of the medaka. About half of these oocytes developed into 7-day-old embryos, and the exogenous gene products were detected in embryos. These experiments show that oviparous animals can be very suitable models to further research at the molecular level into aspects of developmental biology.

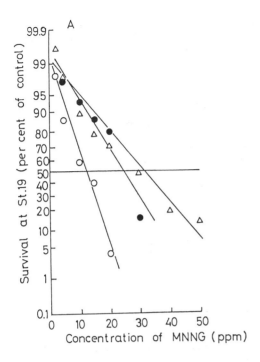
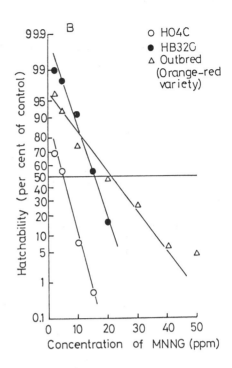

FIGURE 13. Dose-survival relationship at Stage 19 and hatching in different strain embryos after MNNG treatment. Embryos were treated with MNNG at the morula stage for 2 h. (A) Survival at Stage 19; (B) Hatchability. (From Hyodo-Taguchi, Y. and Egami, N., *Zool. Sci.*, 2, 305, 1985. With permission.)

III. THE GOLDFISH

A. GENERAL BIOLOGY

Goldfish is a domesticated form of a crucian carp, *Carassius auratus*. Golden forms of the fish, the ancestral form, were developed at least a thousand years ago. By subsequent selection and breeding, the Chinese and the Japanese produced several domesticated forms with greatly enlarged fins, a short and deformed body, different body colors, and sometimes, enlarged eyes. Now, goldfish is the most popular pet fish in Japan, and more than 20 varieties are cultivated, including the Wakin (common goldfish), Rukin (veiltails), Runchu (eggfish), and Demekin (telescope). The Wakin is generally used for biological experiments. Although such fancy goldfish are delicate and generally rather difficult to rear, small-sized Wakin are easy to keep in the laboratory. They can tolerate a wide range of temperature, from −2 to 34°C (their optimum range is 12 to 26°C). Usually regarded as a small fish, goldfish grow to a length of 5 to 10 cm and a weight of 10 to 50 g, and some attain sexual maturity within the first year of life.

Breeding goldfish is rather difficult in indoor laboratory conditions. In outside aquariums or garden pools, males and females are recognizable at the start of the breeding season because the female becomes swollen with eggs, while the male develops tubercles, known as pearl organs, on the gill covers and pectoral fins (these may be difficult to see unless the fish is viewed from a certain angle). The female lays 500 to 1000 eggs, each 0.5 mm in diameter, between May and August, which are fertilized after laying, the male following the female around all the time she is spawning. The eggs stick to water plants and hatch in 3 to 5 d when the temperature is 21 to 25°C. The larvae, 2 to 3 mm long and tadpole-like in shape, hang on to the water plants for the next 48 h, by which time the yolk sac is used up, the fins have grown, and the fry are able to feed independently. Growth is rather rapid within a few months, but then it gradually slows down; the fish may attain a length of 15

TABLE 4

Acute Toxicity and Tumor Induction by MNNG Treatment at Different Developmental Stages in Different Inbred Strains of Medaka

Strain	Stage of treatment	MNNG (ppm)	Hatchability[a] (%)	1 d	No. of fish surviving at:[b] 1 month	3 months	6 months	1 year	No. of fish with tumor during 1 year (%)
HO4C	Morula[c]	0	101/156 (65)	65(100)	27(42)	19(29)	15(23)	0(0)	0
		2.5	73/172 (42)	98(100)	9(9)	9(9)	8(8)	0(0)	0
		5	52/226 (23)	64(100)	18(28)	18(28)	10(16)	0(0)	0
HO4C	Stage 22[d]	20	385/444 (87)	355(100)	3(1)	3(1)	2(1)	0(0)	—
		40	169/227 (74)	141(100)	2(1)	2(1)	1(1)	0(0)	—
HO4C	Stage 43[e]	0	92/100 (92)	117(100)	31(26)	28(24)	19(16)	12(10)	0
		20	296/336 (88)	284(100)	80(28)	79(28)	57(20)	5(2)	1(1.3)
		40	242/298 (81)	264(100)	74(28)	59(22)	49(19)	0(0)	2(2.7)
HB32C	Morula[c]	0	229/259 (89)	208(100)	69(33)	67(32)	55(26)	33(16)	0
		2.5	143/172 (83)	162(100)	18(11)	18(11)	18(11)	0(0)	0
		5	137/171 (80)	135(100)	39(29)	39(29)	39(29)	34(25)	0
		10	112/146 (77)	89(100)	29(33)	29(33)	28(31)	17(19)	0
		15	87/188 (40)	73(100)	24(33)	20(27)	18(25)	0(0)	0
HB32C	Stage 22[d]	20	484/521 (93)	440(100)	96(22)	94(21)	70(16)	15(3)	0
		40	336/413 (81)	272(100)	50(18)	49(18)	32(12)	14(5)	0
		80	140/174 (81)	114(100)	18(16)	16(14)	3(3)	0(0)	3(17)
HB32C	Stage 34[e]	0	192/198 (97)	192(100)	84(44)	77(40)	59(31)	42(22)	0
		20	252/274 (92)	245(100)	79(32)	79(32)	67(27)	8(3)	4(5.1)
		40	138/153 (90)	138(100)	78(57)	75(54)	38(28)	2(1)	8(10)
		80	182/204 (89)	144(100)	4(3)	1(1)	0(0)	—	—

Note: Embryos at the morula stage were treated with MNNG for 2 h and embryos at Stage 22 and 34 were treated with MNNG for 20 h.[31]

[a] Number of fry hatched per number of embryos treated.
[b] Numbers in parentheses are percentages of survivors.
[c] Morula: 2 h after fertilization at 25°C.
[d] Stage 22: 3 d after fertilization at 25°C (lens formation stage).
[e] Stage 34: 7d after fertilization at 25°C (the last embryonic stage).

to 20 cm and a maximum weight of 500 g, depending on environmental conditions. Pet goldfish can be long-lived, up to 25 years having been recorded, but their mean lifespan is 7 to 8 years.

B. HUSBANDRY

If large-scale outdoor ponds and an indoor facility for breeding are available, it is possible to breed goldfish. In the National Institute of Radiological Sciences (NIRS), goldfish (Wakin) have been bred for over 20 years, and sufficient numbers are available for experimental work. However, in general, goldfish for experimental use are obtained commercially from dealers. These fish vary in age, history, body size, and health, as well as genetic heterogeneity, although the fish obtained at any one time may be rather homogeneous because they usually come from one colony. It is recommended that a sufficient number of fish be obtained for one experiment at one time, and be acclimated to laboratory conditions.

In NIRS, we have 64 outdoor concrete aquariums ($2 \times 1 \times 0.4$ m) for holding the experimental goldfish. Generally, 100 small (3 to 5 cm) fish and 30 moderate-sized (10 to 20 cm) fish are kept in each aquarium. In the laboratory, 10- to 20-l plastic aquariums are used, each holding five or ten goldfish. The aquariums are aerated, and the fish are fed commercial fish pellets or Tetramin®.

C. THE GOLDFISH IN STUDIES OF RADIATION EFFECTS ON CELL RENEWAL SYSTEMS

To analyze the mechanisms of development of radiation damage, since 1962 we have studied mitotic and DNA synthetic activities of various tissues of the goldfish. The fish have several advantages and disadvantages for studies of radiation effects.

The biological effects of radiation in fish are affected by physical, chemical, and biological factors. Fish are poikilotherms, and their body temperature, metabolic rate, and physiological function depend on environmental temperature.[64,65] Using goldfish, it was shown that the renewal rate and length of the generation cycle in the cell renewal system also depend upon temperature, as does the length of survival of fish after irradiation with lethal doses.[66]

Most radiation injuries leading to death of fish are related to the cell renewal systems, with the exception of damage to the central nervous system. The kinetic studies of cell populations were made on the hematopoietic tissue,[67,68] the gills,[69] and the intestine[70,71] of irradiated goldfish held at different temperatures, using autoradiography with ^{3}H-TdR. These results showed that development and recovery processes of radiation-induced injury were based on the kinetics of cell populations in critical tissues.

Cultured cell lines derived from normal and tumor tissue from goldfish were established in several laboratories in Japan.[72-74] It has been reported that fish cells are generally more resistant than mammalian cells to ionizing radiation, are more sensitive to UV light, and are equally sensitive to mutagenic action of certain chemicals.[75] The usefulness of these cultured fish cells in biomedical research is discussed in the following chapter.

D. THE GOLDFISH OPTIC NERVE AS A MODEL SYSTEM FOR FUNCTIONAL REGENERATION

Regeneration of the optic nerve in adult fish and amphibians may result in recovery of visual function.[76,77] This regenerative phenomenon is not observed in higher vertebrates. The goldfish visual pathway has this regenerating ability, as well as considerable qualities of plasticity. Throughout the life of the goldfish there is a symmetrical increase in the number of retinal ganglion cells and an asymmetric growth of the optic tectum to which the optic axons project.[78-80] Plasticity results from the shifting of terminals in the optic tectum to maintain the established retinotopic projection. Plasticity is limited but significant in adult higher vertebrates and is crucial during development. The goldfish visual pathway has therefore emerged as an important model system for the study of optic nerve regeneration as well as for CNS development. The regrowth and regeneration of the goldfish optic nerve have been studied at the molecular level.[81-86]

IV. OTHER SMALL FISHES IN BIOMEDICAL RESEARCH

A. THE PLATYFISH-SWORDTAIL HYBRID MELANOMA SYSTEM

Platyfish (*Xiphophorus maculatus*) and swordtails (*Xiphophorus helleri*) are small viviparous killifish living in freshwater. Certain populations of platyfish are characterized by black spot patterns of macromelanophores on the skin. The macromelanophore genes of the platyfish express malignant phenotypes which produce melanomas when they are introduced into the genome of the swordtail by interspecific hybridizations. This hybrid melanoma system is a useful model for studies on cancer etiology and was introduced to cancer research about 50 years ago by Kosswig[87] and Haüssler[88] in Germany and by Gordon[89] in the U.S. In 1957 Anders and Anders started their studies with this system and have published a large number of reports.[61,90,91]

Platyfish and swordtails have been domesticated for a long time by aquarists, and many color varieties have been produced by hybridization between different populations of the same species or different species of this genus. Several macromelanophore patterns are known in both natural populations and domesticated stocks of platyfish and are determined by macromelanophore genes. The spotted-red platyfish carrying a macromelanophore gene, *Sp,* often are used as one of the parents for hybridization; the varieties of swordtails, the other parent, have no macromelanophores.

The fishes are best kept in well-planted and well-aerated aquariums at 26°C and fed with dried *Daphnia* or flakes for aquarium fish (Tetramin®). Adult platyfish range in size from 14 to 45 mm, while adult swordtails are larger (50 to 100 mm). The age at which sexual maturity is reached varies from 7 weeks to 32 weeks, depending on their genetic background and temperature. Fertilization in *Xiphophorids* is internal. Females give birth to living young which develop inside the ovarian follicle. The gestation period is approximately 21 d, but may depend upon the temperature of the water. At birth, the young are fully formed. Brood intervals also are temperature dependent: at 23°C, most broods are born at intervals between 31 and 39 d, while at 25°C, the interval is 23 to 28 d.[14]

Interspecific hybrids are produced by artificial insemination. To produce the first filial generation (F$_1$), swordtail females and platyfish males are used as parents, since the bigger body size of swordtail females makes artificial insemination easier and the swordtail females ordinarily produce more offspring in a single brood than platyfish females. Brother-and-sister matings of the F$_1$ offspring produce the second generation (F$_2$). Backcross (BC) generations are produced by crossing melanotic hybrids with swordtails.

This hybrid melanoma system was used recently in Japan to study the genetic mechanisms of melanoma formation.[92,93] An age dependence in melanoma development was shown, and the hybrid fish developed pigmentary abnormalities, melanosis, at early ages. The number of fish developing melanomas increased sharply from the age of 5 to 19 months. The cumulative percentage of tumor-bearing fish at 19 months was 22%.

B. THE CENTRAL MUDMINNOW IN THE STUDY OF CHROMOSOME ABERRATIONS

Small fishes have been used to detect genetic effects of various chemical and physical agents, including the production of dominant lethal,[53,55] point mutation,[50-52,54] and cytogenetic damages.[94,95] Among these, cytogenetic methods offer a relatively fast and simple means for detecting genotoxic effects. However, most fishes are not well suited for cytogenetic investigation because they have large numbers of small chromosomes. Kligerman et al. proposed a model system aproach to the study of chromosome damage in fishes.[12,96] Fishes of the genus *Umbra* have excellent karyotypes for clastogenicity studies. The central mudminnow *Umbra limi* has a low number of large chromosomes, the karyotype consisting of 22 large meta- and submetacentric chromosomes. Large numbers of fish can be captured easily and held for study. Tissue from the intestine, kidneys, and gills are suitable for chromosome preparations. The chromosomes of *Umbra limi* responded typically to low-level radiation exposure, and this fish would seem to be an ideal cytogenetic model for the study of induced chromosome aberrations. Moreover, *in vivo* sister chromatid exchange analysis to study potentially dangerous waterborne chemicals was accomplished successfully in the mudminnow system.[97,98]

There are only three species of the genus *Umbra:* one in Europe *(Umbra krameri)* and two in North America *(Umbra pygmaea* of the U.S. Atlantic coastal region and *Umbra limi* of inland waters).[99] The central mudminnow is a small, robust fish with a small mouth and soft-rayed fins. The fish is usually about 5 to 10 cm long but may reach 15 cm. The color is yellowish-brown to green, with narrow stripes running along the flanks and back. Mudminnows usually live along the sides of streams under dense floating vegetation. They feed on a variety of small organisms (plankton, aquatic insects, and sometimes smaller fish). Their breeding season is in early spring, when water temperature is about 13°C. Mature females, from 5 to 10 cm long, spawn from 20 to nearly 1500 eggs, 1.6 mm in diameter. The eggs are laid singly, are adhesive, and stick to the heavy vegetation of the habitat. Eggs hatch in about 6 d, and the young are about 5 mm long. One-year-old fish reach 3 to 9 cm. The fish is hardy and will live for many months in an aquarium at temperatures below 15°C.

Central mudminnows are maintained in the National Institute of Radiological Sciences in Japan. The fish were purchased from the Zoological Center International, Hinsdale, IL. They are kept in heavily planted outdoor pools (80 l in volume) into which groundwater flows continuously. The temperature rises to about 20°C at midsummer and drops to 5°C or below at midwinter (temperatures above 20°C are lethal). Freshwater oligochaetes are given *ad libitum*. A lymphocyte culture technique for human chromosome preparations established by Moorhead et al.[100] was applied to tissues of *Umbra limi*.[101] The advantage of this method lies in the fact that small amounts (0.1 ml) of blood can be repeatedly withdrawn, without harm, from the sinus venosus of the fish after a minimal interval of 2 weeks. The dicentric yields after X-irradiation were significantly lower than these in human lymphocyte chromosomes.

C. THE RAINBOW TROUT AS A SENSITIVE ANIMAL MODEL FOR EXPERIMENTAL CARCINOGENESIS

Liver cancer appeared in domesticated trout in many hatcheries in the U.S. from 1957 to 1960. During the process of searching for the causes of the cancer, it was found that there was a highly potent carcinogen alfatoxin B_1 (AFB_1) in pellets given as food to the rainbow trout (reviewed by Sinnhuber et al.[102]). AFB_1 since has proved to be one of the most potent liver carcinogens known. Subsequent research also has shown that rainbow trout, *Salmo gairdneri,* a member of the family *Salmonidae* (salmon and trout), is one of the animals most sensitive to its carcinogenic effects. Many of the species, subspecies, and strains of trout and salmon have been tested for their sensitivity to alfatoxin. The Mount Shasta strain of rainbow trout is the most sensitive. Wild steelhead trout had a hepatoma incidence of only 6% after 12 months on a diet that contained 8 ppb of AFB, whereas the Mount Shasta strain of domesticated rainbow trout had an incidence of 85 to 100%. In other experiments, levels of AFB_1 as low as 0.4 to 0.5 ppb in the diet of rainbow trout induced a significant incidence of liver cancer.[103,104]

Hendricks et al. developed an innovative system for the detection of hepatocarcinogens.[11,105,106] Embryos of rainbow trout were exposed to a 0.5-ppm aqueous solution of AFB_1 for 30 min, then raised under standard conditions. After 1 year, approximately $2/3$ of the population had developed hepatoma. Sensitivity to AFB_1 increases with increasing age of the embryo, and the model system shows a dose response to increasing concentrations of the carcinogens. The embryonic model system is sensitive to the hepatocarcinogenicity of other AFB_1 metabolites (aflatoxicol, aflatoxin M_1, and aflatoxin G_1), sterigmatocystin, versicolorin A, dimethylnitrosamine, diethylnitrosamine and *N*-methyl-*N'*-nitro-*N*-nitrosoguanidine.

V. SUMMARY AND CONCLUSIONS

To demonstrate the value of small freshwater fish in biomedical research, we reviewed some examples of the use of the medaka, especially the inbred strains. The advantages of goldfish for studies of radiation effects also were discussed. A main conclusion drawn is that inbred strains of the medaka offer many potential advantages as test animals for the detection of mutagenic, carcinogenic, and teratogenic agents in water.

Moreover, genetic uniformity is found in gynogenetic populations of some small fish and in clones of fish, and these animals are useful materials for biomedical research. Uniformity eliminates genetic variability as a source of experimental error, so that there is better comparability of data between experiments and between investigators. The use of genetically controlled small fish as an alternative to mammals may contribute to a reduction in the use of experimental mammals and to exact genetic analysis of complicated phenomena, such as carcinogenesis and developmental abnormality.

The use of the inbred strains of the medaka is particularly suitable since each inbred strain has its own inherent advantages, depending on the nature of the problem to be studied. For instance, for toxicity assays the HO4C strain is valuable because the embryos are extremely sensitive to chemical agents, such as MNNG, and to radiation. The adults also are sensitive to MNNG. In carcinogenicity testing, however, their lack of sensitivity to the lethal effects of agents becomes a disadvantage. Many carcinogens also act as general toxicants, producing tissue damage rather than neoplasms. Because tumor induction follows a postexposure latency, the fish must survive through the general toxic effects of the carcinogen. Therefore, the HB32C strain is a better test animal for carcinogens because this strain is relatively resistant to acute lethal effects; pigment-cell tumors are easily induced after a single treatment with MNNG. Using hybridization and backcrosses of members of different inbred strains of the medaka with different sensitivities to carcinogens, further comparative studies of oncogens can be expected.

The usefulness of the embryos as a model for carcinogenesis in rainbow trout, *Salmo gairdneri,* was stressed by Hendricks. In the inbred strain of the medaka (HB32C), pigment-cell tumors were induced in adult and embryos after treatment with MNNG. In adults, melanomas appeared 3 to 6 months later, whereas following a short period of treatment at embryonic stage, the tumors were produced in adult fish 6 to 10 months later. For long-term experiments, a high mortality before the formation of the first tumor is a disadvantage. Experimental studies of carcinogenesis in mammals require much time and laboratory space and are rather expensive. The inbred strains of the medaka seem to offer economic advantages as laboratory animals for cancer research.

We proposed the use of the dominant lethal test of γ-irradiated males and visible mutation system in multiple recessive tester stocks of the medaka for testing mutagens. However, additional data are required to define the most suitable inbred strain to use for scoring genetic effects. Recently we established cell lines in culture derived from each inbred strain of the medaka: the transplantability of cultured tumor cells in syngeneic hosts may be useful in studies of oncogenic transformation, when combined with molecular techniques. Such cell lines are useful to analyze carcinogenic processes by *in vitro* and *in vivo* experiments.

REFERENCES

1. **Marine, D. and Lenhart, C. H.,** Observations and experiments on the so-called thyroid carcinoma of brook trout *(Salvelinus fontinalis)* and its relation to ordinary goitre, *J. Exp. Med.,* 12, 311, 1910.
2. **Yamamoto, T.,** *Medaka (Killifish) — Biology and Strains,* Keigaku, Tokyo, 1976.
3. **Schultz, R. J. and Schultz, M. E.,** Characteristics of a fish colony of *Poeciliopsis* and its use in carcinogenicity studies with 7, 12-dimethylbenz(a)anthracene and diethylnitrosamine, *Natl. Cancer Inst. Monogr.,* 65, 5, 1985.
4. **Woodhead, A. D., Setlow, R. B., and Pond, V.,** The Amazon molly, *Poecilia formosa,* as a test animal in carcinogenicity studies: chronic exposures to physical agents, *Natl. Cancer Inst. Monogr.,* 65, 45, 1985.
5. **Egami, N., Etoh, H., Tachi, C., and Hyodo, Y.,** Relationship between survival time after exposure to X-rays and dose of the irradiation in the fishes, *Carassius auratus* and *Oryzias latipes, Dobutsugaku Zasshi,* (in Japanese, with English abstr.), 71, 313, 1962.
6. **Aoki, K.,** The effects of whole body X-irradiation on the hematopoietic tissue in the goldfish, *Carassius auratus, Dobutsugaku Zasshi,* (in Japanese, with English abstr.), 72, 283, 1963.
7. **Hyodo, Y.,** Effect of X-irradiation on the intestinal epithelium of the goldfish, *Carassius auratus* I. Histological changes in the intestine of irradited fish, *Annot. Zool. Jpn.,* 37, 104, 1964.

8. **Etoh, H.,** The operation for removal of the kidneys from goldfish, *Kagaku (Tokyo),* (in Japanese), 32, 606, 1962.

9. **Jones, P. S., Tesser, P., Keyser, K. T., Quitschke, W., Samadi, R., Karten, H. J., and Schechter, N.,** Immunohistochemical localization of intermediate filament proteins of neuronal and nonneuronal origin in the goldfish optic nerve: specific molecular markers for optic nerve structures, *J. Neurochem.,* 47, 1226, 1986.

10. **Tesser, P., Jones, P. S., and Schechter, N.,** Elevated levels of retinal neurofilament mRNA accompany optic nerve regeneration, *J. Neurochem.,* 47, 1235, 1986.

11. **Hendricks, J. D.,** The use of rainbow trout *(Salmo gairdneri)* in carcinogen bioassay, with special emphasis on embryonic exposure, in *Phyletic Approaches to Cancer,* Dawe, C. J. et al., Eds., Jpn Sci. Soc. Press, Tokyo, 1981, 227.

12. **Kligerman, A. D.,** The use of aquatic organisms to detect mutagens that cause cytogenetic damage, in *Radiation Effects on Aquatic Organisms,* Egami, N., Ed., University Park Press, Baltimore, 1980, 241.

13. **Babich, H. and Borenfreund, E.,** *In vitro* cytotoxicity of organic pollutants to bluegill sunfish (BF-2) cells, *Environ. Res.,* 42, 229, 1987.

14. **Kallman, K. D.,** The platyfish, *Xiphophorus maculatus,* in *Handbook of Genetics,* Vol. 4, King, R. C., Ed., Plenum Press, New York, 1975, 81.

15. **Ozato, K. and Wakamatsu, Y.,** Multi-step genetic regulation of oncogene expression in fish hereditary melanoma, *Differentiation,* 24, 181, 1983.

16. **Setlow, R. B., Woodhead, A. D., and Hart, R. W.,** Animal model: damage to DNA in the Amazon molly by physical and chemical agents, *Am. J. Pathol.,* 91, 213, 1978.

17. **Egami, N.,** Record of the number of eggs obtained from a single pair of *Oryzias latipes* kept in laboratory aquarium, *J. Fac. Sci. Univ. Tokyo, Sect. 4,* 8, 521, 1959.

18. **Matui, K.,** Illustration of the normal course of development in the fish, *Oryzias latipes* (in Japanese), *Jpn. J. Exp. Morphol.,* 5, 33, 1949.

19. **Gamo, H. and Terajima, I.,** The normal stages of embryonic development of the medaka, *Oryzias latipes* (in Japanese), *Jpn. J. Ichthyol.,* 10, 31, 1963.

20. **Parenti, L. R.,** Phylogenetic aspects of tooth and jaw structure of the medaka, *Oryzias latipes,* and other beloniform fishes, *J. Zool. (London),* 211, 561, 1987.

21. **Egami, N. and Etoh, H.,** Life span data for the small fish, *Oryzias latipes, Exp. Gerontol.,* 4, 127, 1969.

22. **Iriki, S.,** Studies on the chromosome of pisces, On the chromosome of *Aplocheilus latipes, Sci. Rep. Tokyo Bunrika Daigaku Sect. B,* 1, 127, 1932.

23. **Aida, T.,** On the inheritance of color in a freshwater fish, *Aplocheilus latipes* Temminck and Schlegel, with special reference to sex-linked inheritance, *Genetics,* 6, 554. 1921.

24. **Ojima, Y. and Hitotsumachi, S.,** The karyotype of the medaka, *Oryzias latipes, Chromosome Inform. Seu.,* 10, 15, 1969.

25. **Uwa, H. and Ojima, Y.,** Detailed and banding karyotype analyses of the medaka, *Oryzias latipes* in cultured cells, *Proc. Jpn. Acad.,* Ser. B, 57, 39, 1981.

26. **Toyama, K.,** Some examples of medelian characters, *Nihon Ikkushugaku Kaiho,* (in Japanese), 1, 1, 1916.

27. **Ishihara, M.,** Inheritance of body color in *Oryzias latipes* (abstr., in Japanese), *Zool. Mag.,* 28, 177, 194, 1916.

28. **Naruse, K., Ijiri, K., Shima, A., and Egami, N.,** The production of cloned fish in the medaka *(Oryzias latipes), J. Exp. Zool.,* 236, 335, 1985.

29. **Shima, A., Shimada, A., and Egami, N.,** Specific-locus mutation frequencies in the medaka, *Oryzias latipes.* I. Gamma-ray induced mutation frequency at the locus b, *J. Radiat. Res.,* 28, 91, 1987.

30. **Hyodo-Taguchi, Y.,** Establishment of inbred strains of the teleost, *Oryzias latipes, Dobutsugaku Zasshi,* (in Japanese, English abstr.,), 89, 283, 1980.

31. **Hyodo-Taguchi, Y. and Egami, N.,** Establishment of inbred strains of the medaka *Oryzias latipes* and the usefulness of the strains for biomedical research, *Zool. Sci.,* 2, 305, 1985.

32. **Shima, A., Shimada, A., Sakaizumi, M., and Egami, N.,** First listing of wild stocks of the medaka *Oryzias latipes* currently kept by the Zoological Institute, Faculty of Science, University of Tokyo, *J. Fac. Sci. Univ. Tokyo Sect. 4,* 16, 27, 1985.

33. **Egami, N. and Kukita, Y.,** X-ray effects on rejection of transplanted fins in the fish, *Oryzias latipes, Transplantation,* 8, 301, 1969.

34. **Kikuchi, S. and Egami, N.,** Effects of γ-irradiation on the rejection of transplanted scale melanophores in the teleost, *Oryzias latipes, Dev. Comp. Immunol.,* 7, 51, 1983.

35. **Kikuchi, S., Imaizumi, A., and Egami, N.,** The effect of the temperature on the immunologic memory against allograft transplantation in the fish *Oryzias latipes, J. Fac. Sci. Univ. Tokyo Sect. 4,* 15, 325, 1983.

36. **Kallman, K. D. and Gordon, M.,** Genetics of fin transplantation in *Xiphophorin* fishes, *Ann. N.Y. Acad. Sci.,* 73, 599, 1958.

37. **Kallman, K. D.,** An estimate of the number of histocompatibility loci in the teleost *Xiphophorus maculatus, Genetics,* 50, 583, 1964.

38. **Sakaizumi, M., Egami, N., and Moriwaki, K.,** Allozymic variation in wild populations of the fish, *Oryzias latipes, Proc. Jpn. Acad.,* Ser. B, 56, 448, 1980.
39. **Sakaizumi, M., Moriwaki, K., and Egami, N.,** Allozymic variation and regional differentiation in wild populations of the fish *Oryzias latipes, Copeia,* 1983, 311, 1983.
40. **Ohyama, A., Hyodo-Taguchi, Y., Sakaizumi, M., and Yamagami, K.,** Lactate dehydrogenase isozymes of the inbred and outbred individuals of the medaka, *Oryzias latipes, Zool. Sci.,* 3, 773, 1986.
41. **Egami, N.,** Studies on the variation of the number of the anal fin-rays in *Oryzias latipes.* I. Geographical variation in wild populations, *Jpn. J. Ichthyol.,* 3, 87, 1953.
42. **Hyodo-Taguchi, Y.,** A new mutant in the egg character of the medaka *Oryzias latipes, Dobutsugaku Zasshi,* (in Japanese), 88, 185, 1979.
43. **Tomita, H.,** Gene analysis in the medaka *(Oryzias latipes), Medaka,* 1, 7, 1982.
44. **Egami, N., Hyodo-Taguchi, Y., and Etoh, H.,** Recovery from radiation effects on organized cell population in fish at different temperatures, *Proc. Int. Conf. Radiat. Biol. Cancer,* 117, 1967.
45. **Ijiri, K.,** Ultraviolet irradiation of the gametes and embryos of the fish *Oryzias latipes:* effects on their development and the photoreactivation phenomenon, *J. Fac. Sci. Univ. Tokyo Sect. 4,* 14, 351, 1980.
46. **Hamaguchi, S.,** Differential radiosensitivity of germ cells according to their developmental stages in the teleost, *Oryzias latipes,* in *Radiation Effects on Aquatic Organisms,* Egami, N., Ed., University Park Press, Baltimore, 1980, 119.
47. **Hyodo-Taguchi, Y. and Yoshioka, H.,** Effects of X-irradiation on adult fish of the different inbred strains of the fish, *Oryzias latipes, J. Radiat. Res.,* 28, 17, 1987.
48. **Hyodo-Taguchi, Y.,** Effects of UV irradiation on embryonic development of different inbred strains of the *Oryzias latipes, J. Radiat. Res.,* 24, 221, 1983.
49. **Newcombe, H. B. and McGregor, J. F.,** Major congenital malformations from irradiations of sperm and eggs, *Mutat. Res.,* 4, 663, 1967.
50. **Schröder, J. H.,** X-ray induced mutations in the Poeciliid fish, *Lebistes reticulatus* Peters, *Mutat. Res.,* 7, 75, 1969.
51. **Chakrabarti, S., Streisinger, G., Singer, F., and Walker, C.,** Frequency of γ-ray induced specific locus and recessive lethal mutations in mature germ cells of the zebrafish, *Brachydanio rerio, Genetics,* 103, 109, 1983.
52. **Walker, C. and Streisinger, G.,** Induction of mutations by γ-rays in pregonial germ cells of zebrafish embryos, *Genetics,* 102, 125, 1983.
53. **Egami, N., Shimada, A., and Hama-Furukawa, A.,** Dominant lethal mutation rate after γ-irradiation of the fish, *Oryzias latipes, Mutat. Res.,* 107, 265, 1983.
54. **Shimada, A., Shima, A., and Egami, N.,** Production of a multiple recessive tester stock of the medaka *Oryzias latipes, Zool. Sci.,* 3, 1008, 1986.
55. **Shimada, A. and Egami, N.,** Dominant lethal mutations induced by MMS and mitomycin C in the fish *Oryzias latipes, Mutat. Res.,* 125, 221, 1984.
56. **Hyodo-Taguchi, Y. and Matsudaira, H.,** Induction of transplantable melanoma by treatment with *N*-methyl-*N'*-nitro-*N*-nitrosoguanidine in an inbred strain of the teleost *Oryzias latipes, J. Natl. Cancer Inst.,* 73, 1219, 1984.
57. **Ishikawa, T., Shimamine, T., and Takayama, S.,** Histologic and electron microscopy observation on diethylnitrosamine-induced hepatomas in small aquarium fish *(Oryzias latipes), J. Natl. Cancer Inst.,* 55, 909, 1975.
58. **Kyono, Y., Shima, A., and Egami, N.,** Changes in the labeling index and DNA content of liver cells during diethylnitrosamine-induced liver tumorigenesies in *Oryzias latipes, J. Natl. Cancer Inst.,* 63, 71, 1979.
59. **Aoki, K. and Matsudaira, H.,** Induction of hepatic tumors in a teleost *(Oryzias latipes)* after treatment with methylazoxymethate: brief communication, *J. Natl. Cancer Inst.,* 59, 1747, 1977.
60. **Aoki, K. and Matsudaira, H.,** Factors influencing methylazoxymethanol acetate initiation of liver tumors in *Oryzias latipes:* carcinogen dosage and time of exposure, *Natl. Cancer Inst. Monogr.,* 65, 345, 1984.
61. **Anders, F., Schartl, M., Barnekow, A., and Anders, A.,** *Xiphophorus* as an *in vivo* model for studies on normal and defective control of oncogenes, *Adv. Cancer Res.,* 42, 192, 1984.
62. **Hyodo-Taguchi, Y., and Matsudaira, H.,** Higher susceptibility to *N*-methyl-*N'*-nitro-*N*-nitrosoguanidine-induced tumorigenesis in an interstrain hybrid of the fish, *Oryzias latipes* (medaka), *Jpn. J. Cancer Res. (Gann),* 78, 487, 1987.
63. **Ozato, K., Kondoh, H., Inohara, H., Iwamatsu, T., Wakamatsu, Y., and Okada, T. S.,** Production of transgenic fish: introduction and expression of chicken δ-crystallin gene in medaka embryos, *Cell Differ.,* 19, 237, 1986.
64. **Hirano, O. and Matsui, I.,** Note on body-temperature of fish. I. Relation between body-temperature and water-temperature, *J. Shimonoseki Coll. Fish,* (in Japanese), 4, 79, 1955.
65. **Fry, F. E. J. and Hart, J. S.,** The relation of temperature to oxygen consumption in the goldfish, *Biol. Bull.,* 94, 66, 1948.

66. **Hyodo-Taguchi, Y.,** Rate of development of intestinal damage in the goldfish after X-irradiation and mucosal cell kinetics at different temperatures, in *Gastro-intestinal Radiation Injury,* Sullivan, M. F., Ed., Excerpta Medica, Amsterdam, 1968, 120.

67. **Etoh, H.,** Changes in incorporation of ^{3}H-thymidine into hematopoietic cells of goldfish following X-irradiation (in Japanese, English abstr., *Dobutsugaku Zasshi,* 77, 213, 1968.

68. **Etoh, H.,** Changes in ^{3}H-thymidine incorporation into hematopoietic cells of goldfish during recovery period from radiation injury, *Annot. Zool. Jpn.,* 42, 159, 1969.

69. **Kashiwagi, M. and Etoh, H.,** Histological and autoradiographical observations of the effects of X-irradiation on gill epithelium of the goldfish, *Carassius auratus, Annot. Zool. Jpn.,* 43, 93, 1970.

70. **Hyodo, Y.,** Development of intestinal damage after X-irradiation and ^{3}H-thymidine incorporation into intestinal epithelial cells of irradiated goldfish, *Carassius auratus,* at different temperatures, *Radiat. Res.,*26, 383, 1965.

71. **Hyodo-Taguchi, Y.,** Effect of X-irradiation on DNA synthesis and cell proliferation in the intestinal epithelial cells of goldfish at different temperatures with special reference to recovery process, *Radiat. Res.,* 41, 568, 1970.

72. **Suyama, I. and Etoh, H.,** A cell line derived from the fin of the goldfish *Carassius auratus, Dobutsugaku Zasshi,* 88, 321, 1979.

73. **Shima, A., Nikaido, O., Shinohara, S., and Egami, N.,** Continued *in vitro* growth of fibroblast-like cells (RBCF-1) derived from caudal fin of the fish, *Carassius auratus, Exp. Gerontol.,* 305, 1980.

74. **Matsumoto, J., Ishikawa, T., Prince Masahito, and Takayama, S.,** Permanent cell lines from erythrophoromas in goldfish *(Carassius auratus), J. Natl. Cancer Inst.,* 64, 879, 1980.

75. **Mitani, H., Etoh, H., and Egami, N.,** Resistance of a cultured fish cell line (CAF-MM1) to γ-irradiation, *Radiat. Res.,* 89, 334, 1982.

76. **Sperry, R. W.,** Mechanisms of neural maturation, in *Handbook of Experimental Psychology,* Stevens, S. S., Eds., John Wiley & Sons, New York, 1951, 236.

77. **Sperry, R. W.,** Functional regeneration in the optic system, in *Regeneration in the Central Nervous System,* Windle, W. F., Ed., Charles C Thomas, Springfield, IL, 1955, 66.

78. **Johns, P. S. and Easter, S. S.,** Growth of the adult goldfish eye — increase in retinal cell number, *J. Comp. Neurol.,* 176, 331, 1977.

79. **Meyer, R. L.,** Evidence from thymidine labeling for continuing growth of retina and tectum in juvenile goldfish, *Exp. Neurol.,* 59, 99, 1978.

80. **Raymond, P. A., Easter, S. S., Burnham, J. A., and Powers, M. K.,** Postembryonic growth of the optic tectum in goldfish, *J. Neurosci.,* 3, 1092, 1983.

81. **Heacock, A. M. and Agranoff, B. W.,** Enhancement labeling of a retinal protein during regeneration of the optic nerve in goldfish, *Proc. Natl. Acad. Sci. U.S.A.,* 73, 828, 1976.

82. **Schecter, N., Francis, A., Deutsch, D. G., and Gazzaniga, M. S.,** Recovery of tectal nicotinic-cholinergic receptor sites during optic nerve regeneration in goldfish, *Brain Res.,* 166, 57, 1979.

83. **Giulian, D., Buisseaux, H. D., and Cowburn, D.,** Biosynthesis and intraaxonal transport of proteins during neuronal regeneration, *J. Biol. Chem.,* 255, 6494, 1980.

84. **Benowitz, L. I. and Lewis, E. R.,** Increased transport of 44,000- to 49,000-dalton acidic protein during regeneration of the goldfish optic nerve: a two-dimensional gel analysis, *J. Neurosci.,* 3, 2153, 1983.

85. **Quitschke, W. and Schechter, N.,** Specific optic nerve protein during regeneration of the goldfish retinotectal pathway, *Brain Res.,* 258, 69, 1983.

86. **Perry, G. W., Burmeister, D. W., and Grafstein, B.,** Changes in protein content of goldfish optic nerve during degeneration and regeneration following nerve crush, *J. Neurochem.,* 44, 1142, 1985.

87. **Kosswig, C.,** Über Bastarde der Teleostier *Platypoecilus* und *Xiphophorus, Z. Indukt, Abstamm. Vererbungsl.,* 44, 253, 1927.

88. **Haüssler G.,** Über Melanombildungen bei Bastarden von *Xiphophorus maculatus* var robra, *Klin. Wochen.,* 7, 1561, 1928.

89. **Gordon, M.,** The genetics of a viviparous top-minnow *Platypoecilus;* the inheritance of two kinds of melanophores, *Genetics,* 12, 253, 1927.

90. **Anders, F., Schartl, M., and Scholl, E.,** Evaluation of environmental and hereditary factors in carcinogenesis, based on studies in *Xiphophorus,* in *Phyletic Approaches to Cancer,* Dawe, C. J. et al., Eds., Jpn Sci. Soc. Press, Tokyo, 1981, 289.

91. **Anders, A. and Anders, A.,** Etiology of cancer as studies in the platyfish-swordtail system, *Biochim. Biophys. Acta,* 516, 61, 1978.

92. **Wakamatsu, Y.,** Two types of melanomas in a new experimental system of platyfish-swordtail hybrids, *Dev. Growth Differ.,* 22, 731, 1980.

93. **Ozato, K. and Wakamatsu, Y.,** Age-specific incidence of hereditary melanomas in the *Xiphophorus* fish hybrids, *Carcinogenesis,* 2, 129, 1981.

94. **Woodhead, D. S.,** Influence of acute irradiation on induction of chromosome aberrations in cultured cells of the fish *Amica splendens,* in, *Biological and Environmental Effects of Low-Level Radiation,* Vol. 1, IAEA, Vienna, 1976, 67.

95. **Hoaftman, R. N.,** The induction of chromosome aberrations in *Nothobranchius rachowi* (Pisces: Cyprinodontidae) after treatment with ethyl methanesulphate or benz(a)pyren, *Mutat. Res.,* 91, 347, 1981.

96. **Kligerman, A. D., Bloom, S. E., and Howell, W. M.,** *Umbra limi:* a model for the study of chromosome aberrations in fishes, *Mutat. Res.,* 31, 225, 1975.

97. **Kligerman, A. D. and Bloom, S. E.,** Sister chromatid differentiation and exchanges in adult mudminnows *(Umbra limi)* after *in vivo* exposure to 5-bromodeoxyuridine, *Chromosoma,* 56, 101, 1976.

98. **Kligerman, A. D.,** Induction of sister chromatid exchanges in the central mudminnow following *in vivo* exposure to mutagenic agent, *Mutat. Res.,* 64, 205, 1979.

99. **Scott, W. B.,** Central mudminnow *Umbra limi* (Kirtland), in *Freshwater Fishes of Canada,* Scott, W. B. and Crossman, E. J., Eds., Fisheries Research Board of Canada, Bull. 184, Ottawa, 1973, 341.

100. **Moorhead, P. S., Nowell, P. C., Mellman, W. T., Battips, D. M., and Hungerford, D. A.,** Chromosome preparations of leukocyte cultured from human peripheral blood, *Exp. Cell Res.,* 20, 613, 1960.

101. **Suyama, I. and Etoh, H.,** X-ray-induced dicentric yields lymphocytes of the teleost, *Umbra limi, Mutat. Res.,* 107, 111, 1983.

102. **Sinnhuber, R. Q., Hendricks, J. D., Wales, J. H., and Putnam, G. G.,** Neoplasms in rainbow trout, a sensitive animal model for environmental carcinogenesis, *Ann. N.Y. Acad. Sci.,* 298, 389, 1977.

103. **Lee, D. J., Wales, J. H., Ayres, J. L., and Sinnhuber, R. Q.,** Synergism between cyclopropenoid fatty acids and chemical carcinogens in rainbow trout *(Salmo gairdneri), Cancer Res.,* 28, 2312, 1968.

104. **Halver, J. E.,** Aflatoxicosis and trout hepatoma, in *Aflatoxin, Scientific Background, Control and Implications,* Goldblatt, F. A., Ed., Academic Press, New York, 1969, 265.

105. **Hendricks, J. D., Wales, J. H., Sinnhuber, R. Q., Nixon, J. E., Loveland, P. M., and Scanlan, R. A.,** Rainbow trout *(Salmo gairdneri)* embryos: a sensitive animal model for experimental carcinogenesis, *Fed. Proc. Fed. Am. Soc. Exp. Biol.,* 39, 3222, 1980.

106. **Hendricks, J. D., Meyers, T. R., Casteel, J. L., Nixon, J. E., Loveland, P. M., and Bailey, G. S.,** Rainbow trout embryos: advantages and limitations for carcinogenesis research, *Natl. Cancer Inst. Monogr.,* 65, 275, 1984.

Chapter 13

FRESHWATER FISH IN RESEARCH ON RADIATION BIOLOGY

Akihiro Shima

TABLE OF CONTENTS

I. INTRODUCTION

The purpose of this chapter is to illustrate in detail how two or three species of freshwater laboratory fish have been used very successfully to solve problems in biomedical research that are of vital concern to humans.

Fish are the most numerous of the vertebrates, and the number of species is roughly estimated to be 20,000. This fact suggests the diversity of fish not only in number of species but also in other various aspects, such as shape and size, vertical distribution (from high mountain regions down to deep sea), living water temperature (from Antarctic freezing water up to hot springs in tropical areas), and genome size (DNA content per cell ranging from 30 to 3000% of that of humans). Turning to the phyletic point of view, however, fish have often been said to be ancestors to humans. This statement is supported by the accumulating evidence from anatomy, comparative physiology and biochemistry, embryology, genetics, and fossils. Furthermore, current knowledge from all these areas of the biological sciences strongly suggests the unity of the class Pisces.

The species of fish covered in this chapter will be limited to the following freshwater teleosts: medaka *(Oryzias latipes)*, goldfish *(Carassius auratus)*, and Amazon molly *(Poecilia formosa)*. The space allotted to each will be different because these fishes have different degrees of development as laboratory animals, which should be under not only genetic but also environmental control.

II. MEDAKA AND OTHER FRESHWATER FISH IN RESEARCH ON RADIATION BIOLOGY

A. DEVELOPMENT OF FRESHWATER FISH AS LABORATORY ANIMALS

The development of fish as laboratory animals is still behind that of rodents, particularly the mouse. Nevertheless, readers are encouraged to refer to the book written by Yamamoto[1] on the medaka. With regard to the Amazon molly, the articles by Hart et al.[2] and Woodhead et al.[3] give ample information. Unfortunately, no adequate reference on goldfish as laboratory animals written in languages other than Japanese can be found, except for a monograph recently written by Andrews[4] as a fishkeeper's guide.

B. RESPONSES TO IONIZING RADIATION
1. Somatic Aspect: Dose-Survival Relationship and $LD_{50/30}$ Value

As the most convenient and unmistakable end point for the quantitation of the acute effects of radiations on living organisms, individual death has widely been used. When animals are whole-body exposed to grading doses of ionizing radiations, such as X-rays, gamma rays, or neutrons, the probability of individual death increases with increasing doses. Daily counting of individual deaths and close observation of animals of each exposure group, including the control, are continued for, say, 30 d, after exposure. Plotting cumulative percentage mortality (on the ordinate) against days after irradiation (on the abscissa) gives a curve showing dose-mortality relationship. If percentage survival is used instead of percentage mortality, then a dose-survival curve is obtained. Such dose-survival relationship has been studied in most detail using primarily the mouse. At very high doses (greater than the order of 10,000 rad of low linear energy transfer (LET) radiations), almost all animals die within a matter of hours. In this case, the organ responsible for the individual death appears to be the central nervous system. Death at intermediate doses (approximately 500 to 1000 rad) results from damage to the gastrointestinal stem cells. At much lower doses on the order of a few hundred rad, generally about 50% of mice die within 30 d due primarily to the failure of the hematopoietic system.

To present and compare radiation sensitivity using individual death as an end point, the concept of the 50% lethal dose (LD_{50}) has commonly been used. Depending upon a specified length of time when 50% of animals die out, $LD_{50/4}$ (4 d after irradiation) corresponds to the dose responsible for so-called gastrointestinal death in mice, and $LD_{50/30}$ (30 d after irradiation) to hematopoietic death in mice. However, in the case of hematopoietic death in humans, the term $LD_{50/60}$ should be used because of the differences in cell turnover rate of the hematopoietic organ. Among these LD_{50} values, for the purpose of convenience and simplicity, $LD_{50/30}$ is most often used to compare radiation sensitivity among different species of animals.

The distribution of $LD_{50/30}$ values in mammals ranges from 155 rad (sheep) up to 1520 rad (desert mouse).[5] But even in the same species, strain differences (547 to 790 rad) can also be found, as reported by Kondo et al.[6] for 10 inbred mouse strains.

These acute radiation syndromes observed in rodents, above all in mice, were studied by the use of the fish, poikilothermic laboratory animals, from a view point of comparative radiation biology: Egami et al.[7] and Hyodo-Taguchi and Egami[8] examined relationships between X-ray doses and survival time of the medaka *Oryzias latipes* and the goldfish *Carassius auratus* at different temperatures.

First, they[7,8] examined histopathological changes in the fish *Oryzias latipes* irradiated with various doses of X-rays and tried to see if there was any relation between the severity of the injuries in particular organs (critical organs) and radiation doses delivered. The findings obtained from a series of experiments at 22 to 23°C were that (1) in fish exposed to extremely high doses (greater than approximately 100,000 rad), damage to the brain resulted in instantaneous death within a few minutes after irradiation or even during irradiation; (2) for dose ranges from 4000 to 64,000 rad the mean survival time of the fish was around 10 d (irrespective of radiation doses) and the major histopathological findings at death of the fish were injuries in the gastrointestinal tract; and (3) at doses between 2000 and 4000 rad the major cause of death appeared to be the failure in hematopoietic function. All of these histopathological findings in irradiated *Oryzias latipes* well coincided with those observed in irradiated mice. This is one part of the evidence for the validity of the fish as useful alternatives to mammalian models.

Second, and more quantitatively, Egami and Etoh[9] and Etoh and Egami[10] examined the survival time of the fish *Oryzias latipes* as a function of temperature after acute irradiation with X-rays. The control experiments[9] performed at 22 to 23°C indicated that when the logarithm of mean survival time in days (ordinate) was plotted against the logarithm of radiation dose (abscissa), a dose-survival curve consisting of 4 phases (rough classification) was obtained. The curve was almost flat (phase 1) from 250 rad (mean survival time, or T_s = 93.7 d) to 2000 rad (T_s = 73.8 d), followed by the steep decline (phase 2) toward 4000 rad, where T_s was 11.4 d. Again, a plateau (phase 3) was observed from 4000 rad (T_s = 11.4 d) up to 64,000 rad (T_s = 5.87 d), with intermediate T_s values of 8.34 d for 8000 rad, 8.84 d for 16,000 rad, and 8.06 d for 32,000 rad, respectively. Then the curve bent sharply downward, from 64,000 rad toward 128,000 rad (T_s = 0.33 d) (phase 4). Phase 2 corresponds to the period where the critical organ is the hematopoietic tissues. The $LD_{50/30}$ of the medaka *Oryzias latipes* is in this range of doses, say, 2000 to 3000 rad. In phase 3, almost the same T_s were obtained irrespective of the doses, and hence this range is called the ''dose-independent range''. The height of the plateau, that is, the mean survival time, is the reflection of the cell turnover rate in the gastrointestinal tract, which is the major organ responsible for the death of the fish exposed to this range of radiation doses; the slower the turnover rate of the intestinal epithelial cells, the longer the mean survival time.

It should be noted that in mice the mean survival period in this dose-independent phase is about 4 d. When medaka irradiated with X-ray doses within phase 3 (dose-independent range) were kept at 37°C, the temperature commonly considered to be the body temperature of mammals, the mean survival time of the medaka approximated 4 d.[10] The earlier occurrence of death of the irradiated medaka when transferred to a higher temperature (37°C) is basically due to the loss of cell renewal ability of the intestinal epithelium and to the shortening of the migration time of the epithelial cells along the villi. That is, the rate of cell loss from the intestinal villi becomes greater than the rate of cell renewal. These two rates are known to be temperature dependent.[8] In the nonirradiated control fish, the two rates are in a dynamic equilibrium, and the function of the intestinal epithelium is assured as epithelial cells, which are generated from the bottom of the villi (crypts), migrate along the villi and finally detach from the tip of the villi to the lumen of the intestinal tract. However, exposure of the fish to radiation reduces the proliferative ability of the intestinal stem cells in a dose-dependent fashion. Although the recovery process from radiation damage is temperature dependent, the manifestation of the inhibitory effect of radiation is essentially independent of temperature, whereas time for cells to migrate along the villi is temperature dependent but radiation independent. Thus, when irradiated fish are raised at a higher temperature, the migration of cells becomes faster while the supply of cells is reduced, resulting in the imbalance between loss and supply of the cell. This lack of balance between cell loss and renewal eventually resulted in the intestinal denudation and loss of sodium ion from the fish body[7] and in the death of the fish. This is one of the most successful demonstrations that some poikilothermal animals can be a good model for studying the mechanisms of radiation death in mammals.

Shima et al.[11] irradiated the Amazon molly *Poecilia formosa* with γ-rays of ^{60}Co or ^{137}Cs and examined the relationship between dose and mean survival time. The $LD_{50/30}$ values obtained were 6000 rad for clone I and 5000 rad for clone II which had been raised by Woodhead and Setlow in the Brookhaven National Laboratory in New York. The mean survival times (T_s) were as follows: 117 ± 81 d for 3000 rad, 77.5 ± 21.7 d for 4000 rad, 18.7 ± 1.9 d for 8000 rad, 8.64 ± 0.71 d for 16,000 rad, 8.53 ± 1.5 d for 32,000 rad, 2.3 ± 1.2 d for 64,000 rad, and 0.04 d (approximately 1 h) for 100,000 rad, respectively. The comparison of these data with those obtained from the medaka[9] revealed that there were some differences in T_s values in phases 1 and 2, particularly the latter, indicating that the susceptibility of hematopoietic tissues to ionizing radiation may be different between medaka and Amazon molly. However, T_s values in the dose-independent range, which correspond to gastrointestinal death, were almost the same between the 2 species of fish. This is an indication that studying cell population kinetics, particularly in the intestinal epithelium of the fish, can provide useful clues for elucidating mechanism(s) responsible for radiation-induced injury in mammals.

To summarize, the aforementioned results on fish strongly support the view that fish are useful alternatives to mammalian models in the study of acute somatic effects of ionizing radiation on living organisms.

2. Genetic Aspects: Specific-Locus Mutation Frequency

To estimate genetic risk from ionizing radiation in humans, accumulation of knowledge from animal experiments has to be continued; in particular, further search is needed on the dose-response relationship using mutations as an end point.[12] The major portions of such animal data have been obtained by the mouse specific-locus method.[13-16]

The specific-locus method enables the detection of radiation-induced forward mutations in the first filial generation after exposure of one of the parents to radiation. To use this method, the tester stock, which were recessive homozygous at the seven specific loci,[17] were developed in the mouse. Seven such genes were selected as markers that control the easily recognizable phenotypes, among which were coat color, coat pattern, and anomalies of ears and tail. Usually females of the recessive homozygous were used as testers. Male mice, which are homozygous dominant for all seven marker genes (wild type), are irradiated with graded doses of radiation at various dose-rates, and mated to the tester females at various times after irradiation. If no forward mutation occurred in the germ cells of the control as well as the irradiated males, all offspring obtained from crossing these males with the control tester females would show the dominant phenotypes. If the male germ cells carrying forward mutation(s) at any specific loci, however, give rise to progeny with the gametes of the control tester females, those progeny would manifest recessive phenotypes. Thus, we can detect occurrence of mutation(s) at the F_1 generation and mutants obtained can be subjected to further tests like allelism and recessive lethality. According to Searle,[14] the main advantages of the specific-locus method are (1) recessive mutations are detected in the first generation after exposure; (2) recessive lethal mutations at the specific loci are detected as well as recessive visible ones, although there will be some loss of those with dominant deleterious effects; and (3) progeny inheriting the newly arisen specific-locus mutations can normally be recognized with ease, so there is little likelihood of any subjective bias.

By the use of this seven-locus test stock of the mouse, a series of extensive experiments have been performed for over 30 years by several groups of researchers, especially by W. L. and L. B. Russell at the Oak Ridge National Laboratory in Tennessee. The total number of control offspring examined reached more than 800,000 and these enormous data have been called ''Historical control'' to which referred are almost all researches concerning mutation induction in the germ cells. A comprehensive though not so recent review[18] can give basically important information with regard to specific-locus mutations induced by radiation in mice.

According to Kosswig,[19] the first use of fish for studying classical genetics goes back to as early as 1914 when Gerschler crossed *Xiphophorus* and *Platypoecilus*. Since then, fishes have been used as laboratory animals in various fields of the biological sciences. In 1973, Schroeder[20] edited a book entitled *Genetics and Mutagenesis of Fish* (as Proceedings of the Ichthyological Symposium on Genetics and Mutagenesis held from October 13 through 15, 1972, in Neuherberg, Federal Republic of Germany). In this book (seven chapters in all) one chapter was devoted to mutagenesis, and three of four papers on this subject were concerned with radiation genetics of fish. Schroeder[20] reviewed the end points used so far in research on radiation mutagenesis in fish: mutations at dominant X- and Y-linked color genes, dominant lethal mutations and changes of viability, major congenital malformations, exchanges between the sex chromosomes, newly induced recessive mutations, formation of premelanomas, changes of behavioral traits, and induced gynogenesis. In addition to these end points, γ-ray-induced specific-locus and recessive lethal mutations were recently studied in the zebrafish *Brachydanio rerio*.[21] Despite these extensive studies on radiation mutagenesis using fish, no research has been done so far, to the best of the author's knowledge, on a detailed dose-response relationship using the specific-locus mutation method in fish whose results can be compared with mutation induction in mice. Therefore, we decided to tackle this problem by establishing the multiple recessive tester stock in the medaka and by accumulating data on mutation frequencies at specific loci.

a. Establishment of the Multiple Recessive Tester Stock Fish in the Medaka

In the medaka, the number of mutant genes identified was 37 in 1975 (Tomita[22]) but reached 75 at the time of writing this manuscript.[23] All of these mutant genes are concerned with either body color or morphogenetic deformities.

For developing the tester medaka for the specific-locus method, the choice of the marker genes to be used is crucial. We chose the markers primarily based on the following criteria: the mutant phenotypes should be easy to distinguish from their wild types; the mutant phenotypes should be stably expressed throughout the life of the fish; and the multiple recessive homozygosity should not reduce the fecundity and viability of the fish. The six loci we chose were as follows: *b* (colorless melanophores, autosomal, recessive), *Da* (double anal fins, autosomal, incomplete dominant), *gu* (reduced deposition of guanine in iridocytes, autosomal, recessive), *lf* (leucophore-free, autosomal, recessive), *pl* (no pectoral fins throughout the life, autosomal, recessive), and *r* (colorless xanthophores, sex linked, recessive).

Our first attempt to obtain multiple recessive medaka was begun in 1984 using the following five genes as markers: *b, Da, gu, pl,* and *r*. After trial and error, we recently established the multiple recessive tester stock in the medaka *Oryzias latipes (b/b Da/Da gu/gu pl/pl r/r).*[24] The expression of the phenotypic characteristics of each mutant gene, which was homozygously incorporated into one fish together with the remaining four genes also in homozygosity, was not altered by the presence of other four mutant genes. However, considerable reduction was found in the fecundity and viability of the tester fish. Furthermore, dorsoventral spinal curvatures were often observed, particularly in the females of the tester.

Therefore, we decided to change marker genes.[25] Since some preliminary experiments strongly indicated that the presence of the *b* and *r* loci in homozygosity induces the zygote lethality when the *gu* and *pl* loci become homozygous, we deleted the *r* locus. Furthermore, the recognition of the mutant phenotype expression of the *Da* and *pl* loci was rather difficult to do early during the embryonic development. The manifestation of the mutation phenotype of the *lf* locus, however, was evident as early as 3 d after fertilization, and the presence of this locus in homogyzosity affected neither viability nor fecundity of the *(b/b gu/gu)* fish. In addition, no morphogenetic deformities could be found. Thus, we established the new tester stock homozygous for the three genes *(b/b gu/gu lf/lf).*[25] We are now using the females of this tester stock for crossing with irradiated males of the homozygous dominant at the three loci concerned (wild type).

In connection with mutant genes of the medaka, it should be noted that although more than 70 mutant genes are known in this species, no information has yet been available with regard to linkage relationship, particularly relative to centromeres. Therefore, in our laboratory, attempts were made to map 5 visible mutant genes,[26] which were used as markers in the aforementioned tester stock, and 10 polymorphic enzyme loci[27] by the use of the diploid gynogenesis technique established in the medaka.[28] Recent success in gene-centromere mapping of these 15 loci obviously increased the usefulness of the medaka as genetically controlled laboratory animals.[26,27]

b. Induction of Mutations by Radiation at Specific Loci in the Medaka[29]

In parallel with development of the multiple recessive tester stock in the medaka *Oryzias latipes,* it appeared necessary to examine whether or not the medaka per se are valid for the detection of germ cell mutations at the specific locus.

The simplest tester fish available for specific-locus testing is the so-called orange-red variety medaka, which are homozygous recessive at the *b* locus. In the medaka *Oryzias latipes,* melanin formation in the melanophores is under the control of the three autosomal alleles *(B, B',* and *b).*[1,22] The allele *B* is dominant to the other two alleles, *B'* and *b*. The body color of the wild type *(B/B)* fish is brown due to the full melanin formation in the melanophores. The gene *B'* is recessive to *B* but dominant to *b*. The homozygotes at the *B'* locus *(B'/B'),* as well as the heterozygotes *(B'/b),* show the body color of variegated black. In contrast with *B* and *B'* alleles, the melanin formation is remarkably reduced in the *b/b* homozygotes, and hence the body color is orange-red due to the colorless or amelanotic melanophores. In our experience, the recognition of the presence of melanotic melanophores on the yolk sac of embryos can be accomplished as early as 1.5 d after fertilization. The ease of recognition of the mutation phenotype strongly suggests that the orange-red medaka can be a tester suitable for detecting forward mutation at the *B* locus *(B→b).*

Therefore, males of the wild type *(B/B)* were irradiated with 500 or 1000 R of γ-rays at a dose of 100 R/min using a 4000 Ci ^{137}Cs gamma source at the Research Center for Nuclear Science and Engineering, the University of Tokyo. Immediately after exposure, the irradiated as well as the nonirradiated males were mated with nonirradiated tester female fish which were recessive homozygous at the *b* locus. The fertilized eggs thus obtained were plated one by one each into one well of plastic microtiter test plates and observed under a stereoscopic microscope for the presence (no mutation) or absence (mutation) of melanotic melanophores on the yolk sac. All of these procedures were performed in the animal rearing room at 26 to 30°C with illumination control (14 h of light and 10 h of dark). Under these experimental conditions, the average time of hatching was about 7 d after fertilization.

Depending on the time of fertilization after exposure, the maturation stage of male germ cells at the time of exposure was judged as follows: sperm, 1 to 3 d after irradiation; spermatids, 4 to 10 d; and spermatogonia, 30 or more days, respectively.

Figure 1 shows the frequency of specific-locus mutations induced in spermatogenic cells of the medaka exposed to acute γ-irradiation, together with mouse data obtained by Russell.[13,16,17] In the case of the medaka, such fry were judged to be viable that could survive more than 4 d after hatching; hence the term ''viable mutants'' indicates those mutants that survived more than 4 d after hatching, and ''viable mutant frequencies'' can be defined as the number of viable mutants divided by the number of hatched fry that survived more than 4 d after hatching. Although our data are still incomplete, frequencies of viable mutations so far observed in sperm, spermatids, and spermatogonia of the medaka are fairly well in accordance with the comparable results on mice, which have a huge historical accumulation. Also coinciding with the findings of Russell[13,16,17] and other authors[14,15] was the result that only very few mutations could be found in spermatogonia in comparison with postspermatogonial germ cells, such as sperm and spermatids. A plausible explanation for low yields of mutations from spermatogonia is a high repair potential of spermatogonia and/or elimination of gross chromosomal damage (germinal selection) at mitosis and meiosis. These results justify the use of the medaka as a new experimental organism for mutagenicity studies.

Another advantage of using the medaka, the oviparous fish, for detecting germ cell mutation is the fact that the mutant phenotype can be confirmed as early as 1.5 d after fertilization due to the transparent egg membrane. Indeed, almost all the mutants whose mutation phenotype could be recognized very early during development died out before or very soon after hatching. The kind of mutations which are concealed by dominant lethals cannot be detected in viviparous animals such as mice, and therefore no suitable terminology exists for expressing them. Therefore, in an attempt to quantitate the gross frequency of specific-locus mutations, we proposed a new terminology, ''frequencies of total mutants'', which can be defined as the ratio of the number of total mutants to the number of fertilized eggs minus the number of early deaths. For male germ cells of all maturation stages, the frequencies of total mutations were one order of magnitude higher than the corresponding frequencies of viable mutations. These results indicate that the vast majority of the mutants are not viable and eventually die during the development process, resulting in so-called dominant lethals. The big difference in frequencies of total and viable mutations, in other words, the close relationship between frequencies and lethalities of total mutations, strongly indicates that the major damage induced in male germ cells by gamma rays would be chromosomal deletions rather than point mutations. Very recently, we succeeded in developing an experimental procedure which allows us to cultivate the somatic cells of each embryo. We believe that this success is useful not only for rescuing somatic cells of mutant embryos, almost all of which would die before hatching, but also for analyzing the mechanisms of dominant lethals induced by gamma radiation as well as by chemicals. Thus, all these results support the validity of the use of the Medaka for the quantitative study of genetic effects of radiations, particularly for the development of a line of research that could not have been accomplished readily with mice, the viviparous experimental animal.

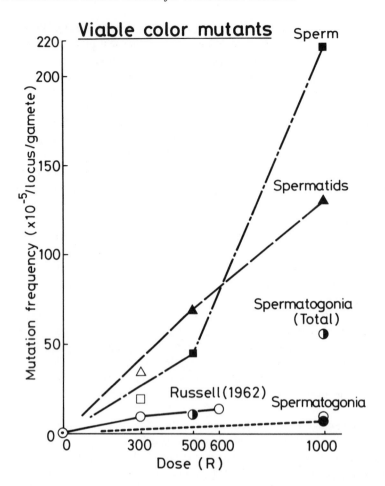

FIGURE 1.   Frequencies of viable mutants induced by acute irradiation with γ-rays in spermatogenic cells of the medaka *Oryzias latipes*. Male fish of the wild type *(B/B)* were exposed to [137]Cs γ-rays at a dose rate of 100 R/min, and mated with tester females which were recessive homozygous at the *b* locus. Mutation frequencies at the *b* locus in offspring which could survive more than 4 d after hatching were scored for sperm (■), spermatids (▲), and spermatogonia (●), respectively. For spermatogonia, frequencies of total mutants (◑) are also shown. Results on mice obtained by Russell[13,16,17] for sperm (□), spermatids (△), and spermatogonia (○) are shown for comparison.

III. SUMMARY

Some biomedical findings obtained using freshwater fishes were summarized to bring to the attention of researchers useful alternatives to mammalian models. Particular emphases were given to the use of fishes in research on radiation biology. Somatic as well as genetic effects of ionizing radiation, which are of vital concern to humans, were successfully studied using the Medaka *Oryzias latipes*. Furthermore, an introduction was made to the research that could not have been readily accomplished using viviparous laboratory animals like mice. The usefulness of cultured cells of fish for study of radiation effects and DNA repair will appear elsewhere.[30]

REFERENCES

1. **Yamamoto, T.**, *Medaka (Killifish) — Biology and Strains*, Keigaku, Tokyo, 1975.
2. **Hart, R. W., Livesey, H. R., and Setlow, R. B.**, Reproduction and potential research use of the gynogenetic teleost *Poecilia formosa, Drum Croaker*, 16, 1, 1976.
3. **Woodhead, A. D., Setlow, R. B., and Hart, R. W.**, Genetically uniform strains of fish as laboratory models for experimental studies of the effects of ionizing radiation, in *Methodology for Assessing Impacts of Radioactivity on Aquatic Ecosystems*, IAEA Technical Reports Ser. No. 190, Vienna, 1979, 317.
4. **Andrews, C.**, *Fancy Goldfishes*, Salamander Books, London, 1987.
5. **Hall, E. J.**, *Radiobiology for the Radiologist*, 2nd ed., Harper & Row, Philadelphia, 1978, 214.
6. **Kondo, K., Nagami, T., and Teramoto, S.**, Differences in hematopoietic death among inbred strains of mice, in *Comparative Cellular and Species Radiosensitivity*, Bond, V. P. and Sugahara, T., Eds., Igaku Shoin, Tokyo, 1969, 20.
7. **Egami, N., Hyodo-Taguchi, Y., and Etoh, H.**, Recovery from radiation effects on organized cell population in fish at different temperatures, in *Proc. Int. Conf. Radiation Biology and Cancer*, Sugahara, T., Ed., Radiation Society of Japan, Tokyo, 1967, 117.
8. **Hyodo-Taguchi, Y. and Egami, N.**, Development of intestinal radiation injury and recovery at different temperatures in fish, in *Comparative Cellular and Species Radiosensitivity*, Bond, V. P. and Sugahara, T., Eds., Igaku Shoin, Tokyo, 1969, 244.
9. **Egami, N. and Etoh, H.**, Dose-survival time relationship and protective action of reserpine against X-irradiation in the fish, *Oryzias latipes, Annot. Zool. Jpn.*, 35, 188, 1962.
10. **Etoh, H. and Egami, N.**, Effect of temperature on survival period of the fish, *Oryzias latipes*, following irradiation with different X-ray doses, *Annot. Zool. Jpn.*, 38, 114, 1965.
11. **Shima, A., Shimada, A., Woodhead, A. D., and Setlow, R. B.**, unpublished data, 1986.
12. United Nations Scientific Committee on the Effects of Atomic Radiation, *Ionizing Radiation: Sources and Biological Effects, 1982 Report to the General Assembly, with Annexes*, United Nations, New York, 1982.
13. **Russell, W. L.**, An augmenting effects of dose fractionation on radiation-induced mutation rate in mice, *Proc. Natl. Acad. Sci. U.S.A.*, 48, 1724, 1962.
14. **Searle, A. G.**, Mutation induction in mice, in *Advances in Radiation Biology*, Vol. 4, Lett, J. T., Adler, H. I., and Zelle, M., Eds., Academic Press, New York, 1974, 131.
15. **Ehling, U. H.**, Specific-locus mutations in mice, in *Chemical Mutagens*, Vol. 5, Hollaender, A. and de Serres, F. J., Eds., Plenum Press, New York, 1978, 233.
16. **Russell, W. L. and Kelly, E. M.**, Mutation frequencies in male mice and the estimation of genetic hazards of radiation in men, *Proc. Natl. Acad. Sci. U.S.A.*, 79, 542, 1982.
17. **Russell, W. L.**, X-ray-induced mutations in mice, *Cold Spring Harbor Symp. Quant. Biol.*, 16, 327, 1951.
18. **Green, E. L. and Roderick, T. H.**, Radiation genetics, in *Biology of the Laboratory Mouse*, 2nd ed., Green, E. L., Ed., Dover, New York, 1968, 165.
19. **Kosswig, C.**, The role of fish in research on genetics and evolution, in *Genetics and Mutagenesis of Fish*, Schroeder, J. H., Ed., Springer-Verlag, Berlin, 1973, 3.
20. **Schroeder, J. H.**, Teleosts as a tool in mutation research, in *Genetics and Mutagenesis of Fish*, Schroeder, J. H., Ed., Springer-Verlag, Berlin, 1973, 91.
21. **Chakrabarti, S., Streisinger, G., Singer, F., and Walker, C.**, Frequency of gamma-ray induced specific locus and recessive lethal mutations in mature germ cells of the zebrafish, *Brachydanio rerio, Genetics*, 103, 109, 1983.
22. **Tomita, H.**, Mutant genes in the Medaka, in *Medaka (Killifish) — Biology and Strains*, Yamamoto, T., Ed., Keigaku, Tokyo, 1975, 251.
23. **Tomita, H.**, personal communication, 1987.
24. **Shimada, A., Shima, A., and Egami, N.**, Establishment of a multiple recessive tester stock in the fish *Oryzias latipes, Zool. Sci.*, 5, 895, 1988.
25. **Shimada, A. and Shima, A.**, unpublished data, 1988.
26. **Naruse, K., Shimada, A., and Shima, A.**, Gene-centromere mapping for 5 visible mutant loci in multiple recessive tester stock of the medaka *(Oryzias latipes), Zool. Sci.*, 5, 489, 1988.
27. **Naruse, K. and Shima, A.**, *Biochemical Genetics*, in press, 1989.
28. **Naruse, K., Ijiri, K., Shima, A., and Egami, N.**, The production of cloned fish in the medaka *(Oryzias latipes), J. Exp. Zool.*, 236, 335, 1985.
29. **Shima, A. and Shimada, A.**, Induction of mutations in males of the fish *Oryzias latipes* at a specific locus after gamma irradiation, *Mutat. Res.*, 198, 93, 1988.
30. **Shima, A., Isa, K., Komura, J., Hayasaka, K., and Mitani, H.**, Somatic cell genetic study of DNA repair in cultured fish cells, in *Proc. 7th Int. Conf. Invertebrate and Fish Tissue Culture*, Kuroda, Y., Kurstak, E., and Maramorosch, K., Eds., Jpn. Sci. Soc. Press, Tokyo/Springer-Verlag, Berlin, 1988, 215.

Chapter 14

THE GOLDFISH VISUAL PATHWAY: INTERMEDIATE FILAMENT PROTEINS IN NERVE GROWTH AND DEVELOPMENT

Nisson Schechter, Paul S. Jones, Wolfgang Quitschke, and Paul Tesser

TABLE OF CONTENTS

I. INTRODUCTION: THE GOLDFISH VISUAL SYSTEM

The retinotectal pathways that are capable of nerve regeneration, as in the goldfish, have served as important model systems for both theoretical and empirical work in problems associated with the development of the nervous system. Regeneration in the visual system of the goldfish has been studied by anatomical,[1-7] behavioral,[5,8,9] and electrophysiological[10-13] techniques. These studies have had a significant influence on our view of neuronal development and have resulted in a central concept for neurobiology: the underlying basic development of the nervous system is genetically predetermined and is expressed as a series of regulated events which result in its observed morphology and unique function. The biochemical analysis of this system has lagged behind the other modes of neurobiological analysis; however, extensive progress has been made recently because of the newer techniques of molecular biology. The goldfish visual system is well suited for investigating biochemical events associated with axonal growth, synapse formation, and the regeneration process. While these studies are germane to understanding the development of the visual systems, they also strongly reflect on the general problem of the regulation of neurogenesis at the molecular level.

Historically, the study of neuronal development has progressed by two general approaches: the examination of events during embryonic development of neural pathways and the analysis of experimentally perturbed systems in adult animals. Each of these approaches has its limitations, especially with regard to biochemical strategies.

A major difficulty with the use of embryonic systems is that many other processes besides the one of interest are undergoing changes concurrently. The high turnover rates for some macromolecules, along with substantial net synthesis in several simultaneously developing neural systems, result in a low signal-to-background ratio for most neurochemical measurements. The probability of detecting specific molecular events such as the expression of critical proteins, even in a well-defined system, is extremely low.

In the adult animal, however, an excellent biochemical response can be elicited, but here the investigator may be dealing with induced degenerative effects or abnormal connections. Some of these induced connections are even behaviorally potent,[14] but in no case has "benefit" been demonstrated for the animal. Moreover, there are several experimental systems where sprouting or other kinds of experimentally induced abnormal axonal growth can be observed.[15-16] Although these systems produce a wealth of anatomical, electrophysiological, and chemical events to study, they fail to offer one critical factor — a function that is beneficial for the animal.

A system that inherently minimizes the disadvantages mentioned previously is the retinotectal visual pathway of goldfish, and therefore it is an appropriate choice for study of the putative molecular determinants of nerve growth and development. The use of the adult fish minimizes unrelated changes in brain growth and tissue development that are present in developing animals. While there is some growth of the eye and brain in the adult goldfish,[17-19] the growth rate is extremely slow compared with the changes in the retina and optic nerve which occur after nerve injury.[20] Also, growth of retinal cell populations correlates with body size[17-18] and can be further minimized simply by housing the fish in small tanks to impede body growth.

The retinotectal system has been well defined anatomically, with regard to the retina, the optic nerve, and the tectum (the tectum being the principal central target of the axons from the optic nerve). These tissues are all anatomically distinct and accessible in goldfish; surgical manipulations are simple, as are dissections for biochemical analysis. Since the axonal fibers from each eye project to the contralateral tectum,[7,21-22] one side of the pathway serves as a control for manipulations of the other. In addition, the eye can be injected with various radiolabeled precursors which will be incorporated into macromolecules and transported to the corresponding tectum and other brain targets by axonal flow mechanisms.[23] Furthermore, the retinotectal system is to a large extent a biochemically closed system, since the axons are uninterrupted in their innervation of the tectum.[24] Precursor pools are small and have rapid turnovers, thereby permitting radiolabeling to high specific activities. The goldfish system also provides excellent behavioral correlates since there are easily measured visual responses that are dependent on the integrity of the retinotectal pathway.

Metabolic activities associated with the outgrowth of optic axons, mainly in the goldfish visual system, have been recently and excellently reviewed.[84] Research in our laboratory[25-28] as well as others[29-34] shows the goldfish visual system to be amenable to biochemical analysis and that specific proteins can be identified, the expression of which correlates with nerve development. In some cases, proteins in nerves of lower vertebrates with a capacity for regeneration have homologous counterparts in mammalian nerves during neurogenesis. For example, the growth-associated proteins (GAPs) which are observed in axotomized peripheral nerves of rats and rabbits[35] are also under intense investigation in lower vertebrates such as fish[36-37] and toads.[38,39] The synthesis of GAPs is induced in neurons of these lower vertebrates after nerve crush.

This review will focus on research performed in our laboratory which concerns itself with the intermediate filaments of the goldfish visual pathway. These structural proteins may be a molecular link between certain aspects of nerve growth and regeneration and may have counterparts involved in mammalian neurogenesis.[40] From this very specific example, we hope to show how with the goldfish visual pathway one can successfully take, at the molecular level, a reductive approach to the analysis of gene expression during neurogenesis.

II. INTERMEDIATE FILAMENT PROTEINS: CLASSIFICATION AND STRUCTURAL CHARACTERISTICS

The major structural components of the cytoskeleton of most eukaryotic cells are the intermediate filaments (IF; 8 to 11 nm), the microtubules (20 to 25 nm), and the microfilaments (4 to 6 nm). The proteins of the IF complex, in contrast to the other two categories of cytoplasmic filaments, are far more diverse and display a distinctive cellular, tissue, and species heterogeneity.[41] Although IFs share common characteristics, such as insolubility in nonionic detergents and similar ultrastructural features, they have differences which have allowed them to be classified by immunological and biochemical criteria and according to their cellular disposition. The 5 categories which have been delineated mainly for higher vertebrates are: desmin (molecular weight, 53 kDa) in muscle cells; keratins (about 20 proteins whose molecular weight ranges between 40 and 70 kDa) in epithelial cells; vimentin (58 kDa) in mesenchymal cells; glial fibrillary acidic protein (GFAP; 50 kDa) in glial-type cells; and the neurofilament "triplet" proteins (68 kDa, 150 kDa, and 200 kDa) in neurons.[41] This classification system is not rigorous, and the list of addenda and exceptions continues to increase. For example, the classification of IF proteins may be somewhat further extended as it has recently been shown that clathrin light chain polypeptides, as well as some of the nuclear lamina proteins, share significant sequence homologies with IF proteins.[49-51] Vimentin has been co-localized with other IFs to specific cell types *in vivo* and in tissue culture.[42-45] It is noteworthy to this review that this IF protein is also transiently expressed within neurons in higher vertebrates during neurogenesis, and then, at later stages, the conventional neurofilament proteins are observed.[46,48]

Amino acid sequence analysis[52,53] and molecular cloning experiments[54-58] show that IF proteins share a common structural organization.[84] These proteins contain a chymotrypsin-resistant 40-kDa core region which has a conserved amino acid sequence. This core is segmented into conserved, α-helical-rich domains, which are separated by short runs of helical-disrupting amino acids. The helical domains are composed, qualitatively, of repeats of seven amino acids referred to as a heptad unit. The first and fourth amino acid are usually nonpolar residues. This results in an arrangement where the nonpolar residues reside along the helix in a manner which allows two helices to associate as a result of the interaction of these amino acids. This yields a coiled-coil structure with two helices coiling around a common axis. This two-chain coiled-coil arrangement gives rise to higher levels of filamentous structure although the mechanism of how this occurs is not certain. The core's carboxyl domain is the most highly conserved for the majority of IF proteins and contains an epitope which is recognized by a monoclonal antibody that reacts with all IFs (the AIF antibody).[59] The diversity associated with IF proteins is generated by two variable domains, one on either side of the core region. For the neurofilament "triplet" proteins, the increase in size for the higher-molecular-weight proteins is achieved by an extension of the carboxy terminal variable domain.[60] An additional level of diversity is imparted to IF proteins, particularly neurofilament proteins, by the preferential phosphorylation[61-63] of the variable regions. Phosphorylation occurs axoplasmically;[64] in addition, somatic phosphorylation has also been observed.[65] Furthermore, it has been suggested that the phosphorylation of IF proteins may be a regulatory event for regional axonal specialization in the CNS.[64] In summary, IF proteins display structural homology as well as diversity. It has been suggested that the diverse variable domains may be involved in special functions associated with the differentiated state of specific cell types, while the conserved regions meet functional requirements common to all cell types where IF proteins are expressed.[40,66]

III. GOLDFISH INTERMEDIATE FILAMENT PROTEINS: GENERAL CHARACTERISTICS

As a group, the neuronal tissues of lower vertebrates display a greater complexity in the composition and the distribution of IF proteins than that observed in higher vertebrates.[40] In the goldfish visual pathway, for example, the neuro- and glial-filament proteins do not match those found in mammalian neural tissue.[67] Although a neurofilament "triplet" is observed, there is a significant downward shift in molecular weight. In a one-dimensional gel of cytoskeletal proteins from goldfish optic nerve, bands are observed at 58, 80, and 145 kDa.[67] The predominant IF proteins in the goldfish pathway are the 58-kDa proteins, which we have designated as the ON proteins (optic nerve proteins). They can be separated into at least four major components by two-dimensional gel electrophoresis with isoelectric points of approximately 5.2 to 5.4 (Figure 1A).

These proteins have all the characteristics of IF proteins:[67] they are insoluble in nonionic detergents, are transported within the slow phase of axonal transport, and can be reconstituted into 10-nm filaments.[67] In addition, they have a chymotrypsin-resistant 40-kDa core and are phosphorylated in the variable extension domains.[66] They also react with the AIF antibody (Figure 1B). Although the ON proteins share this antigenicity with all IF proteins, polyclonal antibodies have been generated which specifically recognize the ON proteins and can distinguish between proteins ON_1/ON_2 and ON_3/ON_4. Furthermore, antibody studies and peptide fragment analysis indicates that the ON proteins are different from mammalian vimentin and GFA.[66]

IV. SITES OF SYNTHESIS AND LOCALIZATION OF PROTEINS ON_1 TO ON_4

Several lines of evidence show that proteins ON_1 and ON_2 are of neuronal origin, whereas proteins ON_3 and ON_4 are of nonneuronal origin. A loss of proteins ON_1 and ON_2 is observed in the optic nerve after optic nerve crush or enucleation.[26] *In vivo* experiments in which ^{35}S-methionine is injected into the eye preferentially label ON_1 and ON_2 in the optic nerve,

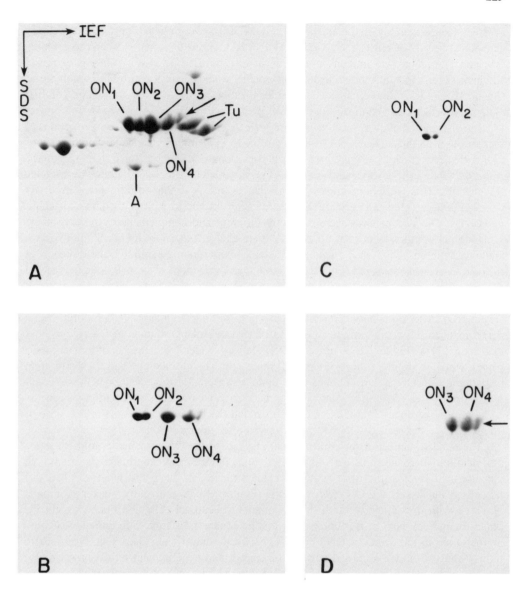

FIGURE 1. Immunoblots of goldfish optic nerve proteins demonstrating the specificities of anti-intermediate filament antibodies. (A) Coomassie blue-stained gel of optic nerve proteins; Tu, tubulins; Ac, actin; (B) An immunoblot probed with the AIF monoclonal antibody; (C) An immunoblot probed with anti-ON_1/ON_2; (D) An immunoblot probed with anti-ON_3/ON_4. A component of lower concentration at the acidic side of ON_4 is indicated by an arrow. (From Jones, P. S., Tesser, P., Keyser, K. T., Quitschke, W., Samadi, R., Karten, H. J., and Schechter, N., *J. Neurochem.*, 47, 1226, 1986. With permission.)

while proteins ON_3 and ON_4 remain unlabeled under these conditions. *Ex vivo* experiments with retinal tissue show incorporation of ^{35}S-methionine into ON_1 and ON_2 and a minimal incorporation into ON_3. In contrast, when optic nerve tissue alone is incubated with the radiolabeled amino acid, proteins ON_3 and ON_4 are labeled, whereas ON_1 and ON_2 are not. One can conclude from these biochemical experiments that proteins ON_1 and ON_2 are synthesized in the retina, presumably the retinal ganglion cells, and transported into the optic nerve.[27] Proteins ON_3 and ON_4 are of nonneuronal origin and are synthesized in glial-type cells associated with the optic nerve.

Histological studies give the biochemical results a visual reality, and provide additional evidence for the neuronal and nonneuronal origins of the two groups of proteins. Polyclonal antibodies were prepared to the 58-kDa goldfish IF proteins ON_1 and ON_2 and to proteins ON_3 and ON_4. The two antibody preparations were purified by affinity chromatography, and their specificities to the two groups of proteins were demonstrated by immunoblotting (Figure 1C and D).[68] Immunohistochemical localization studies were performed on goldfish optic nerve and retina (Figures 2 and 3).

The distribution of the respective proteins were determined in optic nerve cross sections.[68] Incubations with the anti-ON_1/ON_2 antibody resulted in a granular, punctate staining pattern over optic nerve fascicles (Figure 2A). This is consistent with an intraaxonal staining pattern. Anti-ON_3/ON_4 reactivity appeared as a fibrous meshwork within and surrounding the fascicles (Figure 2B). This is a nonneuronal staining pattern and is similar to the pattern observed when optic nerve glial cells in higher vertebrates are visualized with glial-filament-specific probes.[69,70] The two patterns were distinctly different and complementary to each other. Longitudinal sections of normal optic nerve were also incubated with both antibody preparations. With anti-ON_1/ON_2, a pattern of axonal fibers running parallel to the direction of the optic nerve was observed (Figure 2C). Longitudinal sections of normal optic nerve with anti-ON_3/ON_4 revealed a pattern similar to that observed in optic nerve cross sections in that the fibers are seen extending in all directions in and out of the plane of the section (Figure 2D). After optic nerve crush or enucleation, staining with anti-ON_1/ON_2 of the optic nerve distal to the crush zone resulted in a pattern of isolated clumps of reactivity rather than long fibers (Figure 2E). Anti-ON_3/ON_4 staining under the same conditions shows a pattern which is similar to normal nerves (Figure 2F). The morphology and the distribution of proteins ON_3/ON_4 seem unaffected by optic nerve crush or enucleation. Gliosis is apparently not observed here, in contrast to higher vertebrates where significant gliosis occurs with optic nerve injury.[71-73] These data also parallel the biochemical results in that little or no increase in synthesis of ON_3/ON_4 is observed after injury.[26,33]

The results obtained from a histological analysis of the ON proteins in goldfish retina also parallel the conclusions derived from previous biochemical experiments.[74] The retinal ganglion cell layer is the most heavily labeled structure in the goldfish retina when this tissue is reacted with anti-ON_1/ON_2 antibody (Figure 3A). Significantly weaker staining is observed with this antibody in all layers containing neuronal cells and fibers, except the photoreceptor cell bodies. It is noteworthy that no apparent differences in immunoreactivity are observed in the levels of ON_1 and ON_2 between control and experimental (postcrush) retina. The two-dimensional gel patterns of control and experimental retinal proteins also reveal no apparent differences in the concentrations of ON_1 and ON_2. These results suggest that though the levels of synthesis of these proteins increase, there is no accumulation of these proteins in the ganglion cell bodies. Alternatively, the antibodies might not be sensitive enough to distinguish increases in the perikaryal concentrations of these proteins resulting from optic nerve crush.

The predominant retinal labeling with the anti-ON_3/ON_4 antibody is observed in cells residing in the inner nuclear layer which are morphologically similar to the horizontal cells, including axon terminals of the cone horizontal cells (Figure 3B). These results are interesting in light of the results of Drager[44] and Shaw and Weber.[75] They have shown that antibodies that react with vimentin in mammalian optic nerve astrocytes also label horizontal cells in certain adult mammals. The similar cellular localizations of ON_3/ON_4 and vimentin in goldfish and mammals, respectively, may suggest some shared functions between these two proteins in lower and higher vertebrates.

V. RESPONSE OF GOLDFISH "ON" PROTEINS TO NERVE INJURY

The ON proteins were originally of interest because of the observed changes in their

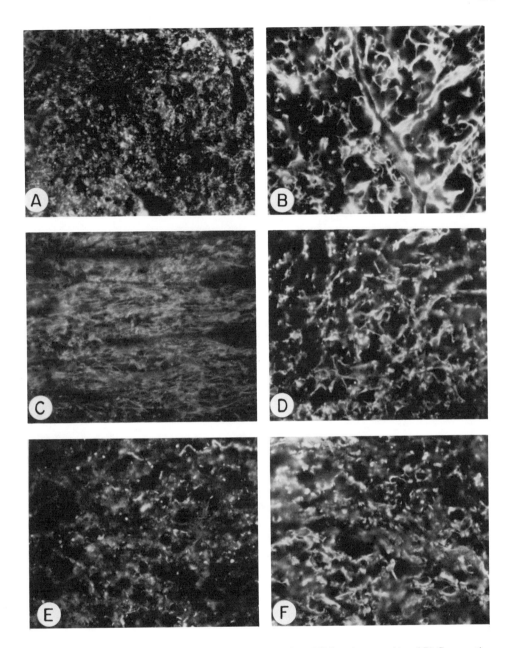

FIGURE 2. Immunohistochemical localization of ON proteins in goldfish optic nerve. (A and B) Cross-sections of normal optic nerve incubated with anti-ON_1/ON_2 and anti-ON_3/ON_4, respectively; (C and D) Longitudinal sections of normal optic nerve treated as in A and B; (E and F) Longitudinal sections of optic nerve 9 d after enucleation, treated as in A and B. Antibody reactivity was visualized by indirect immunofluorescence. (From Jones, P. S., Tesser, P., Keyser, K. T., Quitschke, W., Samadi, R., Karten, H. J., and Schechter, N., *J. Neurochem.*, 47, 1226, 1986. With permission.)

concentrations induced by nerve injury. There are some important questions to be considered: How does the synthesis of the ON proteins respond to optic nerve crush, and in what time frame does the synthesis and accumulation occur within the optic nerve after crush? This was investigated in experiments where the levels of proteins ON_1 to ON_4 were measured within the optic nerve in response to nerve disconnection.[26] In addition, *ex vivo* experiments determined the levels of ^{35}S-methionine which were incorporated into each of the ON

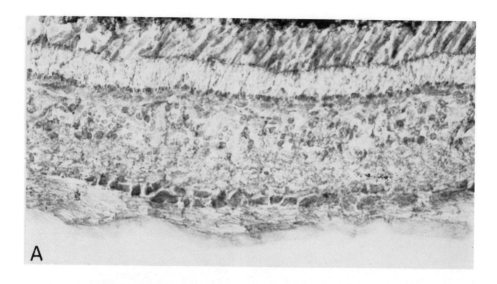

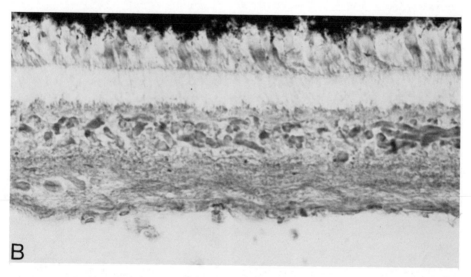

FIGURE 3.   Immunohistochemical localization of the ON proteins in goldfish retina. The pigment layer (the posterior portion of the retina) is at the top, and the retinal ganglion cell and optic fiber layers are at the bottom of the photographs. (A) Anti-ON_1/ON_2, shown 20 d after optic nerve crush, when ON_1 and ON_2 synthesis is at its maximum. There is significant reactivity in the retinal ganglion cell layer, thus confirming the neuronal origin of these proteins. Reactivity is also seen in the optic fiber layer and the inner plexiform layer and occasionally in amacrine cells. (B) Anti-ON_3/ON_4. Incubations of both normal and regenerating retina with anti-ON_3/ON_4 results in the staining of the optic fiber layer and the inner plexiform layer, as well as cells in the inner nuclear layer. These cells appear to be the horizontal cells. (From Jones, P. J. and Schechter, N., *J. Comp. Neurol.*, 266, 112, 1987. With permission.)

proteins. Finally, translation products of retinal RNA were analyzed at various time points after optic nerve crush.

The levels of proteins ON_1 to ON_4 within the optic nerve were investigated at various time points after optic nerve crush or enucleation (Figure 4). The proteins were separated by two-dimensional gel electrophoresis, and a quantitative densitometric analysis[76] was performed on each Coomassie blue-stained spot. The data were expressed as a ratio between experimental and control levels of each protein normalized to the identical amount of protein per gel. Proteins ON_3 and ON_4 remain largely unaffected by both optic nerve crush and eye

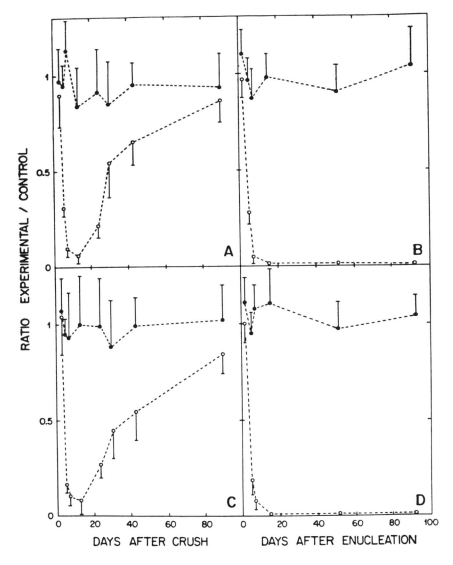

FIGURE 4. The relative concentration of proteins ON_1, ON_2, ON_3, and ON_4 in the optic nerve as a function of time after disconnection expressed as a mean ($\pm$ SD) ratio of experimental to control. (A) ON_4 ($\bullet$-$\bullet$) and ON_2 (o-o) after optic nerve crush; (B) ON_4 ($\bullet$-$\bullet$) and ON_2 (o-o) after enucleation; (C) ON_3 ($\bullet$-$\bullet$) and ON_1 (o-o) after crush; (D) ON_3 ($\bullet$-$\bullet$) and ON_1 (o-o) after enucleation. (From Quitschke, W. and Schechter, N., *Brain Res.*, 258, 69, 1983. With permission.)

removal. Proteins ON_1 and ON_2, however, decrease rapidly in the distal portion of the nerve after either optic nerve crush or eye removal and approach undetectable levels 7 d after disconnection. After eye removal, proteins ON_1 and ON_2 remain undetectable, whereas after optic nerve crush, a gradual restoration of their levels is observed. The process is nearly complete 80 to 90 d after crush when ON_1 and ON_2 approach control levels. If the nerve is analyzed proximal to the crush zone, the levels of ON_1 and ON_2 remain unaffected. ON_1 and ON_2 are only affected in the portion of the nerve distal to the crush. More recently, this observation was also confirmed histologically.[68]

The effects of optic nerve crush on protein synthesis in the retina and optic nerve were investigated *ex vivo*. Tissues were incubated in the presence of ^{35}S-methionine and the amount of incorporation of radioactivity was measured quantitatively. Total protein synthesis in the

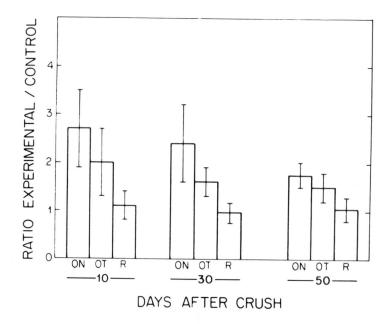

FIGURE 5.   Experiment-to-control ($\pm$ SD) ratio of specific activity (dpm/mg protein) after an *in vitro* incubation for 3 h in the presence of ^{35}S-methionine at 10, 30, and 50 d after optic nerve crush. ON, optic nerve; OT, optic tectum; R, retina. Each bar represents 5 experiments with 5 fish per experiment. The ranges of levels of incorporation were: ON, $5 \times 10^5 - 2 \times 10^6$; OT, $3 \times 10^5 - 1 \times 10^6$; R, $1 \times 10^6 - 4 \times 10^6$ dpm/mg protein. (From Quitschke, W. and Schechter, N., *J. Neurochem.*, 41, 1137, 1983. With permission.)

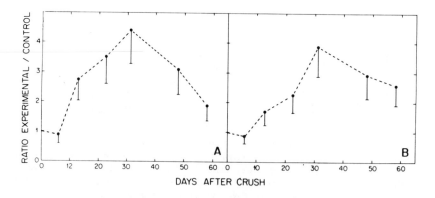

FIGURE 6.   Experimental-to-control ratio of specific activity of proteins ON$_1$ (A) and ON$_2$ (B) determined from gels containing retinal proteins at various times after optic nerve crush. Each point represents the mean ($\pm$ SD) of four to six experiments with five fish per experiment. (From Quitschke, W. and Schechter, N., *J. Neurochem.*, 41, 1137, 1983. With permission.)

retina remains largely constant following optic nerve crush (Figure 5); however, proteins ON$_1$ and ON$_2$ exhibit a significant increase in their synthesis in this tissue with a maximum fourfold rate observed at approximately 30 d after crush (Figure 6). Since ON$_1$ and ON$_2$ synthesis increases significantly in the retina, whereas total protein synthesis does not, it appears that very specific changes in protein expression occur in this tissue. Furthermore, one can conclude that such changes occur in specific cell types in the retina, such as the retinal ganglion cells for ON$_1$ and ON$_2$; because such cell types make up only a fraction of the total retina, one does not detect an increase in total protein synthesis. Within the optic

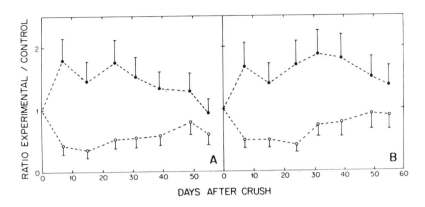

FIGURE 7. Experimental-to-control ratio of specific activity of proteins ON_3 (A) and ON_4 (B) determined from gels containing optic nerve proteins at various times after disconnection. (●), Labeling ratio of ON_3 (A) and ON_4 (B) between experimental and control preparations normalized with regard to protein content per spot: specific activity (ON_3 or ON_4, experimental)/specific activity (ON_3 or ON_4, control). Because the incorporation of radioactivity into total protein also increases as a result of optic nerve crush (Figure 5), an additional ratio (○) is determined that is normalized with regard to total specific activity (dpm/mg) in experimental and control preparations: specific activity (total protein, control)/specific activity (total protein, experimental) × specific activity (ON_3 or ON_4, experimental)/specific activity (ON_3 or ON_4, control). Each point represents the mean (±SD) of four to six experiments with five fish per experiment. (From Quitschke, W. and Schechter, N., *J. Neurochem.*, 41, 1137, 1983. With permission.)

nerve, the amount of specific labeling into ON_3 and ON_4 also increases after nerve injury (Figure 7). A stable incorporation into ON_3 and ON_4 at levels close to 2-fold over control tissue occurs with a gradual decline to normal values by 50 d postcrush. However, within the optic nerve, although the synthesis of proteins ON_3 and ON_4 increases, this must be compared with an increase in total protein synthesis. Figure 7 indicates that while the rate of ON_3 and ON_4 synthesis does increase, this increase is actually a fraction of the increase of the rate of total protein synthesis. Therefore, in this tissue, many changes in protein expression probably occur, with ON_3 and ON_4 representing one such increase in expression within glial cells.

Experiments which examined both qualitative and quantitative changes of *in vitro* translation products were carried out as a way of analyzing the expression of corresponding mRNAs.[77] Figure 8 shows translation products programmed by total RNA and separated by two-dimensional electrophoresis. Approximately 300 spots were visible on the original autoradiogram, and proteins of greater than 100 kDa were synthesized. The boxed region indicates the area of gel that is presented in Figure 9A to D. Significant increases in the synthesis levels of ON_1 and ON_2 at 20 and 35 d after optic nerve crush were observed (Figure 9B and C). Before crush, protein ON_1 is synthesized *in vitro* at low levels; ON_2 is barely detectable. By 20 d, however, they are both synthesized at high levels. After 35 d, both of the proteins return to their precrush levels.

The spots were initially identified as ON_1 and ON_2 based on their comigration with unlabeled optic nerve ON_1 and ON_2 to the same molecular weight and isoelectric point. The anti-ON_1/ON_2 antisera and the general AIF monoclonal antibody were used to confirm the identification of ON_1 and ON_2 synthesized *in vitro*. When anti-ON_1/ON_2 is reacted with [^{35}S]-labeled proteins programmed by RNA from 20-d postcrush retinas, only proteins ON_1 and ON_2 are precipitated (Figure 10A and B). Figure 10A shows the precipitated proteins separated by two-dimensional gel electrophoresis and visualized by autoradiography. The same gel region that is presented in Figure 9A to D is presented here. Figure 10B shows

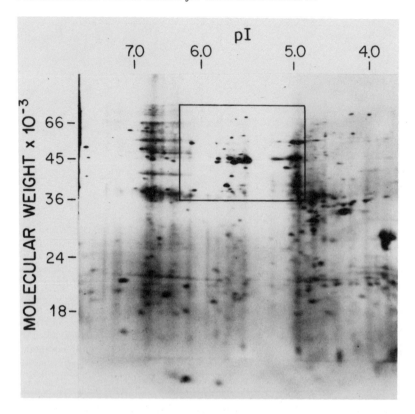

FIGURE 8.   Autoradiogram of two-dimensional gel of total *in vitro* translation products programmed by total RNA from normal goldfish retina; 50 μg of total cellular RNA was isolated from goldfish retinas and translated into proteins in a rabbit reticulocyte lysate system. Approximately 6×10^5 cpm of ^{35}S-methionine-labeled translation products were separated by two-dimensional gel electrophoresis. Autoradiography was carried out for 3 d. The boxed region corresponds to the autoradiograms shown in Figures 9 and 10. (From Tesser, P., Jones, P. S., and Schechter, N., *J. Neurochem.*, 47, 1235, 1986. With permission.)

the same region of an autoradiogram of the proteins precipitated by the AIF monoclonal antibody. Again, ON_1 and ON_2 are precipitated. However, the major protein, which is also indicated in Figure 9A and is present in constant amounts during the regeneration process, is seen in Figure 9A to D. Previous studies on optic nerve proteins did not identify a 38 kDa protein as an intermediate filament protein. This protein may be a retinal intermediate filament protein that is not transported into the optic nerve, or it may turn over rapidly.

The increase in ON_1 and ON_2 synthesis was quantified by scanning with a densitometer[76] autoradiograms of two-dimensional gels that contained identical amounts of radioactivity and that were exposed for the same period of time. The results show that ON_1 synthesis increases at least 10-fold over precrush levels, and ON_2 increases 30-fold over precrush levels (Figure 11). These data demonstrate that the intermediate filament proteins ON_1 and ON_2, which are part of the cytoskeleton of the goldfish optic nerve, are encoded by mRNA synthesized in the retina. The level of mRNA for these proteins increases significantly after optic nerve crush as measured by the *in vitro* translation of retinal RNA characterizations of the products.

It is noteworthy that a slight increase in the synthesis of ON_1 and ON_2 is also observed for RNA from the control eye at 20 and 35 d after crush, although not nearly as great as that observed for the crushed right eye. It has been suggested that a systemic factor affects retinal ganglion cells of both eyes during the regeneration process.[78] Similarly, it may be possible that a tectal factor is retrogradely transported, and this factor may stimulate specific gene expression leading to an increased synthesis of proteins such as ON_1 and ON_2. Although it may be directed toward the regenerating nerve, some of this factor could be diverted and induce synthesis in the noninjured eye.

One observes a rapid regrowth of axons during the regeneration process, but functional vision does not return until approximately 60 d after optic nerve crush.[20,79] The dramatic increase in ON_1 and ON_2 synthesis coincides with the point when axons have reached the tectum, perhaps due to the retrograde transport of tectal factors to retinal ganglion cells. Although the ON proteins are synthesized at high levels early in the regeneration process, they are transported in the slow phase of transport[67] and may reach the tectum between 25 to 40 d after crush, which coincides with the period of synaptogenesis. We speculate that ON_1 and ON_2 have structural characteristics which contribute to the process of synapse maturation and stabilization and are required at the later stages of regeneration.

VI. CONCLUSION

There are several points to consider with regard to the role of intermediate filament proteins in nerve development and regeneration. To date, a precise biological function has not been attributed to this class of proteins, although much structural information is now available. A comparative examination of these proteins between lower and higher vertebrates offers an opportunity to gain some insight to their function. For example, the expression of intermediate filament changes as a function of embryonic development and differentiation. It has been shown that during embryogenesis of the avian and mammalian CNS, vimentin is transiently expressed before or during neuronal differentiation followed by a switch in synthesis to the more conventional neurofilament triplet proteins.[46-48] A possible parallel can be drawn between embryogenesis and the ongoing development of the goldfish retinotectal pathway. As a result of the continued increase in the number of retinal ganglion cells and the asymmetric growth of the optic tectum throughout the life of the adult goldfish, the terminals of the optic axons continually shift positions to maintain the retinotopic projections.[80-82] A similar mechanism may be operating in the frog visual system.[83] It is possible that the retinal ganglion cells from these more plastic systems are organized in such a way that they are in a quasi-embryonic state of constant differentiation. The regenerative capacity may be a consequence of this state. However, this state differs from mammalian embryonic development since the neuron in the lower vertebrate must simultaneously grow and be functionally active. Consequently, any structural element in the lower vertebrate must meet both the plasticity property, which may be fulfilled by vimentin in the embryonic mammalian neuron, and a functional property related to the distinct needs of the neuron, which is fulfilled by the neurofilament proteins in adult mammals. Furthermore, it is also likely that the plasticity and regenerative properties are not only exclusively mediated by the neurons but also by the associated nonneuronal cells. The intermediate filament proteins of lower vertebrates may be associated with both the functional specializations and the plasticity requirements of these neurons. The structure of certain intermediate filament proteins may be well suited to participate in such multiple or dual functions within the cell. Present experiments using molecular cloning techniques are directed toward a comparative structural analysis between the goldfish ON proteins and the intermediate filament proteins of higher vertebrate nerves.

ACKNOWLEDGMENTS

This research was supported by a grant from the National Institutes of Health (EY 05212) to Nisson Schechter. The secretarial assistance of Dorothy Caselles is greatly appreciated.

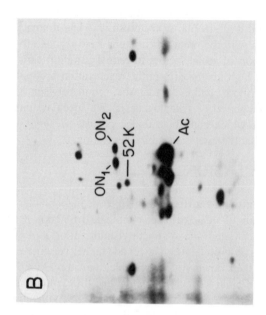

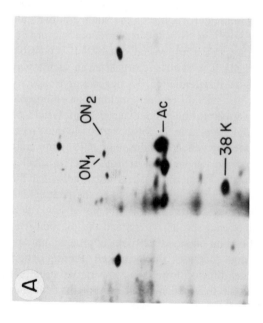

FIGURE 9.   Autoradiograms of the same regions of two-dimensional gels of translation products programmed by RNA isolated from retinas at different time points before and after optic nerve crush. (A) Precrush; (B) At 10 d after crush; (C) At 32 d after crush; (D) At 85 d after crush. The total RNA translated was 50 μg, and 9×10^5 cpm were electrophoresced. All panels are regions of a 48-h exposure. Arrows indicate the proteins designated as intermediate filament proteins ON_1 and ON_2, actin (Ac), a 52 kDa protein not observed on precrush gels, and a 38 kDa protein. (From Tesser, P., Jones, P. S., and Schechter, N., *J. Neurochem.*, 47, 1235, 1986. With permission.)

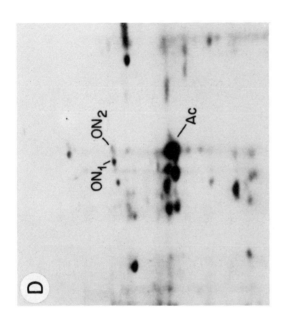

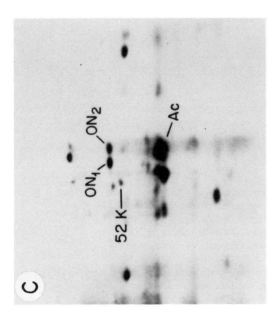

FIGURE 9 continued.

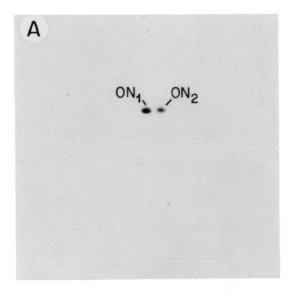

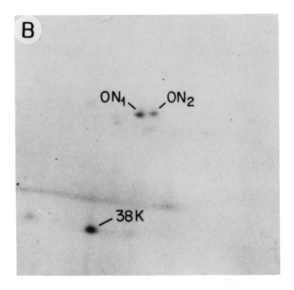

FIGURE 10. Immunoprecipitations of *in vitro* translation products. *In vitro* translation reactions were directed by RNA from 20-d postcrush retinas and the reaction products incubated with either anti-ON$_1$/ON$_2$ antibodies or the AIF monoclonal antibody. Immune complexes were precipitated with Protein A-Sepharose in the case of anti-ON$_1$/ON$_2$ antibodies or with rabbit anti-mouse serum in the case of the AIF monoclonal antibody. (A) Immunoprecipitation employing anti-ON$_1$/ON$_2$ antibodies; (B) Immunoprecipitation employing the AIF monoclonal antibody. (From Tesser, P., Jones, P. S., and Schechter, N., *J. Neurochem.*, 47, 1235, 1986. With permission.)

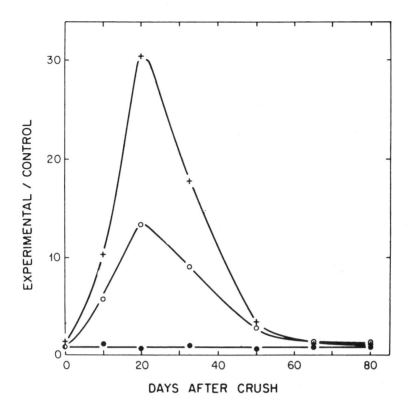

FIGURE 11. Relative changes in the levels of translation products at various time points after optic nerve crush. The ratios of total *in vitro* synthesis levels directed by right-eye retina RNA to left-eye retina RNA for a given time point were determined by comparisons of total TCA-precipitable counts programmed by 50 μg of total retina RNA. Aliquots of 5 μl were removed from each 150 μl *in vitro* translation reaction, TCA precipitated on glass fiber filters, and radioactivity counted by liquid scintillation counting. The ratios of the levels of *in vitro* synthesized ON₁ or ON₂ at a given time point to control levels (precrush levels) were determined by comparing densitometric quantitations of spots corresponding to ON₁ or ON₂ on the original autoradiograms. Ratio of *in vitro* synthesis directed by right-eye retina RNA to left-eye retina RNA (●); level of ON₁ relative to precrush levels (○); level of ON₂ relative to precrush levels (+). (From Tesser, P., Jones, P. S., and Schechter, N., *J. Neurochem.*, 47, 1235, 1986. With permission.)

REFERENCES

1. **Sperry, R.,** Patterning of central synapses in regeneration of the optic nerve in teleosts, *Physiol. Zool.*, 28, 351, 1948.
2. **Sperry, R.,** Chemoaffinity in the orderly growth of nerve fiber patterns and connections, *Proc. Natl. Acad. Sci. U.S.A.*, 50, 703, 1963.
3. **Sperry, R.,** The problem of central nervous reorganization after nerve regeneration and muscle transposition, *Q. Rev. Biol.*, 20, 311, 1945.
4. **Sperry, R. W.,** Mechanisms of neural maturation, in *Handbook of Experimental Psychology,* John Wiley & Sons, New York, 1951, 236.
5. **Aurora, H. L. and Sperry, R. W.,** Color discrimination after optic nerve regeneration in the fish, *Astronotus ocellatus, Dev. Biol.*, 7, 234, 1963.
6. **Attardi, D. G. and Sperry, R. W.,** Preferential selection of central pathways by regenerating optic fibers, *Exp. Neurol.*, 7, 46, 1963.
7. **Murray, M.,** Regenerating retinal fibers into the goldfish optic tectum, *J. Comp. Neurol.*, 168, 175, 1976.

8. **Francis, A., Bengston, C., and Gazzaniga, M. S.,** Intraocular equivalence after optic nerve regeneration in goldfish, *Exp. Neurol.,* 52, 94, 1976.
9. **Weiler, I. J.,** Restoration of visual acuity after optic nerve section and regeneration, in *Astronotus ocellatus, Exp. Neurol.,* 15, 377, 1966.
10. **Maturana, H., Lettvin, J. Y., McCulloch, W. S., and Pitts, W. H.,** Physiological evidence that cut optic nerve fibers in the frog regenerate to their proper places in the tectum, *Science,* 130, 1709, 1959.
11. **Gaze, R.,** Regeneration of the optic nerve in *Xenopus laevis, Q. J. Exp. Physiol.,* 44, 290, 1959.
12. **Jacobson, M. and Gaze, R. M.,** Selection of appropriate tectal connections by regenerating optic fibers in adult goldfish, *Exp. Neurol.,* 13, 418, 1965.
13. **Jacobson, M. and Gaze, R.,** Types of visual response from single units in the optic nerve of the goldfish, *Q. J. Exp. Physiol.,* 49, 199, 1964.
14. **Schneider, G. and Jhaver, S.,** Neuroanatomical correlates of spared or altered function after brain lesions in the newborn hamster, *Plasticity and Recovery of Function in the Central Nervous System,* D. Stein et al., Eds., Academic Press, New York, 1974.
15. **Raisman, G.,** Neuronal plasticity in the septal nuclei of the adult rat, *Brain Res.,* 14, 25, 1969.
16. **Lynch, G., Deadwyler, S., and Cotman, C.,** Post lesion axonal growth produces permanent functional connections, *Science,* 180, 1364, 1973.
17. **Johns, P. R. and Easter, S. S.,** Growth of the adult goldfish eye — increase in retinal cell number, *J. Comp. Neurol.,* 176, 331, 1977.
18. **Johns, P. R.,** Growth of the adult goldfish eye. III. Source of the new retinal cell, *J. Comp. Neurol.,* 176, 343, 1977.
19. **Meyer, R. L.,** Evidence from thymidine labeling for continuing growth of retina and tectum in juvenile goldfish, *Exp. Neurol.,* 59, 99, 1978.
20. **Schechter, N., Francis, A., Deutsch, D. G., and Gazzaniga, M. S.,** Recovery of tectal nicotinic-cholinergic receptor sites during optic nerve regeneration in goldfish, *Brain Res.,* 166, 57, 1979.
21. **Sharma, S. X.,** The retinal projections in the goldfish: an experimental study, *Brain Res.,* 39, 213, 1977.
22. **Ito, H.,** Fine structures of the carp tectum opticum, *J. Hirnforsch.,* 12, 325, 1970.
23. **Grafstein, B.,** Axonal transport communication between soma and synapse, *Adv. Biochem. Psychopharmacol.,* 1, 11, 1969.
24. **Cuenod, M., Marko, P., and Niederer, E.,** Disappearance of particulate tectal protein during optic nerve degeneration in the pigeon, *Brain Res.,* 49, 422, 1973.
25. **Francis, A. and Schechter, N.,** Activity of cholinergic enzymes in the goldfish optic tectum after disconnection: evidence for primary cholinergic fibers, *Neurochem. Res.,* 4, 547, 1979.
26. **Quitschke, W. and Schechter, N.,** Specific optic nerve proteins during regeneration of the goldfish retinotectal pathway, *Brain Res.,* 258, 69, 1983.
27. **Quitschke, W. and Schechter, N.,** *In vitro* protein synthesis in the goldfish retinotectal pathway during regeneration: evidence for specific axonal proteins of retinal origin in the optic nerve, *J. Neurochem.,* 41, 1137, 1983.
28. **Tesser, P., Jones, P. S., and Schechter, N.,** Elevated levels of retinal neurofilament mRNA accompany optic nerve regeneration, *J. Neurochem.,* 47, 1235, 1986.
29. **Heacock, A. M. and Agranoff, B. W.,** Enhanced labeling of retinal protein during regeneration of the optic nerve in goldfish, *Proc. Natl. Acad. Sci. U.S.A.,* 73, 878, 1976.
30. **Giulian, D., Ruisseaux, H. D., and Cowburn, D.,** A study of proteins from ganglion cells of the goldfish retina, *J. Biol. Chem.,* 255, 6486, 1980.
31. **Benowitz, L. I., Shashoua, V. E., and Yoon, M. G.,** Specific changes in rapidly transported proteins during regeneration of the goldfish optic nerve, *J. Neurosci.,* 1, 300, 1981.
32. **Koenig, E. and Adams, P.,** Local protein synthesizing activity in axonal fields regenerating *in vitro, J. Neurochem.,* 39, 386, 1982.
33. **Perry, G. W., Burmeister, D. W., and Grafstein, B.,** Changes in protein content of goldfish optic nerve during degeneration and regeneration following nerve crush, *J. Neurochem.,* 44, 1142, 1985.
34. **Stein-Izsak, C., Harel, A., Solomon, A., Belkin, M., and Schwartz, M.,** Alterations in mRNA translation products associated with regenerative responses in the retina, *J. Neurochem.,* 45, 1754, 1985.
35. **Skene, J. H. P. and Willard, M.,** Axonally transported proteins associated with axon growth in rabbit central and peripheral nervous system, *J. Cell Biol.,* 89, 96, 1981.
36. **Benowitz, L. I. and Lewis, E. R.,** Increased transport of 44,000- to 49,000-dalton acidic proteins during regeneration of the goldfish optic nerve: a two-dimensional gel analysis, *J. Neurosci.,* 3, 2153, 1983.
37. **Perrone-Bizzozero, N. I. and Benowitz, L. I.,** Expression of a 48-kilodalton growth-associated protein in the goldfish retina, *J. Neurochem.,* 48, 644, 1987.
38. **Skene, J. H. P. and Willard, M.,** Changes in axonally transported proteins during axon regeneration in toad retinal ganglion cells, *J. Cell Biol.,* 89, 86, 95, 1981.
39. **Skene, J. H. P. and Willard, M.,** Characteristics of growth-associated polypeptides in regenerating toad retinal ganglion cell axons, *J. Neurosci.,* 1, 419, 1981.

40. **Quitschke, W., Jones, P. S., and Schechter, N.**, A survey of intermediate filament proteins in optic nerve and spinal cord: evidence for differential expression, *J. Neurochem.*, 44, 1465, 1985.

41. **Lazarides, E.**, Intermediate filaments: a chemically heterogeneous developmentally regulated class of proteins, *Ann. Rev. Biochem.*, 51, 219, 1982.

42. **Connell, N. D. and Rheinwald, J. G.**, Regulation of the cytoskeleton in mesothelial cells: reversible loss of keratin and increase in vimentin during rapid growth in culture, *Cell*, 34, 245, 1983.

43. **Czernobilsky, B., Moll, R., Levy, R., and Franke, W. W.**, Co-expression of cytokeratin and vimentin filament in mesothelial, granulosa and rete ovarii cells of the human ovary, *Eur. J. Cell Biol.*, 37, 175, 1985.

44. **Drager, U. L.**, Coxsistence of neurofilaments and vimentin in a neurone of adult mouse retina, *Nature (London)*, 303, 169, 1983.

45. **Schnitzer, J., Franke, W. W., and Schachner, M.**, Immunocytochemical demonstration of vimentin in astrocytes and ependymal cells of developing and adult mouse nervous system, *J. Cell Biol.*, 90, 1981.

46. **Tapscott, S. S., Bennett, G. S., Toyama, Y., Kleinbart, F., and Holtzer, H.**, Intermediate filament proteins in the developing chick spinal cord, *Dev. Biol.*, 86, 40, 1981.

47. **Bignami, A., Raju, T., and Dahl, D.**, Localization of vimentin, the nonspecific intermediate filament protein in embryonal glia and in early differentiating neurons, *Dev. Biol.*, 91, 286, 1982.

48. **Cochard, P. and Paulin, D.**, Initial expression of neurofilaments and vimentin in the central and peripheral nervous system of the mouse embryo *in utero*, *J. Neurosci.*, 4, 2080, 1984.

49. **McKeon, F. D., Kirschner, M. W., and Caput, D.**, Homologies in both primary and secondary structure between nuclear envelope and intermediate filament proteins, *Nature (London)*, 319, 463, 1986.

50. **Fisher, D. Z., Chaudhary, N., and Blobel, G.**, cDNA sequencing of nuclear lamins A and C reveals primary and secondary structural homology to intermediate filament proteins, *Cell Biol.*, 83, 6450, 1986.

51. **Jackson, A. P., Seow, H. F., Holmes, N., Drickamer, K., and Parham, P.**, Clathrin light chains contain brain-specific insertion sequences and a region of homology with intermediate filaments, *Nature (London)*, 326, 154, 1987.

52. **Geisler, N., Kaufmann, E., and Weber, K.**, Protein chemical characterization of three structurally distinct domains along the protofilament unit of desmin 10 nm filaments, *Cell*, 30, 277, 1982.

53. **Geisler, N., Plessman, U., and Weber, K.**, Related amino acid sequence in neurofilaments and non-neuronal intermediate filaments, *Nature (London)*, 296, 448, 1982.

54. **Quax, W., Egberts, W. V., Hendriks, W., Quax-Jeuken, Y., and Bloemendal, H.**, The structure of the vimentin gene, *Cell*, 35, 215, 1983.

55. **Zehner, Z. E. and Paterson, B. M.**, Characterization of the chicken vimentin gene: single copy gene producing multiple mRNAs, *Proc. Natl. Acad. Sci. U.S.A.*, 80, 911, 1983.

56. **Lewis, S. A., Balcarek, J. M., Krek, V., Shelanski, M., and Cowan, N. J.**, Sequence of a cDNA clone encoding mouse glial fibrillary acidic protein: structural conservation of intermediate filaments, *Proc. Natl. Acad. Sci. U.S.A.*, 81, 2743, 1984.

57. **Capetanaki, Y. G., Ngai, J., and Lazarides, E.**, Characterization and regulation in the expression of a gene coding for the intermediate filament protein desmin, *Proc. Natl. Acad. Sci. U.S.A.*, 81, 6909, 1984.

58. **Lewis, S. A. and Cowan, N. J.**, Genetics, evolution, and expression of the 68,000-mol-wt neurofilament protein: isolation of a cloned cDNA probe, *J. Cell Biol.*, 100, 843, 1985.

59. **Pruss, R. M., Mirsky, R., Raff, M. C., Thorpe, R., Dowding, A. S., and Anderton, B. H.**, All classes of intermediate filaments share a common antigenic determinant defined by a monoclonal antibody, *Cell*, 27, 419, 1981.

60. **Geisler, N., Kaufman, E., Fischer, S., Plessman, U., and Weber, K.**, Neurofilament architecture combines structural principles of intermediate filaments with carboxy-terminal extensions increasing in size between triplet proteins, *EMBO J.*, 2, 1295, 1983.

61. **Shecket, G. and Lasek, R. S.**, Neurofilament protein phosphorylation: species generality and reaction characteristics, *J. Biol. Chem.*, 257, 4788, 1982.

62. **Julien, J. P. and Mushynski, W. E.**, Multiple phosphorylation sites in mammalian neurofilament poly-peptides, *J. Biol. Chem.*, 257, 10467, 1982.

63. **Sternberger, L. A. and Sternberger, N. H.**, Monoclonal antibodies distinguish phosphorylated and non-phosphorylated forms of neurofilaments *in situ*, *Proc. Natl. Acad. Sci. U.S.A.*, 80, 6126, 1983.

64. **Nixon, R. A., Brown, B. A., and Marotta, C. A.**, Posttranslational modification of a neurofilament protein during axoplasmic transport: implications for regional specialization of CNS axons, *J. Cell Biol.*, 94, 150, 1982.

65. **Nixon, R. A., Lewis, S. E., and Marotta, C. A.**, Posttranslational modification of a neurofilament proteins by phosphate during axoplasmic transport in retinal, ganglion cell neurons, *J. Neurosci.*, 7, 1145, 1987.

66. **Quitschke, W. and Schechter, N.**, Homology and diversity between intermediate filament paroteins of neuronal and nonneuronal origin in goldfish optic nerve, *J. Neurochem.*, 46, 545, 1986.

67. **Quitschke, W. and Schechter, N.**, 58,000 dalton intermediate filament proteins of neuronal and nonneu-ronal origin in the goldfish visual pathway, *J. Neurochem.*, 42, 569, 1984.

68. **Jones, P. S., Tesser, P., Keyser, K. T., Quitschke, W., Samadi, R., Karten, H. J., and Schechter, N.,** Immunohistochemical localization of intermediate filament proteins of neuronal and nonneuronal origin in the goldfish optic nerve: specific molecular markers for optic nerve structures, *J. Neurochem.,* 47, 1226, 1986.

69. **Schachner, M., Smith, C., and Schoonmaker, G.,** Immunological distinction between neurofilament and glial fibrillary acidic proteins by mouse antisera and their immunohistological characterization, *Dev. Neurosci.,* 1, 1, 1978.

70. **Yen, S. H. and Fields, K. L.,** Antibodies to neurofilament, glial filament and fibroblast intermediate filament proteins bind to different cell types of the nervous system, *J. Cell Biol.,* 88, 115, 1981.

71. **Bignami, A. and Dahl, D.,** The astroglial response to stabbing. Immunofluorescence studies with antibodies to astrocyte-specific proteins (GFA) in mammalian and submammalian vertebrates, *Neuropathol. Appl. Neurobiol.,* 2, 99, 1976.

72. **Reier, P. J. and Webster, H. deF.,** Regeneration and remyelination of Xenopus tadpole optic nerve fibres following transection or crush, *J. Neurocytol.,* 3, 591, 1974.

73. **Stensaas, L. J. and Feringa, E. R.,** Axon regeneration across the site of injury in the optic nerve of the newt *Triturus pyrrhogaster, Cell Tissue Res.,* 179, 501, 1977.

74. **Jones, P. J. and Schechter, N.,** Distribution of specific intermediate filament proteins in the goldfish retina, *J. Comp. Neurol.,* 266, 112, 1987.

75. **Shaw, G. and Weber, K.,** The intermediate filament complement of the retina: a comparison between different mammalian species, *Eur. J. Cell Biol.,* 33, 95, 1984.

76. **Quitschke, W. and Schechter, N.,** A non-computerized scanning method for determining relative protein quantities and synthesis rates on two-dimensional electrophoretic gels, *Anal. Biochem.,* 124, 231, 1982.

77. **Tesser, P., Jones, P. S., and Schechter, N.,** Elevated levels of retinal neurofilament mRNa accompany optic nerve regeneration, *J. Neurochem.,* 47, 1235, 1986.

78. **Barron, K. D., McGuinness, C. M., Misantone, L. J., Zanakis, M. F., Grafstein, B., and Murray, M.,** RNA content of normal and axotomized retinal ganglion cells of rat and goldfish, *J. Comp. Neurol.,* 236, 265, 1985.

79. **Grafstein, B. and Murray, M.,** Transport of protein in goldfish optic nerve during regeneration, *Exp. Neurol.,* 25, 494, 1969.

80. **Johns, P. R. and Easter, S. S.,** Growth of the adult goldfish eye — increase in retinal cell number, *J. Comp. Neurol.,* 176, 331, 1977.

81. **Meyer, R. L.,** Evidence from thymidine labeling for continuing growth of retina and tectum in juvenile goldfish, *Exp. Neurol.,* 59, 99, 1978.

82. **Raymond, P. A., Easter, S. S., Jr., Burnham, J. A., and Powers, M. K.,** Postembryonic growth of the optic tectum in goldfish, *J. Neurosci.,* 3, 1092, 1983.

83. **Reh, T. A. and Constantine-Paton, M.,** Qualitative and quantitative measures of plasticity during the normal development of the *Rana pipiens* retinotectal projection, *Dev. Brain Res.,* 10, 187, 1983.

84. **Franke, W. W.,** Nuclear lamins and cytoplasmic intermediate filament proteins: a growing multigene family, *Cell,* 48, 3, 1987.

Chapter 15

AMPHIBIANS: A RICH SOURCE OF BIOLOGICAL DIVERSITY

F. Harvey Pough

TABLE OF CONTENTS

I. INTRODUCTION

The animals grouped as amphibians (anurans, salamanders, and caecilians) represent the end points of some 250 million years of evolution. Anurans (frogs, toads, tree frogs, and other tailless amphibians) have been particularly successful; the 3500 species of anurans occur on all continents except Antarctica and live in habitats that extend from tropical forests through temperate deserts to seashores. The taxonomic and geographic diversity that characterizes amphibians is accompanied by variation in every aspect of their biology — development, morphology, physiology, behavior, and ecology. Many perturbations of biological processes that are experimentally induced for biomedical research are normal components of the biology of amphibians.

For example, amphibians show greater interspecific variation in maximal rates of oxygen consumption than do mammals or birds, and these variations are associated with interspecific variation in the structural and biochemical characteristics of their tissues. Furthermore, ontogenetic change in rates of oxygen consumption by amphibians can be as great as interspecific variation. Amphibians include species in which pulmonary gas exchange can be turned on or off at will, and species that lack lungs entirely. The reproductive modes of amphibians extend from eggs that hatch into aquatic larvae through terrestrial eggs to viviparity and include diverse forms of parental care. The genetic structure of amphibian species encompasses polyploids and forms that may be gynogenetic. These kinds of variation are rich sources of material for research.

Exploiting the opportunities for biomedical research offered by the diversity of amphibians requires familiarity with their evolution and life history, as well as an appreciation of the magnitude of interspecific and intraspecific variation. Evolutionary hypotheses can be framed only in the context of the phylogeny of the group studied.[1,2] Biomedical researchers too often report that they studied "the frog", an organism that does not exist. No 2 of the 3500 species of frogs are exactly alike, and even closely related species may respond quite differently to experimental manipulations. Furthermore, maintenance conditions for different species of amphibians are quite distinct and must be based on the ecological characteristics of each species.

In this chapter, I shall review the current understanding of the phylogenetic relationships of amphibians to provide a context in which to consider the ways in which amphibians can be used in biomedical research. Next I shall summarize information about some aspects of the biology of amphibians that have particular significance for biomedical research, with emphasis on studies of comparative and environmental physiology. I shall conclude with suggestions for methods of housing amphibians in the laboratory.

General information about the biology of amphibians can be found in Duellman and Trueb,[3] Halliday and Adler,[4] Frost,[5] Pough et al.,[6] Carroll,[7] Llinás and Precht,[8] Taylor and Guttman,[9] Vial,[10] Moore,[11] Lofts,[12] Taylor,[13] Feder,[14] and the *Catalogue of American Amphibians and Reptiles*.[15]

II. PHYLOGENETIC RELATIONSHIPS AND THE DIVERSITY OF AMPHIBIANS

The three orders of modern amphibians — salamanders (Urodela), frogs (Anura), and caecilians (Gymnophiona) — are often grouped in the subclass Lissamphibia ("smooth amphibians" — i.e., lacking scales). The question of whether modern amphibians have a common origin has been debated for the past half century. Currently a common origin of the three orders in the Paleozoic appears likely,[16-18] but hypotheses of two or even three independent origins have also been proposed.[7,19,20] The oldest fossils known that may represent modern amphibians are isolated vertebrae of Permian age (about 250 million years ago) that appear to include both salamander and anuran types. The earliest modern amphibian fossil that is relatively complete may be *Triadobatrachus,* a possible anuran from the early Triassic (about 225 million years ago). Fossils later than *Triadobatrachus* come from the Jurassic and Cretaceous (180 to 63 million years ago) and are as specialized as modern forms; in most cases they can be tentatively assigned to modern families.

Thus, the diversity of living amphibians is very ancient. Frogs and salamanders have been separated from each other for as long as they have been separated from the evolutionary lineages leading to birds or mammals. Consequently the anatomy and physiology of amphibians cannot necessarily be assumed to show primitive conditions of avian or mammalian anatomy and physiology.[21] Instead, all three lineages represent the current end points of millions of years of evolutionary change, and all are mosaics that combine character states that are primitive for tetrapods with specializations (derived characters) that are unique to particular evolutionary lineages. For example, the limbs, hands, and feet of humans retain the primitive tetrapod character state — one bone in the proximal portion of the limb, two bones in the distal portion, and five digits. All amphibians are more specialized than humans in the form of their limbs: no amphibian has more than four toes on the front foot, and anurans and caecilians are still more specialized — the tibia and fibula of anurans are fused, and caecilians have lost their limbs entirely. The same situation applies to the gas exchange system of amphibians and humans: humans retain nearly exclusive reliance on the lungs for gas exchange as the earliest tetrapods probably did, whereas amphibians have specializations of the skin and cardiovascular system that facilitate cutaneous gas exchange.[22,23] Assumptions about evolutionary progress and the direction of evolutionary change must be carefully considered; the living amphibians are as modern as any other kind of living organism.

A. GENERAL CHARACTERISTICS OF AMPHIBIANS

All living adult amphibians are carnivorous, and little morphological specialization is associated with different dietary habits within each order. Amphibians eat almost anything they are able to catch, kill, and swallow. In aquatic forms the tongue is broad, flat, and relatively immobile, but many terrestrial amphibians protrude the tongue from the mouth to capture prey.

1. Salamanders

The salamanders (order Urodela or Caudata) have the most generalized body form and locomotion of the living amphibians (Figure 1). Salamanders are elongate, and all but a very few species of completely aquatic salamanders have four functional limbs. The approximately 350 species of salamanders are almost entirely limited to the northern hemisphere; their southernmost occurrence is in northern South America. North and Central America have the greatest diversity of salamanders — more species of salamanders are found in Tennessee than in all of Europe and Asia combined. Paedomorphosis (the retention of larval characteristics, including larval tooth and bone patterns, the absence of eyelids, retention of a functional lateral line system, and in some cases retention of external gills) is widespread among salamanders, and several families of aquatic salamanders are constituted solely of such paedomorphic forms (Table 1).

The largest living salamanders are the Japanese and Chinese giant salamanders *(Andrias),* which reach lengths of a meter or more. The related North American hellbenders *(Cryptobranchus)* grow to 60 cm. All are members of the family Cryptobranchidae and are paedomorphic and permanently aquatic. Another large aquatic salamander frequently used in biomedical research, the mudpuppy *(Necturus),* is paedomorphic and does retain external gills. Mudpuppies occur in lakes and streams in eastern North America. Two genera of large, eel-like paedomorphic salamanders that live in southeastern North America have been used in biomedical research: *Siren* has external gills and retains only front legs, whereas *Amphiuma* lacks gills and has four tiny legs.

FIGURE 1.   The red-spotted newt (*Notophthalmus viridescens*). (Top) Eggs laid in water hatch into aquatic larvae with external gills. (Center) After several months the larvae metamorphose into a terrestrial form, the red eft. The skin of efts contains tetrodotoxin, and their bright orange color is aposematic. (Bottom) After a period of terrestrial life that may last as long as 14 years the efts transform into aquatic adults. Adult newts lack gills. The eft stage of the life history of newts is characteristic of montane populations; some populations in the coastal plain omit the eft stage, transforming directly from larvae to adults or (in a few places) becoming neotenic adults that retain external gills (Photographs copyright by the author.)

TABLE 1
Families of Living Amphibians

Order Urodela (or Caudata) — the salamanders (ca. 350 species)

Suborder Sirenoidea (includes the following family)

Sirenidae — elongate aquatic salamanders with gills and lacking the pelvic girdle and hindlimbs (three species in North America)

Suborder Cryptobranchoidea (includes the following families)

Hynobiidae — small to medium terrestrial or aquatic salamanders (100—200 mm) with external fertilization of the eggs and aquatic larvae (32 species in Asia)

Cryptobranchidae — very large (750 mm) to enormous (1520 mm) aquatic salamanders with external fertilization of the eggs (one species in North America; two in Asia)

Suborder Salamandroidea (includes the following families)

Proteidae — paedomorphic aquatic salamanders with external gills (six species in North America and Europe)

Dicamptodontidae — small to large (100—350 mm) semiaquatic salamanders with aquatic larvae (three species in North America)

Amphiumidae — very large (100 mm) elongate aquatic salamanders lacking gills (three species in North America)

Salamandridae — small to medium terrestrial and aquatic salamanders (49 species in Europe and Asia, 4 in North America)

Ambystomatidae — small to large terrestrial salamanders with aquatic larvae (30 species in North America)

Plethodontidae — very small (27 mm) to medium aquatic or terrestrial salamanders, some with aquatic larvae others with direct development (219 species in North, Central, and South America; 1 species in Europe)

Order Anura — The frogs, toads, tree frogs, etc. (ca. 3500 species) (includes the following families)

Leiopelmatidae — small (50 mm) semiaquatic frogs with internal fertilization (three species in New Zealand; one in North America)

Discoglossidae — small to medium (50—85 mm) frogs with aquatic larvae (14 species in Europe and Asia)

Rhinophrynidae — a burrowing frog with aquatic larvae (one species in Central America)

Pipidae — specialized aquatic frogs; *Xenopus, Hymenochirus,* and some species of *Pipa* have aquatic larvae; other species of *Pipa* have eggs that develop into juvenile frogs (26 species in South America and Africa)

Pelobatidae — short-legged terrestrial frogs with aquatic larvae (83 species in North America, Asia, Europe, and northern Africa)

Pelodytidae — small (50 mm) terrestrial frogs with aquatic larvae (two species in Europe and Asia)

Myobatrachidae — very small (20 mm) to large (115 mm) aquatic and terrestrial frogs with diverse reproductive modes (99 species in Australia, Tasmania, and New Guinea)

Heleophrynidae — medium frogs that live in mountain streams and have aquatic larvae (three species in extreme southern Africa)

Sooglossidae — small terrestrial frogs that lay eggs on land. The eggs hatch into juvenile frogs or into nonfeeding tadpoles that are carried on the back of an adult (three species in the Seychelles Islands)

Leptodactylidae — very small (12 mm) to enormous (250 mm) frogs from all habitats and with diverse modes of reproduction (710 species in southern North America, Central and South America, and the West Indies)

Bufonidae — very small (20 mm) to enormous (250 mm) mostly terrestrial toads. Most have aquatic larvae, but some species of *Nectophrynoides* are viviparous (335 species in North, Central, and South America, Africa, Europe, and Asia)

Brachycephalidae — very small terrestrial frogs that probably have direct development (two species in southeastern Brazil)

Rhinodermatidae — small terrestrial frogs that lay eggs on land. The tadpoles of *Rhinoderma rufum* are transported to water, whereas those of *R. darwini* complete development in the vocal sacs of the male (two species in southern Chile and Argentina)

Pseudidae — aquatic frogs with enormous tadpoles (250 mm for *Pseudis paradoxa*) that metamorphose into medium-sized (70 mm) adults (four species in South America)

Hylidae — most hylids are arboreal, but a few species are aquatic or terrestrial. Most species have aquatic larvae, but the marsupial frogs show several variations, including direct development of juvenile frogs (630 species in North, Central, and South America, Europe, Asia, and Australia)

Centrolenidae — mostly small arboreal frogs with aquatic larvae that live in streams (65 species in Central and South America)

Dendrobatidae — small terrestrial frogs, many of which are brightly colored and extremely toxic. Terrestrial eggs hatch into tadpoles that are transported to water by an adult (117 species in Central and South America)

Ranidae — medium (50 mm) to enormous (300 mm) aquatic or terrestrial frogs. Most have aquatic tadpoles, but several genera show direct development (667 species in North, Central, and South America, Europe, Asia, and Africa)

TABLE 1 (continued)
Families of Living Amphibians

Hyperoliidae — small to medium mostly arboreal frogs with aquatic larvae (206 species in Africa, Madagascar, and the Seychelles Islands)

Rhacophoridae — very small to large mostly arboreal frogs. Some species have filter-feeding aquatic larvae, whereas others lay eggs in holes in trees and have larvae that do not feed (186 species in Africa, Madagascar, and Asia)

Microhylidae — small to medium terrestrial or arboreal frogs. Many have aquatic larvae, but some species have nonfeeding tadpoles and others have direct development (279 species in North, Central, and South America, Asia, Africa, and Madagascar)

Order Gymnophiona — the caecilians (approximately 170 species) (includes the following families)

Ichthyophiidae — moderately large (to 500 mm) terrestrial caecilians with aquatic larvae (41 species in southeast Asia)

Rhinatrematidae — small (to 300 mm) terrestrial caecilians, believed to have aquatic larvae (nine species in South America)

Scolecomorphidae — moderately large terrestrial caecilians, possibly viviparous (seven species in Africa)

Caeciliidae — very small (100 mm) to very large (1.5 m) terrestrial caecilians with both oviparous and viviparous species, no aquatic larval stage (91 species in Central and South America, Africa, India, and the Seychelles Islands)

Typhlonectidae — large (to 750 mm) aquatic caecilians with viviparous reproduction (19 species in South America)

Modified from Pough, F. H., Heiser, J. B., and McFarland, W. N., Vertebrate Life, 3rd ed., Macmillan, New York, 1989. With permission.

Terrestrial salamanders like the North American mole salamanders *(Ambystoma)* and the European salamanders *(Salamandra* and *Triturus)* have aquatic larvae that lose their gills at metamorphosis. The most fully terrestrial salamanders, the lungless plethodontids, include species in which the young hatch from eggs as miniatures of the adult and lack an aquatic larval stage.

2. Anurans

In contrast to the limited number of species of salamanders and their restricted geographic distribution, the anurans include nearly 3500 species and have a nearly worldwide distribution. Specializations of body form can be used to distinguish many different kinds of anurans (Figure 2), and the difficulty is finding names for them — the diversity of anurans exceeds the number of common names that are used to distinguish them (Table 1). Animals called frogs usually have long legs and move by jumping. Many species in the family Ranidae *(Rana* and several other genera) have this body form, and very similar jumping frogs are found in other families. Stout-bodied terrestrial anurans that make short hops instead of long leaps are often called toads. They usually have blunt heads, heavy bodies, relatively short legs, and little webbing between the toes. This body form is represented by members of the family Bufonidae, and very similar body forms are found in other families, including the spadefoot toads *(Scaphiopus,* family Pelobatidae) of North America and the horned frogs *(Ceratophrys* and *Odontophrynus,* family Leptodactylidae) of South America. Arboreal frogs usually have large heads and eyes, and often slim waists and long legs. Arboreal frogs in many different families (Hylidae, Centrolenidae, Rhacophoridae, Leptodactylidae, and others) move by quadrupedal walking and climbing as much as by leaping. Many arboreal species have enlarged toe disks and are called tree frogs. Specialized aquatic anurans in the family Pipidae (including the African clawed frogs *Xenopus, Hymenochirus, Pseudohymenochirus,* and the South American *Pipa)* and fully aquatic members of other families are dorsoventrally flattened and have thick waists, powerful hind legs, and large hind feet with extensive webs. They also have well-developed lateral line systems and sensory structures at the tips of their fingers.

FIGURE 2. Diverse body forms of anurans. (Top left) The North American frog *Rana utricularia* (a member of the *Rana pipiens* species group) is a long-legged terrestrial anuran that moves in long jumps. (Top right) The North American toad (*Bufo americanus*) is a short-legged, terrestrial anuran that moves by short hops. (*Bottom left*) The Central American red-eyed tree frog (*Agalychnis callidryas*) is an arboreal species that walks and climbs. The toes have expanded tips. (Bottom right) The South American aquatic frog *Telmatobius culeus* lives in Lake Titicaca in the Andes (3812 m above sea level). The extensive folds of skin increase its respiratory surface area. (Photographs copyright by the author.)

Aquatic species of anurans use suction feeding to engulf food in the water, but most semiaquatic and terrestrial species have highly specialized sticky tongues that can be flipped out to trap prey and carry it back to the mouth.

3. Caecilians

The third group of living amphibians is the least known and does not even have an English common name. These are the caecilians (order Gymnophiona), legless burrowing or aquatic amphibians that occur in tropical habitats around the world (Figure 3). The eyes of caecilians are covered by skin or even by bone, but the retinas of many species have the layered organization that is typical of vertebrates and appear to be functional as photoreceptors. Conspicuous dermal folds (annuli) encircle the bodies of caecilians. The primary annuli overlie vertebrae and myotomal septa and reflect body segmentation. Many species of caecilians have dermal scales in pockets in the annuli; scales are not known in the other groups of living amphibians. A second unique feature of caecilians is a pair of protrusible tentacles between the eye and nostril. Some structures that are associated with the eyes of other vertebrates have become associated with the tentacles of caecilians. One of the eye muscles (the retractor bulbi) has become the retractor muscle for the tentacle, a second eye muscle (the levator bulbi) moves the tentacle sheath, and the Harderian gland lubricates the channel of the tentacle. It is likely that the tentacle is a sensory organ that allows chemical substances to be transported from the animal's surroundings to the Jacobson's organ on the roof of the mouth. Caecilians feed on elongate prey like earthworms and (for the aquatic species) fishes.

FIGURE 3.   A caecilian, *Ichthyophis kohtaoensis,* from Thailand. (Photograph copyright by the author.)

III. DERIVED FEATURES OF AMPHIBIAN BIOLOGY

The position of the orders of modern amphibians at the end of long, independent evolutionary lineages has led to many differences in the biology of the groups. Indeed, the surprise is that enough similarities remain after so long a separation to allow us to recognize the three orders as a subclass. The separation of some families of living amphibians traces back to the Mesozoic, not long after the origin of mammals and contemporaneous with the origin of birds, and the differences between families of amphibians are correspondingly more profound than the differences between families of birds or mammals, which are more recently derived. The diversity of amphibians must be considered in the selection of appropriate research species. In this section I shall summarize some of the major patterns of diversity in characteristics of the biology of amphibians that are important for biomedical research.

A. REPRODUCTION AND LIFE HISTORY

Amphibians show greater diversity in reproductive modes and life history patterns than any other group of tetrapods. In fact, the range of reproductive modes among the 4000 species of amphibians far exceeds that of any other group of vertebrates except for fishes, which outnumber amphibian species nearly 100 to 1. The generalization that amphibians return to water to lay eggs that hatch into aquatic larvae and subsequently metamorphose into terrestrial adults is incorrect. That pattern, which is probably primitive for tetrapods, characterizes a majority of the species of anurans that live in temperate regions, where most biologists also live. However, tropical species of anurans employ a variety of different reproductive modes, and even in temperate regions many salamanders produce eggs that hatch directly into terrestrial juveniles, entirely bypassing an aquatic stage.

Most species of amphibians lay eggs: the eggs may be deposited in water or on land, and they may hatch into aquatic larvae or into miniatures of the terrestrial adults. The adults of some species of frogs carry eggs attached to the surface of their bodies. Others carry their eggs in pockets in the skin of the back or flanks, in the vocal sacs, or even in the stomach. In still other species the females retain the eggs in the oviducts and give birth to living young. Many amphibians have no parental care of their eggs or young, but in many other species a parent remains with the eggs and sometimes with the hatchlings, transports tadpoles from the nest to water, and in a few species an adult even feeds the tadpoles.

1. Salamanders

Most groups of salamanders use internal fertilization of the eggs; two primitive families, the Hynobiidae and Cryptobranchidae, retain external fertilization. Internal fertilization among salamanders is accomplished not by an intromittent organ but by the transfer of a packet of sperm (the spermatophore) from the male to the female. In most cases, salamanders that mate in water also lay their eggs in water. The eggs may be laid singly or in a mass of transparent gelatinous material. The eggs hatch into gilled aquatic larvae which, except in paedomorphic forms, transform into terrestrial adults. Some families, especially the lungless salamanders (Plethodontidae), include species that have dispensed partly or entirely with an aquatic larval stage. A few salamanders give birth to living young. The European salamander *(Salamandra salamandra)* produces 20 or more small larvae, each about one twentieth the length of an adult. The larvae are released in water and have an aquatic stage that lasts about 3 months. The closely related alpine salamander *(S. atra)* gives birth to one or two fully developed young about one third the adult body length. A female alpine salamander produces as many eggs as a European salamander, but only one egg in each oviduct develops; the remaining eggs break up into a mass that provides food for the developing embryo. The embryos are nourished by secretions from the walls of the oviducts after the yolk is exhausted.[24,25]

a. Paedomorphosis

The terms that describe organisms that reproduce while still in the juvenile form have been the subjects of a controversy that is not yet resolved.[26] Paedomorphosis (''juvenile form'') can be achieved by delayed somatic development (in which case it is called neoteny) or by precocious reproductive development (progenesis). Neoteny appears to be the usual mechanism of paedomorphosis among amphibians.[3] Neoteny is common among salamanders — four families are characterized by neoteny (Table 1), and all families have at least some populations that are neotenic. A neotenic salamander, the axolotl *(Ambystoma mexicanum)*, has been important in biomedical studies for more than a century.[27] Most studies of the endocrinological basis of neoteny have employed this species and the related tiger salamander, *A. tigrinum*. Obligate neotenes appear to be insensitive to the effects of thyroid hormone, whereas facultative neotenes fail to synthesize the levels of thyroid hormones required for metamorphosis.[28-30]

2. Anurans

Fertilization is external in most anurans; the male uses his forelegs to clasp the female in the pectoral region (axillary amplexus) or pelvic region (inguinal amplexus). Amplexus may be maintained for several hours, days, and in one species for months before the female lays eggs. Internal fertilization has been demonstrated in the Puerto Rican coquí *(Eleutherodactylus coqui)* and may be widespread among frogs that lay eggs on land.[31] Fertilization must also be internal for the few species of anurans that give birth to living young.

Anurans show even greater diversity in their modes of reproduction than urodeles, and similar reproductive habits have evolved independently in different groups. The primitive anuran reproductive pattern is thought to consist of laying large numbers of small eggs.[32] This pattern is retained in two of the most widespread genera of anurans, *Rana* and *Bufo*. For example, toads may lay as many as 10,000 eggs in one clutch, and the female puts about half the energy in her body into the eggs. In temperate regions this reproductive pattern is almost universal among anurans, but in tropical areas as many as 80% of the species of anurans have other patterns.

One method of increasing the proportion of eggs that hatch successfully is to give them protection from predators, and a number of methods of accomplishing this have evolved. Many arboreal frogs lay their eggs in the branches of trees overhanging water. The eggs undergo embryonic development out of the reach of aquatic egg predators, and when the tadpoles hatch they drop into the water and take up an aquatic existence. Other frogs achieve the same result by constructing foam nests that float on the water surface. The female emits a copious mucous secretion that she beats into a foam with her hind legs, and the eggs are laid in the foam mass. When the tadpoles hatch, they drop through the foam into the water.

Although these methods reduce egg mortality, the tadpoles are subjected to predation and competition. Many anurans avoid both problems by finding or constructing breeding sites free from competitors and predators. Some tree frogs, for example, lay their eggs in the water that accumulates in bromeliads — epiphytic tropical plants that grow in trees and are morphologically specialized to collect rainwater. A large epiphyte may hold several liters of water, and the frogs pass through egg and larval stages in that protected microhabitat. Many tropical frogs lay eggs on land near water. The eggs or tadpoles may be released from the nest sites when pond levels rise after a rainstorm. Other frogs construct pools in the mud banks beside streams. These volcano-shaped structures are filled with water by rain or seepage and provide a favorable environment for the eggs and tadpoles. Some frogs have eliminated the tadpole stage entirely. These species lay large eggs on land that develop directly to little frogs. This pattern is characteristic of about 20% of all anuran species.

Parental care is widespread among anurans.[33] Even among the forms with a primitive breeding pattern, the males of some species maintain territories and attack almost any small animal that intrudes. Adults of many species of frogs guard the eggs; in some cases it is the male, in others the female, and in many cases it is not known which sex is involved because external sex identification often is difficult with anurans. Some of the frogs that lay their eggs over water remain with them. Some species sit beside the eggs, others rest on top of them (Figure 4). Removing the guarding frog frequently results in the eggs being eaten by predators or desiccating and dying before hatching.[34-36]

Some of the poison-dart frogs (genera *Colostethus, Dendrobates,* and *Phyllobates* in the family Dendrobatidae) of the American tropics deposit their eggs on the ground, and one of the parents remains with the eggs until they hatch into tadpoles. The tadpoles adhere to the adult and are transported to water (Figure 5). Females of the Panamanian frog *Colostethus inguinalis* carry their tadpoles for more than a week. The tadpoles increase in size during this period and have food in their stomachs, suggesting that they detach from their mother to feed and then return to her.[37] Females of another Central American poison-dart frog, *Dendrobates pumilio,* release their tadpoles into small pools of water in the leaf axils of plants, and then return at intervals to the pools to deposit unfertilized eggs that the tadpoles eat.[38,39]

Other anurans, instead of remaining with the eggs, carry the eggs with them. The male of the European midwife toad *(Alytes obstetricans)* gathers the egg strings about his hind legs as the female lays them. He carries them with him until they are ready to hatch, at which time he releases the tadpoles into water. The male of the terrestrial Darwin's frog *(Rhinoderma darwinii)* of Argentina snaps up the eggs the female lays and carries them in his vocal pouches, which extend back to the pelvic region. The embryos pass through metamorphosis in the vocal sacs and emerge as fully developed froglets. Males are not alone

FIGURE 4. The Puerto Rican coqui (*Eleutherodactylus coquí*). (Top) A male coqui resting on top of a clutch of eggs in the axil of a palm frond. (Bottom) The embryos, which can be seen through the egg capsules, are white. They pass through metamorphosis in the eggs and emerge as miniatures of the adults. (Photographs copyright by the author.)

FIGURE 5.   Transport of tadpoles by adult frogs.(Top) A female *Colostethus inguinalis* carries an entire brood of tadpoles. They grow during the week or more that they remain on her back. (Photograph copyright by the author.) (Bottom) Female *Dendrobates pumilio* usually transport their tadpoles individually to the leaf axils of plants, then return at intervals of several days to deposit trophic eggs that the tadpoles eat. (Photograph copyright by Douglas Brust.)

in caring for eggs. The females of a group of tree frogs carry the eggs on their back, either in an open oval depression, a closed pouch, or individual pockets.[40,41] Progesterone induces incubatory changes in the brooding pouch of the frog *Gasthrotheca riobambae*.[42] The eggs develop into miniature frogs before they leave their mother's back. A similar specialization is seen in the completely aquatic Surinam toad, *Pipa*. In the breeding season the skin of the female's back thickens and softens. In egg laying the male and female in amplexus swim in vertical loops in the water. On the upward part of the loop the female is above the male and releases a few eggs that fall onto his ventral surface. He fertilizes them and, on the downward loop, presses them against the female's back. They sink into the soft skin and a cover forms over each egg, enclosing it in a small capsule. The eggs develop through metamorphosis in the capsules.

Tadpoles of the two species of the Australian frog genus *Rheobatrachus* are carried in the stomach of the female frog.[43] The female swallows eggs or newly hatched larvae and retains them in her stomach through metamorphosis. *R. silus* was the first species in which this behavior was described, and it shows extensive morphological and physiological modifications of the stomach. These changes include distension of the proximal portion of the stomach, separation of individual muscles cells from the surrounding connective tissue, and inhibition of hydrochloric acid secretion, perhaps by prostaglandin released by the tadpoles. In 1984 a second species of gastric brooding frog, *R. vitellinus*, was discovered in Queensland.[44] Strangely, this species lacks the extensive structural changes in the stomach that characterize the gastric brooding of *R. silus*. The striking differences between the two species suggest the surprising possibility that this bizarre reproductive mode might have evolved independently in the species, and offers intriguing opportunities for comparisons of the control mechanisms involved.

Only a few anurans are viviparous. The females of some African bufonids *(Nectophrynoides)* retain the eggs in the oviducts and give birth to baby toads, and the golden coquí *(Eleutherodactylus jasperi,* a Puerto Rican leptodactylid) has a similar mode of reproduction.[45-48] Nutrients are supplied to the fetuses of *Nectophrynoides occidentalis* by secretions from the walls of the oviducts.[49]

a. Tadpoles

Although many species of frogs have evolved reproductive modes that bypass an aquatic larval stage, a life history that includes a tadpole has certain advantages. A tadpole is a completely different animal from an adult anuran, both morphologically and ecologically. Tadpoles of most species of anurans are filter-feeding herbivores, whereas all adult anurans are carnivores that catch prey individually. Because of these differences, tadpoles can exploit resources that are not available to adult anurans.[50]

Toad *(Bufo americanus)* and cascade frog tadpoles *(Rana cascade)* are able to distinguish siblings from nonsiblings, and they associate preferentially with siblings.[51-55] Recognition is probably accomplished by olfaction, and toad tadpoles can distinguish full siblings from maternal half siblings, and they can distinguish maternal half siblings from paternal half siblings.

b. Metamorphosis

The morphological specializations of tadpoles are entirely different from those of adult frogs, and the transition from tadpole to frog involves a very complete metamorphosis in which tadpole structures are broken down and their chemical constituents are rebuilt into the structures of adult frogs. Extensive changes occur at the molecular and tissue level.[56-64] The timing of metamorphosis is flexible and reflects a balance of costs and benefits.[65-67]

3. Caecilians

The reproductive biology of caecilians is as specialized as their body form and ecology.[68-70] Internal fertilization is accomplished by a male intromittent organ that is protruded from the cloaca. Some species of caecilians lay eggs, and the female may coil around the eggs, remaining with them until they hatch. Initial growth of the fetuses is supported by yolk contained in the egg at the time of fertilization, but this yolk is exhausted long before embryonic development is complete. In *Typhlonectes* the fetuses have absorbed all of the yolk in the eggs by the time they are 30 mm long. The fetuses obtain energy by scraping material from the walls of the oviducts with specialized embryonic teeth. Under the influence of progesterone the epithelium of the oviduct proliferates and forms thick beds surrounded by ramifications of connective tissue and capillaries. As the fetuses exhaust their yolk supply, these beds begin to secrete a thick, white, creamy substance that has been called uterine milk. The fetuses emerge from their egg membranes, uncurl, and align themselves lengthwise in the oviducts, where they apparently bite the walls of the oviduct, stimulating secretion and stripping some epithelial cells and muscle fibers that they swallow with the uterine milk. Differences in the details of fetal dentition in different species of caecilians suggest that this specialized form of fetal nourishment may have evolved independently in different phylogenetic lines. (Analogous methods of supplying energy to fetuses are known in some elasmobranch fishes.)

Gas exchange appears to be achieved by apposition of fetal gills to the walls of the oviducts. Both the gills and the oviductal wall are highly vascularized, and it seems likely that exchange of gases, and possibly of small molecules such as metabolic substrates and waste products, takes place across the adjacent gill and oviduct. The gills are absorbed before birth, and cutaneous gas exchange may be important for fetuses late in development.

B. TEMPERATURE RELATIONS AND THERMOREGULATION

Amphibians are ectotherms, that is, they obtain the heat needed to raise their body temperatures to levels that permit activity from external sources. Under natural conditions amphibians thermoregulate by selecting suitable microclimates, and control of body temperature is an important aspect of their biology.[71-74] Thermoregulation and hydroregulation interact for these permeable-skinned animals.[75-78]

The pervasive effects of temperature described in the following sections emphasize the importance of controlling temperature both during maintenance of amphibians and during experimental treatments.

1. Effects of Temperature

The consequences of changes in body temperature on physiological systems of anurans are well documented. Studies of anurans have demonstrated temperature-dependent changes in rates of biochemical reactions, in the conduction velocity of nerves, in the isometric twitch tension and power output of limb muscles, and changes in rates of aerobic and anaerobic metabolism.[79-91] Organismal function of anurans also is temperature-sensitive. Locomotion (hopping and swimming speed or jump distance) of several species of anurans increased with increasing temperatures to 27 to 28°C, and then decreased at higher temperatures.[92-95] Low temperatures inhibit digestion of prey and may reduce the rate and efficiency of energy assimilation.[96-98] Ectothermal vertebrates, including amphibians, show behavioral fever, selecting higher body temperatures in a thermal gradient when they have bacterial infections, and this response can also be elicited by injection of prostaglandin E_1 into the hypothalamus.[99-101]

2. Thermal Acclimation and Seasonal Cycles

Amphibians undergo extensive changes at the organismal, tissue, cellular, and molecular levels in response to temperature acclimation, including changes in temperature tolerance, body temperatures selected in a temperature gradient, metabolic rate, heart rate, stroke volume, responsiveness to catecholamines, gas exchange, blood oxygen affinity, acid-base balance, enzyme activity, cell membrane composition, lipid metabolism, and ion balance.[102-128]

The phenomenon of heat hardening (increased resistance to high temperatures appearing soon after an initial exposure to a thermal stress[129]) has been demonstrated in the laboratory and in the field for amphibians.[102,130-133] Dehydration[133] and exposure to ionizing radiation[134] also increased the resistance of amphibians to heat. These phenomena may be associated with the synthesis of heat shock proteins, a group of evolutionary conservative proteins that are widespread among eukaryotes. Their synthesis is induced by processes that interfere with electron transport or oxidative metabolism.[135,136] Heat shock proteins have been demonstrated in salamanders following exposure to high temperatures,[137] but induction of heat shock protein and induction of thermal tolerance were independent processes.[138]

The effects of temperature and the capacity for acclimation change seasonally and in response to temperature and photoperiod.[72,97,98,103,119,120,123,139,140,267] This lability must be considered in housing animals and in planning experiments.

C. WATER RELATIONS

Amphibians have a glandular skin that lacks external scales and is highly permeable to water.[141-143] Both the permeability and glandularity of the skin have been of major importance in shaping the ecology and evolution of amphibians. Mucous glands are distributed over the entire body surface and secrete mucopolysaccharide compounds. The primary function of the mucus is to keep the skin moist and permeable.[144] Experimentally produced interference with mucous gland secretion can lead to lethal overheating in frogs undergoing normal basking activity.[145]

The internal osmotic pressure of amphibians is approximately two thirds that characteristic of most other vertebrates.[146] The primary reason for the dilute body fluids of amphibians is low sodium content — approximately 100 mM compared with 150 mM in other vertebrates. Amphibians can withstand the loss of a substantially higher proportion of their body water than can other vertebrates, and daily fluctuations in body water content may exceed 20% of body mass during normal behavior.[95,147-150]

The main sites of exchange of water and solutes between the body and the environment are the skin, kidneys, and bladder.[151,152] Arginine vasotocin (AVT) increases water and sodium permeability of these sites, and its effect is more pronounced in terrestrial than in aquatic species. AVT reduces glomerular filtration rate and stimulates reabsorption of water from the tubules and bladder of terrestrial amphibians. It stimulates sodium uptake by aquatic anurans, but antidiuretic responses are usually absent for these species.

1. Uptake and Storage of Water

Amphibians do not drink water. Because of the permeability of their skins, species that live in aquatic habitats face a continuous osmotic influx of water that they must balance by producing urine. Species in arid habitats rarely encounter enough liquid water in one place to drink it, and if they should find a puddle, they can quickly absorb water through their skins. The impressive specializations of terrestrial amphibians are ones that facilitate rehydration from limited sources of water. The pelvic patch, an area of highly vascularized skin in the pelvic region, is responsible for a very large portion of an anuran's cutaneous water absorption.[153-157] Amphibians can absorb water from substrates against gradients of water potential as large as -1000 kPa.[158] The ability of different species of amphibians to withdraw water from soil roughly correlates with the aridity of their habitat. The control of salt and water balance of amphibians in vivo and in vitro has been reviewed by Koefoed-Johnsen.[152]

The urinary bladder plays an important role in the water relations of terrestrial amphibians, especially anurans.[146,152] Amphibian kidneys produce urine that is hyposmotic to the blood, so the urine in the bladder is dilute. Amphibians can reabsorb water from urine to replace water they lose by evaporation, and terrestrial amphibians have larger bladders than aquatic species. Storage capacities of 20 to 30% of the body mass of the animal are common for terrestrial anurans, and some species have still larger bladders: The Australian desert frogs *Notaden nicholsi* and *Neobatrachus wilsmorei* (Myobatrachidae) can store urine equivalent to about 50% of their body mass, and a bladder volume of 78.9% of body mass has been reported for the Australian leptodactylid frog *Helioporus eyrei*.[159] Accurate determination of the body mass of a frog requires that the bladder be emptied by inserting a fine cannula; the weight so measured is called the standard mass.[160]

2. Nitrogen Excretion and Retention

Fully aquatic species of amphibians like *Xenopus laevis* and *Necturus maculosus* are ammonotelic, whereas semiaquatic and terrestrial species are ureotelic.[161] Two groups of tree frogs are primarily uricotelic — two African rhacophorids in the genus *Chiromantis*, and four South American hylids in the genus *Phyllomedusa*.[162-167]

A transition from ammonotelism to ureotelism occurs during larval development and coincides with an increase in the activity of the ornithine-urea cycle enzymes in the liver.[168] Ureotelism appears to be the rule for terrestrial eggs.[169-171] The accumulation of urea may allow the eggs to draw water by osmosis from the ventral surface of an attending parent (Figure 4). A whitish, crystalline material that accumulates in the eggs late in the development of *Batrachyla taeniata* (a South American leptodactylid) may be uric acid.[172]

The concentration of urea in tissues and plasma increases from 3-fold to more than 20-fold during water stress.[167,173] The osmotic pressure resulting from these high concentrations of urea facilitates withdrawal of water from the soil[174,175] or counteracts the effects of high external salinities.[173,176-181] Some amphibians accumulate salt and other metabolites to balance high external fluid concentrations.[182-187] Acid media, including bodies of water affected by acid precipitation, inhibit sodium uptake by amphibians.[188]

3. The Effects of Dehydration

The ability of amphibians to sustain metabolism and activity is remarkably resistant to the effects of dehydration, greatly exceeding that of any other group of vertebrates. Some terrestrial species of anurans can tolerate water losses equivalent to 30% of their initial body mass, at least for periods of hours, and even under these extreme conditions they retain locomotor capacity and neuromuscular coordination.[95,189-192] The effects of dehydration on the tissues and organ systems of anurans have been characterized: dehydration has been reported to produce a large increase in the concentration of potassium in muscle tissues and smaller increases in sodium and chloride concentrations. Muscle water content may be conserved at the expense of plasma volume.[193-197] The velocities of muscular contraction and relaxation are reduced as tissue water is lost, maximum tension decreases, and recovery time increases.[193,198] Rates of glomerular filtration and urine production decrease during dehydration, and reabsorption of water from the kidney tubules and the bladder increases.[151,152,199] Dehydration also can affect the acid-base status of amphibians: the marine toad *(Bufo marinus)* maintained a stable arterial pH accompanied by acidification of the urine during dehydration, but slow dehydration in air led to respiratory acidosis for the green toad *(B. viridis)*.[195,200,201] Changes in neurotransmitter amino acids and physical and chemical changes in the visual and olfactory system have been noted during exposure of western toads *(B. boreas)* to hypersaline conditions.[202-208]

Dehydration of American toads *(B. americanus)*, leopard frogs *(Rana pipiens)*, and the Puerto Rican frog *Eleutherodactylus coqui* reduced oxygen consumption during activity.[150,190] Dehydration decreased maximum heart rate, stroke volume, and maximum rate of oxygen consumption of bullfrogs *(Rana catesbeiana)*, African clawed frogs *(Xenopus laevis)*, and marine toads *(Bufo marinus)*. These effects were attributed partly to increases in hematocrit and blood viscosity, partly to osmotic effects on the tissues (including skeletal and cardiac muscle), and partly to vasoconstriction.[209-214]

4. Waterproofing

Many amphibians have specialized mechanisms that reduce evaporative water loss. These methods include formation of a cocoon from as many as 60 layers of shed skin[215-221] and spreading a layer of dermal secretions over the body.[222] Inherently low rates of evaporative water loss are characteristic of many arboreal frogs,[223,224] and several species of anurans have rates of evaporative water loss an order of magnitude lower than those of most amphibians. These species include two African rhacophorids in the genus *Chiromantis,* four South American hylids in the genus *Phyllomedusa,* and four African hyperoliids in the genus *Hyperolius.*[156,157,163-167,225,226] The rates of cutaneous evaporation of these frogs are similar to those of lizards, birds, and mammals.

D. AEROBIC AND ANAEROBIC METABOLISM

The range of aerobic metabolic capacities that exists among amphibians exceeds that known for birds or mammals, and some relationships of structure and function are apparent among amphibians. Rates of oxygen consumption apparently become limiting to anurans more often than to mammals. Some species of anurans may routinely reach and exceed their limits of aerobic metabolism, and they derive a significant portion of their energy requirements from anaerobiosis.[82,227-231] Cutaneous and buccopharyngeal gas exchange is important for all amphibians, and plethodontid salamanders are entirely lungless.[22,23,232] Lunged salamanders *(Ambystoma laterale* and *A. tigrinum)* had higher rates of oxygen consumption and higher aerobic scopes during exercise than did lungless salamanders of equivalent size *(Desmognathus ochrophaeus* and *D. quadramaculatus),* and the lunged salamanders sustained exercise longer.[233] Elimination of pulmonary and buccopharyngeal gas exchange by a lunged salamander *(Ambystoma tigrinum)* reduced steady-state oxygen consumption during locomotion and locomotor stamina.[234]

1. Structure and Function

Work with mammals has not conclusively demonstrated rate-limiting steps for organismal oxygen consumption at the level of the lung or the lung-blood barrier, the capillaries, or the mitochondria.[235] However, correlations of mass-specific rates of oxygen consumption appear with tissue structure, blood, and muscle biochemistry.[236,237] Similar generalizations hold for anurans: neither lung volume nor the capillarity of gas exchange surfaces could be shown to be limiting to rates of oxygen uptake in a comparison of several species of anurans,[238] but there was a correlation between the rate of oxygen consumption and blood oxygen capacity, heart mass, and blood oxygen capacity.[239-241] Short-term requirements for increased oxygen delivery during activity appear to be met by increased extraction of oxygen.[124,242] Ontogenetic increases in rates of oxygen consumption during forced activity are associated with simultaneous increases in heart mass, hemoglobin concentration, or hematocrit in several species of anurans.[243] Interspecific variation in the aerobic capacities of anurans is great: some species increase standard rates of oxygen consumption 7-fold during activity, whereas other species show increases as large as 26-fold.[230] Anaerobic metabolism shows similar variation: the increase in whole-body lactate during 4 min of intense activity in 17 species of anurans ranged from 3- to 12-fold.[230]

Substantial differences in aerobic capacity exist among adult individuals of a species. For example, in a sample of 16 adult male toads *(Bufo americanus)* mass-specific metabolic rates during activity ranged from 0.79 to 1.54 ml O_2/g·h, and repeatable individual variation accounted for 89% of the total variance in aerobic capacity.[244] Individual variation in ventricle mass and hemoglobin concentration were positively correlated with variation in the rate of oxygen consumption during forced activity for bullfrogs *(Rana catesbeiana).*[241] The aerobic capacity of some species of anurans increases after metamorphosis, whereas other species (in some cases congeneric species) show no postmetamorphic change in aerobic capacity;

ontogenetic changes in aerobic capacity are associated with increases in heart mass, hematocrit, and hemoglobin concentration.[243] The magnitude of the ontogenetic change can be as great as the differences among species.

2. Muscle Fiber Types and Metabolic Capacities

Five types of muscle fibers have been distinguished in anurans — three Fast Twitch (FT) and two Tonic (T).[245] Fast-twitch fibers predominate in the anurans that have been studied.[246] Even the rectus abdominus, which has a graded response to stimulus, contains only 20% tonic fibers, and these are located in the ventral midline.[247,248]

Lannergren and Smith,[249] Smith and Ovalle,[245] and Putnam and Bennett[250] reported similar proportions of the various fiber types in the muscles of *Bufo, Rana,* and *Xenopus,* whereas Sperry[251] found that *Bufo* had a higher proportion of oxidative fibers than did *Rana* and *Xenopus.*

Among mammals, differences in fiber composition of muscles are correlated with performance at the level of tissues and organs. That correlation between fiber type and muscle performance is not apparent among anurans.[250] However, some functional properties of anuran muscles do appear to be related to patterns of organismal performance. Muscles from *Bufo boreas, Rana pipiens,* and *Xenopus laevis* had similar proportions of fibers but quite different functional properties. Muscles from *Bufo* had slower contraction times that those from *Rana* and *Xenopus* and accumulated less lactate during fatiguing stimulus. These functional differences accord with differences in organismal performance of these species of anurans.

3. Biochemical Characteristics of Muscle Tissues

In contrast to the failure of muscle fiber types of anurans to correlate with organismal performance, the activities of aerobic and glycolytic enzymes vary among species of anurans roughly in proportion to aerobic and anaerobic metabolism at the organismal level.[250,252,253] Differences in the biochemistry of muscles can be correlated with differences in organismal oxygen consumption for the spring peeper, *Hyla crucifer.* Vocalizing male frogs consume oxygen at rates 50% above those they achieve during forced locomotion, and the trunk muscles used in vocalization have correspondingly higher oxidative capacity than the limb muscles. The trunk muscles of male spring peepers have higher oxidative capacities than the trunk muscles of females, but the oxidative capacities of the limb muscles are the same for the two sexes.[254,255]

E. POLYPLOIDY

Formation of polyploids is a relatively common evolutionary event among amphibians.[256-258] Approximately 260 species of anurans have been karyotyped, and 29 bisexual species of polyploid anurans have been described, distributed among 11 genera in 5 families (Table 2). In addition, polyploidy can be induced by various means during the early embryonic development of diploid species.[259] Several of the naturally occurring polyploid species are associated with cryptic diploid or lower-order polyploid species that are very similar to the polyploids in external morphology, behavior, and ecology. The grey tree frogs, *Hyla chrysoscelis* (diploid) and *H. versicolor* (tetraploid), are the best-studied polyploid anurans in North America.[260] Electrophoretic analysis suggests that the species diverged less than 500,000 years ago.[261-263] Cells of *H. versicolor* contain about 1.9 times the amount of DNA found in *H. chrysoscelis,* and 1.6 times the total nuclear DNA.[264] The presence of four nucleoli in the majority of cells[265] suggests that *H. versicolor* retains the activity of much of its rRNA, and this hypothesis is confirmed by studies of rRNA gene number.[266] Cell volumes of erythrocytes, sperm, and toe-pad epidermis of *H. versicolor* are approximately 1.8 times those of *H. chrysoscelis.*[267,268] Polyploidy potentially could affect a wide

TABLE 2
Polyploid Amphibians

Family	Species	Ploidy	Distribution	Ref.
Anura				
Bufonidae	*Bufo danatensis*	4n = 44	Turkmen	309
	B. viridis	2n = 22/4n = 44	Kirghiza	310
	B. sp. D	4n = 40	Ethiopia	311
	B. asmare	4n = 40	Ethiopia	312
Hylidae	*Hyla chrysoscelis/*	2n = 24/	North America	258
	H. versicolor	4n = 48		
	Phyllomedusa burmeisteri	2n + 26/4n = 52	Brazil	313
Leptodacytlidae	*Ceratophrys aurita*	8n = 104	Brazil	314
	C. ornata	2n = 26/8n = 104	Argentina	315
	Odontophrynus americanus	2n = 22/4n = 44	Argentina	316, 317
	Pleuroderma bibroni	4n = 44	Uruguay	318
	P. kriegi	4n = 44	Argentina	318
	Neobatrachus sudelli	4n = 48	Australia	319
	N. sutor	4n = 48	Australia	319
Pipidae	*Xenopus tropicalis*	2n = 20	West Africa	320
	X. borealis/mulleri	4n = 36	Southeast Africa	320, 321
	X. clivii	4n = 36	Ethiopia	320, 321
	X. fraseri	4n = 36	Cameroon	320, 321
	X. gilli	4n = 36	Cape Town	320, 321
	X. laevis	4n = 36	South Africa	320, 321
	X. epitropicalis	4n = 40	Africa	322
	X. amieti	8n = 72	Uganda	323
	X. vestitus	8n = 72	Uganda	320, 324
	X. wittei	8n = 72	Uganda	323
	X. sp. n.	8n = 72	Uganda	320, 324
	X. ruwenzoriensis	10n = 104	Uganda	320, 324
Ranidae	*Dicroglossus occipitalis*	2n = 26/4n = 52	Liberia	311
	Pyxicephalus delalandii	2n = 26/4n = 52	South Africa	311
	Rana esculenta	3n = 39	Europe	273, 274
Urodela				
Ambystomatidae	*Ambystoma tremblayi*	3n = 42	North America	275
	A. platineum	3n = 42	North America	275
	A. laterale × A. texanum	3n = 42	North America	325
	A. platineum × A. texanum	4n = 56	North America	277
Sirenidae	*Pseudobranchus striatus*	4n = 64	North America	326
	Siren intermedia	4n = 46	North America	326, 327
	S. lacertina	4n = 52	North America	326, 327

Note: All of the species listed are bisexual except for the triploids. Some species have both diploid and tetraploid populations. The polyploid status of the Sirenidae is conjectural.

Modified from Reference 280.

range of tissue and cellular processes because of its effect on cell size. However, the resting metabolic rates, aerobic capacities, and anaerobic capacities of the *H. versicolor* and *H. chrysoscelis* are indistinguishable despite the differences in cell size between the species.[269-272]

The European frog *Rana esculenta* is a hybrid between *R. ridibunda* and *R. lessonae*.[273,274] Female *R. esculenta* produce both large and small ova; the large ova are diploid with one *ridibunda* and one *lessonae* genome, and the small ova are haploid with one *ridibunda* genome. Salamanders of the *Ambystoma jeffersonianum* complex of eastern North America include two bisexual diploid species *(A. jeffersonianum* and *A. laterale)* and two all-female triploid species *A. platineum* and *A. tremblayi).*[275-277] It has been assumed that ova of the triploid species must be activated by sperm from one of the diploid species to develop (gynogenesis) but that the sperm made no genetic contribution to the embryo. This hypothesis has recently been challenged: Bogart[278] has reported that the meiotic mechanism used by triploid females allows low levels of gene exchange to occur, and Lowcock[279] has traced gene exchange in hybrid complexes of *Ambystoma.*

African clawed frogs (species of *Xenopus*) are commonly used in biomedical studies; the genus *Xenopus* apparently spans a range from 2 to 10 n.[280]

F. TOXINS

Although there is evidence that the secretions of the mucous glands of some species of amphibians are irritating or toxic to predators, the chemical defense system is located primarily in the poison glands. These glands are concentrated on the dorsal surfaces of the animal, and defense postures of both anurans and urodeles present the glandular areas to potential predators. A wide variety of irritating and, in some cases, exceedingly toxic compounds are produced by amphibians.[281-291] The compounds produced by different groups reflect their phylogenetic relationship, and skin toxins have been helpful in classification of some taxa.

IV. MAINTENANCE OF AMPHIBIANS IN THE LABORATORY

The diversity of amphibians and the range of ecological and behavioral specializations found in the group make it impossible to generalize about their requirements in captivity. The most important consideration is that an investigator be fully informed about the biology and ecology of the species being studied.[292] Techniques for care of amphibians in the laboratory will be found in Stewart,[292] Nace et al.,[293] anonymous,[294] Culley,[295] Zimmerman,[296] and Frost.[303] A large zoo may be a helpful source of advice; many zoos now have breeding programs for amphibians, especially endangered species.

A. HOUSING

The body temperatures of amphibians are normally slightly below environmental temperatures as a result of evaporative cooling, and this effect should be considered in setting room temperature. Species of amphibians from temperate regions are best kept at body temperatures between 15 and 20°C, whereas tropical species should have body temperatures between 20 and 25°C. A daily temperature cycle of 5°C would duplicate conditions the animals encounter in nature and is probably desirable, but no experimental data are available to test that assumption.

Water loss is a critical problem for amphibians, especially when they are held in temperature-controlled conditions in which air is first cooled and then heated to a pre-set temperature — a method that creates low relative humidities with correspondingly high water vapor pressure deficits. Terrestrial cages should be covered with solid lids to raise the humidity.

1. Aquatic and Semiaquatic Species

Some amphibians are entirely aquatic and live well in aquariums or holding tanks. Fully aquatic forms include the larvae of salamanders and anurans, neotenic salamanders such as the axolotl *(Ambystoma mexicanum),* adult African clawed frogs *(Xenopus* and the smaller genus *Hymenochirus),* Surinam toads *(Pipa),* aquatic caecilians, and the aquatic salamanders *Siren, Pseudobranchus, Amphiuma, Necturus, Proteus, Cryptobranchus,* and *Andrias.*

Most other species of frogs, including the species of *Rana* often used in biomedical research, are at least semiterrestrial and are best kept in cages that allow them to move between dry substrate and water. A glass aquarium tilted up at one end is usually suitable for this purpose, and trays with dry substrate and water dishes are also effective. The cool flow-through water system commonly employed to house frogs from the *Rana pipiens* group used in teaching laboratories is satisfactory only when animals are purchased in the fall as they are about to enter hibernation. Frogs can be held near torpor in dim light at 4°C for several months and induced to emerge in breeding condition by gradually increasing temperature and photoperiod. At other times of year these conditions disrupt the annual rhythm of the frogs with unpredictable effects on physiological processes and organismal function.

Cannibalism is normal for salamander larvae and for the tadpoles of some species of frogs including *Ceratophrys, Pyxicephalus,* and some species of *Dendrobates* and *Scaphiopus.* These species must be reared individually: disposable plastic glasses are convenient containers for the purpose. Cannibalism is less common among adult amphibians, although large salamanders sometimes eat smaller cagemates, and some frogs (including species of *Ceratophrys, Pyxicephalus,* and *Leptodactylus pentadactylus*) normally prey on other frogs. Both crowding and large disparities of body size among cagemates should be avoided.

2. Terrestrial Species

Many species of amphibians are terrestrial as adults. Some of these return to water to breed but spend the rest of the year in terrestrial habitats where they encounter pools of water only after rains. Many species commonly used in biomedical research fall into this category, including all species of toads *(Bufo),* spadefoot toads *(Scaphiopus),* and the leopard frogs *(Rana pipiens* species complex) as well as many other species of *Rana.* These anurans are easily housed in cages with a substrate of bark chips (available from garden supply stores) and water dishes. Hiding places are helpful; plastic boxes or other easily cleaned objects of suitable size can be used. Rough or sharp edges should be avoided.

Fully terrestrial amphibians lay eggs that hatch into miniatures of the adults. Many species of frogs, including the neotropical genus *Eleutherodactylus,* belong in this category, as do many species of plethodontid salamanders and some caecilians. These animals often do best in terraria. Small semifossorial animals (salamanders, caecilians, some frogs) bury themselves in moist sphagnum moss. Arboreal species may prefer terraria with plants in which they can climb and hide.

B. FOOD AND FOOD SUPPLEMENTS

All amphibians are carnivorous, and many species respond most readily to live, or at least moving, food. Nutritional requirements of amphibians are poorly defined, and providing a variety of food types is a wise precaution. Feeding only one type of food leads to apparent nutritional deficiencies.[297-300] The conditions under which food items are reared affect their nutritional quality; instructions for rearing a variety of food species can be found in Nace and Buttner,[299] Master,[301] anonymous,[302] and Frost,[303] Crickets and other insects used to feed captive amphibians should be dusted with a mixture of a powdered vitamins and bone meal.

1. Aquatic Amphibians

Aquatic salamanders (adults and larvae), caecilians, and anurans *(Xenopus, Hymenochirus, Pipa)* will eat a variety of invertebrates that are often available from tropical fish supply stores, including *Tubifex* worms and brine shrimp *(Artemia).* Some aquatic amphibians apparently respond to the odor of food and will eat freeze-dried fish or frog food, trout pellets, and small pieces of liver.

2. Terrestrial Amphibians

Terrestrial amphibians usually require live prey and eat earthworms, crickets, fruit flies, *Tenebrio* larvae (mealworms), *Tribolium* (flour beetle) larvae, and larval and adult wax moths *(Galleria* and *Achroea)*. Large anurans and salamanders will eat new-born mice and rats. Collecting insects with a sweep net or a light trap can be a convenient way to supply a varied diet. A method of using falling drops of water to move dead insects so that anurans will seize them has been described by Buttner.[300]

3. Larvae

Larval salamanders eat the same sorts of food as aquatic salamanders. Most anuran tadpoles are herbivorous. They will eat lettuce or spinach softened by boiling, many kinds of dry fish food, and pelleted rabbit or mouse food suspended in a mixture of agar and gelatin.[303] Carnivorous tadpoles can be fed brine shrimp *(Artemia)*.

C. DISEASES

Little is known about the diagnosis or treatment of diseases of amphibians — references can be found in Reichenbach-Klinke and Elkan,[304] Elkan and Reichenbach-Klinke,[305] Cosgrove,[306] Marcus,[307] and Hoff et al.[308]

ACKNOWLEDGMENT

This work was supported by the New York State Agriculture Experiment Station, Project No. 412.

REFERENCES

1. **Felsenstein, J.,** Phylogenies and the comparative method, *Am. Nat.,* 125, 1, 1985.
2. **Huey, R. B.,** Phylogeny, history, and the comparative method, in *New Directions in Ecological Physiology,* Feder, M. E., Bennett, A. F., Burggren, W. W., and Huey, R. B., Eds., Cambridge University Press, Cambridge, 1988, chap. 4.
3. **Duellman, W. E. and Trueb, L.,** *Biology of Amphibians,* McGraw-Hill, New York, 1986.
4. **Halliday, T. R. and Adler, K., Eds.,** *The Encyclopedia of Reptiles and Amphibians,* George Allen and Unwin Ltd., London, 1986.
5. **Frost, D. R.,** *Amphibian Species of the World: a Taxonomic and Geographical Reference,* Assoc. Syst. Collections, Allen Press, Lawrence, KS, 1985.
6. **Pough, F. H., Heiser, J. B., and McFarland, W. N.,** *Vertebrate Life,* 3rd ed., Macmillan, New York, 1989, ch. 11.
7. **Carroll, R. L.,** *Vertebrate Paleontology and Evolution,* W. H. Freeman, San Francisco, 1987.
8. **Llinás, R. and Precht, W., Eds.,** *Frog Neurobiology,* Springer-Verlag, Berlin, 1976.
9. **Taylor, D. H. and Guttman, S. I., Eds.,** *The Reproductive Biology of Amphibians,* Plenum Press, New York, 1977.
10. **Vial, J. L., Ed.,** *Evolutionary Biology of the Anurans,* University of Missouri Press, Columbia, 1973.
11. **Moore, J. A., Ed.,** *Physiology of the Amphibia,* Vol. 1, Academic Press, New York, 1964.
12. **Lofts, B., Ed.,** *Physiology of the Amphibia,* Vol. 2, Academic Press, New York, 1974.
13. **Taylor, E. H.,** *The Caecilians of the World,* University of Kansas Press, Lawrence, 1968.
14. **Feder, M. E.,** Integrating the ecology and physiology of plethodontid salamanders, *Herpetologica,* 39, 291, 1983.
15. **Various authors,** *Catalogue of American Amphibians and Reptiles,* Society for the Study of Amphibians and Reptiles, 1971 *et seq.*
16. **Parsons, T. and Williams, E.,** The relationship of modern Amphibia: a reexamination, *Q. Rev. Biol.,* 38, 26, 1963.
17. **Trueb, L. and Cloutier, R.,** Historical constraints on lissamphibian osteology, *Am. Zool.,* 27, 33A, 1987.
18. **de Queiroz, K. and Cannatella, D. C.,** The monophyly and relationships of the Lissamphibia, *Am. Zool.,* 27, 60A, 1987.

19. **Nieuwkoop, P. D. and Satasurya, L. A.,** Embryological evidence for a possible polyphyletic origin of recent amphibians, *J. Embryol. Exp. Morphol.,* 35, 159, 1976.
20. **Gardiner, B. G.,** Gnathostome vertebrae and the classification of the Amphibia, *Zool. J. Linn. Soc.,* 79, 1, 1983.
21. **Pough, F. H.,** The advantages of ectothermy for tetrapods, *Am. Nat.,* 115, 92, 1980.
22. **Johansen, K. and Burggren, W.,** Cardiovascular function in the lower vertebrates, in *Hearts and Heart-like Organs,* Vol. 1, Bourne, G., Ed., Academic Press, New York, 1980, chap. 3.
23. **Feder, M. E. and Burggren, W. W.,** Cutaneous gas exchange in vertebrates: design, patterns, control and implications, *Biol. Rev.,* 60, 1, 1985.
24. **Vilter, V. and Vilter, A.,** Sur la gestation de la salamandre noir des Alpes, *Salamandra atra* Laur., *C. R. Seances Soc. Biol. Paris,* 157, 464, 1960.
25. **Vilter, V. and Vilter, A.,** Sur l'evolution des corps jaunes ovariens chez *Salamandra atra* Laur. des Alpes vaudoises, *C. R. Soc. Biol.,* 158, 457, 1964.
26. **Gould, S. J.,** *Ontogeny and Phylogeny,* Harvard University Press, Cambridge, MA, 1977.
27. **Smith, H. M. and Smith, R. B.,** *Synopsis of the Herpetofauna of Mexico. I. Analysis of the Literature on the Mexican Axolotl,* Lundberg, Augusta, WV, 1971.
28. **Norris, D. O. and Platt, J. E.,** T_3 and T_4 induced rates of metamorphosis in immature and sexually mature larvae of *Ambystoma tigrinum* (Amphibia, Caudata), *J. Exp. Zool.,* 189, 303, 1974.
29. **Taurog, A.,** The effect of TSH and long acting thyroid stimulator on the thyroid ^{131}I metabolism on metamorphosis in the Mexican axolotl *(Ambystoma mexicanum), Gen. Comp. Endocrinol.,* 24, 257, 1974.
30. **Taurog, A., Oliver, C., Eskay, R. L., Porter, J. C., and McKenzie, J. M.,** The role of TRH in the neoteny of the Mexican axolotl *(Ambystoma mexicanum), Gen. Comp. Endocrinol.,* 24, 267, 1974.
31. **Townsend, D. S., Stewart, M. M., Pough, F. H., and Brussard, P. F.,** Internal fertilization in an oviparous frog, *Science,* 212, 469, 1981.
32. **Salthe, S. N. and Duellman, W. E.,** Quantitative constraints associated with reproductive mode in anurans, in *Evolutionary Biology of the Anurans,* Vial, J. L., Ed., University of Missouri Press, Columbia, 1973, 229.
33. **Wells, K. D.,** Parental behavior in male and female frogs, in *Natural Selection and Social Behavior: Recent Research and New Theory,* Alexander, R. D. and Tinkle, D. W., Eds., Chiron Press, Newton, MA, 1971, 184.
34. **Villa, J. and Townsend, D. S.,** Viable frog eggs eaten by phorid fly larvae, *J. Herpetol.,* 17, 278, 1983.
35. **Taigen, T. L., Pough, F. H., and Stewart, M. M.,** Water balance of terrestrial anuran (*Eleutherodactylus coqui*) eggs: importance of parental care, *Ecology,* 65, 248, 1984.
36. **Townsend, D. S., Stewart, M. M., and Pough, F. H.,** Male parental care and its adaptive significance in a neotropical frog, *Anim. Behav.,* 32, 421, 1984.
37. **Wells, K. D.,** Evidence for growth of tadpoles during parental transport in *Colostethus inguinalis, J. Herpetol.,* 14, 428, 1980.
38. **Weygoldt, P.,** Complex brood care and reproductive behavior in captive poison-arrow frogs, *Dendrobates pumilio* O. Schmidt, *Behav. Ecol. Sociobiol.,* 7, 329, 1980.
39. **Brust, D. G.,** Maternal brood care by *Dendrobates pumilio:* a frog that feeds its young, *Am. Zool.,* 27, 74A, 1987.
40. **Duellman, W. E. and Maness, S. J.,** The reproductive behavior of some hylid marsupial frogs, *J. Herpetol.,* 14, 213, 1980.
41. **del Pino, E. M.,** Morphology of the pouch and incubatory integument in marsupial frogs (Hylidae), *Copeia,* 1980, 10, 1980.
42. **del Pino, E. M.,** Progesterone induces incubatory changes in the brooding pouch of the frog *Gastrotheca riobambae* (Fowler), *J. Exp. Zool.,* 227, 159, 1983.
43. **Tyler, M. J., Ed.,** *The Gastric Brooding Frog,* Croom Helm, London, 1983.
44. **Leong, A. S.-Y., Tyler, M. J., and Shearman, D. J. C.,** Gastric brooding: a new form in a recently discovered Australian frog of the genus *Rheobatrachus, Aust. J. Zool.,* 34, 205, 1986.
45. **Xavier, F.,** Action modératrice de la progestérones sur la croissance des embryons chez *Nectophrynoides occidentalis,* Angel, *C. R. Acad. Sci. Ser. D,* 270, 2115, 1970.
46. **Xavier, F.,** An exceptional reproductive strategy in Anura: *Nectophrynoides occidentalis,* Angel (Bufonidae), an example of adaptation to terrestrial life by viviparity, in *Major Patterns in Vertebrate Evolution,* Hecht, M. K., Goody, P. C., and Hecht, B. M., Eds., Plenum Press, New York, 1977, 545.
47. **Wake, M. H.,** The reproductive biology of *Eleutherodactylus jasperi* (Amphibia, Anura, Leptodactylidae) with comments on the evolution of live-bearing systems, *J. Herpetol.,* 12, 121, 1978.
48. **Wake, M. H.,** The reproductive biology of *Nectophrynoides malcolmi* (Amphibia, Bufonidae) with comments on the evolution of reproductive modes in the genus *Nectophrynoides, Copeia,* 1980, 193, 1980.
49. **Xavier, F.,** Le cycle des voies génitales femelles de *Nectophrynoides occidentalis* Angel, amphibien anoure vivipare, *Z. Zellforsch. Mikrosk. Anat.,* 140, 509, 1973.

50. **Wassersug, R. J.,** The adaptive significance of the tadpole stage with comments on the maintenance of complex life cycles in anurans, *Am. Zool.,* 15, 405, 1975.

51. **Waldman, B. and Adler, K.,** Toad tadpoles associate preferentially with siblings, *Nature (London),* 282, 611, 1979.

52. **Waldman, B.,** Sibling recognition in toad tadpoles, the role of experience, *Z. Tierpsychol.,* 56, 341, 1981.

53. **Waldman, B.,** Sibling association among schooling toad tadpoles, field evidence and implications, *Anim. Behav.,* 30, 700, 1982.

54. **Blaustein, A. R. and O'Hara, R. K.,** Genetic control of sibling recognition?, *Nature (London),* 290, 246, 1981.

55. **O'Hara, R. K. and Blaustein, A. R.,** An investigation of sibling recognition in *Rana cascade* tadpoles, *Anim. Behav.,* 29, 1121, 1981.

56. **Deuchar, E. M.,** *Biochemical Aspects of Amphibian Development,* Methuen, London, 1966.

57. **Deuchar, E. M.,** Xenopus: *The South African Clawed Frog,* John Wiley & Sons, London, 1975.

58. **Freiden, E.,** Biochemical adaptation and anuran metamorphosis, *Am. Zool.,* 1, 115, 1961.

59. **Freiden, E.,** Biochemistry of amphibian metamorphosis, in *Metamorphosis, a Problem in Developmental Biology,* 1st ed., Etkin, W. and Gilbert, L. I., Eds., Appleton-Century-Crofts, New York, 1968, 349.

60. **Dodd, M. H. I. and Dodd, J. M.,** The biology of metamorphosis, in *Physiology of the Amphibia,* Vol. 2, Lofts, B. A., Ed., Academic Press, New York, 1976, 467.

61. **Broyles, R. H.,** Changes in the blood during amphibian metamorphosis, in *Metamorphosis, a Problem in Developmental Biology,* 2nd ed., Etkin, W. and Frieden, E., Eds., Plenum Press, New York, 1981, 461.

62. **Atkinson, B. G.,** Biological basis of tissue regression and synthesis, in *Metamorphosis, a Problem in Developmental Biology,* 2nd ed., Etkin, W. and Frieden, E., Eds., Plenum Press, New York, 1981, 379.

63. **Fox, H.,** Cytological and morphological changes during amphibian metamorphosis, in *Metamorphosis, a Problem in Developmental Biology,* 2nd ed., Etkin, W. and Frieden, E., Eds., Plenum Press, New York, 1981, 327.

64. **Fox, H.,** *Amphibian Morphogenesis,* Humana Press, Clifton, NJ, 1984.

65. **Wilbur, H. M. and Collins, J. P.,** Ecological aspects of amphibian metamorphosis, *Science,* 182, 1302, 1973.

66. **Smith-Gill, S. J. and Bervan, K. A.,** Predicting amphibian metamorphosis, *Am. Nat.,* 113, 563, 1979.

67. **Werner, E. E.,** Amphibian metamorphosis: growth rate, predation risk, and optimal size at transformation, *Am. Nat.,* 128, 319, 1986.

68. **Wake, M. H.,** The reproductive biology of caecilians: an evolutionary perspective, in *The Reproductive Biology of Amphibians,* Taylor, D. H. and Guttman, S. I., Eds., Plenum Press, New York, 1977, 73.

69. **Wake, M. H.,** Fetal maintenance and its evolutionary significance in the Amphibia: Gymnophiona, *J. Herpetol.,* 11, 379, 1977.

70. **Wake, M. H.,** A perspective on the systematics and morphology of the Gymnophiona (Amphibia), *Mem. Soc. Zool. Fr.,* No. 43, 21, 1986.

71. **Brattstrom, B. H.,** Amphibian temperature regulation studies in the field and laboratory, *Am. Zool.,* 19, 345, 1979.

72. **Feder, M. E. and Lynch, J. F.,** Effect of elevation, latitude, season, and microhabitat on field body temperatures of neotropical and temperate zone salamanders, *Ecology,* 63, 1657, 1982.

73. **Feder, M. E., Lynch, J. F., Shaffer, H. B., and Wake, D. B.,** Field body temperatures of tropical and temperate zone salamanders, *Smithson. Herpetol. Inf. Serv. Publ.,* 52, 1, 1982.

74. **Lillywhite, H. B., Licht, P., and Chelgren, P.,** The role of behavioral thermoregulation in the growth energetics of the toad, *Bufo boreas, Ecology,* 54, 375, 1973.

75. **Tracy, C. R.,** Water and energy relations of terrestrial amphibians: insights from mechanistic modeling, in *Perspectives in Biophysical Ecology,* Gates, D. M. and Schmerl, R. B., Eds., Springer-Verlag, New York, 1975, 325.

76. **Tracy, C. R.,** A model of the dynamic exchanges of water and energy between a terrestrial amphibian and its environment, *Ecology,* 46, 293, 1976.

77. **Bundy, D. and Tracy, C. R.,** Behavioral responses of American toads *(Bufo americanus)* to stressful thermal and hydric environments, *Herpetologica,* 33, 455, 1977.

78. **Carey, C.,** Factors affecting body temperatures of toads, *Oecologia (Berlin),* 35, 197, 1978.

79. **Bishop, L. G. and Gordon, M. S.,** Thermal adaptation of metabolism in anuran amphibians, in *Molecular Mechanisms of Temperature Adaptation,* Prosser, C. L., Ed., American Association for the Advancement of Science, Washington, D.C., 1967, 267.

80. **Meyer, J. R. and Hegman, J. P.,** Environmental modification of sciatic nerve conduction velocity in *Rana pipiens, Am. J. Physiol.,* 220, 1383, 1971.

81. **Seymour, R. S.,** Physiological correlates of forced activity and burrowing in the spadefoot toad, *Scaphiopus hammondi, Copeia,* 1973, 435, 1973.

82. **Bennett, A. F. and Licht, P.,** Anaerobic metabolism during activity in amphibians, *Comp. Biochem. Physiol. A,* 48, 319, 1974.

83. **Meyer, J. R., Hegman, J. P., and Dingle, H.,** Environmental modification of *Rana pipiens* sciatic nerve function, *Am. J. Physiol.,* 227, 854, 1974.

84. **Carey, C.,** Aerobic and anaerobic energy expenditure during rest and activity in montane *Bufo b. boreas* and *Rana pipiens, Oecologia (Berlin),* 39, 213, 1979.

85. **Rall, J. A.,** Effects of temperature on tension, tension-dependent heat, and activation heat in twitches of frog skeletal muscle, *J. Physiol. (London),* 291, 265, 1979.

86. **Miller, K. and Hutchison, V. H.,** Aerobic and anaerobic scope for activity in the giant toad, *Bufo marinus, Physiol. Zool.,* 53, 170, 1980.

87. **Putnam, R. W. and Bennett, A. F.,** Thermal dependence of behavioural performance of anuran amphibians, *Anim. Behav.,* 29, 502, 1981.

88. **Renaud, J. M. and Stevens, E. D.,** Effect of acclimation temperature and pH on contraction of frog sartorius muscle, *Am. J. Physiol.,* 240, R301, 1981.

89. **Renaud, J. M. and Stevens, E. D.,** The interactive effects of temperature and pH on the isometric contraction of toad sartorius muscle, *J. Comp. Physiol.,* 145, 67, 1981.

90. **Rome, L. C.,** The effect of long-term exposure to different temperatures on the mechanical performance of frog muscle, *Physiol. Zool.,* 56, 33, 1983.

91. **Rome, L. C. and Kushmerick, M. J.,** Energetics of isometric contractions as a function of muscle temperature, *Am. J. Physiol.,* 244, (*Cell Physiol.,* 13), C100, 1983.

92. **Huey, R. B. and Stevenson, R. D.,** Integrating thermal physiology and ecology of ectotherms: a discussion of approaches, *Am. Zool.,* 19, 357, 1979.

93. **Hirono, M. and Rome, L. C.,** Jumping performance of frogs *(Rana pipiens)* as a function of muscle temperature, *J. Exp. Biol.,* 108, 429, 1984.

94. **Miller, K.,** Effect of temperature on sprint performance in the frog *Xenopus laevis* and the salamander *Necturus maculosus, Copeia,* 1982, 695, 1982.

95. **Preest, M. R. and Pough, F. H.,** Interactive effects of body temperature and hydration state on locomotor performance of a toad, *Bufo americanus, Am. Zool.,* 27, 5A, 1987.

96. **Bobka, M. S., Jaeger, R. G., and McNaught, D. C.,** Temperature-dependent assimilation efficiencies of two species of terrestrial salamanders, *Copeia,* 1981, 417, 1981.

97. **Jørgensen, C. B.,** External and internal control of patterns of feeding, growth and gonadal function in a temperate zone anuran, the toad *Bufo bufo, J. Zool. (London),* 210, 211, 1986.

98. **Larsen, L. O.,** Feeding in adult toads; physiology, behaviour, ecology, *Vidensk. Medd. Dan. Naturhist. Foren. Khobenhavn,* 145, 97, 1984.

99. **Myhre, K., Cabanac, M., and Myhre, G.,** Fever and behavioral temperature regulation in the frog *Rana esculenta, Acta Physiol. Scand.,* 101, 219, 1977.

100. **Muchlinski, A. E.,** The energetic cost of the fever response in three species of ectothermic vertebrates, *Comp. Biochem. Physiol. A,* 81, 577, 1985.

101. **Ritchart, J. P. and Hutchison, V. H.,** The effects of ATP and cAMP on the thermal tolerance of the mudpuppy, *Necturus maculosus, J. Therm. Biol.,* 11, 47, 1986.

102. **Hutchison, V. H. and Maness, J. D.,** The role of behavior in temperature acclimation and tolerance in ectotherms, *Am. Zool.,* 19, 367, 1979.

103. **Layne, J. R. and Claussen, D. L.,** Seasonal variation in the thermal acclimation of critical thermal maxima (CTMax) and minima (CTMin) in the salamander *Eurycea bislineata, J. Therm. Biol.,* 7, 29, 1982.

104. **Keen, W. H. and Shroeder, E. E.,** Temperature selection and tolerance in three species of *Ambystoma* larvae, *Copeia,* 1975, 523, 1975.

105. **Feder, M. E. and Pough, F. H.,** Increased temperature resistance produced by ionizing radiation in the newt *Notophthalmus v. viridescens, Copeia,* 1975, 658, 1975.

106. **Hutchison, V. H. and Hill, L. G.,** Thermal selection in the hellbender, *Cryptobranchus allegheniensis,* and the mudpuppy, *Necturus maculosus, Herpetologica,* 32, 327, 1976.

107. **Feder, M. E.,** Environmental variability and thermal acclimation of metabolism in neotropical and temperate zone salamanders, *Physiol. Zool.,* 51, 7, 1978.

108. **Feder, M. E.,** Environmental variability and thermal acclimation of metabolism in tropical anurans, *J. Therm. Biol.,* 7, 23, 1982.

109. **Burggren, W. W. and Wood, S. C.,** Respiration and acid-base balance in the salamander, *Ambystoma tigrinum:* influence of temperature acclimation and metamorphosis, *J. Comp. Physiol.,* 144, 241, 1981.

110. **Lagerspetz, K. Y. H.,** Effect of temperature acclimation on the microsomal ATPases of the frog brain, *J. Therm. Biol.,* 2, 27, 1977.

111. **Lagerspetz, K. Y. H., Harri, M. N. E., and Okslahti, R.,** The role of the thyroid in the temperature acclimation of the oxidative metabolism in the frog *Rana temporaria, Gen. Comp. Endocrinol.,* 22, 169, 1974.

112. **Lagerspetz, K. Y. H. and Skytta, M.,** Temperature compensation of sodium transport and ATPase activity in frog skin, *Acta Physiol. Scand.,* 106, 151, 1979.

113. **Shertzer, R. H., Hart, R. G., and Pavlick, F. M.,** Thermal acclimation in selected tissues of the leopard frog, *Rana pipiens, Comp. Biochem. Physiol. A*, 51, 327, 1975.

114. **Harri, M. N. E.,** The rate of metabolic temperature acclimation in the frog, *Rana temporaria, Physiol. Zool.*, 46, 148, 1973.

115. **Harri, M. N. E.,** Neural control of temperature adaptation in *Rana temporaria*, in *Effects of Temperature on Ectothermic Organisms*, Wieser, W., Ed., Springer-Verlag, Berlin, 1974, 35.

116. **Harri, M. N. E.,** The relation between thyroid activity and responsiveness to cold acclimation in the frog, *Rana temporaria, Comp. Gen. Pharmacol.*, 5, 305, 1974.

117. **Harri, M. N. E. and Hedenstam, R.,** Calorigenic effect of adrenaline and noradrenaline in the frog, *Rana temporaria, Comp. Biochem. Physiol.*, 41A, 409, 1972.

118. **Harri, M. N. E. and Tirri, R.,** Lowered sensitivity to acetylcholine in hearts from cold-acclimated rats and frogs, *Acta Physiol. Scand.*, 90, 509, 1974.

119. **Harri, M. N. E. and Talo, A.,** Effect of season and temperature acclimation on the heart rate-temperature relationship in the isolated frog's heart, *(Rana temporaria), Comp. Biochem. Physiol. A*, 52, 409, 1975.

120. **Harri, M. N. E. and Talo, A.,** Effect of season and temperature acclimation on the heart rate-temperature relationship in the frog, *Rana temporaria, Comp. Biochem. Physiol. A*, 52, 469, 1975.

121. **Baranska, J. and Wlodawer, P.,** Influence of temperature on the composition of fatty acids and lipogenesis in frog tissues, *Comp. Biochem. Physiol.*, 28, 553, 1969.

122. **Ballantyne, J. S. and George, J. C.,** An ultrastructural and histological analysis of cold acclimation in vertebrate skeletal muscle, *J. Therm. Biol.*, 3, 109, 1978.

123. **Weathers, W. W.,** Circulatory responses of *Rana catesbeiana* to temperature, season, and previous thermal history, *Comp. Biochem. Physiol. A*, 51, 43, 1975.

124. **Weathers, W. W.,** Influence of temperature acclimation on oxygen consumption, haemodynamics and oxygen transport in bullfrogs, *Aust. J. Zool.*, 24, 321, 1976.

125. **DeCosta, J., Alonso-Bedate, M., and Fraile, A.,** Temperature acclimation in amphibians: changes in lactate dehydrogenase activities and isoenzyme patterns in several tissues from adult *Discoglossus pictus pictus* (Otth.) tadpoles, *Comp. Biochem. Physiol. B*, 70, 331, 1981.

126. **Enig, M., Ramsay, J., and Eby, D.,** Effect of temperature on pyruvate metabolism in the frog: the role of lactate dehydrogenase isozymes, *Comp. Biochem. Physiol. B*, 53, 145, 1976.

127. **Tsugawa, K.,** Direct adaptation of cells to temperature: similar changes of LDH isozyme patterns by *in vitro* and *in situ* adaptations in *Xenopus laevis, Comp. Biochem. Physiol. B*, 55, 259, 1976.

128. **Tsugawa, K.,** Thermal dependence in kinetic properties of lactate dehydrogenase from the African clawed toad, *Xenopus laevis, Comp. Biochem. Physiol. B*, 66, 459, 1980.

129. **Alexandrov, V. Ya.,** Cytophysical and cytoecological investigations of resistance of plant cells towards the action of high and low temperatures, *Q. Rev. Biol.*, 39, 35, 1964.

130. **Maness, J. D. and Hutchison, V. H.,** Acute adjustment of thermal tolerance in vertebrate ectotherms following exposure to critical thermal maxima, *J. Therm. Biol.*, 5, 225, 1980.

131. **Hutchison, V. H. and Murphy, K.,** Behavioral thermoregulation in the salamander *Necturus maculosus* after heat shock, *Comp. Biochem. Physiol. A*, 82, 391, 1985.

132. **Rutledge, P. S., Spotila, J. R., and Easton, D. P.,** Heat hardening in response to two types of heat shock in the lungless salamanders *Eurycea bislineata* and *Desmognathus ochrophaeus, J. Therm. Biol.*, 12, 235, 1987.

133. **Pough, F. H.,** Natural daily temperature acclimation of eastern red efts, *Notophthalmus v. viridescens* (Rafinesque) (Amphibia: Caudata), *Comp. Biochem. Physiol. A*, 47, 71, 1974.

134. **Feder, M. E.,** Biochemical and metabolic correlates of thermal acclimation in the rough-skinned newt, *Taricha granulosa, Physiol. Zool.*, 56, 513, 1983.

135. **Ashburner, M. and Bonner, J. J.,** The induction of gene activity in *Drosophila* by heat shock, *Cell*, 17, 241, 1979.

136. **Ashburner, M.,** The effects of heat shock and other stress on gene activity: an introduction, in *Heat Shock from Bacteria to Man*, Ashburner, M., Schlesinger, M., and Tissieres, A., Eds., Cold Spring Harbor Press, Cold Spring Harbor, NY, 1982, 1.

137. **Rutledge, P. S., Easton, D. P., and Spotila, J. R.,** Heat shock protein from the lungless salamanders *Eurycea bislineata* and *Desmognathus ochrophaeus, Comp. Biochem. Physiol. B*, 88, 13, 1987.

138. **Easton, D. P., Rutledge, P. S., and Spotila, J. R.,** Heat shock protein induction and induced thermal tolerance are independent in adult salamanders, *J. Exp. Zool.*, 241, 263, 1987.

139. **Koskela, P. and Pasanen, S.,** Effects of thermal acclimation on seasonal liver and muscle glycogen content in the common frog, *Rana temporaria* L., *Comp. Biochem. Physiol.*, 50, 723, 1975.

140. **Lagerspetz, K. Y. H.,** Interactions of season and temperature acclimation in the control of metabolism in the Amphibia, *J. Therm. Biol.*, 2, 223, 1977.

141. **Farquhar, M. G. and Palade, G. E.,** Functional organization of the amphibian skin, *Proc. Natl. Acad. Sci. U.S.A.*, 51, 569, 1964.

142. **Lindeman, B. and Vôute, C.,** Structure and function of the epidermis, in *Frog Neurobiology,* Llinás, R. and Precht, W., Eds., Springer-Verlag, Berlin, 1976.

143. **Whitear, M.,** A functional comparison between the epidermis of fish and of amphibians, *Symp. Zool. Soc. London,* 39, 291, 1977.

144. **Lillywhite, H. B. and Licht, P.,** A comparative study of integumentary mucous secretions in amphibians, *Comp. Biochem. Physiol. A,* 51, 937, 1975.

145. **Lillywhite, H. B.,** Thermal modulation of cutaneous mucus discharge as a determinant of evaporative water loss in the frog, *Rana catesbeiana, Z. Vgl. Physiol.,* 73, 84, 1971.

146. **Shoemaker, V. H. and Nagy, K. A.,** Osmoregulation in amphibians and reptiles, *Ann. Rev. Physiol.,* 39, 449, 1977.

147. **Dole, J. W.,** Summer movements of adult leopard frogs, *Rana pipiens* Schreber, in northern Michigan, *Ecology,* 46, 236, 1965.

148. **Dole, J. W.,** The role of substrate moisture and dew in the water economy of leopard frogs, *Rana pipiens, Copeia,* 1967, 141, 1967.

149. **Lee, A. K.,** Water economy of the burrowing frog, *Helioporus eyrei* (Gray), *Copeia,* 1968, 741, 1968.

150. **Pough, F. H., Taigen, T. L., Stewart, M. M., and Brussard, P. F.,** Behavioral modification of evaporative water loss by a Puerto Rican frog, *Ecology,* 64, 244, 1983.

151. **Bentley, P. J.,** *Endocrines and Osmoregulation. A Comparative Account of the Regulation of Water and Salt in Vertebrates,* Springer-Verlag, Berlin, 1971.

152. **Koefoed-Johnsen, V.,** Control mechanisms in amphibians, in *Mechanisms of Osmoregulation in Animals: Maintenance of Cell Volume,* Gilles, R., Ed., John Wiley & Sons, New York, 1979, chap. 6.

153. **Bentley, P. J. and Main, A. R.,** Zonal differences in permeability of the skin of some anuran Amphibia, *Am. J. Physiol.,* 2, 361, 1972.

154. **Roth, J. J.,** Vascular supply to the ventral pelvic region of anurans as related to water balance, *J. Morphol.,* 140, 443, 1973.

155. **Christensen, C. N.,** Adaptations in the water economy of some anuran Amphibia, *Comp. Biochem. Physiol. A,* 47, 1035, 1974.

156. **Kobelt, F. and Linsenmair, K. E.,** Adaptations of the reed frog *Hyperolius viridiflavus* (Amphibia, Anura, Hyperoliidae) to its arid environment. I. The skin of *Hyperolius viridiflavus nitidulus* in wet and dry season conditions, *Oecologia (Berlin),* 68, 533, 1986.

157. **Geise, W. and Linsenmair, K. E.,** Adaptations of the reed frog *Hyperolius viridiflavus* (Amphibia, Anura, Hyperoliidae) to its arid environment. II. Some aspects of the water economy of *Hyperolius viridiflavus nitidulus* under wet and dry season conditions, *Oecologia (Berlin),* 68, 542, 1986.

158. **van Berkum, F., Pough, F. H., Stewart, M. M., and Brussard, P. F.,** Altitudinal and interspecific differences in the rehydration abilities of Puerto Rican frogs *(Eleutherodactylus), Physiol. Zool.,* 55, 130, 1982.

159. **Main, A. R. and Bentley, P. J.,** Water relations of Australian burrowing frogs and tree frogs, *Ecology,* 45, 379, 1964.

160. **Ruibal, R.,** The adaptive value of bladder water in the toad, *Bufo cognatus, Physiol. Zool.,* 35, 218, 1962.

161. **McClanahan, L. L.,** Nitrogen excretion in arid-adapted amphibians, in *Environmental Physiology of Desert Organisms,* Hadley, N. F., Ed., Dowden, Hutchinson & Ross, Stroudsburg, PA, 1975.

162. **Shoemaker, V. H., Balding, D., Ruibal, R., and McClanahan, L. L., Jr.,** Uricotelism and low evaporative water loss in a South American frog, *Science,* 175, 1018, 1971.

163. **Loveridge, J. P.,** Observations on nitrogenous excretion and water relations of *Chiromantis xerampelina* (Amphibia, Anura), *Arnoldia,* 5, 1, 1970.

164. **Loveridge, J. P.,** Strategies of water conservation in southern African frogs, *Zool. Afr.,* 11, 319, 1976.

165. **Shoemaker, V. and McClanahan, L.,** Evaporative water loss, nitrogen excretion and osmoregulation in phyllomedusine frogs, *J. Comp. Physiol.,* 100, 331, 1975.

166. **Balinsky, J. B., Chemaly, S. M., Currins, A. E., Lee, A. R., Thompson, R. L., and Van der Westhuizen, D. R.,** A comparative study of the enzymes of urea and uric acid metabolism in different species of Amphibia, and in the adaptation to the environment of the tree frog *Chiromantis xerampelina* Peters, *Comp. Biochem. Physiol. B,* 54, 549, 1976.

167. **Drewes, R. C., Hillman, S. S., Putnam, R. W., and Sokol, D. M.,** Water, ion and nitrogen balance in the treefrog, *Chiromantis petersi,* with comments on the structure of the integument, *J. Comp. Physiol.,* 116, 257, 1977.

168. **Cohen, P. P.,** Biochemical aspects of metamorphosis: transition from ammonotelism to ureotelism, *Harv. Lect. Ser.,* 60, 119, 1966.

169. **Candelas, G. C. and Gomez, M.,** Nitrogen excretion in tadpoles of *Leptodactylus albilabris* and *Rana catesbeiana, Am. Zool.,* 3, 521, 1963.

170. **Martin, A. A. and Cooper, A. K.,** The ecology of terrestrial anuran eggs, genus *Crinia* (Leptodactylidae), *Copeia,* 1972, 163, 1972.

171. **Shoemaker, V. H. and McClanahan, L. L., Jr.,** Nitrogen excretion in the larvae of a land-nesting frog *(Leptodactylus bufonius), Comp. Biochem. Physiol. A,* 44, 1149, 1973.

172. **Cei, J. M. and Capurro, L. F.,** Biología y desarollo de *Eupsophus taeniatus* Girard, *Invest. Zool. Chil.,* 4, 150, 1958.

173. **Balinsky, J. B.,** Adaptation of nitrogen metabolism to hyperosmotic environment in Amphibia, *J. Exp. Zool.,* 215, 335, 1981.

174. **Ruibal, R., Tevis, L., Jr., and Roig, V.,** The terrestrial ecology of the spadefoot toad *Scaphiopus hammondi, Copeia,* 1969, 571, 1969.

175. **McClanahan, L. L.,** Changes in body fluids in burrowed spadefoot toads as a function of soil water potential, *Copeia,* 1972, 209, 1972.

176. **Gordon, M. S., Schmidt-Nielsen, K., and Kelly, H. M.,** Osmotic regulation in the crab-eating frog *(Rana cancrivora), J. Exp. Biol.,* 38, 659, 1961.

177. **Gordon, M. S. and Tucker, V. A.,** Osmotic regulation in the tadpoles of the crab-eating frog *(Rana cancrivora), J. Exp. Biol.,* 42, 437, 1965.

178. **Schmidt-Nielsen, K. and Lee, P.,** Kidney function in the crab-eating frog *(Rana cancrivora), J. Exp. Biol.,* 39, 167, 1962.

179. **Dicker, S. E. and Elliott, A. B.,** Water uptake by the crab-eating frog, *Rana cancrivora,* as affected by osmotic gradients and by neurohypophyseal hormones, *J. Physiol. (London),* 207, 119, 1970.

180. **Gasser, K. W. and Miller, B. T.,** Osmoregulation of larval blotched tiger salamanders, *Ambystoma tigrinum melanostictum,* in saline environments, *Physiol. Zool.,* 59, 643, 1986.

181. **Degani, G.,** Urea tolerance and osmoregulation in *Bufo viridis* and *Rana ridibunda, Comp. Biochem. Physiol. A.,* 82, 833, 1985.

182. **Gordon, M. S.,** Osmotic regulation in the green toad *(Bufo viridis), J. Exp. Biol.,* 39, 261, 1962.

183. **Gordon, M. S.,** Intracellular osmoregulation in skeletal muscle during salinity adaptation in two species of toads, *Biol. Bull.,* 2, 218, 1965.

184. **Katz, U.,** NaCl adaptation in *Rana ridibunda* and a comparison with the euryhaline toad *Bufo viridis, J. Exp. Biol.,* 63, 763, 1975.

185. **Katz, U.,** Salt induced changes in sodium transport across the skin of the euryhaline toad *Bufo viridis, J. Physiol. (London),* 247, 537, 1975.

186. **Liggins, G. W. and Grigg, G. C.,** Osmoregulation of the cane toad, *Bufo marinus,* in salt water, *Comp. Biochem. Physiol. A,* 82, 613, 1985.

187. **Degani, G., and Nevo, E.,** Osmotic stress and osmoregulation of tadpoles and juveniles of *Pelobates syriacus, Comp. Biochem. Physiol. A,* 83, 365, 1986.

188. **Freda, J.,** The influence of acidic pond water on amphibians: a review, *Water Air Soil Pollut.,* 30, 439, 1986.

189. **Beuchat, C. A., Pough, F. H., and Stewart, M. M.,** Response to simultaneous dehydration and thermal stress in three species of Puerto Rican frogs, *J. Comp. Physiol. B,* 154, 579, 1984.

190. **Gatten, R. E.,** Activity metabolism of anuran amphibians: tolerance to dehydration, *Physiol. Zool.,* 60, 576, 1987.

191. **Sherman, E. and Stadlen, S. G.,** The effect of dehydration on rehydration and metabolic rate in a lunged and lungless salamander, *Comp. Biochem. Physiol. A,* 85, 483, 1986.

192. **Stefanski, M., Gatten, R. E., Jr., and Pough, F. H.,** Activity metabolism of salamanders: tolerance to dehydration, *J. Herpetol.,* 23, 45, 1989.

193. **Shoemaker, V. H.,** The effects of dehydration on electrolyte concentrations in a toad, *Bufo marinus, Comp. Biochem. Physiol.,* 13, 261, 1964.

194. **Hillman, S. S.,** Some effects of dehydration on internal distributions of water and solutes in *Xenopus laevis, Comp. Biochem. Physiol. A,* 61, 303, 1978.

195. **Boutilier, R. G., Randall, D. J., Shelton, G., and Toews, D. P.,** Acid-base relationships in the blood of the toad, *Bufo marinus.* II. The effects of dehydration, *J. Exp. Biol.,* 82, 345, 1979.

196. **Degani, G. and Warburg, M. R.,** Changes in concentrations of ions and urea in both plasma and muscle tissue in a dehydrated hylid anuran, *Comp. Biochem. Physiol. A,* 77, 357, 1984.

197. **Hillman, S. S., Zygmunt, A., and Baustian, M.,** Transcapillary fluid forces during dehydration in two amphibians, *Physiol. Zool.,* 60, 339, 1987.

198. **Howarth, J. V.,** The behavior of frog muscle in hypertonic solutions, *J. Physiol. (London),* 144, 167, 1958.

199. **Gallardo, R., Pang, P. K. T., and Sawyer, W. H.,** Neural influences on bullfrog renal functions, *Proc. Soc. Exp. Biol. Med.,* 165, 233, 1980.

200. **Katz, U.,** The effect of dehydration on the *in vivo* acid-base status of the blood in the toad *Bufo viridis, J. Exp. Biol.,* 88, 403, 1980.

201. **Tufts, B. L. and Toews, D. P.,** Renal function and acid-base balance in the toad *Bufo marinus* during short-term dehydration, *Can. J. Zool.,* 64, 1054, 1986.

202. **Baxter, C. F.,** Intrinsic amino acid levels and the blood-brain barrier, in *Progress in Brain Research,* 29, Lajtha, A. and Ford, D., Eds., Elsevier, Amsterdam, 1968, 429.

203. **Baxter, C. F. and Baldwin, R. A.,** A functional role for amino acids in the adaptation of tissues from the nervous system to alterations in envrionmental osmolarity, in *Amino Acids as Chemical Transmitters,* Fohnum, F., Ed., Plenum Press, New York, 1979, 599.

204. **Baxter, C. F., Baldwin, R. A., Tachiki, K. H., Rose, B. B., and Dole, J. W.,** The effects of an altered plasma osmolality upon behavior and cellular levels of amino acids in the nervous system, *Int. Soc. Neurochem. Abstr.,* 7, 214, 1979.

205. **Baxter, C. F., Dole, J. W., Tachiki, K. H., Rose, B. B., and Baldwin, R. A.,** Amphibian feeding responses as related to temporal and regional changes in amino acid patterns of the central nervous system, *Trans. Am. Soc. Neurochem.,* 8, 601, 1982.

206. **Tachiki, K. H. and Baxter, C. F.,** Role of carbon dioxide fixation: blood aspartate and glutamate in the adaptation of amphibian brain tissues to a hyperosmotic internal environment, *Neurochem. Res.,* 5, 993, 1980.

207. **Tachiki, K. H., Baxter, C. F., Dole, J. W., and Rose, B. B.,** Amino acids and feeding behavior during long term osmotic adaptation, *Trans. Am. Soc. Neurochem.,* 12, 122, 1981.

208. **Dole, J. W., Rose, B. B., and Baxter, C. F.,** Hyperosmotic saline environment alters feeding behavior in the western toad, *Bufo boreas, Copeia,* 1985, 645, 1985.

209. **Hillman, S. S.,** The roles of oxygen delivery and electrolyte levels in the dehydrational death of *Xenopus laevis, J. Comp. Physiol. B,* 129, 169, 1978.

210. **Hillman, S. S.,** Physiological correlates of differential dehydration tolerance in anuran amphibians, *Copeia,* 1980, 125, 1980.

211. **Hillman, S. S.,** The effects of *in vivo* and *in vitro* hyperosmolality on skeletal muscle performance in the amphibians *Rana pipiens* and *Scaphiopus couchi, Comp. Biochem. Physiol. A,* 73, 709, 1982.

212. **Hillman, S. S.,** Inotropic influence of dehydration and hyperosmolal solutions on amphibian cardiac muscle, *J. Comp. Physiol.,* 154, 325, 1984.

213. **Hillman, S. S., Withers, P. C., Hedrick, M. S., and Kimbal, P. B.,** The effects of erythrocythemia on blood viscosity, maximal systemic oxygen transport capacity and maximal rates of oxygen consumption in an amphibian, *J. Comp. Physiol.,* 155, 577, 1985.

214. **Hillman, S. S.,** Dehydrational effects on cardiovascular and metabolic capacity in two amphibians, *Physiol. Zool.,* 60, 608, 1987.

215. **Lee, A. K. and Mercer, E. H.,** Cocoon surrounding desert dwelling frogs, *Science,* 157, 87, 1967.

216. **Reno, H. W., Gehlbach, F. R., and Turner, R. A.,** Skin and aestivational cocoon of the aquatic amphibian *Siren intermedia, Copeia,* 1972, 625, 1972.

217. **McClanahan, L. L., Shoemaker, V. H., and Ruibal, R.,** Structure and function of the cocoon of a ceratophryd frog, *Copeia,* 1976, 179, 1976.

218. **McClanahan, L. L., Ruibal, R., and Shoemaker, V. H.,** Rate of cocoon formation and its physiological correlates in a ceratophryd frog, *Physiol. Zool.,* 56, 430, 1983.

219. **Loveridge, J. P. and Crayé, G.,** Cocoon formation in two species of southern African frogs, *S. Afr. J. Sci.,* 75, 18, 1979.

220. **Loveridge, J. P. and Withers, P. C.,** Metabolism and water balance of active and cocooned African bullfrogs, *Pyxicephalus adspersus, Physiol. Zool.,* 54, 203, 1981.

221. **Ruibal, R. and Hillman, S. S.,** Cocoon structure and function in the burrowing hylid frog, *Pternohyla fodiens, J. Herpetol.,* 15, 403, 1981.

222. **Blaylock, L. A., Ruibal, R., Platt-Aloia, K.,** Skin structure and wiping behavior of phyllomedusine frogs, *Copeia,* 1976, 283, 1976.

223. **Seibert, E. A., Lillywhite, H. B., Wassersug, R. J.,** Cranial coossification in frogs: relationship to rate of evaporative water loss, *Physiol. Zool.,* 47, 261, 1974.

224. **Wygoda, M. L.,** Low cutaneous evaporative water loss in arboreal frogs, *Physiol. Zool.,* 57, 329, 1984.

225. **Withers, P. C., Hillman, S. S., Drewes, R. C., and Sokol, O. M.,** Water loss and nitrogen excretion in sharp-nosed frogs *(Hyperolius nasutus:* Anura Hyperoliidae), *J. Exp. Biol.,* 97, 335, 1982.

226. **Withers, P. C., Louw, G., and Nicholson, S.,** Water loss, oxygen consumption, and colour change in "waterproof" reed frogs, *S. Afr. J. Sci.,* 78, 30, 1982.

227. **Bennett, A. F. and Licht, P.,** Relative contributions of anaerobic and aerobic energy production during activity in Amphibia, *J. Comp. Physiol.,* 87, 351, 1973.

228. **Hutchison, V. H. and Miller, K.,** Anaerobic capacity of amphibians, *Comp. Biochem. Physiol. A,* 63, 213, 1979.

229. **Putnam, R. W.,** The basis for differences in lactic acid content after activity in different species of anuran amphibians, *Physiol. Zool.,* 52, 509, 1979.

230. **Taigen, T. L., Emerson, S. B., and Pough, F. H.,** Ecological correlates of anuran exercise physiology, *Oecologia (Berlin),* 52, 49, 1982.

231. **Gatten, R. E.**, The uses of anaerobiosis by amphibians and reptiles, *Am. Zool.*, 25, 945, 1985.
232. **Feder, M. E., Full, R. J., and Piiper, J.**, Elimination kinetics of acetylene and Freon 22 in resting and active lungless salamanders, *Respir. Physiol.*, 72, 229, 1988.
233. **Full, R. J., Anderson, B. D., Finnerty, C. M., and Feder, M. E.**, Exercising with and without lungs. I. The effects of metabolic cost, maximal oxygen transport, and body size on terrestrial locomotion in salamander species, *J. Exp. Biol.*, 138, 471, 1988.
234. **Feder, M. E.**, Exercising with and without lungs. II. Experimental elimination of pulmonary and buccopharyngeal gas exchange in individual salamanders *(Ambystoma tigrinum)*, *J. Exp. Biol.*, 138, 487, 1988.
235. **Weibel, R. and Taylor, C. R., Eds.**, Design of the mammalian respiratory system, *Respir. Physiol.*, 44, 1, 1981.
236. **Scheur, J. and Tipton, C. M.**, Cardiovascular adaptations to physical training, *Annu. Rev. Physiol.*, 39, 221, 1977.
237. **Clausen, J. P.**, Effect of physical training on cardiovascular adjustments to exercise in man, *Physiol. Rev.*, 57, 779, 1977.
238. **Hillman, S. S. and Withers, P. C.**, An analysis of respiratory surface area as a limit to activity metabolism in anurans, *Can. J. Zool.*, 57, 2100, 1979.
239. **Hillman, S. S.**, Cardiovascular correlates of maximal oxygen consumption rates in anuran amphibians, *J. Comp. Physiol. B*, 128, 169, 1976.
240. **Hillman, S. S.**, The effect of anaemia on metabolic performance in the frog, *Rana pipiens*, *J. Exp. Zool.*, 211, 107, 1980.
241. **Walsberg, G. E., Lea, M. S., and Hillman, S. S.**, Individual variation in maximum aerobic capacity: cardiovascular and enzymatic correlates in *Rana catesbeiana*, *J. Exp. Zool.*, 239, 1, 1986.
242. **Becker, B., Rathscheck, H., Muller, H. K., and Schroeder, W.**, Sauerstoffdruk, Sauerstoffverbrauch und Durchblutung in der Extremitatenmuskulatur des wachen Frosches, *Pfluegers Arch. Ges. Physiol. Menschen Tiere*, 321, 15, 1970.
243. **Pough, F. H. and Kamel, S.**, Post-metamorphic change in activity metabolism of anurans in relation to life history, *Oecologia*, 65, 138, 1984.
244. **Wells, K. D. and Taigen, T. L.**, Reproductive behavior and aerobic capacities of male American toad *(Bufo americanus)*: Is behavior constrained by physiology?, *Herpetologica*, 40, 292, 1984.
245. **Smith, R. S. and Ovalle, W. K.**, Varieties of fast and slow extrafusal muscle fibers in amphibian hind limb muscles, *J. Anat.*, 116, 1, 1973.
246. **Lannergren, J.**, Structure and function of twitch and slow fibers in amphibian skeletal muscle, in *Basic Mechanics of Ocular Motility and Their Clinical Implication*, Lennerstrand, G. and Bach-y-Rita, R., Eds., Pergamon Press, New York, 1975, 63.
247. **Forrester, T. and Schmidt, H.**, An electrophysiological investigation of the slow muscle fiber system in the frog rectus abdominus muscle, *J. Physiol.(London)*, 207, 447, 1970.
248. **Uhrik, B. and Schmidt, H.**, Distribution of slow muscle fibers in the frog rectus abdominus muscle, *Pfluegers Arch. Ges. Physiol. Menschen Tiere*, 225, 627, 1973.
249. **Lannergren, J. and Smith, R. S.**, Types of muscle fibers in toad skeletal muscle, *Acta Physiol. Scand.*, 68, 263, 1966.
250. **Putnam, R. W. and Bennett, A. F.**, Histochemical, enzymatic, and contractile properties of skeletal muscle of three anuran amphibians, *Am. J. Physiol.*, 244, (*Reg. Int., Comp. Physiol.*, 13) R558, 1973.
251. **Sperry, D. G.**, Fiber type composition and postmetamorphic growth of anuran hindlimb muscles, *J. Morphol.*, 170, 321, 1981.
252. **Bennett, A. F.**, Enzymatic correlates of activity metabolism in anuran amphibians, *Am. J. Physiol.*, 226, 1149, 1974.
253. **Baldwin, J., Friedman, G., and Lillywhite, H.**, Adaptation to temporary muscle anoxia in anurans: activities of glycolytic enzymes in muscles from species differing in their ability to produce lactate during exercise, *Aust. J. Zool.*, 25, 15, 1977.
254. **Taigen, T. L. and Wells, K. D.**, Energetics of vocalization by an anuran amphibian *(Hyla versicolor)*, *J. Comp. Physiol. B*, 155, 163, 1985.
255. **Taigen, T. L., Wells, K. D., and Marsh, R. L.**, The enzymatic basis of high metabolic rates in calling frogs, *Physiol. Zool.*, 58, 719, 1985.
256. **Olmo, E. and Morescalchi, A.**, Evolution of the genome and cell size in salamanders, *Experientia*, 31, 804, 1975.
257. **Olmo, E. and Morescalchi, A.**, Genome and cell sizes in frogs: a comparison with salamanders, *Experientia*, 34, 44, 1978.
258. **Horner, H. A. and MacGregor, H. C.**, C values and cell volume: their significance in the evolution and development of amphibians, *J. Cell Sci.*, 63, 135, 1983.
259. **Fankhause, G.**, The effects of changes in chromosome number on amphibian development, *Q. Rev. Biol.*, 20, 20, 1945.
260. **Wasserman, A. O.**, Polyploidy in the common tree toad, *Hyla versicolor* Le Conte, *Science*, 167, 385, 1970.

261. **Maxson, L., Pepper, E., and Maxson, R. D.,** Immunological resolution of a diploid-tetraploid species complex of treefrogs, *Science,* 197, 1012, 1977.
262. **Ralin, D. B.,** "Resolution" of diploid-tetraploid treefrogs, *Science,* 202, 335, 1978.
263. **Ralin, D. B., Romano, M. A., and Kilpatrick, C. W.,** The tetraploid treefrog *Hyla versicolor:* evidence for a single origin from the diploid *H. chrysoscelis, Herpetologica,* 39, 212, 1983.
264. **Bachmann, K. and Bogart, J. P.,** Comparative cytochemical measurements in the diploid-tetraploid species pair of hylid frogs *Hyla chrysoscelis* and *H. versicolor, Cytogenet. Cell Genet.,* 15, 186, 1975.
265. **Cash, M. N. and Bogart, J. P.,** Cytological differentiation of the diploid-tetraploid species pair of North American treefrogs (Amphibia, Anura, Hylidae), *J. Herpetol.,* 12, 555, 1978.
266. **Toivonen, I. A., Crowe, D. T., Detrick, R. J., Klemann, S. W., and Vaughn, J. C.,** Ribosomal RNA gene number and sequence divergence in the diploid-tetraploid species pair of North American hylid treefrogs, *Biochem. Genet.,* 21, 299, 1983.
267. **Ralin, D. B.,** Evolutionary aspects of mating call variation in a diploid-tetraploid species complex of treefrogs (Anura), *Evolution,* 31, 721, 1977.
268. **Green, D. M.,** Size differences in adhesive toe-pad cells of treefrogs of the diploid-polyploid *Hyla versicolor* complex, *J. Herpetol.,* 14, 15, 1980.
269. **Goniakowska, L.,** The respiration of erythrocytes of some amphibians *in vitro, Bull. Acad. Pol. Sci.,* 18, 793, 1970.
270. **Goniakowska, L.,** Metabolism, resistance to hypotonic solutions, and ultrastructure of erythrocytes of five amphibian species, *Acta Biol. Cracov., Ser. Zool.,* 16, 113, 1973.
271. **Monnickendam, M. A. and Balls, M.,** The relationship between cell sizes, respiration rates and survival of amphibian tissues in long-term organ cultures, *Comp. Biochem. Physiol. A,* 44, 871, 1973.
272. **Kamel, S., Marsden, J. E., and Pough, F. H.,** Diploid and tetraploid grey treefrogs *(Hyla chrysoscelis* and *H. versicolor)* have similar metabolic rates, *Comp. Biochem. Physiol. A,* 82, 217, 1985.
273. **Berger, L.,** Systematics and hybridization in the *Rana esculenta* complex, in *The Reproductive Biology of the Amphibians,* Taylor, D. H. and Guttman, S. I., Eds., Plenum Press, New York, 1977, 367.
274. **Graf, J.-D. and Polls, M.,** Evolutionary genetics of the *Rana esculenta* hybrid complex, in *Evolution and Ecology of Unisexual Vertebrates,* Dawley, R. M. and Bogart, J. P., Eds., New York State Museum, Albany, in press, 1989.
275. **Uzzell, T. M., Jr.,** Natural triploidy in salamanders related to the *Ambystoma jeffersonianum* complex (Amphibia, Caudata), *Copeia,* 1964, 257, 1964.
276. **Sessions, S. K.,** Cytogenetics of diploid and triploid salamanders in the *Ambystoma jeffersonianum* complex, *Chromosoma,* 84, 599, 1982.
277. **Morris, M. A. and Brandon, R. A.,** Gynogenesis and hybridization between *Ambystoma platineum* and *A. texanum* in Illinois, *Copeia,* 1984, 324, 1984.
278. **Bogart, J. P.,** A mechanism for gene exchange via all-female hybrids, in *Evolution and Ecology of Unisexual Vertebrates,* Dawley, R. M. and Bogart, J. P., Eds., New York State Museum, Albany, in press, 1989.
279. **Lowcock, L. A.,** Biogeography of hybrid complexes of *Ambystoma:* interpreting unisexual-bisexual genetic data in space and time, in *Evolution and Ecology of Unisexual Vertebrates,* Dawley, R. M. and Bogart, J. P., Eds., New York State Museum, Albany, 1988 (in press).
280. **Marsden, J. E.,** Single Locus Segregation Patterns of Allozymes Encoding Loci in the Tetraploid Treefrog *Hyla versicolor* (Anura, Amphibia, Hylidae), Master's thesis, Cornell University, Ithaca, NY, 1985.
281. **Lutz, B.,** Venomous toads and frogs, in *Venomous Animals and Their Venoms,* Vol. 2, Bücherl, W. and Buckley, E., Eds., Academic Press, New York, 1971, chap. 37.
282. **Deulofeu, V. and Rúveda, E. A.,** The basic constituents of toad venom, in *Venomous Animals and Their Venoms,* Vol. 2, Bücherl, W. and Buckley, E., Eds., Academic Press, New York, 1971, chap. 38.
283. **Daly, J. W. and Witkop, B.,** Chemistry and pharmacology of frog venoms, in *Venomous Animals and Their Venoms,* Vol. 2, Bücherl, W. and Buckley, E., Eds., Academic Press, New York, 1971, chap. 39.
284. **Meyer, K. and Linde, H.,** Collection of toad venoms and chemistry of the toad venom steroids, in *Venomous Animals and Their Venoms,* Vol. 2, Bücherl, W. and Buckley, E., Eds., Academic Press, New York, 1971, chap. 40.
285. **Habermehl, G.,** Toxicology, pharmacology, chemistry, and biochemistry of salamander venom, in *Venomous Animals and Their Venoms,* Vol. 2, Bücherl, W. and Buckley, E., Eds., Academic Press, New York, 1971, chap. 41.
286. **Myers, C. W. and Daly, J. W.,** Preliminary evaluation of skin toxins and vocalizations in taxonomic and evolutionary studies of poison-dart frogs (Dendrobatidae), *Bull. Am. Mus. Nat. Hist.,* 157, 173, 1976.
287. **Myers, C. W., Daly, J. W., and Malkin, B.,** A dangerously toxic new frog *(Phyllobates)* used by Embera Indians of western Colombia, with discussion of blowgun fabrication and dart poisoning, *Bull. Am. Mus. Nat. Hist.,* 161, 307, 1978.
288. **Daly, J. W., Brown, G. B., Mensah-Dwumah, M., and Myers, C. W.,** Classification of skin alkaloids from neotropical poison-dart frogs (Dendrobatidae), *Toxicon,* 16, 163, 1978.

289. **Flier, J., Edwards, M. W., Daly, J. W., and Myers, C. W.,** Widespread occurrence in frogs and toads of skin compounds interacting with the oubain site of Na$^+$, K$^+$-ATPase, *Science,* 208, 503, 1980.

290. **Daly, J. W., Hignet, R. J., and Myers, C. W.,** Occurrence of skin alkaloids in non-dendrobatid frogs from Brazil (Bufonidae), Australia (Myobatrachidae) and Madagascar (Mantellinae), *Toxicon,* 22, 905, 1984.

291. **Daly, J. W., Spande, T. F., Whittaker, N., Highet, R. J., Feigl, D., Nishimori, N., Tokuyama, T., and Myers, C. W.,** Alkaloids from dendrobatid frogs: structures of two ω-hydroxy congeners of 3-butyl-5-propylindolisidine and occurrence of 2,5-disubstituted pyrrolidines and a 2,6-disubstituted piperidine, *J. Nat. Prod.,* 49, 265, 1986.

292. **Stewart, K. W.,** Amphibians, in *Guide to the Care and Use of Experimental Animals,* Vol. 2, Canadian Council on Animal Care, Ottawa, 1980-1984, chap. 2.

293. **Nace, G. W., Culley, D. D., Emmons, M. B., Gibb, E. L., Hutchison, V. H., and McKinnell, R. G.,** *Amphibians: Guidelines for the Breeding, Care and Management of Laboratory Animals,* Subcommittee on Amphibian Standards, Institute of Laboratory Animal Resources (BAS/NRC), Washington, D.C., 1974.

294. **Anon.,** *Guidelines for Use of Live Amphibians and Reptiles in Field Research,* American Society of Ichthyologists and Herpetologists, The Herpetologists' League, Society for the Study of Amphibians and Reptiles, 1987.

295. **Culley, D. D.,** Culture and management of the laboratory frog, *Lab Anim.,* 5, 30, 1976.

296. **Zimmerman, H.,** *Tropical Frogs,* TFH Publ., Neptune, NJ, 1979.

297. **Modzelewski, E. H., Jr. and Culley, D. D., Jr.,** Growth responses of the bullfrog, *Rana catesbeiana,* fed various live foods, *Herpetologica,* 30, 397, 1974.

298. **Lehman, G.,** The effects of four arthropod diets on the body and organ weights of the leopard frog, *Rana pipiens,* during vitellogenesis, *Growth,* 42, 505, 1978.

299. **Nace, G. W. and Buttner, J. K.,** Culture of crickets: a food for amphibians and reptiles, *Axolotl Newsl.,* 12, 23, 1983.

300. **Buttner, J. K.,** Conditioning laboratory frogs to accept non-living food, *Lab. Anim.,* May/June, 40, 1984.

301. **Master, C. O.,** *Encyclopedia of Live Foods,* T. F. H. Publ., Neptune, NJ, 1975.

302. Anon., Carolina Arthropods Manual, Carolina Biological Supply Co., Burlington, NC, 1982.

303. **Frost, J. S.,** A time-efficient, low cost method for the laboratory rearing of frogs, *Herpetol. Rev.,* 13, 73, 1982.

304. **Reichenbach-Klinke, H. and Elkan, E.,** *The Principal Diseases of Lower Vertebrates,* Academic Press, London, 1965.

305. **Elkan, E. and Reichenbach-Klinke, H.,** *Color Atlas of the Diseases of Fishes, Amphibians and Reptiles,* T. F. H. Publ., Neptune, NJ, 1974.

306. **Cosgrove, G. E.,** Amphibian diseases, in *Current Veterinary Therapy, VI, Small Animal Practice,* Kirk, R. W., Ed., W. B. Saunders, Philadelphia, 1977.

307. **Marcus, L. C.,** *Veterinary Biology and Medicine of Captive Amphibians and Reptiles,* Lea & Febiger, Philadelphia, 1981.

308. **Hoff, G. L., Frye, F. L., and Jacobsen, E. R., Eds.,** *Diseases of Amphibians and Reptiles,* Plenum Press, New York, 1984.

309. **Pisanetz, Ye M.,** New polyploid species of *Bufo danatensis* Pizanetz, sp. nov., from the Turkmen SSR, *Sopov Akad. Nauk Ukr. Rsp. Ser. B Heol. Biol. Nauky,* No. 3, 277, 1978.

310. **Mazik, E. Yu., Kadyrova, B. K., and Tokosunov, A. T.,** Peculiarities of the karyotype of the green toad *(Bufo viridis)* of Kirghizia, *Zool. Zh.,* 55, 1740, 1976.

311. **Bogart, J. P. and Tandy, M.,** Polyploid amphibians: three more diploid-tetraploid cryptic species of frogs, *Science,* 193, 334, 1976.

312. **Tandy, M., Bogart, J. P., Largen, M. J., and Feener, D. J.,** A tetraploid species of Bufo (Anura, Bufonidae) from Ethiopia, *Monit. Zool. Ital. Suppl.,* 17, 1, 1982.

313. **Batistic, R. F., Soma, M., Becak, M. L., and Becak, W.,** Further studies on polyploid amphibians: a diploid population of *Phyllomedusa burmeisteri, J. Hered.,* 66, 160, 1975.

314. **Becak, M. L., Becak, W., and Rabello, M. N.,** Further studies on polyploid amphibians (Ceratophrydidae). I. Mitotic and meiotic aspects, *Chromosoma,* 22, 192, 1967.

315. **Saez, F. A. and Brum, N.,** Citogenetica de anfibios anuraos de America de Sud. Los chromosomas de *Odontophyrnus americanus y Ceratophrys ornata, An. Fac. Med. Univ. Repub. Montivideo,* 44, 414, 1959.

316. **Becak, M. L., Becak, W., and Rabello, M. N.,** Cytological evidence of constant tetraploidy in the bisexual South American frog *Odontophrynus americanus, Chromosoma,* 19, 188, 1966.

317. **Becak, M. L. and Becak, W.,** Further studies on polyploid amphibians. III. Meiotic aspects of the interspecific triploid hybrid: *Odontophyrnus cultripes* (2n = 22) × *O. americanus* (4n = 44), *Chromosoma,* 31, 377, 1970.

318. **Barrio, A. P., Rinaldi de Cheri, P.,** Estudios citogeneticos sobre el genero *Pleuroderma* y sus consecuencias evolutivas (Amphibia, Anura, Leptodactylidae), *Physis (Florence),* 30, 309, 1970.

319. **Mahoney, M. J. and Robinson, E. S.,** Polyploidy in the Australian leptodactylid frog genus *Neobatrachus, Chromosoma,* 81, 199, 1980.

320. **Tymowska, J. and Fischberg, M.,** Chromosome complements of the genus *Xenopus, Chromosoma,* 44, 335, 1973.

321. **Kobel, H. R.,** Evolutionary trends in *Xenopus* (Anura: Pipidae), *Ital. J. Zool.,* 8, 118, 1981.

322. **Tymowska, J. and Fischberg, M.,** A comparison of the karyotype, constitutive heterochromatin, and nucleolar organizer regions of the new tetraploid species *Xenopus epitropicalis* Fischberg and Picard with those of *Xenopus tropicalis* Gray (Anura, Pipiade), *Cytogenet. Cell Genet.,* 34, 149, 1982.

323. **Kobel, H. R., DuPasquier, L., Fischberg, M., and Gloor, H.,** *Xenopus amieti* sp. nov. (Anura: Pipidae) from the Cameroons: another case of tetraploidy, *Rev. Suisse Zool.,* 87, 919, 1980.

324. **Fischberg, M. and Kobel, H. R.,** Two new polyploid *Xenopus* species from western Uganda, *Experientia,* 34, 1012, 1978.

325. **Downs, F. L.,** Unisexual *Ambystoma* from the Bass Islands of Lake Erie, *Occas. Pap. Mus. Zool. Univ. Michigan,* 685, 1, 1978.

326. **Morescalchi, A. and Olmo, E.,** Sirenids: a family of polyploid urodeles?, *Experientia,* 30, 491, 1974.

327. **Morescalchi, A.,** Chromosome evolution in the caudate Amphibia, in *Evolutionary Biology,* Vol. 8, Dobzhansky, T., Hecht, M. K., and Steere, W. C., Eds., Plenum Press, New York, 1975, 338.

Chapter 16

AMPHIBIAN MODELS IN ENDOCRINE RESEARCH

David O. Norris

TABLE OF CONTENTS

I. INTRODUCTION

Amphibians have not been used extensively as model systems for the investigation of basic endocrinological phenomena. However, the limited work performed to date suggests that they should be utilized more extensively in the future. Amphibians have several advantages over mammals as laboratory animals. The requirements for maintenance and rearing of amphibians[1-7] and for culture of amphibian tissues and organs[8] are not so stringent as for mammals. Many processes appear to be organized more simply in amphibians than in the comparable mammalian systems, and in some cases it is easier to separate the individual components. Furthermore, various processes that may occur simultaneously in the standard laboratory mammals (for example, the appearance of different spermatogenetic stages) may be separated in time in amphibians, making it possible to study regulatory factors in isolation.

In this chapter, I shall review briefly some of the major endocrinological model systems that have been utilized and which should continue to yield valuable information of importance to both comparative and clinically oriented physiologists. Most references will be those of major review articles. In addition, I shall suggest an array of basic studies that might be developed as models for investigating other endocrinological events as they occur in amphibians. There is, of course, one caveat recently stressed by Licht[9] to be considered. We must be very cautious in making extrapolations from nonmammalian systems to mammalian systems and vice versa before the systems are sufficiently elucidated to warrant such extrapolations. Nevertheless, understanding how nonmammalian systems function and are regulated may provide considerable insight into mammalian systems.

II. HISTORICAL ROLE OF AMPHIBIANS IN BIOMEDICAL RESEARCH

Endocrine research on amphibians has resulted in numerous discoveries that have contributed directly to biomedical research. For example, the importance of the pituitary gland was first demonstrated experimentally in 1916 by Allen[10] and Smith,[11] who independently studied the effects of removal of the embryonic precursor of the pituitary from early amphibian embryos. They each observed changes in pigmentation (lightening of the skin due to loss of melanotropin, MSH) and reduced body growth (lack of growth hormone, GH). In addition Smith noted the reduced development of the thyroid gland following hypophysectomy, linking the thyroid to a dependence on the pituitary gland.

Alleviation of diabetic ketoacidosis and high circulating glucose levels following hypophysectomy of an experimentally pancreatectomized animal was accomplished first by Houssay.[12] Houssay performed his experiments with the toad *Bufo arenarum*. It was later that the importance of this technique for mammalian studies was fully appreciated, and such a doubly-operated (pancreatectomized and hypophysectomized) laboratory animal became known as a "Houssay animal".

A pregnancy test replacing the use of rabbits was devised in female *Xenopus laevis*.[13,14] This test used oviposition following injection of urine from a pregnant female as evidence for the presence of human chorionic gonadotropin (hCG) in the urine. This test was later replaced by the spermiation assay in male toads[15] or frogs.[16] This latter bioassay for pregnancy was more economical than the female assay and remained in use until development of antibody-based assays for hCG occurred in the 1970s.

The stimulatory action of thyroid hormones on development was first shown by Gudernatsch,[17] who fed horse thyroid glands to amphibian larvae (*Rana temporaria* and *R. esculenta*) and observed that they underwent metamorphosis. This classical observation was the first confirmed action for thyroid hormones in vertebrates.

III. AMPHIBIAN MODELS IN ENDOCRINE RESEARCH

A. THE AMPHIBIAN URINARY BLADDER: MODEL FOR ALDOSTERONE ACTION

The use of the amphibian urinary bladder as an *in vitro* model for studying the action of the adrenal mineralocorticoid aldosterone arose from studies in the 1950s which demonstrated that these organs could adsorb fluid, especially under conditions of dehydration, and return this fluid to the blood.[18,19] Furthermore, the urinary bladder was capable of actively transporting sodium ions from the mucosal (urinary) side to the serosal (blood) surface under the influence of the neurophypophyseal octapeptide hormone arginine vasopressin.[20,21] The movement of sodium ions was followed by water uptake. Although these events were shown to occur in several species of frogs and toads, the urinary bladder of *Bufo marinus* became the principal *in vitro* system for studying the influence of hormones, especially aldosterone, on ion and water movement. Although the mammalian urinary bladder is capable of limited sodium transport similar to that described for the amphibian,[22] the amphibian bladder has become a useful model for the action of aldosterone on kidney tubules.[23]

Studies on the amphibian urinary bladder have examined the effects of aldosterone on the mitochondrial-rich cell in the urinary bladder mucosa and have focused on characterization of the aldosterone receptor, the stimulation of a new mRNA synthesis, the ensuing synthesis of new protein, and the nature of the resulting aldosterone-induced proteins.[23,24]

B. THE AMPHIBIAN OVARY: MODEL FOR OOCYTE GROWTH, OOCYTE MATURATION, AND OVULATION

The anuran amphibian ovarian follicle has proved to be an excellent *in vitro* model for studying oocyte maturation and ovulation. Several studies in the 1930s and 1940s showed that ovulation could be induced *in vitro* with certain steroids and pituitary extracts.[25-27] This technique was adopted again during the 1960s (e.g., Wright,[28] Schuetz[29]) and has led to extensive, detailed studies on the roles of pituitary gonadotropins, steroids, adenylate cyclase, and prostaglandins in the course of oocyte maturation (germinal vesicle breakdown), and the ovulatory process.[30-33] In addition, the *in vitro* ovulation technique can be employed as a sensitive, rapid bioassay for pituitary gonadotropins[34,35] and for hypothalamic gonadotropin-releasing hormone (GnRH).[36,37]

This method for *in vitro* ovulation is very simple and works equally well in urodele amphibians.[38] Each ovary can be excised and cut into small pieces. An ovarian fragment is placed in a small culture dish or test tube containing physiological saline or an organ culture medium. The sensitivity of the system may be enhanced by prior injection (priming dose) *in vitro* of pituitary gonadotropin.[39] Test substances can be added to the culture medium, and ovulations can be observed between 12 and 48 h. This system has been modified to study events in naked oocytes after removal of the layer of follicle cells that normally surrounds the oocyte.[31,33]

The anuran amphibian has proven an excellent system for the evaluation of oocyte growth, especially with respect to the interaction of liver and ovary in the process of vitellogenesis.[40-42] This system is also useful in studying the mechanism of action by estradiol on a target tissue and the subsequent stimulation of protein synthesis.

C. THE AMPHIBIAN TAIL FIN: MODEL FOR HORMONE ACTION AND INTERACTION

Tissues of the amphibian tail have provided another useful *in vitro* system for studying the mechanisms of thyroid hormone action.[43] This approach was first employed by Derby[44,45] to analyze tail fin regression during metamorphosis of anuran tadpoles. Most recently, it has been used to examine interactions of prolactin and thyroid hormones[46,47] as well as involvement of cAMP[48] and prostaglandins[49] in thyroid hormone action on tadpole tails. It is also a useful assay for the pharmacological study of the biological activity of thyroid hormone analogs.[50]

Platt and his colleagues[51-53] have since extended this system to the urodele larval tail fin. The interactions of thyroid hormones, prolactin, and neurophyphophysial octapeptides have been studied with this method.

D. *IN VIVO* MODEL FOR INDUCTION OF BONE AND CARTILAGE

An interesting system was developed recently to study the actions of thyroid hormones on induction of bone and cartilage during premetamorphic development of *Bombina orientalis* tadpoles.[54,55] This system uses plastic pellets containing minute quantities of thyroid hormones implanted into the brain region of small tadpoles. These initial studies demonstrate that cartilage and bone formation are dissociated events and respond independently to thyroid hormones. These new findings may have important biomedical implications.

E. THE AMPHIBIAN CORPUS LUTEUM

Well-vascularized postovulatory follicles (corpora lutea) were first described in a frog by Swammerdam in 1738,[56] although little importance was attached to these structures until recent years. The corpora lutea of both oviparous and viviparous amphibians[57,58] may be used more extensively in the future for studying corpora lutea formation, function, and regression. The most thorough studies have been conducted by Xavier[58] in the viviparous frog *Nectophrynoides occidentalis*. This work has shown interesting parallels between formation, regulation and regression in this species and mammals. Even among the oviparous species, the life of the corpus luteum can be extended by treatment with gonadotropic preparations,[59] and it has been shown to be steroidogenic.[57,58]

F. PERIFUSION SYSTEMS: MODELS FOR REGULATION STUDIES

A promising approach to which amphibian tissues are readily adaptable is the perifusion (or superfusion) system. This type of system involves placing an organ in a chamber through which a physiological solution is passed at a constant flow rate. Regulatory chemicals or tissue extracts can be introduced at intervals and the effluent from the chamber collected and analyzed by radioimmunoassay, high-performance liquid chromatography, or other sensitive method. Two or three tissues can be arranged in a series of perifused chambers so that the secretory response of one tissue can be examined in terms of the effects it produces on the second tissue in line. This can be especially useful where it is not possible to assay a hormone from tissue 1 but where tissue 2 responds to the effluent from tissue 1 by producing another secretory substance that can be measured easily.

In amphibians, perifusion systems are being used to examine pituitary responsiveness to hypothalamic peptides, such as thyrotropin-releasing hormone (TRH)[60] and GnRH,[61] and the responsiveness of adrenal tissue to corticotropin[62-64] and of testes[65] or ovaries[37] to gonadotropic preparations. Such systems may be especially helpful for studying seasonal refractoriness in the pituitary-gonadal axis and may have direct application to seasonally breeding mammals as well.

IV. POTENTIAL AMPHIBIAN MODELS

Urodele amphibians have not been employed frequently in endocrine research. The Mexican axolotl *Ambystoma mexicanum* has been used sparingly for endocrine research, although it has been used for many other kinds of physiological investigations. One disadvantage of the axolotl for some endocrine studies is that it has been selected and bred for decades for its ability to become sexually mature as a larva without undergoing the typical metamorphosis from a larva to a terrestrial type of salamander. The retention of larval characteristics in a sexually breeding animal is often referred to as paedogenesis.[66] Furthermore, the type of paedogenesis that occurs in several species of salamander may be termed neoteny: a condition where somatic development is retarded and sexual maturation occurs on a normal schedule.[67] Among these neotenous salamanders there occur two types of neotenes. Animals of the first type are called obligate neotenes. These species, such as the mudpuppy *Necturus* spp., never undergo metamorphosis to a terrestrial form and remain aquatic. They retain the external gills and fish-like body characteristic of larval salamanders. Obligate neotenes are insensitive to thyroid hormones that normally induce metamorphosis in amphibians. The second type of neotene is characterized by the facultative neotenic species which can be induced to undergo metamorphosis following treatment with thyroid hormones. Within the facultative neotenes are species that rarely or never undergo metamorphosis in nature as well as species, such as the tiger salamander *Ambystoma tigrinum,* that frequently undergo spontaneous metamorphosis. Although the tiger salamander is distributed throughout much of North America, it is mainly in the southwestern United States that neoteny commonly occurs.

To date, the bullfrog *Rana catesbeiana,* the leopard frog *R. pipiens,* and the clawed frog *Xenopus laevis* have been the most heavily utilized amphibians for endocrine research. In an evolutionary sense this may be unfortunate since at least morphologically and probably physiologically, salamanders are more similar to the primitive amphibians that gave rise to modern amphibians and also to the reptiles. The tiger salamander, in part because of its facultatively neotenic life history, is an excellent animal for a wide variety of endocrine investigations. It is readily available over much of the U.S. as well as in portions of southern Canada and northern Mexico. Techniques established for axolotls[68] could be used to breed colonies of tiger salamanders. Larvae or adults are easy to maintain in the laboratory, where they can be trained to eat a variety of foods. Because of their opportunistic feeding patterns,[69] they will consume almost any organism that is small enough to swallow. Hence, they eat a variety of insects, earthworms, crustaceans, and larval amphibians. They can be maintained in the laboratory with mealworms, crickets, or earthworms available from commercial suppliers. Very young tiger salamander larvae have specific salinity and temperature requirements,[2] but older larvae can be exposed to a wide range of salinity and temperature without experiencing mortality. These older larvae can be maintained readily in dechlorinated tap water at 15 to 20°C. Furthermore, they can be held at low temperatures (2 to 4°C) for long periods of time without apparent ill effects. When metamorphosed animals are desired, larvae can be induced to metamorphose by placing them on long photoperiods.[70-72]

Larvae or adults can be anesthetized with tricane methane sulfonate (MS-222) before surgical procedures, such as hypophysectomy or gonadectomy. Survival of operated animals is excellent, and infection of surgical wounds is rare. Large larvae (20 to 50 g or more) can be reared and used to make surgical operations easier.

Tiger salamander larvae provide useful material for developmental and/or reproductive studies. Such investigations might involve differentiation of immature reproductive structures or the maturation of the hypothalamic neuroendocrine system and the induction of metamorphosis. Furthermore, larvae are useful for studies of sexual maturation, especially young larvae from neotenic populations. The neotenes as well as metamorphosed adults are seasonal breeders and are good animals for examining the environmental and physiological bases for seasonality.

A. DEVELOPMENTAL STUDIES

1. Reproductive Structures

The tiger salamander was first used in the 1930s for studying the role of androgens in gonaduct differentiation (see Norris[73]). The primordia of oviducts are present initially in both male and female larvae. These oviducts increase in size in response to either androgens or estrogens.[73] It is possible to induce formation of male-type cloacal glands in male or female larvae with either testosterone or dihydrotestosterone.[74,75] In males, these glands contribute to the formation of the spermatophore used to transfer spermatozoa from the male to the female. Although androgens and estrogens synergize in stimulation of oviduct growth,[76] they apparently are antagonistic on cloacal tissues.[77] Detailed studies of amphibian oviducts and cloacal tissues may provide new understanding into the process of sexual differentiation and the interactions of steroids. At least one of these cloacal glands has its origin from the urogenital sinus and may be homologous to the mammalian prostate gland.

2. Neuroendocrine Differentiation and Metamorphosis

Metamorphosis is a dramatic event in the life history of many amphibians, and some special insights may be gained from pursuing studies of this process in a facultative neotene such as the tiger salamander. The onset of metamorphosis is controlled by the hypothalamo-hypophyseal-thyroid axis as recently summarized by Dent[78] and Norris and Dent.[79] This system appears to be totally independent of maturation and/or activation of the hypothalamo-hypophyseal-gonadal axis responsible for sexual maturation, since these events are always separated in time. The larvae either undergo metamorphosis and later become sexually mature or become sexually mature at about the normal time, retaining the possibility of activating metamorphosis at a later date. Thus, facultative neotenes provide a natural experiment for examining the effects of environmental factors on activation of certain neural and neuroendocrine pathways.

Low circulating levels of thyroid hormones in larvae are believed to control maturation of the nervous system, which in turn amplifies the production of thyroid hormones through activation of the hypophalamo-hypophysial-thyroid axis (see Norris and Dent[79]). The distribution of monoaminergic neurotransmitters (norepinephrine, dopamine, serotonin) and neuropeptides (immunoreactive TRH, GnRH, etc.) are known for amphibians,[79] and we[80,81] have begun to examine changes in the metabolism of monoaminergic neurotransmitters in the tiger salamander brain during metamorphosis.

B. REPRODUCTIVE ENDOCRINOLOGY

Seasonal fluctuations in androgens and reproductive events have been reported for male neotenic tiger salamanders.[82] The seasonal nature of the spermatogenetic cycle of tiger salamanders and other species may be ideal for the study of regulation of specific stages (e.g., Moore[83]) because the spermatogenetic cysts in an entire region of the testis are synchronized; that is, all exhibit the same stage of development.

In females, many events of the reproductive cycle are known,[84] and the process of oogenesis and vitellogenesis could be examined in a similar manner to the anuran studies described above (Section III.B). Ovulation can be induced readily *in vivo*[38,85,86] and *in vitro*.[38] The responsiveness of larval tiger salamander oviducts to neurohypophysial peptides (arginine vasotocin) has been reported.[87] This *in vitro* system could be exploited to examine the interactions of steroids with neurophypophysial factors on oviduct contractions as well as possible interactions of neurotransmitters, prostaglandins, various second messengers (cAMP, inositol phosphate), and other hormones.

V. SOURCES FOR AMPHIBIANS

Amphibians for research may be obtained from commercial suppliers or from fish bait dealers (larval tiger salamanders), but one should be concerned about the condition of these animals by the time they reach the laboratory. Data concerning such factors as collection site, collection procedures, conditions and duration of capture, and nutritional and/or re-

285

productive state are rarely available; such animals may not be suitable for many kinds of studies. Furthermore, indiscriminate collecting practices has virtually eliminated amphibians in some locales and continues to threaten others. Together with the constant reduction in aquatic and wetland habitats resulting from expanding human populations, the switch from mammalian models for endocrine and other types of physiological research to amphibians may prove to be detrimental to the continued survival of favorite amphibian species. Consequently, the widespread use of any amphibian model for research must be accompanied by conscientious development of facilities for the proper breeding and rearing of these animals under controlled conditions.

VI. SUMMARY

The use of amphibians for endocrine research has led to many important discoveries that have had direct application to the biomedical community. Several *in vitro* model systems have been discussed, including the urinary bladder, ovarian follicle, tail fin, and various types of perifusion systems. Three *in vitro* systems also were identified that have *in vitro* potential as well: vitellogenesis and oocyte growth, development of cartilage and bone, and the corpus luteum. Finally, the use of a facultative neotene, the tiger salamander, was discussed as a potential system for investigating a variety of basic problems.

REFERENCES

1. **Bragg, A.,** Notes on the behavior of toads in captivity, *Wasmann J. Biol.,* 14, 301, 1956.
2. **Cohen, N.,** A Method for mass rearing *Ambystoma tigrinum* during and after metamorphosis in a laboratory environment, *Herpetologica,* 24, 86, 1968.
3. **Gibbs, E. L., Nace, G. W., and Emmons, M. B.,** The live frog is almost dead, *Bioscience,* 21, 1027, 1971.
4. **Nace, G. W. and Richards, C. M.,** Living frogs. II. Care, *Carol. Tips,* 35, 41, 1972.
5. **Nace, G. W. and Richards, C. M.,** Living frogs. III. Tadpoles, *Carol. Tips,* 35, 45, 1972.
6. **Marcus, L. C.,** *Veterinary Biology and Medicine of Captive Amphibians and Reptiles,* Lea & Febiger, Philadelphia, 1981.
7. **Claussen, D. L. and Layne, J. R., Jr.,** Growth and survival of juvenile toads, *Bufo woodhousei,* maintained on four different diets, *J. Herpetol.,* 17, 107, 1983.
8. **Monnickendam, M. A. and Balls, M.,** Amphibian organ culture, *Experientia,* 29, 1, 1973.
9. **Licht, P.,** Suitability of the mammalian model in comparative reproductive endocrinology, in *Comparative Endocrinology: Developments and Direction,* Ralph, C. L., Ed., Alan R. Liss, New York, 1986, 95.
10. **Allen, B.,** The results of extirpation of the anterior lobe of the hypophysis and of the thyroid of *Rana pipiens* larvae, *Science,* 44, 755, 1916.
11. **Smith, P. E.,** Experimental ablation of the hypophysis in the frog embryo, *Science,* 44, 280, 1916.
12. **Houssay, B. A.,** Hypophyseal functions in the toad *Bufo arenarum* Hensel, *Q. Rev. Biol.,* 24, 1, 1949.
13. **Bellerby, C. W.,** A rapid test for the diagnosis of pregnancy, *Nature (London),* 133, 494, 1934.
14. **Shapiro, H. A. and Zwarenstein, H.,** A rapid test for pregnancy on *Xenopus laevis, Nature (London),* 133, 339, 1934.
15. **Galli-Mainini, C.,** Pregnancy tests using the male toad, *J. Clin. Endocrinol.,* 7, 653, 1947.
16. **Wiltberger, P. B. and Miller, D. F.,** The male frog, *Rana pipiens,* as a new test animal for early pregnancy, *Science,* 107, 198, 1948.
17. **Gudernatsch, J. F.,** Feeding experiments on tadpoles. I. The influence of specific organs given as food on growth and differentiation. A contribution to the knowledge of organs with internal secretion, *Arch. Entwicklungsmech. Org.,* 35, 457, 1913.
18. **Ewer, R. W.,** The effect of pituitrin on fluid distribution in *Bufo regularis* Reuss, *J. Exp. Biol.,* 29, 173, 1952.
19. **Sawyer, W. H. and Schisgall, R. M.,** Increased permeability of the frog bladder to water in response to dehydration and neurohypophysial extracts, *Am. J. Physiol.,* 187, 312, 1956.

20. **Leaf, A., Anderson, J., and Page, L. B.,** Active sodium transport by the isolated toad bladder, *J. Gen. Physiol.*, 41, 657, 1958.
21. **Bentley, P. J.,** The effects of neurohypophysial extracts on water transfer across the walls of the isolated urinary bladder of the toad *Bufo marinus, J. Endocrinol.*, 17, 201, 1958.
22. **Brand, P. H. and Higgins, J. T., Jr.,** Effect of adrenal hormones on water and electrolyte metabolism, in *The Adrenal Gland,* Mulrow, P. J., Ed., Elsevier, New York, 1986, 201.
23. **Morris, D. J.,** The metabolism and mechanism of action of aldosterone, *Endocr. Rev.*, 2, 234, 1981.
24. **Bentley, P. J. and Scott, W. N.,** The actions of aldosterone, in *General, Comparative, and Clinical Endocrinology of the Adrenal Cortex,* Vol. 2, Chester Jones, I. and Henderson, I. W., Eds., Academic Press, New York, 1978, 498.
25. **Rugh, R.,** Ovulation in the frog. I. Pituitary relations in induced ovulation, *J. Exp. Zool.*, 71, 149, 1935.
26. **Zwarenstein, H.,** Experimental induction of ovulation with progesterone, *Nature (London)*, 139, 112, 1937.
27. **Langan, W. B.,** Ovulatory responses of *Rana pipiens* to mammalian gonadotropic factors and sex hormones, *Proc. Soc. Exp. Biol. Med.*, 47, 59, 1941.
28. **Wright, P. A.,** Influence of estrogen on induction of ovulation *in vitro* in *Rana pipiens, Gen. Comp. Endocrinol.*, 1, 381, 1961.
29. **Schuetz, A. W.,** Action of hormones on germinal vesicle breakdown in frog *(Rana pipiens)* oocytes, *J. Exp. Zool.*, 166, 347, 1967.
30. **Maller, J. L.,** Oocyte maturation in amphibians, in *Development Biology, A Comprehensive Synthesis,* Vol. 1, *Oogenesis,* Browder, L. W., Ed., Plenum Press, New York, 1985, 289.
31. **Schuetz, A. W.,** Local control mechanisms during oogenesis and folliculogenesis, in *Developmental Biology, A Comprehensive Synthesis,* Vol. 1, *Oogenesis,* Browder, L. W., Ed., Plenum Press, New York, 1985, 3.
32. **Schuetz, A. W.,** Hormonal dissociation of ovulation and maturation of oocytes: ovulation of immature amphibian oocytes by prostaglandin, *Gamete Res.*, 15, 99, 1986.
33. **Nagahama, Y.,** Endocrine control of oocyte maturation, in *Hormones and Reproduction in Fish, Amphibians and Reptiles,* Norris, D. O. and Jones, R. E., Eds., Plenum Press, New York, 1987, 171.
34. **Licht, P.,** Luteinizing hormone (LH) in the reptilian pituitary gland, *Gen. Comp. Endocrinol.*, 22, 463, 1974.
35. **Licht, P., Farmer, S. W., and Papkoff, H.,** The nature of the pituitary gonadotropins and their role in ovulation in a urodele amphibian *(Ambystoma tigrinum), Life Sci.*, 17, 1049, 1975.
36. **Hubbard, G. M. and Licht, P.,** *In vitro* study of the direct ovarian effects of gonadotropin releasing hormone (GnRH) in the frogs *Rana pipiens* and *Rana catesbeiana, Gen. Comp. Endocrinol.*, 60, 154, 1985.
37. **Hubbard, G. M. and Licht, P.,** Effects of cycloheximide on *in-vitro* testosterone secretion from *Rana catesbeiana* ovaries, *Comp. Biochem. Physiol.*, 84A, 401, 1986.
38. **Norris, D. O. and Duvall, D.,** Hormone-induced ovulation in *Ambystoma tigrinum:* influence of prolactin and thyroxine, *J. Exp. Zool.*, 216, 175, 1981.
39. **Thorton, V. F.,** A bioassay for progesterone and gonadotropins based on the meiotic division of *Xenopus* oocytes *in vitro, Gen. Comp. Endocrinol.*, 16, 599, 1971.
40. **Wallace, R. A.,** Oocyte growth in nonmammalian vertebrates, in the *Vertebrate Ovary,* Jones, R. E., Ed., Plenum Press, New York, 1978, 469.
41. **Wallace, R. A.,** Vitellogenesis and oocyte growth in nonmammalian vertebrates, in *Developmental Biology, A Comprehensive Synthesis,* Vol. 1, *Oogenesis,* Browder, L. W., Ed., Plenum Press, New York, 1985, 127.
42. **Ho, S.-M.,** Endocrinology of vitellogenesis, in *Hormones and Reproduction in Fishes, Amphibians and Reptiles,* Norris, D. O. and Jones, R. E., Eds., Plenum Press, New York, 1987, 145.
43. **Atkinson, B. G.,** Biological basis of tissue regression and synthesis, in *Metamorphosis,* 2nd ed., Gilbert, L. I. and Frieden, E., Eds., Plenum Press, New York, 1981, 397.
44. **Derby, A.,** An *in-vitro* quantitative analysis of the response of tadpole tissues to thyroxine, *J. Exp. Zool.*, 168, 147, 1968.
45. **Derby, A.,** An *in-vitro* quantitative analysis of the response of tadpole tissue to hormone treatment, in *Hormones in Development,* Hamburgh, M. and Barrington, E. J. W., Eds., Appelton-Century-Crofts, New York, 1971, 261.
46. **Derby, A.,** The effect of prolactin and thyroxine on tail resorption in *R. pipiens: in vivo* and *in vitro, J. Exp. Zool.*, 193, 15, 1975.
47. **Ray, L. B. and Dent, J. N.,** Observation on the interaction of prolactin and thyroxine in the tail of the bullfrog tadpole, *Gen. Comp. Endocrinol.*, 64, 36, 1986.
48. **Ray, L. B. and Dent, J. N.,** Investigations on the role of cAMP in regulating the resorption of the tail fin from tadpoles of *Rana catesbeiana, Gen. Comp. Endocrinol.*, 64, 44, 1986.

49. **Mobbs, I. G., King, V. A., and Wassersug, R. J.,** Prostaglandin E_2 does not inhibit metamorphoses of tadpole tails in tissue culture, *Exp. Biol.,* 47, 151, 1988.

50. **Frieden, E.,** The dual role of thyroid hormones in vertebrate development and calorigenesis, in *Metamorphosis,* 2nd ed., Gilbert, L. I. and Frieden, E., Eds., Plenum Press, New York, 1981, 545.

51. **Platt, J. E. and LiCause, M. J.,** Effects of oxytocin in larval *Ambystoma tigrinum:* acceleration of induced metamorphosis and inhibition of the antimetamorphic action of prolactin, *Gen. Comp. Endocrinol.,* 41, 84, 1980.

52. **Platt, J. E., Christopher, M. A., and Sullivan, C. A.,** The role of prolactin in blocking thyroxine-induced differentiation of tail tissue in larval and neotenic *Ambystoma tigrinum, Gen. Comp. Endocrinol.,* 35, 402, 1978.

53. **Platt, J. E., Brown, G. B., Erwin, S. A., and McKinley, K. T.,** Antagonistic effects of prolactin and oxytocin on tail fin regression and acid phosphatase activity in metamorphosing *Ambystoma tigrinum, Gen. Comp. Endocrinol.,* 67, 247, 1986.

54. **Hanken, J. and Hall, B. K.,** Skull development during anuran metamorphosis. II. Role of thyroid hormones in osteogenesis, *Anat. Embryol.,* 178, 219, 1988.

55. **Hanken, J. and Summers, C. H.,** Skull development during anuran metamorphosis. III. Role of thyroid hormones in chondrogenesis, *J. Exp. Zool.,* 24, 156, 1988.

56. **King, H. D.,** The follicle sacs of the amphibian ovary, *Biol. Bull. (Woods Hole, Mass.),* 3, 245, 1902.

57. **Saidapur, S. K.,** Structure and function of postovulatory follicles (corpora lutea) in the ovaries of non-mammalian vertebrates, *Int. Rev. Cytol.,* 75, 243, 1982.

58. **Xavier, F.,** Functional morphology and regulation of the corpus luteum, in *Hormones and Reproduction in Fishes, Amphibians and Reptiles,* Norris, D. O. and Jones, R. E., Eds., Plenum Press, New York, 1987, 241.

59. **Pancharatna, M. and Saidapur, S. K.,** The luteotrophic effect of homoplastic pituitary pars distalis homogenate, PMSG, and HCG on the corpora lutea of the hypophysectomized frog, *Rana cyanophlyctis* (SCHN), *Gen. Comp. Endocrinol.,* 53, 375, 1984.

60. **Leroux, P., Tonon, M. C., Saulot, P., Jegou, S., and Vaudry, H.,** *In vitro* study of frog *(Rana ridibunda* Pallas) neurointermediate lobe secretion by use of a simplified perifusion system. II. Lack of action of thyroxine on TRH-induced α-MSH secretion, *Gen. Comp. Endocrinol.,* 51, 323, 1983.

61. **Porter, D. A. and Licht, P.,** The cellular basis of the calcium dependence of GnRH-stimulated gonadotropin release from frog, *Rana pipiens,* pituitaries, *J. Exp. Zool.,* 240, 353, 1986.

62. **Leboulenger, F., Delarue, C., Tonon, M. C., Jegou, S., and Vaudry, H.,** *In vitro* study of frog *(Rana ridibunda* Pallas) interrenal function by use of a simplified perifusion system. I. Influence of adrenocorticotropin upon corticosterone release, *Gen. Comp. Endocrinol.,* 36, 327, 1978.

63. **Leboulenger, F., Belanger, A., Delarue, C. P., Netchitailo, P., Perroteau, I., Roullet, M., Jegou, S., Tonon, M. C., and Vaudry, H.,** *In vitro* study of frog *(Rana ridibunda* Pallas) interrenal function by use of a simplified perfusion system. V. Influence of adrenocorticotropin upon progesterone production, *Gen. Comp. Endocrinol.,* 45, 465, 1981.

64. **Delarue, C., Nethcitailo, P., Leboulenger, F., Perroteau, I., Escher, E., and Vaudry, H.,** *In vitro* study of frog *(Rana ridibunda* Pallas) interrenal function by use of a simplified perifusion system. VII. Lack of effect of somatostatin on angiotensin-induced corticosteroid production *Gen. Comp. Endocrinol.,* 54, 333, 1984.

65. **Boujard, D. and Joly, J.,** The dynamics of the steroidogenic response of perifused *Xenopus* testis explants to gonadotropins, *Gen. Comp. Endocrinol.,* 51, 405, 1983.

66. **Gould, S. J.,** *Ontogeny and Phylogeny,* Belknap Press, Cambridge, MA, 179, 1977.

67. **Gould, S. J.,** *Ontogeny and Phylogeny,* Belknap Press, Cambridge, MA, 321, 1977.

68. **Lawrence, L. M.,** *Axolotl Newsletter,* No. 9, University of Indiana Axolotl Colony, Bloomington, IN, 1980.

69. **Norris, D. O.,** Seasonal changes in diet of paedogenetic tiger salamanders, *Ambystoma tigrinum mavortium, J. Herpetol.,* 23, in press, 1989.

70. **Norris, D. O., Duvall, D., Greendale, K., and Gern, W. A.,** Thyroid function in pre- and postspawning neotenic tiger salamanders *(Ambystoma tigrinum), Gen. Comp. Endocrinol.,* 35, 512, 1977.

71. **Norman, M. F. and Norris, D. O.,** Effects of metamorphosis on the *in-vitro* sensitivity of thyroid glands from the tiger salamander *Ambystoma tigrinum,* to bovine thyrotropin, *Gen. Comp. Endocrinol.,* 67, 77, 1987.

72. **Norman, M. F., Carr, J. A., and Norris, D. O.,** Adenohypophysial-thyroid activity of the tiger salamander, *Ambystoma tigrinum,* as a function of metamorphosis and captivity, *J. Exp. Zool.,* 242, 55, 1987.

73. **Norris, D. O.,** Regulation of male gonaducts and sex accessory structures, in *Hormones and Reproduction in Fishes, Amphibians and Reptiles,* Norris, D. O. and Jones, R. E., Eds., Plenum Press, New York, 1987, 327.

74. **Norris, D. O. and Austin, H. B.,** Testosterone and prolactin effects on induction and differentiation of cloacal glands in larval *Ambystoma tigrinum, Am. Zool.,* 26, 2A, 1986.

75. **Norris, D. O., Austin, H. B., and Hijazi, A.,** Induction of cloacol and dermal skin glands of tiger salamander larvae *(Ambystoma tigrinum)* : effects of testosterone and prolactin, *Gen. Comp. Endoncrinol.,* 73, 194, 1989.

76. **Norris, D. O., Carr, J. A., and Featherston, R. J.,** Action and interactions of estradiol, testosterone and dihydrotestosterone on oviducts and cloacal tissues of larval *Ambystoma tigrinum, J. Wyo. Acad. Sci.,* in press, 1988.

77. **Norris, D. O. and Moore, F. L.,** Antagonism of testosterone-induced cloacal development by estradiol-17B in immature larval tiger salamanders *(Ambystoma tigrinum), Herpetologica,* 31, 255, 1975.

78. **Dent, J. N.,** Hormonal interaction in amphibian metamorphosis, *Am. Zool.,* 28, 297, 1988.

79. **Norris, D. O. and Dent, J. N.,** Neuroendocrine aspects of amphibian metamorphosis, in *Development, Maturation and Senescence of the Neuroendocrine System,* Schreibman, M. P. and Scanes, C. G., eds., Academic Press, San Diego, CA, in press, 1989.

80. **Carr, J. A., Norris, D. O., Desan, P. H., Smock, T., and Norman, M. F.,** Quantitative measurement of brain amines and metabolites during metamorphosis of tiger salamander larvae, *Physiologist,* 30, 142, 1987.

81. **Norris, D. O., Carr, J. A., Desan, P. H., Smock, T., and Norman, M. F.,** Levels of biogenic monoamines and their metabolites in the tiger salamander brain during metamorphosis, in preparation.

82. **Norris, D. O., Norman, M. F., Pancak, M. K., and Duvall, D.,** Seasonal variations in spermatogenesis, testicular weights, vasa deferentia, and androgen levels in neotenic male tiger salamanders, *Ambystoma tigrinum, Gen. Comp. Endocrinol.,* 60, 51, 1985.

83. **Moore, F. L.,** Spermatogenesis in larval *Ambystoma tigrinum:* positive and negative interactions of FSH and testosterone, *Gen. Comp. Endocrinol.,* 26, 525, 1975.

84. **Norris, D. O.,** Seasonal variations in ovaries and oviducts of neotenic female tiger salamanders, *Ambystoma tigrinum mavortium,* in preparation.

85. **Ketterer, D. and Forbes, W. R.,** Induction of spawning in the Mexican axolotl *(Ambystoma mexicanum)* by lutenizing hormone, *J. Endocrinol.,* 55, 457, 1972.

86. **Humphrey, R. R.,** Factors influencing ovulation in the Mexican axolotl as revealed by induced spawning, *J. Exp. Zool.,* 199, 209, 1977.

87. **Guillette, L. J., Jr., Norris, D. O., and Norman, M. R.,** Response of amphibian *(Ambystoma tigrinum)* oviduct to arginine vasotocin and acetylcholine : *in vitro* influence of steroid hormone pretreatment *in vivo, Comp. Biochem. Physiol.,* 80C, 151, 1982.

Chapter 17

REPTILE MODELS FOR BIOMEDICAL RESEARCH

Neil Greenberg, Gordon M. Burghardt, David Crews, Enrique Font, Richard E. Jones, and Gerald Vaughan

TABLE OF CONTENTS

I. INTRODUCTION

Reptiles possess considerable interest as exemplars of alternative tactics for solving ecologically important problems of survival and efficient energetics. They also are contemporary representatives of the evolutionary precursors of our own species. But beyond these traditional zoological concerns, recent research informed by the ethological approach to problems of causation indicates that several unique qualities of reptiles may provide models useful for research on a diverse array of problems of biomedical interest. These include developmental, endocrinological, neurological, and cellular aspects of stress-related and affective disorders, feeding, reproductive dysfunction, and even Parkinson's disease.

A. THE ETHOLOGICAL APPROACH

Ever since our understanding of the endocrine aspects of social behavior was advanced by workers such as Lehrman[1] and Hinde,[2] the clarity of their central insight — the essential reciprocity of internal and external cues in the control of behavior (see Figure 1) — has illuminated many other areas. An appreciation for such interactions is the essence of the ethological approach, which has become integrated into many areas of biomedical thinking since three of its earliest proponents, Konrad Lorenz, Niko Tinbergen, and Karl von Frisch, were awarded a Nobel Prize for medicine in 1973.

The correlation of environmental (including social) factors with internal (including genetic and physiological) effects promises to provide new insight into the causation of some of the most complex behavioral patterns, including psychiatric problems.[4,5] For example, many disorders of behavior would be better understood if we had a clearer understanding of the manner in which autonomic and somatic units of physiology and behavior are integrated, brought under the control of external stimuli, and reciprocally mediated in emotional arousal, particularly at higher neural levels.[6-8] Insight into such areas would provide a significant contribution to the development of testable hypotheses involving the mutual influences of the social environment and physiology in reproductive medicine and psychosocial disorders. Recent research has, in fact, addressed aspects of this challenge particularly amenable to study in reptiles: the relationship between higher neural function and hormones associated with stress, aggression, and reproductive behavior.

This review will sample several areas of active research that exploit unique qualities of various reptiles with a degree of validity made possible largely by an ethologically informed appreciation for specific physiological and behavioral adaptations to the natural environments of their respective subjects.[9] Specific paradigms and programs to be described include chemosensory aspects of behavioral dysfunction, the effects of physiological stress on aggressiveness and reproductive behavior, developmental aspects of reproductive function, ovulatory cycling, affective disorders involving the basal forebrain, drug-induced neuropathies of central neural dopaminergic function, and the cellular physiology of dermal chromatophores and their prospect for clarifying hormone function.

Most reptile research that suggests interesting biomedical models centers on a relatively few well-known, easily cared for species: the ubiquitous green anole (*Anolis carolinensis*), garter snakes (genus *Thamnophis*), and the remarkable all-female species of parthenogenetic whiptail (genus *Cnemidophorus*). Each of these research programs has provided findings that exemplify the manner in which reptiles can illuminate areas of biomedical interest.

II. CHEMOSENSORY ASPECTS OF BEHAVIORAL DYSFUNCTION

Although mammals, including human beings, depend largely on the visual sensory modality, research has revealed that chemical senses are also of great importance. For

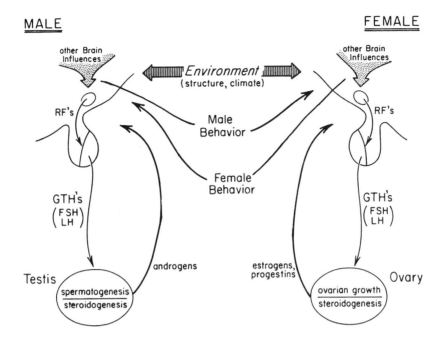

FIGURE 1. Interaction of the environment, hormones, and behavior in the regulation of vertebrate reproduction. GTH's are gonadotrophic hormones and RF's are hypothalamic releasing factors. (Adapted from Crews, D., in *Behavior and Neurology of Lizards*, Greenberg, N. and MacLean, P.D., Eds., NIMH, MD, 1978, 149.)

example, congenitally anosmic children (Kallmann's syndrome) show delayed sexual maturity and some never achieve normal reproductive function.[10,11] Further, 25% of adults developing anosmia report a loss of sexual interest.[12]

In most vertebrates, olfactory and vomeronasal chemical senses are difficult to distinguish, but reptiles offer a preparation with which to study the separation of these two systems in the detection and mediation of chemical signals.[13-18] Squamate reptiles, particularly snakes, have well-developed olfactory systems. In some lizards and all snakes the vomeronasal or Jacobson's organ in the roof of the mouth communicates with the accessory olfactory bulb of the brain and is intimately involved in feeding, sexual, social, and other behavior systems.[13,19,20] While the vomeronasal organ is present in many mammals and is important in reproductive behavior, in snakes it seems to play a particularly critical role. Further, because the squamate vomeronasal organ is morphologically distinct and separate from the nasal olfaction system, its function is much more easily evaluated. Consequently, reptiles provide a model system to work out those ways in which vomeronasal input operates that are not amenable to study in mammals. For example, the accessibility of the vomeronasal nerve and the clear-cut behavioral responses to pheromones make the garter snake a useful preparation for studies of the role of chemosensory systems in the regulation of reproduction.[15] Even chemosensory aspects of nonspecific arousal, as manifest in so well an integrated behavioral pattern as locomotor exploration, is amenable to ethological dissection in reptiles: in this paradigm, two forms of lizard tongue flicking, associated respectively with vomeronasal function and possible gustatory function, are differentially affected by exposing subjects to a mild stressor, thereby inducing autonomic activation.[21] Most research on reptilian chemosensory function, however, has dealt with feeding. In this regard, among the most studied species are garter snakes (*Thamnophis*), a widespread and frequently common North American genus.

A. FEEDING

From birth on, garter snakes also rely primarily on chemical cues in food choice and discrimination.[13,19] Some species also exhibit a remarkable refusal of food during their mating period despite the fact that they have been hibernating for 6 to 8 months,[14] an anorexia that may reveal aspects of the relationship between feeding, reproductive behavior, and hypothalamic neuropeptides.[22] While the role of chemical cues in mammalian ingestive behavior is an area of active research[23] and can involve obesity and other disorders, isolation rearing of snakes can uncover developmental plasticity and pathways. For example, certain neonatal experiences can alter the food preferences of garter snakes (genus *Thamnophis*) sharply;[24] other changes may be age or size related.[19] In one study, garter snakes were often found to prefer earthworms to fish, both natural foods, even though they would grow much better on a fish diet.[25] Chemically, there is much similarity between the components on the surface of fish and of worms that trigger attack in garter snakes.[26] Thus we have a convenient preparation to trace the relationships among congenital chemical-food cue preferences, nutrition, and experience at the chemical, physiological, and behavioral levels.

III. RECIPROCAL INFLUENCES OF PHYSIOLOGICAL STRESS AND REPRODUCTIVE BEHAVIOR

The stressful aspects of the social environment are known to have profound effects on the physiology of many species,[27-29] including man,[30] as well as inducing or contributing to numerous pathologies in animals, which can provide invaluable models for medical research. Often, very specific stress-associated disorders can be delineated in animals including, for example, renal insufficiency in tree shrews,[31] retarded growth rate in blennid fish,[32] and a host of reproductive dysfunctions.[33] As Selye has made abundantly clear,[29] psychological stressors that involve the primary need of survival can induce release of hormones by endocrine glands that can, in turn, lead to a broad spectrum of dysfunctions of secondary systems, including disordered personal and social relationships.

The manner in which the physiological stress response can influence reproduction in many taxa was reviewed by Greenberg and Wingfield,[33] who found that stress cannot be presumed to be invariably deleterious but is best viewed as a tactic for adaptive physiological responses beyond the scope of everyday needs. A subtle but significant aspect of this has been addressed in studies on the green anole lizard *Anolis carolinensis* that have shown how social stress arising from the observation of aggression between males may effectively suppress a female's reproductive activity.[34,35] Such studies are directly relevant to disorders associated with female reproductive function.

The mutual influences of the social environment and stress endocrinology are also clear in the male-male interactions of *A. carolinensis* that result in the establishment of social-dominance relationships.[36] The social behavior of wild-caught animals reared in the laboratory, however, raises problems of experimental validity. Fortunately, many reptiles, and *A. carolinensis* in particular, will display the relevant elements of their social behavioral repertoire readily. For conducting studies of social organization, an effective compromise between the external validity of field work and the internal validity possible in the controlled laboratory setting can be achieved by utilizing naturalistic laboratory habitats adjusted to elicit benchmark elements of natural behavior observed in the field. Gradual simplification of the environment will then permit selective emphasis of behavioral patterns that, while spontaneously emitted, are not usually amenable to observation or are seen only rarely under field conditions.[37]

A. BODY COLOR AS AN INDEX OF THE PHYSIOLOGICAL STRESS RESPONSES

The body color changes seen in *A. carolinensis*, well known in the field and easily

293

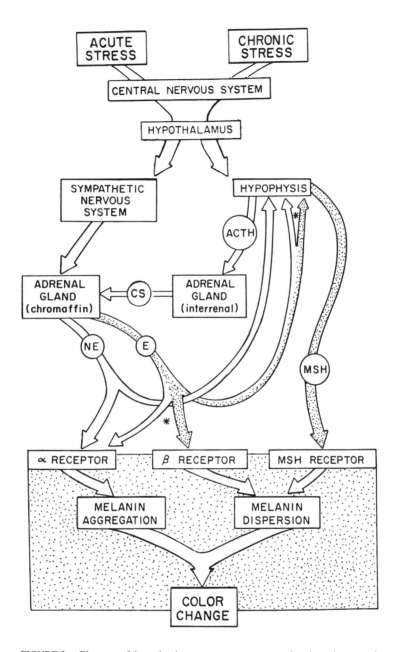

FIGURE 2. Elements of the endocrine response to an acute or chronic environmental stressor and their effect on melanin aggregation and dispersion within a dermal melanophore in *A. carolinensis*. E and NE represent epinephrine and norepinephrine, respectively; the asterisk indicates a biphasic effect of E whereby low levels stimulate and high levels inhibit body color darkening; CS, corticosterone, can stimulate additional synthesis of E from NE by facilitating a crucial methylating enzyme; ACTH, adrenal corticotrophic hormone; MSH, melanocyte-stimulating hormone (melanotropin). (Adapted from Reference 36.)

reproduced in the laboratory, provide us with an excellent model for the examination of several hypotheses related to stress physiology and other autonomic aspects of aggression and status. *A. carolinensis* may be unique among reptiles in that its dermal chromatophores are known to be free of sympathetic innervation.[38] Thus, body color is, under the exclusive influence of circulating hormones (Figure 2). Melanocyte-stimulating hormone (MSH) or relatively low levels of epinephrine (E) will cause darkening of melanophores, while nor-epinephrine (NE) or high levels of E will cause its lightening.

The hormones that stimulate melanophores (E, NE, and MSH) are all sensitive to environmental stressors. While the catecholamines are well known to be involved in a great diversity of physiological stress-adaptive responses, MSH has only recently been implicated in physiological as well as behavioral responses to stressors.[39] Indeed, MSH is sometimes regarded as a neurotrophic factor for central cholinergic neurons, and its deficiency has been associated with clinical aspects of Alzheimer's disease.[40] The hormone associated with a chronic stress response, the glucocorticoid, corticosterone (CS), does not usually affect color directly but can alter the rate of synthesis of E fron NE by facilitating a key enzyme in the process. Further, because one site of the body (the ''eye spot'') is specifically devoid of hormone receptors that might balance or mask the effect of E, we possess a specific *in situ* ''bioassay'' of adrenal ''medullary'' (chromaffin cell) activation.

In the laboratory as in the field, aggressive interactions of males cause rapid shifts in color, with losers frequently appearing brown at the conclusion of a fight. If formerly isolated individuals are required to continue to cohabit a single vivarium after an aggressive encounter, changes indicative of social dominance are apparent and typical body color generally shifts: those that had become socially dominant developed a slightly lighter (more green) color, while subordinates generally become darker (more brown).

These findings implicating color as a significant aspect of status, when considered in concert with the sensitivity of chromatophores to hormones that also respond to environmental stressors, suggest that subordinates were experiencing chronic stress or at least an upward adjustment in the baseline levels of chronic stress-related hormones. To test this, the circulating levels of the corticosterone were determined for dominants and subordinates by radioimmunoassay.[41] The data indicate that CS, which is classically associated with chronic stress, was significantly elevated in subordinates. Androgens are generally regarded as the hormones crucial to the expression of aggression of the type that structures dominant-subordinate relationships between two cohabiting adult male lizards. Recent evidence has, however, strongly implicated the stress hormones of the pituitary-adrenal axis in the control of aggression.[42] To examine the possible interactions of these two endocrine systems in dominance phenomena, pairs were constituted in which one or both animals were castrated. Although aggressive, castrates become subordinate to intacts but do not exhibit darker body color.[41] Similarly, when both members of the pair are castrates, a status relationship, while slightly weaker, is still formed, but the subordinate does not become darker in body color. If darkening is in fact associated with chronic stress we would not expect elevated blood CS in castrated subordinates, and indeed, radioimmunoassay indicates that this is the case.

These data indicate that stress endocrinology is deeply involved in the aggressive and social status relationships seen in *A. carolinensis,* that reproductive status can alter the stress response, and that body color may prove to be a valuable *in vivo* bioassay of these physiological aspects of their social behavior.

IV. BEHAVIORAL, GONADAL, AND MORPHOLOGICAL SEXUALITY

A complete understanding of the identity and nature of the intrinsic functional association of the elements comprising sexuality requires that one separate conceptually the behavioral patterns related to reproduction from the usually concomitant morphological and physiological bases of reproduction. This distinction between the behavioral and the physiological or morphological aspects of sexual behavior is difficult to study since we presently define behavioral sex by gonadal sex. Indeed, certain behaviors are regarded as secondary sex characteristics. Natural experiments exist, however, that call into question the basic premise that gonadal sex and behavioral sex are intrinsically coupled. That is, there exist all-female parthenogenetic whiptail lizards, *Cnemidophorus uniparens*, that regularly and reliably exhibit behavior patterns remarkably similar to the courtship and copulatory behavior of both sexes of their diploid ancestors. Because representatives of both the bisexual ancestral and the unisexual descendant species still exist, comparison of the mechanisms regulating "sexual" behaviors in both parthenogenetic species and their direct evolutionary ancestors should yield information on the evolution of neuroendocrine mechanisms of controlling species-typical and gender-typical behaviors.[43]

A. DEVELOPMENTAL DISORDERS OF REPRODUCTIVE FUNCTION

How do the neural substrates underlying sexually dimorphic behaviors develop? It is evident that the brain of diploid vertebrates is initially bisexual, changing during development under the influence of the hormonal milieu. Hormones during the perinatal period are not the sole determiner of psychosexual differentiation, however. Testicular feminization (Tfm) is a heredity defect present in a number of mammalian species in which genetic males develop as phenotypic females. Although Tfm rats have an inherited target organ insensitivity to androgen, they will exhibit mounting and introsmission behavior, but not sexual receptivity, when treated with exogenous hormones.[44] This suggests that the gender (genotype) of the individual is a major variable independent of gonadal hormones that must be considered in concepts about the organization of brain tissues that mediate sexually dimorphic behaviors.

The unisexual whiptail lizard *C. uniparens* is a useful animal model with which to investigate developmental aspects of sexuality because it controls for both genetic variation and gender (genotype). These species are isogenic because reproduction is parthenogenetic. There is no male genotype, and no males ever occur. Thus, since the gonads of parthenogenetic lizards always develop as ovaries, any change in primary and secondary sex structures in the unisexual lizards are easy to identify and study. Thus, this animal can be used to address the fundamental question of the inductive influences of the internal environment on sexual differentiation in the absence of male genotype.

Until recently it was assumed that sex in all amniote vertebrates (reptiles, birds, and mammals are all amniotes) is determined by genetic mechanisms. We now know, however, that in many reptiles, the phenomenon of temperature-dependent sex determination raises particularly interesting questions about the role of the environment in the organization of the brain. If incubation temperature determines sex ratios in the leopard gecko, we would predict that incubation temperature may also produce a gradient in psychosexual differentiation. Indeed, preliminary experiments show that females from eggs incubated at a high (mostly male-producing) temperature are more masculine than are females from eggs incubated at a low (female-producing) temperature.

B. EXPERIENCE VS. BIOLOGY

There is considerable debate today about the extent to which gender identity in humans is determined by social context or by biology, such as genetic and, consequently, hormonal factors. This is further complicated by the well-known reciprocal interaction between hormonal state and behavior. Abnormalities in either realm often lead to difficulties in psychosexual adjustment. In the parthenogenetic whiptail lizard, there are no males in the species, yet each individual alternately exhibits male and female sexual behavioral patterns. Thus, it is possible to examine the behavioral components of sexuality free of the complicating elements of physiological sex.[45,46] In addition to providing an animal model for the physiological bases and functional significance of gender-typical behaviors, this animal provides a unique probe into the organizational influences of the internal and external environments on sexual differentiation.[43]

V. MODELS OF OVULATORY CYCLING

In higher primates, only one of the two ovaries ovulates a single egg each menstrual cycle. Although some authors have concluded that ovulation occurs from each ovary randomly in some primates,[47,48] other observations suggest that ovulation usually alternates between ovaries,[49,50] a fact only recently confirmed in women.[51]

How and when a follicle is selected for the next ovulation is controversial. The dominant follicle or corpus luteum may secrete a substance that locally inhibits response of other follicles in the same ovary, thus ensuring that the next ovulation is in the contralateral ovary.[52] Or, progesterone secreted by the corpus luteum may locally inhibit the response of other follicles in that ovary to circulating gonadotropins.[52,53] Possibly, the dominant follicle secretes a protein(s) that inhibits response of other follicles to gonadotropins.[54] However, there is evidence that the dominant follicle does not inhibit the response of other follicles to gonadotropin, but rather that it secretes estrogen, which suppresses circulating follicle-stimulating hormone (FSH) levels below those necessary for growth of other follicles.[55,56]

How does the dominant follicle maintain its growth under these low gonadotropin conditions? It seems probable that access to circulating gonadotropin controls follicular selection within one ovary[57] and that the dominant follicle responds to a level of gonadotropin too low for growth of other follicles, perhaps because of its specialized microvasculature[58,59] or its unique gonadotropin receptor concentrations.[59] How a follicle maintains its dominance may not, however, be relevant to how its dominance arises. Why doesn't a dominant follicle occur in both ovaries? What are the mechanisms of interovarian communication that result in alternation of ovulation?

"Perhaps the most intriguing question on the dynamics of the primate ovarian cycle has not yet been satisfactorily studied; that is, how do we account for the selection of a single dominant follicle despite perfusion of both ovaries with seemingly identical exposure to pituitary gonadotropins?"[60] There is still no satisfactory answer to that question. Because of the relative difficulty of experimentation on primates, other vertebrate species that exhibit alternating, unilateral ovulation can serve as "model systems" to understand this phenomenon. Even though the structure of the ovarian follicular wall of other vertebrates differs from that of mammals, the hormonal control of ovarian function is similar and homologous among vertebrates.[58]

The female lizard *Anolis carolinensis* may possibly be unique among reptiles in that it shows an alternating ovulatory pattern comparable to that in human beings (see Figure 3). All of the many species in the tropical lizard genus *Anolis* exhibit alternating unilateral ovulation of a single egg during the breeding season.[61] Each ovary of *A. carolinensis* contains a series of about 11 ovarian follicles, a few of which are atretic. The follicles in each ovary are arranged in a step-wise size hierarchy, with the smallest follicle being next to a germinal bed and the largest being vitellogenic; all of the follicles except the largest are previtellogenic. Furthermore, there is symmetry in follicular size between ovaries of all ranks of previtellogenic follicles. Only the vitellogenic follicles, one on each side, exhibit asymmetry, the one in the larger ovary being about 6 to 8 mm in diameter and the one in the smaller ovary being about 2.5 mm in diameter.[62] The smaller ovary also contains a secretory corpus luteum.[63] What is important to understand is that each ovary has an approximately 16- to 28-d ovulatory cycle, but because each ovary is 180° out of phase, ovulation of single egg in each female occurs every 8 to 14 d.[64,65]

The endocrinology of ovarian follicular growth and steroid secretion in *Anolis* is very similar to that in mammals. That is, circulating pituitary gonadotropin (similar to mammalian FSH[66]) stimulates all phases of follicular growth, vitellogenesis, steroid hormone secretion, and ovulation.[66,67] Maintenance of the follicular size hierarchy in each ovary may be the result of differences in follicular access to circulating gonadotropins; i.e., access to gonad-

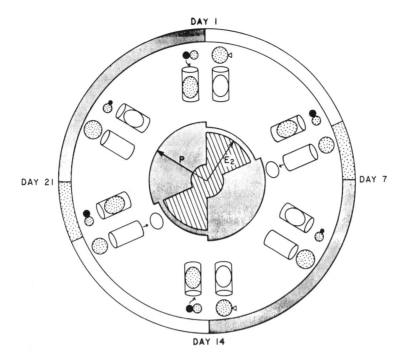

FIGURE 3. A diagrammatic summary of two ovulatory cycles in the lizard *A. caro-linensis*. The vitellogenic follicles are represented as stippled circles and secretory corpora lutea as solid circles; the small triangle represents a regressing corpora luteum; the cylinders represent oviducts with either shelled eggs (open ovals) or unshelled eggs (stippled ovals). In the center of the circle, the relative plasma concentrations of estrogen (hatched) and progesterone (shaded) are represented. The outer ring indicates the presence of full sexual receptivity (shaded), developing receptivity (stippled), or nonreceptivity (clear). The arrows indicate ovulation and oviposition. (Adapted from Reference 63.)

otropin is directly proportional to rank in the hierarchy, with the largest follicle appropriating most of the gonadotropin because of its hypervascularity[68] and its vessel permeability.[57]

As ovaries of *A. carolinensis* grow each spring, there is a large vitellogenic follicle (IL) in one ovary (randomly, right or left), while there is a small previtellogenic follicle (IS) in the contralateral ovary. As the season progresses, IL will be the first to ovulate in the spring (mid-April). If one surgically removes the previtellogenic follicle (IL) and then allows vernal ovarian recrudescence to occur,[70] the old follicle IS (in the contralateral ovary) will grow at a supernormal rate and be the first to ovulate, at a time when the removed IL would have ovulated. Furthermore, removal of IL does not result in supernormal growth of other follicles in the same ovary. Removal of follicle IS alone has no effect, and removal of both IL and IS stimulates growth of smaller follicles in each ovary. The importance of these findings is that the largest (dominant) follicle exerts an inhibition on the largest follicle in the contra-lateral ovary but does not inhibit smaller follicles in the ipsilateral ovary. Also, administration of pharmacological levels of FSH destroys the alternation, the result being symmetrical growth of the large vitellogenic follicles in each ovary.[67,69] This implies some form of interovarian communication, and the present hypothesis is that the dominant follicle somehow either "robs" circulating gonadotropin or prevents response or access of follicle IS to gonadotropin. A careful search for an anatomical basis for shunting of blood back and forth between ovaries did not reveal an obvious vascular mechanism.[71]

Whether the corpus luteum of *Anolis* inhibits follicles in the contralateral ovary as it does in primates is not known. It is possible that local concentrations of progesterone from the corpus luteum inhibit follicular growth in the same ovary, since progesterone is secreted by the *Anolis* corpus luteum[63,72] and this steroid inhibits FSH-stimulated follicular growth in *A. carolinensis*.[73] Presence of a shelled egg in one of the two oviducts of *Anolis* also could somehow (neurogenically or by secretion of an inhibitory substance) inhibit follicular growth in the ipsilateral ovary, but this could not explain asynchronous follicular growth before the first ovulation each spring.

VI. AFFECTIVE DISORDERS INVOLVING THE BASAL FOREBRAIN

The lack of a developed associative cortex and the nearly complete decussation of the optic tracts have shown lizards to be ideal for the study of the role of higher neural structures in behavior.[74,75] These investigations have yielded much information on the functional organization of the areas of the basal forebrain that underlie behavioral patterns such as aggression[36] and courtship.[76]

A. STRIATAL FUNCTION IN REPTILES
Several nuclei within the reptilian basal forebrain appear to correspond to the mammalian basal ganglia, the behavioral functions of which have only relatively recently been appreciated. These structures are now known to be associated with characteristic perceptual and cognitive deficits[77] and may play a profound role in attention structure and sensory gating.[78]

B. STRIATAL CONTROL OF SPECIES-TYPICAL STEREOTYPED BEHAVIOR
The highly stereotyped, species-typical behavioral patterns characteristic of the social displays in many species have provided robust data to illuminate striatal function. MacLean[79] demonstrated that lesions of the globus pallidus in the area where fibers converge to form the ansa lenticularis, or of the ansa itself, impaired a species-typical display in the squirrel monkey, probably by interrupting the pallidal projection to the tegmental area. The nature and specificity of MacLean's findings encouraged an investigation of the paleostriatum of a lizard, the green anole *Anolis carolinensis,* predicated on a close understanding of the units of behavior spontaneously emitted in naturalistic settings.[74]

The investigations of the anoline forebrain utilized a forebrain atlas and stereotaxic technique[80] to place lesions at selected sites and to correlate the sites of damage with subsequent deficits in the expression of stereotypies associated with aggression and reproduction.

Taking advantage of the almost complete decussation of the optic tracts in this species, lesions were stereotactically placed in one hemisphere only, and visual stimuli were restricted to either the lesioned or the intact hemisphere by means of a rubber eye-patch, effectively allowing subjects to function as their own controls. When lesions to the paleostriatum of lizards involved the lateral forebrain bundle, deficits in the species-typical challenge display response to conspecific males occurred,[81] but general arousal and displays that were under less specific stimulus-control such as courtship[76] were unaffected. Greenberg[74,81] characterized such lizards as unable to recognize appropriate stimulus input, a difficulty sometimes characterized as "social agnosia".

C. ANATOMICAL CORRESPONDENCES
The utility of a neurobehavioral model is determined largely by the demonstration that the model and the system for which it is a putative simplification possess corresponding neural organization. This is revealed by detailed analyses of neurocytological structure and chemistry. For example, acetylcholinesterase histochemistry and histofluorescence of catecholamines in the striatum of *Anolis*[82] when considered in concert, would seem to indicate

that its ventral portion could, in part, be homologized to mammalian caudate-putamen on the basis of their similar histochemistry. In support of this hypothesis, the projection indicated by terminal degeneration seen after treatment with a toxin selective for dopaminergic neurons, MPTA, is seen to largely overlap the histofluorescent, cholinesterase-sensitive areas.[83] These areas probably correspond to the terminus of the mesostriatal pathway in mammals, the caudate and putamen. This refines the earlier view that the reptilian ventral striatum is homologous in part to the globus pallidus of mammals.[84] In fact, one element of ventral striatum, the nucleus magnocellularis, is likely paleostriatal. Finally, there is some evidence that the dorsolateral striatum of reptiles may actually correspond to parts of the mammalian pallidum.[84,85]

Anatomical insights can be a powerful guide to the development of functional hypotheses. For example, spermatogenic activity (and thus perhaps reproductive behavior) appears correlated with the intensity of monoamine-indicating fluorescence in the paraventricular hypothalamus of the lizard *Lacerta*.[86] Work on other taxa has made it clear that cells can simultaneously secrete multiple hormones and neurotransmitters capable of orchestrating multiple responses in targets by means of differential concentrations, ratios of multiple agents, regulation of postsecretional metabolism, and/or modulation of receptor responses to specific agents;[87] further, specific activities can differentially deplete dopamine from striatal nuclei.[88]

The repertoire of stereotyped behavioral patterns that can be elicited under controlled conditions and thus are amenable to neurobehavioral investigation has recently expanded. Not only aggression and reproductive phenomena but also specific behavioral units of social dominance[41] and exploratory behavior[21] have been shown to be sensitive to specific stimulus control in concert with known psychoactive stress-sensitive and gonadal hormones. Such phenomena may well be linked to altered arousal or attentional processes not unlike those identified by Schneider[78] as central to striatal-related behavioral disorders in man. In this regard it is interesting that Iverson[89] noted that motivational and motor arousal dysfunction appeared associated with the mesolimbic and mesostriatal systems, respectively.

The striatal complex is a prominant candidate to provide insight into higher neural influences on behavioral dysfunctions because within the repertoire of reptilian behavioral patterns there is, at best, a modest representation of the learning abilities that characterize many mammals. The complex, often reciprocal relationships between neural, endocrine, and behavioral elements are correspondingly simplified and will eventually yield to techniques that can correlate behavioral patterns with physiological variables and their interactions.

VII. CENTRAL DOPAMINE DYSFUNCTION

The "shaking palsy" described early in 1817 by James Parkinson includes an array of symptoms, most prominent of which are a mild resting tremor of a limb, muscular rigidity, and bradykinesia. There is also an impairment of intellectual and cognitive processes ("bradyphrenia"), with an associated depression sometimes preceding the more overt neurological symptoms. Autonomic dysfunctions are also present, and physical or psychological stress can alter the clinical profile in one of two ways: "freezing", an exacerbation or precipitation of neurological deficits, or "paradoxical kinesia", a transient remission of bradykinesia when confronted with a life-threatening emergency.[90] Several of these effects can occur in lizards following administration of a dopaminergic neurotoxin, MPTP.

A. NEUROTOXIC EFFECTS OF MPTP

The meperidine analog *N*-methyl-4-phenyl-1,2,3,6-tetrahydropyridine (MPTP) produces symptoms similar to idiopathic Parkinson's disease,[91] primarily by cytotoxic effects on the dopaminergic substantia nigra (SN) pars compacta, although some studies also report toxic effects in locus ceruleus and the ventral tegmental area.[92] The mode of MPTP cytotoxicity

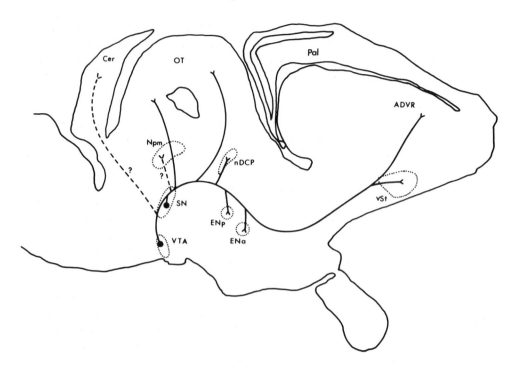

FIGURE 4.   Putative dopaminergic axis in the lizard *A. carolinensis* as indicated by the distribution of degenerating axons and terminals subsequent to administration of the selective neurotoxin, MPTP. ADVR, anterior dorsal ventricular ridge; Cer, cerebellum; ENa and ENp, entopeduncular nuclei anterior and posterior, respectively; nDCP, nucleus of the dorsal posterior commissure; Npm, nucleus profundus mesencephalii; OT, optic tectum; Pal, pallium; SN, substantia nigra; vSt, ventral striatum; VTA, ventral tegmental area. (Adapted from Reference 82.)

is derived from its selective incorporation into neurons through their dopamine re-uptake systems. While the primary site of cell death is the pars compacta of the SN, the adjacent ventral tegmental area and other sites are also often affected. The selectivity appears sensitive to both the age of the animal (less selective in older subjects[91]) and the amount of MPTP administered. In fact at low levels, many SN cells may survive while conspicuous mesostriatal axonopathies appear and tyrosine hydroxylase immunoreactivity in the striatum decreases.[93]

Following systemic administration of MPTP to *Anolis*, the degeneration-sensitive cupric silver method of de Olmos et al.[94] was used to construct maps of the sites of toxic damage. This procedure revealed cytopathological changes at several loci ranging from the forebrain to the cervical cord (Figure 4). Degenerative changes in striatal terminals occur in areas that also show catecholamine histofluorescence and AChE reactivity,[82,95] thus supporting the putative homology of the reptilian striatal afferents and the mammalian mesostriatal pathway.

B. BEHAVIORAL CHANGES FOLLOWING MPTP TREATMENT

Acute behavioral changes in MPTP-treated animals included indications of a physiological stress response and, in particular, color changes, including the formation of a postorbital darkening (the "eyespot"), and nuchal crest erection, both indications of adrenal activation.[36] These effects and a pronounced hypokinesia remitted in all but 8 individuals, all of which received in excess of 50 mg/kg of the drug. These individuals also developed akinesia, postural rigidity, episodic convulsions, and occasionally manifested stereotyped head and neck movements. These individuals died in less than 24 h. Most individuals, however, survived and showed no further symptoms of physiological stress.[95] This survival is consistent with the idea of functional recovery of involved neural tissue; the individuals that did not survive may have suffered a crisis of adaptation due to massive and persevering adrenal activation.[96] Most subjects that received smaller doses, while showing clear indications of neurological damage, displayed no behavioral abnormalities; several, however, showed episodic rigidity and diminished spontaneous behavior. In tests of species-typical aggressive behavior, such subjects performed appropriate stereotyped behavior but at notably low intensities.[115]

VIII. DEVELOPMENTAL-COMMUNICATIVE DISORDERS AND LEARNING

Reptiles generally lack the intense postnatal care found in birds and mammals. Even when present, neonate reptiles are highly precocious and capable of carrying out feeding, defensive, maintenance, and other responses independently.[97] This means that the development of behavior, including adult behavior, can be traced and experimentally manipulated in the absence of the parental care system, some elements of which, especially lactation and thermoregulation, are so profound in mammals. A specific application of this difference has been applied to the origins of play,[98] but many examples can be found in other areas, such as reproduction, aggression, nutrition, chemoreception, and learning.

While reptiles are not noted for their intelligence, they in fact can learn many traditional laboratory tasks, particularly if ecologically relevant reinforcers are utilized.[99] As Burghardt has clearly pointed out, the problem is ours in understanding and responding to their thermal, nutritional, physical, and social needs.[100,101] Reptiles such as crocodilians have learned many complex tasks indeed, but for practical laboratory work the smaller lizards are more tractable, common, and easy to work with. While the highly visual iguanid and agamid lizards have been used most, skinks, geckos, and teiids are often good species and represent a diversity of sensory, dietary, social, and ecological characteristics. Although ecological adaptive needs such as light, heat, humidity, visual complexity, and structure of the habitat must be accommodated, traditional laboratory designs (operant conditioning or the classical maze) can be adapted for determining the role of carefully circumscribed phenomena such as the function of specific neural sites or the effects of environmental toxins.

IX. CELLULAR PHYSIOLOGY OF BODY COLOR IN *ANOLIS CAROLINENSIS*

Perhaps the most remarkable characteristic of the small iguanid lizard *A. carolinensis* is the rapid variation in skin color it exhibits in response to various social and environmental stressors. Chromatic responses, ranging from dark brown to brilliant green, are elicited by light or by systemic hormonal agents rather than neuronal control.[38] Skin color is determined by the interaction of three types of cells.[102] Blue light, produced by thin layer interference, is reflected from crystalline plates within iridophores to pass through a more superficial layer of yellow-pigment-containing xanthophores resulting in the animals' tonic green skin color. Melanophores lie below these two cell types. Brown coloration is produced by dispersion of melanin within melanophores into dendritic processes that terminate above and obscure iridophores and xanthophores. Thus, dispersion and reaggregation of melanin, a chromomotor response, are directly responsible for the variation in color.

Most reptilian species possess the ability to alter their coloration by redistributing pigment. The fact, however, that skin of *A. carolinensis* responds to visible light and to hormones in systemic circulation, but is not innervated, lends it valuable advantage as a model system for the investigation of the concerted action of hormones and in certain problems in photomedicine. Isolated skin is easily maintained in culture for several days with little degradation of its functional characteristics. Since the reaction to either light or hormonal stimulation is a change in color, photometric measurement of variation in albedo provides a simple and convenient indicator of the progress of cellular response.

Two different systems (primary and secondary) for inducing chromomotor responses have been identified.[103] The primary mechanism is a direct response to light. The secondary is response to hormonal stimulation. Exposed *in vitro* to visible light at an intensity of 100 W/m² or less, skin of *A. carolinensis* darkens within 2.5 min to lose 50% or more in reflectivity as pigment is dispersed.[104] This primary (photic) response has been observed in every major group of animals,[105] including primates. In human skin, a redistribution of pigment within keratinocytes in response to UV-A radiation has been described.[106] This response, commonly called the immediate pigment-darkening response, may provide a significant protective skin tanning within minutes of exposure to sunlight. Since significant synthesis of new melanin does not occur until much later following exposure to the sun, immediate pigment redistribution may be important in providing partial interim protection from sunlight. Increased knowledge of possible means of modulating exposure to the UV light in sunlight may have biomedical applications beyond the obvious defense against carcinogenic effects of UV light; indeed, because neural tissue, like melanocytes, is derived from neural crest and therefore may possess antigenic properties much like those of melanocytes, it has been suggested that the association of increased risk of developing multiple sclerosis with increasing latitude may be associated with the fact that UV light can aid in the induction of suppressor T-cells specific for melanocyte-associated antigens. If multiple sclerosis is, as suspected, an autoimmune disease against neural tissue, suppressor cells may confer some measure of protection.[107]

Knowledge of the mechanism of photic response gained from the *A. carolinensis* model may be instrumental in finding methods to stimulate or control the similar response in human skin as a method of reducing UV-A damage. Further, the transduction mechanism for the dermal photic response depends upon calcium and appears to involve changing levels of cyclic GMP.[108] Because of these and other similarities to signal transduction in the retinal rod, the dermal photic response, typified in the skin of *A. carolinensis,* could serve as an easily manipulated model system for the study of signal transduction in retinal cells.

A. HORMONAL INTERACTIONS IN REGULATING CELL FUNCTION

It has become increasingly obvious that cells respond to a variety of hormones. It is as important to be aware of the action of hormones in concert as it is to understand them as solo agents. Melanophores in *A. carolinensis* skin possess alpha-2 as well as beta-2 adrenergic and melanocyte-stimulating hormone (MSH) receptors (except in a patch of skin behind the eye which lacks alpha-2 receptors).[109] The action of MSH bound to its receptor is to activate adenyl cyclase and increase intracellular levels of cyclic AMP, resulting in pigment dispersion and brown skin coloration. Chromomotor response to MSH is so predictable that accurate *in vitro* bioassays for the hormone have been devised that employ estimates of color change in small pieces of *Anolis* skin.[110] Low concentrations of epinephrine, acting through the beta-2 receptor, also stimulate adenyl cyclase. Epinephrine at higher concentrations activates the alpha-2 receptor. This leads to inhibition of adenyl cyclase, reduces cyclic AMP, and overrides the actions of MSH and beta-2 agonists.

Adrenergic receptors play a central role in the regulation of a variety of physiological processes. Adrenergic agonists and antagonists are used in the clinical management of medical conditions such as hypertension, cardiac arrythmias, angina pectoris, thyrotoxicosis, asthma, and allergies, among others. It is imperative to understand the actions and interactions of agonists with each other and with their various antagonists. The melanophore system in *A. carolinensis,* in many instances, can be used to make such determinations. Propranolol, for instance, is a beta antagonist used in the control of certain forms of hypertension. It exerts its action by binding the beta adrenergic receptor, blocking the effects of catacholamines in increasing blood pressure. Although MSH has a separate receptor on the cell surface, propranolol, in addition to blocking the effects of epinephrine, interferes with the action of MSH on the *A. carolinensis* chromatophore.[111] Interactions of this type, easily observed in *A. carolinensis* chromomotor response, are difficult to demonstrate in most model systems.

In summary, the dermal melanophore system in the skin of *A. carolinensis* constitutes a rapid, sensitive, and replicable model system that may provide valuable insight relative to a variety of problems of biomedical importance. These range from elucidation of the mechanisms of movement of organelles within the cell[104] to mechanisms of signal transduction in photoreception, the protection of skin from the harmful effects of UV radiation, and the interactions of hormonal agonists and antagonists.

X. CONCLUSIONS

Many reptile species are particularly well suited for the study of the physiological causes as well as consequences of behavior. As fully terrestrial tetrapods, many physiological systems are fundamentally like our own, yet the interpretation of those elements of physiology that articulate directly with behavior are relatively uncomplicated by virtue of the stereotypy of reptilian behavior and their modest problem solving ability.

Further, many species, including small lizards, are often very tolerant of surgical interventions and will heal quickly and completely in aseptic conditions. Choosing species known for adapting well to captivity can facilitate the study of many individuals housed in compact, yet adequate quarters. While live insect food is necessary for many, the amount needed is far less than an equivalent mass (and far fewer individuals) of, for example, insectivore mammals. Some species are essentially vegetarian; for example, green iguanas (*Iguana iguana*) have served in many physiological investigations. However, only recently has adequate information on optimal captive housing conditions been available,[112] allowing increased replicability and internal validity in studies of behavioral and physiological interactions. In addition, many reptiles are small, inexpensive, easily cared for, and robust subjects for many kinds of manipulations. For many species we possess considerable baseline data on morphology, cytology, neurology, endocrinology, and behavior, frequently subjected to comprehensive review in the continuing series *Biology of the Reptilia,* edited by Carl Gans in collaboration with many reptile specialists.[116]

In general, husbandry and medical treatment for reptiles have improved at a great rate in recent years. Besides the specialized literature, works such as those of Murphy and Collins[9] and Frye[113] offer recent summaries and useful guidelines. As with all animals, experienced reptile experts should be available for consultation and emergency assistance. Many problems are easily avoidable; for example, parasites (especially mites) introduced from wild-caught specimens are a common problem but easily controlled once proper conditions are in place.

Specific areas of biomedical research have even extended to psychiatry. Stereotyped elements of reptilian behavior, particularly dysfunctions of social and communicative behavior, have become increasingly studied in the laboratory study since it has become clear that under controlled laboratory conditions predicated on field observations, several reptile species display the relevant if not the entire behavioral repertoire, including the development of complex social organizations.[37,74] And for many species, social stimuli can be correlated with physiological stress responses, changes in reproductive function, and behavior, often more directly than can comparable dependent variables in mammals.

A. THE EVOLUTIONARY PERSPECTIVE

Beyond an intrinsic interest in reptiles, work with these descendants of the stock which gave rise to modern birds and mammals engenders an evolutionary point of view that provides us with an implicit sense of the flexibility of all organisms. The utility of the evolutionary perspective often seems remote, but concern with it is neither trivial nor sterile, although the details by which it is established are often obscure and the controversy it engenders frequently generates more heat than light. The understanding of putative anatomical, physiological, or behavioral correspondences between animals is central to the development of useful models and requires that the logical problems of analogy, homology, and homoplasy be clearly engaged.[114] Further, the evolutionary perspective has already proved invaluable in suggesting alternative solutions to many of the ecological problems that reptiles share with other terrestrial vertebrates, and it can be expected to provide us with many future animal models to use in solving problems not amenable to experimentation in conventional research species.

ACKNOWLEDGMENTS

Work on chemical control of feeding was supported in part by National Science Foundation (NSF) grants to Gordon Burghardt; research on reproductive behavior was supported in part by NSF grants to Richard Jones and National Institutes of Health grants to David Crews. Work on the central neural function of *Anolis carolinensis* was made possible by a University of Tennessee Faculty Research Grant and by NSF grant BNS-8406028 to Neil Greenberg and a Cole Neuroscience Foundation grant to Robert Switzer; we are grateful for Edward F. O'Connor's thoughtful assistance with the histofluorescence procedures, and to Richard Heikkila for providing the dopamine neurotoxins.

REFERENCES

1. **Lehrman, D. S.,** Interaction of internal and external environments in the regulation of the reproductive cycle of the ring dove, in *Sex and Behavior,* Beach, F. A., Ed., John Wiley & Sons, New York, 1965, 355.
2. **Hinde, R. A.,** Interaction of internal and external factors in the integration of canary reproduction, in *Sex and Behavior,* Beach, F. A., Ed., John Wiley & Sons, New York, 1965, 381.
3. **Crews, D.,** Integration of internal and external stimuli in the regulation of lizard reproduction, in *Behavior and Neurology of Lizards,* Greenberg, N. and MacLean, P. D., Eds., National Institutes of Mental Health, Rockville, MD, 1978, 149.
4. **White, N. F.,** *Ethology and Psychiatry,* University of Toronto Press, Toronto, 1974.
5. **McGuire, M. T. and Fairbanks, L. A.,** *Ethological Psychiatry, Psychopathology in the Context of Evolutionary Biology,* Grune & Stratton, New York, 1977, 230.
6. **Euler, U. S. and Folkow, B.,** The effect of stimulation of autonomic areas in the cerebral cortex upon the adrenaline and noradrenaline secretion from the adrenal gland in the cat, *Acta Physiol. Scand.,* 42, 313, 1958.
7. **Matsui, H.,** Adrenal medullary secretory response to pontine and mesencephalic stimulation in the rat, *Neuroendocrinology,* 33, 84, 1981.
8. **Matsui, H.,** Adrenocortical secretion in response to diencephalic stimulation in the rat, *Neuroendocrinology,* 38, 164, 1984.
9. **Murphy, J. B. and Collins, J. T.,** *Reproductive Biology and Diseases of Captive Reptiles,* Soc. Stud. Amphib. Rept. Contrib. to Herpetol., Soc. Stud. Amphib. Rept., Oxford, Ohio, 1980, 1.
10. **Kallman, F. J., Schoenfeld, W. A., and Barrera, S. E.,** The genetic aspects of primary eunuchoidism, *Am. J. Ment. Defic.,* 48, 203, 1944.
11. **Soules, M. R. and Hammond, C. B.,** Female Kallman's syndrome: evidence for a hypothalamic luteinizing hormone-releasing deficiency, *Fertil. Steril.,* 33, 82, 1980.
12. **Henkin, R. I. P., Schecter, J., Friedewald, W. T., Demets, D. L., and Raff, M.,** A double-blind study of the effects of zinc sulfate on taste and smell dysfunction, *Am. J. Med. Sci.,* 272, 285, 1976.
13. **Burghardt, G. M.,** Chemical perception in reptiles, in *Communication by Chemical Signals,* Johnston, J. W., Jr., Moulton, D. G., and Turk, A., Eds., Appleton-Century-Crofts, New York, 1970, 241.
14. **Crews, D.,** Hormonal control of male and female sexual behavior in the garter snake *(Thamnophis sirtalis parietalis),* Horm. Behav., 7, 451, 1976.

15. **Crews, D.,** Control of male sexual behavior in the Canadian red-sided garter snake, in *Hormones and Behavior in Higher Vertebrates,* Balthazart, J., Prove, E., and Gilles, R., Eds., Springer-Verlag, Berlin, 1983, 398.

16. **Halpern, M.,** Nasal chemical senses in snakes, in *Advances in Vertebrate Neuroethology,* Ewert, J., Capranica, R., and Ingle, D., Eds., Plenum Press, New York, 1983, 141.

17. **Halpern, M. and Kubie, J.,** The role of the ophidian vomeronasal system in species-typical behavior, *Trends Neurosci.,* 7, 472, 1984.

18. **Mason, R. and Crews, D.,** Pheromonal mimicry in garter snakes, in *Chemical Signals in Vertebrates,* Duval, D. and Muller-Schwartz, D., Eds., Plenum Press, New York, 1986.

19. **Burghardt, G. M.,** Behavioral and stimulus correlates of vomeronasal functioning in reptiles: feeding, grouping, sex, and tongue use, in *Chemical Signals in Vertebrates and Aquatic Invertebrates,* Müller-Schwarze, D. and Silverstein, R., Eds., Plenum Press, New York, 1980, 275.

20. **Halpern, M.,** The organization and function of the vomeronasal system, *Ann. Rev. Neurosci.,* 10, 325, 1987.

21. **Greenberg, N.,** Exploratory behavior and stress in the lizard *Anolis carolinensis, Z. Tierpsychol.,* 70, 89, 1985.

22. **Mrosovsky, N. and Sherry, D. F.,** Animal anorexias, *Science,* 207, 837, 1980.

23. **Finger, T. E. and Silver, W. L.,** *Neurobiology of Taste and Smell,* John Wiley & Sons, New York, 1987.

24. **Burghardt, G. M.,** Behavioral ontogeny in reptiles: whence, whither, and why, in *The Development of Behavior: Comparative and Evolutionary Aspects,* Burghardt, G. M. and Bekoff, M., Eds., Garland STPM, New York, 1978, 149.

25. **Scudder-Davis, R. M. and Burghardt, G. M.,** Diet and growth in juveniles of the garter snakes *Thamnophis sirtalis internalis* and *T. radix radix, Growth,* 51, 74, 1987.

26. **Burghardt, G. M., Goss, S., and Schell, F. M.,** Comparison of earthworm and fish-derived chemicals in eliciting prey attack by garter snakes *(Thamnophis), J. Chem. Ecol.,* 14, 855, 1988.

27. **Christian, J. J.,** Endocrine factors in population regulation, in *Biosocial Mechanisms of Population Regulation,* Cohen, M. N., Malpass, R. S., and Klein, H. G., Eds., Yale University Press, New Haven, CT, 1980, 55.

28. **Selye, H.,** *The Stress of Life,* McGraw-Hill, New York, 1956.

29. **Selye, H.,** *Stress in Health and Disease,* Butterworths, Boston, 1976.

30. **Kiritz, S. and Moos, H.,** Physiological effects of social environments, *Psychosom. Med.,* 36, 96, 1974.

31. **Holst, F.,** Renal failure as the cause of death in *Tupaia belanger* exposed to persistent social stress, *J. Comp. Physiol.,* 78, 236.

32. **Wirtz, P.,** Physiological effects of visual contact to a conspecific in *Blennius pholis, J. Comp. Physiol.,* 101, 237, 1975.

33. **Greenberg, N. and Wingfield, J.,** Stress and reproduction: reciprocal relationships in, *Reproductive Endocrinology of Lower Vertebrates,* Norris, D. O. and Jones, R. E., Eds., Plenum Press, New York, 1987, 461.

34. **Crews, D.,** Psychobiology of reptilian reproduction, *Science,* 189, 1059, 1975.

35. **Crews, D.,** Interrelationships among ecological, behavioral and neuroendocrine processes in the reproductive cycle of *Anolis carolinensis* and other reptiles, in *Advances in the Study of Behavior,* Vol. 11, Rosenblatt, J. S., Hinde, R. A., Beer, C. G., and Busnel, M. C., Eds., Academic Press, New York, 1980, 1.

36. **Greenberg, N. and Crews, D.,** Physiological ethology of aggression in amphibians and reptiles, in *Hormones and Aggressive Behavior,* Svare, D. B., Ed., Plenum Press, New York, 1983, 469.

37. **Greenberg, N.,** Ethological consideration in the experimental study of lizard behavior, in *Behavior and Neurology of Lizards,* Greenberg, N. and MacLean, P. D., Eds., NIMH, Rockville, MD, 1978, 203.

38. **Kleinholz, L. H.,** Studies in reptilian color changes. III. Control of the light phase and behavior in isolated skin, *J. Exp. Biol.,* 15, 492, 1938a.

39. **Greenberg, N., Chen, T., and Vaughan, G. L.,** Melanotropin levels are altered by acute and chronic social stress in lizards, *Soc. Neurosci. Abstr.,* 12, 834, 1986.

40. **Anderson, B.,** Is alpha-MSH deficiency the cause of Alzheimer's disease?, *Med. Hypoth.,* 19, 279, 1986.

41. **Greenberg, N., Chen, T., and Crews, D.,** Social status, gonadal state, and the adrenal stress response in the lizard, *Anolis carolinensis, Horm. Behav.,* 18, 11, 1984.

42. **Leshner, A. I.,** *An Introduction to Behavioral Endocrinology,* Oxford University Press, New York, 1978.

43. **Crews, D., Gustafason, J. E., and Tokarz, R. R.,** Psychobiology of parthenogenesis in reptiles, in *Lizard Ecology,* Huey, R., Pianka, E., and Schoener, T., Eds., Harvard University Press, Cambridge, MA, 1983, 205.

44. **Shapiro, B. H., Levine, D. C., and Adler, N. T.,** The testicular feminized rat: a naturally occurring model of androgen independent brain masculinization, *Science,* 209, 418, 1980.

45. **Crews, D.,** On the origin of sexual behavior, *Psychoneuroendocrinology,* 7, 259, 1982.

46. **Crews, D.,** Functional association in behavioral endocrinology, in *Masculinity/Femininity: Concepts and Definitions,* Reinisch, J. M., Rosenblum, L. A., and Sanders, S. A., Eds., Oxford University Press, Oxford, 1986.

47. **Wallach, E. E., Virutamasen, P., and Wright, K. H.,** Menstrual cycle characteristics and side of ovulation in the rhesus monkey, *Fertil. Steril.,* 24, 715, 1973.

48. **Clark, J. R., Dierschke, D. J., and Wolf, R. C.,** Hormonal regulation of ovarian folliculogenesis in rhesus monkeys. I. Concentration of serum luteinizing hormone and progesterone during laparoscopy and patterns of follicular development during successive menstrual cycles, *Biol. Reprod.,* 18, 779, 1978.

49. **Dukelow, W. R.,** Ovulatory cycle characteristics in *Maca fascicularis, J. Med. Primatol.,* 6, 33, 1977.

50. **Hodgen, G. D.,** The dominant ovarian follicle, *Fertil. Steril.,* 38, 281, 1982.

51. **Gougeon, A. and Lefevre, B.,** Histological evidence of alternating ovulation in women, *J. Reprod. Fertil.,* 70, 7, 1984.

52. **Goodman, A. L. and Hodgen, G. D.,** The ovarian triad of the primate menstrual cycle, *Recent Prog. Horm. Res.,* 39, 1, 1983.

53. **Goodman, A. L. and Hodgen, G. D.,** Antifolliculogenic action of progesterone despite hypersecretion of FSH in monkeys, *Am. J. Physiol.,* 243, E387, 1982.

54. **DiZerga, G. S. and Hodgen, G. D.,** The interovarian progesterone gradient: a spatial and temporal regulator of folliculogenesis in the primate ovarian cycle, *J. Clin. Endocrinol. Metab.,* 54, 495, 1982.

55. **DiZerga, G. S., Marrs, R. P., Roche, P. C., Campeau, J. D., and Kling, O. R.,** Human granulosa cell secretion of protein(s) which suppress follicular response to gonadotropins, *J. Clin. Endocrinol. Metab.,* 56, 147, 1983.

56. **Zeleznik, A. J., Hutchinson, J. S., and Schuler, H. M.,** Interference with the gonadotropin-suppressing actions of estradiol in macaques overrides the selection of a single preovulatory follicle, *Endocrinology,* 117, 991, 1985.

57. **Jones, R. E.,** Control of follicular selection, in *The Vertebrate Ovary — Comparative Biology and Evolution,* Jones, R. E., Ed., Plenum Press, New York, 1978a, 763.

58. **Jones, R. E.,** Evolution of the vertebrate ovary — an overview, in *The Vertebrate Ovary — Comparative Biology and Evolution,* Jones, R. E., Ed., Plenum Press, New York, 1978b, 827.

59. **Zeleznick, A. J., Schuler, H. M., and Reichert, L. E.,** Gonadotropin-binding sites in the rhesus monkey ovary: role of the vasculature in the selective distribution of human chorionic gonadotropin to the preovulatory follicle, *Endocrinology,* 109, 356, 1981.

60. **DiZerga, G. S. and Hodgen, G. D.,** Folliculogenesis in the primate ovary, *Endocrinol. Rev.,* 2, 27, 1981.

61. **Smith, H. M., Sinelnik, G., Fawcett, J. D., and Jones, R. E.,** A survey of the chronology of ovulation in anoline lizard genera., *Trans. Kans. Acad. Sci.,* 75, 107, 1973.

62. **Jones, R. E., Fitzgerald, K., Duvall, D., and Banker, D.,** On the mechanisms of alternating and simultaneous ovulation in lizards, *Herpetologica,* 35, 132, 1979.

63. **Jones, R. E., Guillette, L. J., Jr., Summers, C. H., Tokarz, R. R., and Crews, D.,** The relationship among ovarian condition, steroid hormones, and estrous behavior in *Anolis carolinensis, J. Exp. Zool.,* 227, 145, 1983a.

64. **Hamlett, G. W. D.,** Notes on breeding and reproduction in the lizard *Anolis carolinensis, Copeia,* 1952, 183, 1952.

65. **Andrews, R. M.,** Oviposiion frequency of *Anolis carolinensis, Copeia,* 259, 1985.

66. **Licht, P., Papkoff, H., Farmer, S., Muller, C., Tsui, H. W., and Crews, D.,** Evolution of gonadotropin structure and function, *Recent Prog. Horm. Res.,* 169, 1977.

67. **Jones, R. E., Tokarz, R., LaGreek, F., and Fitzgerald, K.,** Endocrine control of clutch size in reptiles. VI. Patterns of FHS-induced ovarian stimulation in adult *Anolis carolinensis, Gen. Comp. Endocrinol.,* 30, 101, 1976.

68. **Gerrard, A. M., Jones, R. E., and Roth, J. J.,** Thecal vascularity in ovarian follicles of different size and rank in the lizard *Anolis carolinensis, J. Morphol.,* 141, 227, 1973.

69. **Licht, P.,** Effect of mammalian gonadotropins (ovine FSH and LH) in female lizards, *Gen. Comp. Endocrinol.,* 14, 98, 1970.

70. **Jones, R. E., Gerrard, A. M., and Roth, J. J.,** Endocrine control of clutch size in reptiles. II. Compensatory follicular hypertrophy following partial ovariectomy in *Anolis carolinensis, Gen. Comp. Endocrinol.,* 20, 550, 1973.

71. **Jones, R. E., Summers, C. H., Austin, H. B., Smith, H. M., and Gleeson, T. T.,** Ovarian, oviductal, and adrenal vascular connection in female lizards (genus *Anolis), Anat. Rec.,* 206, 247, 1983.

72. **Guilette, L. J., Jr. and Fox, S. L.,** Effect of deluteinization on plasma progesterone and gestation in the lizard *Anolis carolinensis, Comp. Biochem. Physiol. A,* 80, 303, 1985.

73. **Jones, R. E.,** Endocrine control of clutch size in reptiles. IV: Estrogen-induced hyperemia and growth of ovarian follicle in the lizard *Anolis carolinensis, Gen. Comp. Endocrinol.,* 25, 211, 1975.

74. **Greenberg, N.,** A neuroethological study of display behavior in the lizard *Anolis carolinensis, Am. Zool.,* 17, 191, 1977.

75. **Crews, D. and Greenberg, N.,** Function and causation of social signals in lizards, *Am. Zool.,* 21, 273, 1981.

76. **Greenberg, N., Scott, M., and Crews, D.,** Role of the amygdala in the reproductive and aggressive behavior of the lizard, *Anolis carolinensis, Physiol. Behav.,* 32, 147, 1984.
77. **Teuber, H. L.,** Complex functions of basal ganglia, in *The Basal Ganglia,* Yahr, P., Ed., Raven Press, New York, 1976, 151.
78. **Schneider, J. S.,** Basal ganglia role in behavior: importance of sensory gating and its relevance to psychiatry, *Biol. Psychiatry,* 19(12), 1693, 1984.
79. **MacLean, P. D.,** Effects of lesions of the globus pallidus on species-typical display behavior of squirrel monkeys, *Brain Res.,* 149, 175, 1978.
80. **Greenberg, N.,** A forebrain atlas and stereotaxic technique for the lizard, *Anolis carolinensis, J. Morphol.,* 174, 217, 1982.
81. **Greenberg, N., MacLean, P. D., and Ferguson, L. F.,** Role of the paleostriatum in species-typical display of the lizard, *Anolis carolinensis, Brain Res.,* 172, 229, 1979.
82. **Greenberg, N., Font, E., and Switzer, R. C., III,** The reptilian striatum revisited, in *The Forebrain in Reptiles: Current Concepts of Structure and Function,* Shwerdtfeger, W. K. and Smeets, W. J., Eds., Karger-Verlag, Basel, 1988, 162.
83. **Northcutt, R. G.,** Forebrain and midbrain organization in lizards and its phylogenetic significance, in *Behavior and Neurology of Lizards,* Greenberg, N. and MacLean, P. D., Eds., NIMH, Rockville, MD, 1978, 11.
84. **Baker-Cohen, K. F.,** Comparative enzyme histochemical observations on submammalian brains, *Rev. Anta. Morphol. Exp.,* 40(7), 17, 1968.
85. **Brauth, S. E. and Kitt, C. A.,** The paleostriatal system of *Caiman crocodilus, J. Comp. Neurol.,* 189, 1980.
86. **Marshall, C.,** Hypothalamic monoamines in lizards *(Lacerta), Cell Tissue Res.,* 205, 95, 1980.
87. **O'Donohue, T. L. and Dorsa, D. M.,** The opiomelanotropinergic neuronal and endocrine systems, *Peptides,* 3, 353, 1982.
88. **Freed, C. R. and Yamamoto, B. K.,** Regional brain dopamine metabolism, a marker for the speed, direction, and posture of moving animal, *Science,* 292, 62, 1985.
89. **Iverson, S. D.,** Striatal function and stereotyped behavior, in *Psychobiology of the Striatum,* Cools, A. R., Lohman, A. H. M., and Van Den Bercken, J. H. L., Eds., Elsevier, New York, 1972, 99.
90. **Zigmond, M. J., Stricker, E. M., and Berger, T. W.,** Parkinsonism: insights from animal models utilizing neurotoxic agents, in *Animal Models of Dementia,* Alan R. Liss, New York, 1987.
91. **Marsden, C. D. and Jenner, P. G.,** The significance of 1-methyl-4-phenyl-1,2,3,6-tetrahydropyridine, in *Selective Neuronal Death,* CIBA Symp., 126, 239, 1987.
92. **Langston, J. W. and Irwin, I.,** MPTP: current concepts and controversies, *Clin. Neuropharmacol,* 9, 485, 1986.
93. **Kitt, C. A., Cork, L. C., Eideberg, E., Tong, T. H., and Price, D. L.,** Injury of catecholaminergic neurons after acute exposure to MPTP, *Ann. N.Y. Acad. Sci.,* 495, 730, 1987.
94. **de Olmos, J. S., Ebbesson, S. O. E., and Heimer, L.,** Silver methods for impregnation of degenerating axons, in *Neuroanatomical Tract-Tracing Methods,* Heimer and Robards, Plenum Press, New York, 1981, 117.
95. **Font, E., Switzer, R. C., III, and Greenberg, N.,** MPTP-induced neuropathology and behavior in the lizard *Anolis carolinensis,* unpublished data, 1988.
96. **Barbeau, A., Dallaire, L., Buu, N. T., Poirier, J., and Rucinska, E.,** Comparative behavioral, biochemical and pigmentary effects of MPTP, MPP⁺ and paraquat in *Rana pipiens, Life Sci.,* 37, 1529, 1985.
97. **Burghardt, G. M.,** Precocity, play, and the ectotherm-endotherm transition, in *Handbook of Behavioral Neurobiology,* Vol. 9, Blass, E. M., Ed., 1988, 107.
98. **Burghardt, G. M.,** On the origins of play, in *Play in Animals and Humans,* Smith, P. K., Ed., Blackwell Scientific, London, 1984, 5.
99. **Brattstrom, B. H.,** Learning studies in lizards, in *Behavior and Neurology of Lizards,* Greenberg, N. and MacLean, P. D., Eds., NIMH, Rockville, MD, 1978, 173.
100. **Burghardt, G. M.,** Learning processes in reptiles, in *Biology of the Reptilia,* Vol. 7, Gans, C. and Tinkle, D., Eds., Academic Press, New York, 1977, 555.
101. **Burghardt, G. M.,** The ontogeny, evolution, and stimulus control of feeding in humans and reptiles, in *The Chemical Senses and Nutrition,* Kare, M. R. and Maller, O., Eds., Academic Press, New York, 1977, 253.
102. **Bagnara, J. T., Taylor, J. D., and Hadley, M. E.,** The dermal chromatophore unit, *J. Cell Biol.,* 38, 67, 1968.
103. **Weber, W.,** Photosensitivity of chromatophores, *Am. Zool.,* 23, 495, 1983.
104. **Vaughan, G. L.,** Photosensitivity in the skin of the lizard, *Anolis carolinensis, Photochem. Photobiol.,* 49, 109, 1987.
105. **Wolken, J. J. and Mogus, M. A.,** Extraocular photosensitivity, *Photochem. Photobiol.,* 29, 189, 1979.

106. **Lavker, R. M. and Kaidbey, K. H.**, Redistribution of melanosomal complexes within keratinocytes following UV-A irradiation: a possible mechanisms for cutaneous darkening in man, *Arch. Dermatol.*, 272, 215, 1982.

107. **Sharpe, R. J.**, The low incidence of multiple sclerosis in areas near the equator may be due to ultraviolet light induced suppressor cells to melanocyte antigens, *Med. Hypoth.*, 19, 319, 1986.

108. **Vaughan, G. L. and Tallent, M. K.**, Pigment movement associated with the dermal photic response is inhibited by 8-bromo-cGMP, *Photochem. Photobiol.*, 1988, 47.

109. **Hadley, M. E. and Goldman, J. M.**, Physiological color changes in reptiles, *Am. Zool.*, 9, 489, 1969.

110. **Tilders, F. H. H., van Delft, A. M. L., and Smelike, P. G.**, Reintroduction and evaluation of an accurate, high capacity bioassay for melanocyte stimulating hormone using the skin of *Anolis carolinensis in vitro*, *J. Endocrinol.*, 66, 165, 1975.

111. **Vaughan, G. L. and Greenberg, N.**, Propranolol, a beta-adrenergic antagonist, retards response to MSH in the skin of *Anolis carolinensis, Physiol. Behav.*, 40, 555, 1987.

112. **Burghardt, G. M. and Rand, A. S., Eds.**, *Iguanas of the World: Their Behavior, Ecology, and Conservation*, Noyes Press, Park Ridge, NJ, 1982.

113. **Frye, F. L.**, *Biomedical and Surgical Aspects of Captive Reptile Husbandry*, Veterinary Medicine Publishing, Edwardsville, KS, 1981.

114. **Hailman, J. P.**, Homology: logic, information, and efficiency, in *Evolution, Brain, and Behavior: Persistent Problems*, Masterton, R. B., Hodos, W., and Herison, H., Eds., Lawrence Erlbaum, Hillsdale, NJ, 1976, 181.

115. **Greenberg, N.**, unpublished data.

116. **Gans, C., Ed.**, *Biology of the Reptilia*, Vols. 1 to 13, Academic Press, New York, 1969—1982; Vols. 14 and 15, John Wiley & Sons, New York, 1985; Vol. 16, Alan R. Liss, New York, 1988.

Chapter 18

SULFHYDRYL-RICH HEMOGLOBINS IN REPTILES: A DEFENSE AGAINST REACTIVE OXYGEN SPECIES?

Evaldo Reischl

TABLE OF CONTENTS

I. SULFHYDRYL-RICH HEMOGLOBINS IN VERTEBRATES AND REACTIVE OXYGEN SPECIES

A. INTRODUCTION

Vertebrate hemoglobins, with the exception of those from the primitive Cyclostomata, are tetrameric molecules with a molecular weight of about 64,500 and composed of four subunits of two different kinds, called α and β for human normal adult hemoglobin, whose subunit composition is designated $\alpha_2^A \beta_2^A$.[1] Some rare exceptions to this pattern of symmetry have been observed, where the tetrameric molecule seems to have more than two kinds of polypeptide chains.[2]

The number of sulfhydryl (-SH) groups in tetrameric vertebrate hemoglobins ranges from only 2 per tetramer in cattle and some other ruminants to 18 per tetramer in the Nile crocodile *(Crocodylus niloticus)*.[3] Normal adult human hemoglobin has six -SH groups per tetramer.[3]

It has long been known, from ultracentrifugation studies, that hemoglobins from reptiles and amphibians form polymers of their tetrameric hemoglobins,[4] and these originate through intermolecular disulfide (-S-S-) bridges and often start to form immediately after erythrocyte hemolysis.[5-7] Hemoglobins polymerized by disulfide bridges can often be detected by blurred electropherograms of untreated samples. The pattern may become neat again, with well-defined zones, after reaction with a disulfide bridge reducing agent, like dithio-erythritol or 2-mercaptoethanol, and stabilized if the reducing treatment is followed by alkylation of the free -SH groups. The same considerations are true for the ion exchange chromatography profiles.[7] Such studies, together with molecular filtration and ultracentrifugation experiments, have shown the occurrence of these polymerizing hemoglobin systems in numerous animal species.[5-11]

The stereochemical position of the -SH groups, whose sequence position is known, can be inferred by using three-dimensional data from human and horse hemoglobins, given by Fermi and Perutz,[12] and from the evidence that the globin chain configuration is the same in all hemoglobins.[13] The usefulness of this approach is confirmed by the fact that when residues, assigned to be totally external, are present, polymerization by -S-S- bridges is nearly always found. If -SH groups are assigned internally, no such polymers are found. In case of assignment to a surface crevice, sometimes polymers have been known to occur, like those in the elasmobranch *Squalus acanthias*,[3,8,12] and in other instances they do not occur, at least under common conditions. The last case is true for birds in general, where in 19 species, for whose hemoglobins complete amino acid sequences have been determined,[3] all have 10 -SH per tetramer in their major hemoglobin component, 4 -SH groups that can be assigned to surface crevices, the other 6 being internal: the 10 -SH groups always occupy the same sequence position.[3] In this paper, those molecules with 8 to 18 -SH groups per tetramer that may or not polymerize by interhemoglobin disulfide bridges are called -SH-rich hemoglobins.

B. THE BIOLOGICAL MEANING OF -SH-RICH HEMOGLOBINS

An indication of the possible biological meaning of -SH-rich polymerizing hemoglobins occurred to us from the observation that such systems are frequently found in animals known from ecophysiological observations to be resistant to hypoxia, such as freshwater turtles, crocodiles, and some species of amphibians.[14-17] These animals are capable of an amphibian way of life and are resistant to hypoxia; their ability to shuttle frequently from hypoxia to normoxia, a situation that could increase the production of reactive oxygen, seems to be important.[16] Another group of animals found to have -SH-rich hemoglobins are the cartilaginous fish or elasmobranchs.[3,8] Interestingly, although not having a dual way of life, shuttling between the atmosphere and water like the reptiles and amphibia mentioned, these animals are reported to be submitted, at times, to an internal hypoxia, for instance during exercise.[18] Further observations seem to reveal a greater generalization, although the number of species to argue from is scarce, that -SH-rich hemoglobins occur in animal species that may become involved in a great power output at times in their normal daily activity, such as the prosimian *Galago crassicaudatus,* the "bush baby", reported to be capable of spectacular jumping activity,[17] and cats, like the house cat and *Pantera p. orientales,* capable of bursts of fast running. Among mammals with low body mass and hence with a high mass-specific metabolic rate,[19] three species are known to have hemoglobins with -SH groups externally located: the mouse (*Mus musculus*), a rat (*Rattus norvegicus*), and the guinea pig (*Cavia a. porcellus*).[3,5]

Birds, known to have a very high metabolic rate, especially during flight,[20] have their hemoglobin's four external -SH groups on surface crevices, and polymers are not commonly observed. Nevertheless, these -SH groups could still interact with low-molecular-weight thiols, like the important tripeptide glutathione. As I shall discuss, all the situations just mentioned might indicate that -SH-rich hemoglobins are an important and a widespread participant in the natural strategies developed to protect organisms from highly reactive oxygen forms, thus the previously proposed association of hypoxia and -SH-rich hemoglobins being just a particular case of augmented reactive oxygen species.[14-16]

C. REACTIVE O_2 SPECIES

Important reactive oxygen species, such as superoxide anion radical (O_2^-), hydrogen peroxide (H_2O_2), and hydroxyl radical ($OH^.$), can originate through various mechanisms: about 5% of the electrons flowing through mitochondria are accepted to take part in the generation of activated O_2 species, by the univalent reduction of O_2, instead of the tetravalent reduction normally observed that leads to formation of water.[21-23] An apparently paradoxical situation, whose explanation was first proposed by Fridowich[24] and now seems confirmed, is that reinflux of O_2 to tissues after hypoxia or ischemia generates a burst of reactive oxygen species. This is probably due to the apport of the electron acceptor O_2 to the cells that accumulated reduced intermediates during the period of O_2 lack; their relatively uncontrolled oxidation, generating the reactive oxygen species, is on the basis of the posthypoxic tissue damage.[24] We think it can be accepted that O_2 reinflux to severely hypoxic animals, in nature, could put them under similar menaces as those faced by hypoxic and ischemic tissues and organs. This should especially happen in this way if no barriers, such as myoglobin-rich musculature or high O_2-affinity hemoglobins, are present. A higher specific metabolic rate can also be expected to generate a greater flux of reactive oxygen species, perhaps from the same leaks that operate under more moderate metabolic rates.

II. -SH-RICH HEMOGLOBINS IN REPTILES

A. INTRODUCTION

Reptiles are a class of tetrapod vertebrates that, like birds and mammals, present all structures necessary for living fully adapted to land.[26] From this characteristic of the group, we shall see that among the reptiles some species have an extraordinary ability to live and dwell underwater: for instance, some freshwater turtles are facultative anaerobes.[27] Nowadays, living reptiles include chelonians, which include turtles and tortoises; rhynchocephalians, with only one remaining species, the tuatara; Squamata, which comprise lizards and snakes; and the crocodilians.[26]

B. OCCURRENCE OF -SH-RICH HEMOGLOBINS IN REPTILES
1. Turtles

From an extensive study on 54 species and subspecies carried out by Sullivan and Riggs,[7] it was shown that -SH-rich hemoglobins are a common feature of numerous turtle species.

Among those species, only in 18 no polymerizing hemoglobins were detected, and, very significantly, the nonpolymerizing hemoglobins include 7 of the 8 testudinid (terrestrial quelonians: tortoises) species examined; 6 species are described to be semiterrestrial or, if aquatic, to dwell in a more terrestrial niche in their habitat.[28] The two species of sea turtles examined, as well as the two species of the family Pelomedusidae, also do not form heavy hemoglobins. These two groups include animals, in which adult representatives may achieve great body masses and hence have low specific metabolic rates.[19] All the active underwater swimmers and foragers, such as the Chelydridae and Kinosternidae, have polymerizing hemoglobins. The aggregates formed are very large and tend to precipitate on storage.[7] The formation of large aggregates indicates that these hemoglobins have externally positioned, highly reactive -SH groups, able to interact extensively.

2. Snakes and Lizards

The hemoglobins of lizards and snakes seem generally not to form polymers by disulfide bridges. Schwantes studied 24 species and subspecies of snakes from 3 families and found no indication of polymer formation.[29] In two species of water snakes, some disulfide polymers could be detected after oxidation to methemoglobin and storage.[14,30] These observations suggest that water snakes do not form polymers easily. In this respect, they could be similar to birds.

Electrophoretic studies on several species of lizards, performed by Guttman, gave no suggestion of polymer formation;[31,32] the species studied apparently included only land-adapted animals, but a report on the marine iguana *(Amblyrhynchus cristatus)* clearly shows that this lizard has hemoglobins that make disulfide polymers.[11] Another iguana, the green leguan *(Iguana iguana),* had its hemoglobins completely sequenced.[33] Although not forming polymers, it has two cysteinyl residues that are on surface crevices. *I. iguana,* differently from the marine iguana, which actively feeds underwater on algae, uses water only to escape from external menance and is able to remain submerged for about 5 h on this occasion.[34] Apparently the marine iguana should shuttle more often from hypoxia to normoxia than the green leguan. Electrophoretic and sequence studies on a varanid lizard *(Varanus griseus)* showed the presence of polymerizing hemoglobins.[35] The lizards from this group are reported to be very active animals that can achieve high rates of O_2 consumption.[19]

3. Crocodilians

This group of reptiles consists of animals that have a typical amphibious life-style.[36] Some time ago their hemoglobins were shown to have a unique pattern of regulation by the bicarbonate anion.[37] Their hemoglobins form polymers by disulfide bridges,[37] and the molecular basis of this is widely documented by the complete amino acid sequences available for three species of this group. They have the -SH-richest hemoglobins detected thus far. *C. crocodylus* has 14 -SH per tetramer, 6 in surface crevices, and 4 totally external residues; *C. niloticus* has 18 -SH per tetramer, with 6 in surface crevices and 6 externally positioned; *A. mississipiensis* has 16 -SH per tetramer, 6 on surface crevices, and 4 externally located.[3]

C. A PHYSIOLOGICAL RATIONALE FOR THE EXTERNALLY LOCATED -SH GROUPS IN REPTILIAN HEMOGLOBINS

High resistance to hypoxia seems to be a characteristic of reptiles, aquatic or not.[38] The molecular and cellular basis to survive to hypoxia and anoxia can be summarized by the concept of metabolic arrest and membrane channel arrest recently reviewed by Hochachka.[39] Reptiles that rely on this physiological characteristic in their daily activity may be under an augmented threat of reactive oxygen, since this could be produced on reinflux of oxygen. Permanently or transiently high specific metabolic rates could be another source of reactive oxygen species, as discussed previously.

Now we would like to focus on some specific mechanisms in which -SH-rich hemoglobin could participate to resist oxygen toxicity. This could happen by avoiding the accumulation of excess oxidized glutathione in erythrocytes, by formation of mixed disulfides with the hemoglobin -SH groups. This would circumvent the necessity to export glutathione from the cell, as proposed by Beutler, since oxidized glutathione can have disturbing effects on cellular metabolism.[40] The oxidized glutathione is produced by the glutathione peroxidase reaction, an important reaction to protect the cell from oxidative stress.[41] Disulfide bridge formation, however, could directly give up reducing potential to combat oxidizing situations by reactions such as: $-SH + -SH \rightleftharpoons -S-S- + 2\ e^- + H^+$. Eldjarn has shown that such a system may be excellent electron bufer *in vitro*.[42] Extensive polymerization of turtle hemoglobins has been observed *in vivo* followed by extensive formation of methemoglobin; the reversal of these reactions has also been observed intracellularly.[43]

Among the reptiles, the species that would be expected to have greater need of -SH-rich hemoglobins should be the ones that must recover from hypoxia often daily, as imposed by their ecophysiology. This could be the case of active underwater swimmers and foragers, like some freshwater turtles and the crocodilians. As previously stated, their hemoglobins polymerize extensively, indicating very -SH-rich molecules.

Chrysemys picta bellii, a freshwater turtle from the family Emididae, which is known to have an extraordinary resistance to hypoxia and is able to survive for a half year under nearly anoxic conditions,[27] has hemoglobins that do not form polymers. However, the primary structure of the three polypeptide chains that form the two hemoglobins present in their erythrocytes shows that eight -SH per tetramer are present, and the cysteinyl residue at position 126 (H_4) is assigned to a surface crevice of the hemoglobin molecule and could form mixed disulfides with glutathione, although, possibly for steric reasons, it does not allow the formation of intertetramer disulfide bridges easily.[33] *C. p. bellii's* hemoglobins were shown to be functionally of the type with high intrinsic O_2 affinity and regulable by organic phosphates.[44] Such hemoglobins could be of value in buffering the O_2 reinflux after long anoxic periods faced by these animals in winter in their natural habitat.[24] These functional characteristics make *C. p. bellii* much different from *Phrynops hilarii,* a South American freshwater turtle whose hemoglobins show low intrinsic oxygen affinity and are very modestly affected by organic phosphates.[45] Such functional properties would be of no help in buffering the O_2 posthypoxic reinflux to tissues. The hemoglobins of *P. hilarii* start to form disulfide polymers immediately after their erythrocyte hemolysis; titration studies showed them to have about 10 -SH per tetramer.[15] Studies on amino acid composition show that one hemoglobin component has 12 -SH per tetramer and the other, 10 -SH per tetramer.[46] The primary structure of one of the components was determined.[47] This chain, α^D, has an externally positioned cysteinyl residue: $\alpha - 81(F2)$. *P. hilarii,* as indicated by observations performed in laboratory and in nature, is a typical freshwater turtle, able to stay underwater in winter for long periods, even for months.

Possession of -SH-rich hemoglobins by an animal species could be an alternative strategy to resist oxygen toxicity. The possession of muscles very rich in myoglobin, as reported for several mammalian divers, like whales,[48] which do not have -SH-rich hemoglobins, could be another alternative strategy. High muscle myoglobin, besides its traditional functions, would certainly be an important barrier, in the oxygen cascade, to avoid a sudden burst of O_2 on reoxygenation at the mitochondrial level. Furthermore, in these organisms, the O_2 stored could be enough to avoid a real tissue hypoxia. Reptiles seem generally to have very low levels of myoglobin in their muscles and would be devoid of such a protection. The possession of O_2 storing capacity or of aquatic respiration through the gut or cloaca could also avoid the piling up of reduced intermediates that presumably are those that, under uncontrolled oxidation, give rise to univalent reduced O_2. In addition, the oxygenation characteristics of the hemoglobins, in an animal's blood, as discussed previously for two freshwater turtles, should also be an important factor in the antioxidant economy of those organisms.

III. BIOMEDICAL ASPECTS

A. ACTIVATED OXYGEN AND DISEASE

It has become clear that activated forms of oxygen are involved in a plethora of important disease states in human beings, such as those related to ischemia of the heart, brain, and gastrointestinal tract,[49-51] as well as to those related to the inflammatory process.[52] The participation of oxygen radicals in carcinogenesis is another large area of medical interest and was recently reviewed by Cerutti.[23] The aging process and life-span of different animal species, and the threat of ionizing radiation to tissues, also are related to processes involving activated oxygen.[53,54] The instances of diseases mentioned and several others lead to an increase in the intra- or extracellular production of reactive oxygen species; a list of such instances is presented by Del Maestro.[54] Physiological defense mechanisms may also involve the production of reactive oxygen intermediates, like the burst of O_2^- generated by phagocytosing neutrophils.[52]

The superoxide anion radical (O_2^-), hydrogen peroxide (H_2O_2), and the hydroxyl radical (OH·) are reactive O_2 species produced in metabolism.[54,55] From these, the hydroxyl radical is the most reactive, being an extremely powerful oxidant.[22] The short-lived and reactive hydroxyl radical can be produced from hydrogen peroxide and ferrous iron, the Fenton reaction. In addition, free hemoglobin is also able to participate in such reactions, and binding to haptoglobin inhibits the process.[56]

A very strongly oxidizing species, like OH·, can initiate the chain process of lipid peroxidation,[55,57] and lipid peroxides can destroy the cell membrane locally or give rise to products that can produce their effects farther away.[57] Other basic damaging reactions of active oxygen include direct reactions with the genetic material, oxidation of thiol enzymes, and direct action on membrane functional units.[57]

B. PROTECTIVE MECHANISMS

Since all aerobic organisms are under the continuous threat of such a multitude of ill effects of oxygen-reactive derivatives, a number of protective mechanisms and agents were also provided by nature to ensure the maintenance of the living processes. They include mechanisms to maintain oxygen tension at convenient low levels in the cell, such as microvascular regulation of blood circulation, intercapillary distances, affinity and storage characteristics of cytochrome oxidase for oxygen and its reaction intermediates.[58] Additionally, aerobic cells possess specialized enzymes to handle reactive oxygen products like superoxide dismutase, catalase, and glutathione peroxidase,[58] as well as antioxidant substances, such as vitamin E, adequate for hydrophobic environments, and ascorbic acid, cysteine, and reduced glutathione that function more adequately as scavengers in hydrophilic compartments or regions of the cell. The concerted action of hydrophobic and hydrophilic scavengers may be necessary.[59]

Our observations, and the association of occurrence of -SH-rich hemoglobins in animal species with ecophysiological requirements that might enhance the production of activated oxygen,[14-16] seem to be an indication that the presence of -SH-rich hemoglobins is another defense mechanism to be added to the list of redox buffering, especially antioxidant defense, of the cellular machinery.

C. -SH-RICH HUMAN HEMOGLOBINS

Normal human hemoglobin has six thiol groups, with two of them at the position β-93, accessible to the exterior of the molecule.[12] The stereochemistry of these -SH groups does not allow the formation of posthemolysis polymers, but they are able to form mixed disulfides with glutathione *in vitro*[60] but apparently not *in vivo*.[61]

Of interest is the existence of six variants from the major normal adult human hemoglobin: Hb Porto Alegre β-9 Ser → Cys,[62] Hb Ta-Li β-83 Gly → Cys,[63] Hb Rainier β-145 Tyr → Cys,[64] Hb Nunobiki α-141 Arg → Cys,[65] Hb Nigeria α-81 Ser → Cys,[66] and Hb Mississippi β-44 Ser → Cys,[67] all of them showing two additional cysteinyl residues. From these variants can be seen the extreme importance of the stereochemical position of the -SH groups to the kind of reaction they may be involved. All six mutations have the novel -SH groups externally positioned. In Hb Porto Alegre, Ta-Li, and Mississippi, disulfide polymers are formed, whereas in Hb Rainier β-145 Tyr → Cys, an intramolecular disulfide bridge is formed,[64] and in Hb Nunobiki α-141 Arg → Cys, the extra -SH seem to oxidize to -So$^-$ and -SO^{--}, without forming intermolecular −S−S− bridges.[65]

Bearers of Hb Ta-Li and Porto Alegre are reported to be healthy normal individuals,[68] and perhaps we should see if these variants do not endow their carriers with an advantage. Studies on Hb Porto Alegre are of special interest, since this variant was detected five times in apparently unrelated familiar groups, all reported to be from Iberic origin; this finding could indicate that these hemoglobins could be of relatively common occurrence in Spain and Portugal.[69]

If such an evolutionary fixation occured, it would indicate that some adaptative advantage exists in the possession of such a hemoglobin. A possible advantage could be the reinforcement of the antioxidant capacity of Hb Porto Alegre carriers, as indicated by the augmented glutathione content and glutathione reductase activity in their erythrocytes.[70] An increase in low-molecular-weight thiol content also was observed in animal species with additional -SH groups on their hemoglobins that we call -SH-rich hemoglobins: in the marsupial *Didelphis virginiana* and in the freshwater turtle *Phrynops hilarii*.[16,71] Besides an augmented antioxidant capacity, such erythrocytes could possess a superior general detoxifying capacity, since glutathione and other thiols are known to be involved actively in detoxification besides redox buffering, and such a detoxifying role for erythrocytes has specifically been suggested.[40] The metabolic mechanisms involved in the concomitant increase in glutathione reductase activity and low-molecular-weight thiol concentration in erythrocytes bearing -SH-rich hemoglobins are not clear, and may be of great interest.

D. ARTIFICIALLY AUGMENTING THE PROTECTIVE MECHANISMS

Recently, several instances of protection to organs from the action of free radicals and active oxygen by scavenging agents, artificially introduced in biological systems, have been reported.[25,72,73] Ways of augmenting glutathione concentration intracellularly have been reported and may be an important strategy to face oxidative stresses.[74,75] Perhaps, in the not too remote future, it will be possible to endow human beings with permanently modified proteins, aiming to correct inborn errors or to achieve a better overall health status of the individual. Human beings with -SH-rich hemoglobins, like Ta-Li and Porto Alegre, or with one molecule with both β-83 and β-9 positions with cysteinyl residues, instead of the normal residues, could have improved antioxidant and detoxifying capacity, and this may be of value as a measure of corrective and or preventive medicine.

Animals with polymerizing hemoglobins, through externally positioned -SH groups, may be valuable in the study of such problems, and their occurrence in several taxa, including mice, rats, and guinea pigs, traditional experimental animals, is certainly of interest. Although reptiles might be useful for particular studies, they may not be the ideal animal for the majority of circumstances, because they are relatively rare animals. Also, the necessary anesthetic procedures are not always easy to apply, due to their impermeable body surfaces and great apneic capacity.[76,77]

ACKNOWLEDGMENTS

Support for this work from the Brazilian agencies *FINEP* and *CNPq*; and from the Alexander von Humboldt Foundation, FRG, are acknowledged.

REFERENCES

1. **Bunn, H. F. and Forget, B. G.,** *Hemoglobin: Molecular, Genetic and Clinical Aspects,* W. B. Saunders, Philadelphia, 1986, chap. 2.
2. **Mied, P. A. and Powers, D. A.,** Hemoglobins of the killifish *Fundulus heteroclitus, J. Biol. Chem.,* 253, 3521, 1978.
3. **Kleinschmidt, T. and Sgouros, J. G.,** Hemoglobin sequences, *Biol. Chem. Hoppe-Seyler,* 368, 579, 1987.
4. **Svedberg, T. and Hedenius, A.,** The sedimentation constants of respiratory proteins, *Biol. Bull.,* 66, 191, 1934.
5. **Riggs, A.,** Polymerization of vertebrate hemoglobins: mechanism and effect on properties, in *Int. Symp. Comparative Hemoglobin Structure,* Thessaloniki, 1966.
6. **Riggs, A., Sullivan B., and Agee, J. R.,** Polymerization of frog and turtle hemoglobins, *Proc. Natl. Acad. Sci. U.S.A.,* 51, 1127, 1964.
7. **Sullivan, B. and Riggs, A.,** Structure, function and evaluation of turtle hemoglobins. I. Distribution of heavy hemoglobins, *Comp. Biochem. Physiol.,* 23, 437, 1967.
8. **Fyhn, U. E. H. and Sullivan, B.,** Elasmobranch hemoglobins: dimerization and polymerization in various species, *Comp. Biochem. Physiol. B,* 50, 119, 1975.
9. **Sullivan, B.,** Reptilian hemoglobins, in *Chemical Zoology,* Vol. 9, *Amphibia and Reptilia,* Florkin, M. and Bradley, T. S., Eds., Academic Press, New York, 1974, chap. 14.
10. **Reischl, E.,** The hemoglobins of the fresh-water teleost *Hoplias malabarica* (Bloch, 1794): heterogeneity and polymerization, *Comp. Biochem. Physiol. B,* 55, 255, 1976.
11. **Higgins, P. J.,** Immunochemical identity of the high and low molecular weight forms of Galapagos marine iguana hemoglobin, *Comp. Biochem. Physiol. B,* 59, 129, 1978.
12. **Fermi, G. and Perutz, M. F.,** *Atlas of Molecular Structures in Biology, Vol. 2, Haemoglobin and Myoglobin,* Phillips, D. C. and Richards, F. M., Eds., Clarendon Press, Oxford, 1981.
13. **Perutz, M. F., Kendrew, J. C., and Watson, H. C.,** Structure and function of haemoglobin. II. Some relations between polypeptide chain configuration and amino acid sequence, *J. Mol. Biol.,* 13, 669, 1965.
14. **Reischl, E., Diefenbach, C. O. da C., and Tondo, C. V.,** Ontogenetic variation of hemoglobins from South American Chelonia (Reptilia), *Comp. Biochem. Physiol. B,* 62, 539, 1979.
15. **Reischl, E., Hampe, O. G., and Crestana, R. H.,** Time course of SH group disappearance from the haemoglobins of the turtle *Phrynops hilarii, Comp. Biochem. Physiol. B,* 77, 207, 1984.
16. **Reischl, E.,** High sulphydryl content in turtle erythrocytes: Is there a relation with resistance to hypoxia? *Comp. Biochem. Physiol. B,* 85, 723, 1986.
17. **Grzimek, B.,** *Grzimeks Tierleben, Enzyklopädie des Tierreiches,* Kindler Verlad, Zurich, 1968-1972.
18. **Bushnell, P. G., Lutz, P. L., Steffensen, J. F., Oikari, A., and Gruber, S. H.,** Increases in arterial blood oxygen during exercise in the lemon shark *(Negaprion brevirostris), J. Comp. Physiol.,* 147, 41, 1982.
19. **Bartholomew, G. A.,** in *Animal Physiology: Principles and Adaptations,* 3rd ed., Gordon, M. S., with Bartholomew, G. A., Grinnel, A. D., Jorgensen, C. B., and White, F. N., Macmillan, New York, 1977, chaps. 3, 8.
20. **Butler, P. J.,** Exercise in non-mammalian vertebrates: a review, *J. R. Soc. Med.,* 78, 739, 1985.
21. **Fridowich, I.,** Superoxide radical and superoxide dismutases, in *Oxygen and Living Processes: An Interdisciplinary Approach,* Gilbert, D. L., Ed., Springer-Verlag, New York, 1981, chap. 13.
22. **Halliwell, B. and Gutteridge, J. M. C.,** Oxygen toxicity, oxygen radicals, transition metals and disease, *Biochem. J.,* 219, 1, 1984.
23. **Cerutti, P. A.,** Prooxidant states and tumor promotion, *Science,* 227, 375, 1985.
24. **Fridowich, I.,** Hypoxia and oxygen toxicity, *Adv. Neurol.,* 26, 255, 1979.
25. **Schlafer, M., Kane, P. F., and Kirsh, M. M.,** Superoxide dismutase plus catalase enhances the efficacy of hypothermic cardioplegia to protect globally ischemic, reperfused heart, *J. Thor Cardiovasc. Surg.,* 83, 830, 1982.
26. **Hildebrand, M.,** *Analysis of Vertebrate Structure,* John Wiley & Sons, New York, 1974.

27. **Ultsch, G. R. and Jackson, D. C.**, Long-term submergence at 3°C of the turtle *Chrysemys picta belli* in normoxic and severely hypoxic water. I. Survival, gas exchange and acid-base status, *J. Exp. Biol.*, 96, 11, 1982.

28. **Bury, R. B.**, Population ecology of fresh-water turtles, in *Turtles: Perspectives and Research*, Harless, M. and Morlock, H., Eds., John Wiley & Sons, New York, 1979, chap. 26.

29. **Schwantes, A. R.**, Hemoglobinas e haptoglobinas em serpentes *(Squamata, Reptilia)*, Ph.D. thesis, Universidade Federal do Rio Grande do Sul, Porto Alegre, RS, 1972.

30. **Sullivan, B.**, Oxygenation properties of snake hemoglobin, *Science*, 157, 1308, 1967.

31. **Guttmann, S. I.**, An electrophoretic study of the hemoglobins of the sand lizards, *Callisaurus, Cophosaurus, Holbrookia* and *Uma.*, *Comp. Biochem. Physiol.*, 34, 569, 1970.

32. **Guttmann, S. I.**, An electrophoretic analysis of the hemoglobins of old and new world lizards, *J. Herpetol.*, 5(1-2), 11, 1971.

33. **Rücknagel, K. P.**, Reptilian — Hämoglobine: Primärstruktunr der hämoglobine und expression der globingene, Ph.D. thesis, Ludwig-Maximilians-Universität, Munich, 1984.

34. **Moberly, W. R.**, The metabolic responses of the common iguana, *Iguana iguana*, to walking and diving, *Comp. Biochem. Physiol.*, 27, 21, 1968.

35. **Ruchnagel, K. P.**, Untersuchungen anhand der Primärstruktur der Hämoglobine von *Chrysemys picta belli, Iguana iguana* and *Varanus griseus*, Diplomarbeit, Ludwig-Maximilians-Universität, Munich, 1981.

36. **Lewis, L. Y. and Gatten, R. E., Jr.**, Aerobic metabolism of American alligators, *Alligator mississippiensis* under standard conditions and during voluntary activity, *Comp. Biochem. Physiol. A*, 80, 441, 1985.

37. **Jelkmann, W. and Bauer, C.**, Oxygen binding properties of caiman blood in the absence and presence of carbon dioxide, *Comp. Biochem. Physiol. A*, 65, 331, 1979.

38. **Jackson, D. C.**, Metabolic depression and oxygen depletion in the diving turtle, *J. Appl. Physiol.*, 24, 503, 1968.

39. **Hochachka, P. W.**, Assessing metabolic strategies of surviving O_2 lack: role of metabolic arrest coupled with channel arrest, *Mol. Physiol.*, 8, 331, 1985.

40. **Beutler, E.**, Active transport of glutathione disulphide from erythrocytes, in *Functions of Glutathione: Biochemical, Physiological, Toxicological and Clinical Aspects*, Larsson, A., Orrenius, L., Holmgren, A., and Mannervik, B., Eds., Raven Press, New York, 1983, 65.

41. **Flohé, L., Günzler, W. A., and Ladenstein, R.**, Glutathione peroxidase, in *Glutathione: Metabolism and Function*, Arias, I. M. and Jakoby, W. B., Eds., Raven Press, New York, 1976, 115.

42. **Eldjarn, L.**, Some biochemical effects of S-containing protective agents and the development of suitable SH/SS systems for *in vitro* studies of such effects, *Prog. Biochem. Pharmacol.*, 1, 173, 1965.

43. **Sullivan, B. and Riggs, A.**, Hemoglobin: reversal of oxidation and polymerization in turtle red cells, *Nature (London)*, 204, 1098, 1964.

44. **Weber, R.**, personal communication, 1986.

45. **Reischl, E., Höhn, M., Jaenicke, R., and Bauer, C.**, Bohr effect, electron spin resonance spectroscopy and subunit dissociation of the hemoglobin components from the turtle *Phrynops hilarii*, *Comp. Biochem. Physiol. B*, 78, 251, 1984.

46. **Reischl, E., Kelinschmidt, T., Rücknagel, K. P., and Braunitzer, G.**, unpublished data, 1986.

47. **Rücknagel, K. P., Reischl, E., and Braunitzer, G.**, Expession von α^D — genen bei Schildkröten, *Chrysemys picta belli* und *Phrynops hilarii* (Testudines), *Hoppe-Seyler Z. Physiol. Chem.*, 365, 1163, 1984.

48. **Castelini, M. A. and Somero, G. N.** Buffering capacity of vertebrate muscle: correlation with potentials for anaerobic function, *J. Comp. Physiol.*, 143, 191, 1981.

49. **Meerson, F. Z., Kagan, V. E., Kozlov, Yu. P., Belkina, L. M., and Arkhipenko, Yu. V.**, The role of lipid peroxidation in pathogenesis of ischemic damage and the antioxidant protection of the heart, *Basic Res. Cardiol.*, 77, 465, 1982.

50. **Petkau, A.**, Concluding remarks: a prospective view of active oxygen in medicine, *Can. J. Physiol. Pharmacol.*, 60, 1425, 1982.

51. **Parks, D. A., Bulkley, G. B., and Neil Granger, D.**, Role of oxygen-derived free radicals in digestive tract diseases, *Surgery*, 94, 415, 1983.

52. **McCord, J. M. and Roy, R. S.**, The pathophysiology of superoxide: roles in inflammation and ischemia, *Can. J. Physiol. Pharmacol.*, 60, 1346, 1982.

53. **Tolmasoff, J. M., Ono, T., and Cutler, R. G.**, Superoxide dismutase: correlation with life-span and specific metabolic rate in primate species, *Proc. Natl. Acad. Sci. U.S.A.*, 77, 2777, 1980.

54. **Del Maestro, R. F.**, An approach to free radicals in medicine and biology, *Acta Physiol. Scand. Suppl.*, 492, 153, 1980.

55. **Halliwell, B.**, Oxygen radicals: a common sense look at their nature and medical importance, *Med. Biol.*, 62, 71, 1984.

56. **Sadrzadeh, S. M. H., Graf, E., Panter, S. S., Hallaway, P. E., and Eaton, J. W.**, Hemoglobin: a biologic Fenton reagent, *J. Biol. Chem.*, 259, 14354, 1984.

57. **Slater, T. F.**, Free-radical mechanisms in tissue injury, *Biochem. J.*, 222, 1, 1984.
58. **Chance, B., Sies, H., and Boveris, A.**, Hydroperoxide metabolism in mammalian organs, *Physiol. Rev.*, 59, 527, 1979.
59. **Wayner, D. D. M., Burton, G. W., and Ingold, K. U.**, The antioxidant efficiency of vitamin C is concentration-dependent, *Biochim. Biophys. Acta*, 884, 119, 1986.
60. **Huisman, T. H. J. and Dozy, A. M.**, Studies on the heterogeneity of hemoglobin. V. Binding of hemoglobin with oxidized glutathione, *J. Lab. Clin. Med.*, 60, 302, 1962.
61. **Srivastava, S. K., Awasthi, Y. C., and Beutler, E.**, Useful agents for the study of glutathione metabolism in erythrocytes, *Biochem. J.*, 139, 289, 1974.
62. **Bonaventura, J. and Riggs, A.**, Polymerization of hemoglobin of mice and man: structural basis, *Science*, 158, 800, 1967.
63. **Blackwell, R. Q., Liu, C.-S., and Wang, C.-L.**, Hemoglobin Ta-Li: β83 Gly→Cys, *Biochim. Biophys. Acta*, 243, 467, 1971.
64. **Hayashi, A., Stamatoyannopoulos, G., Yoshida, A., and Adamson, J.**, Haemoglobin Rainier: β 145 (HC2) tyrosine→cysteine and haemoglobin Bethesda: β 145 (HC2) tyrosine→histidine, *Nature (London) New Biol.*, 230, 264, 1971.
65. **Shimasaki, S.**, A new hemoglobin variant, hemoglobin Nunobiki [α141(HC3) Arg→Cys], *J. Clin. Invest.*, 75, 695, 1985.
66. **Honig, G. R., Schamsuddin, M., Mason, G. R., Vida, L. N., Tremaine, L. M., Tarr, G. E., and Shahidi, N.**, Hemoglobin Nigeria (α-81 Ser→Cys): a new variant associated with α-Thalassemia, *Blood*, 55, 131, 1980.
67. **Adams, J. G., III, Morrison, W. T., Barlow, R. L., and Steinberg, M. H.**, Hb Mississippi [β-44(CD3) Ser→Cys]: a new variant with anomalous properties, *Hemoglobin*, 11, 435, 1987.
68. **Tondo, C. V., Salzano, F. M., and Rucknagel, D. L.**, Hemoglobin Porto Alegre, a possible polymer of normal hemoglobin in a Caucasian Brazilian family, *Am. J. Hum. Genet.*, 14, 401, 1962.
69. **Seid-Akhavan, M., Ayres, M., Salzano, F. M., Winter, W. P., and Rucknagel, D. L.**, Two more examples of Hb Porto Alegre, $\alpha_2\beta_2{}^9$Ser → Cys in Belem, Brazil, *Hum. Hered.*, 23, 175, 1973.
70. **Tondo, C. V.**, Increased erythrocyte glutathione reductase activity in a hemoglobin Porto Alegre (β9 Ser → Cys) carrier, *Biochem. Biophys. Res. Commun.*, 105, 1381, 1982.
71. **Bethlenfalvay, N. C., Waterman, M. R., Lima, J. E., and Waldrup, T.**, Comparative aspects of methemoglobin formation and reduction in opossum *(Didelphis virginiana)* and human erythrocytes, *Comp. Biochem. Physiol. A*, 75, 635, 1983.
72. **Gardner, T. J., Stewart, J. R., Casale, A. S., Downey, J. M., and Chambers, D. E.**, Reduction of myocardial ischemic injury with oxygen-derived free radical scavengers, *Surgery*, 94, 423, 1983.
73. **Ghanayem, B. I., Boor, P. J., and Ahmed, A. E.**, Acrylonitrile-induced gastric mucosal necrosis: role of gastric glutathione *J. Pharmacol. Exp. Ther.*, 232, 570, 1985.
74. **Puri, R. N. and Meister, A.**, Transport of glutathione, as γ-glutamylcysteinylglycl ester, into liver and kidney, *Proc. Natl. Acad. Sci. U.S.A.*, 80, 5258, 1983.
75. **Wellner, V. P., Anderson, M. E., Puri, R. N., Jensen, G. L., and Meister, A.**, Radioprotection by glutathione ester: transport of glutathione ester into human lymphoid cells and fibroblasts, *Proc. Natl. Acad. Sci. U.S.A.*, 81, 4732, 1984.
76. **Bonath, K.**, *Narkose der Reptilian, Amphibien und Fishe*, Verlag Paul Parey, Berlin 1977, chap. 1.
77. **Maxwell, J. H.**, Anesthesia and surgery, in *Turtles: Perspectives and Research*, Harless, M. and Morlock, H., Eds., John Wiley & Sons, New York, 1979, chap. 7.

Chapter 19

THE DOMESTIC FOWL AS A BIOMEDICAL RESEARCH ANIMAL

John A. Brumbaugh and William S. Oetting

TABLE OF CONTENTS

I. INTRODUCTION

The vertebrate class *Aves* has played a prominent role in biological research. To provide a comprehensive review of the literature would be an enormous task filling many volumes. To survey biomedical applications only would still fill several volumes and is definitely beyond the scope of this chapter. The purpose of this chapter is to demonstrate the suitability of avian species for biomedical research using the domestic fowl *(Gallus domesticus)* as a typical model.

The domestic fowl has been so widely used as a biomedical research animal that we are again faced with limiting the scope of our discussion. Prominent discoveries and investigations using the domestic fowl will be discussed. The topics covered reflect the interests of the authors and were selected to show a broad range of applications. Some special examples using turkeys and Japanese quail will also be presented.

This chapter is divided into seven sections. The next section, historical highlights, presents some of the major scientific breakthroughs provided by experiments using the fowl. The section on life cycle and production requirements surveys the fundamental biology of the chicken and describes equipment and facilities. The section on advantages and disadvantages points out the suitability of birds for certain investigations and their lack of suitability for other purposes, which is frequently related to their intrinsic biology. The current models section provides selected examples of ongoing research in a variety of areas in which the fowl is a common research animal. Future applications are discussed in another section, while the last section lists some of the comprehensive resources available for those who wish to begin using the fowl in their research.

II. HISTORICAL HIGHLIGHTS

Punnett[1] in his 1923 edition of *Heredity in Poultry,* has the following dedication: "To William Bateson, whose experiments with poultry offered the first demonstration of Mendelian heredity in the animal kingdom." He is referring to a paper by Bateson and Saunders[2] (1902) and obviously believed that chickens were first used to demonstrate Mendelism in animals. Another paper, published by Cuenot,[3] using mice, also appeared in 1902, so the point is debatable. Regardless of which paper appeared first, chickens were one of the first animal forms to be used in genetic research.

The contribution that the chick has made to embryological research is universally known. Lillie[4] in the preface to the first edition of *Development of the Chick* (1908) refers to the chick as the "never failing resource of the embryologist. . . ." In the current models section, we shall see how much the chick is used in developmental studies.

The fact that some cancers can have a viral etiology was first reported in the fowl in 1908 and 1909 by Ellerman and Bang[5,6] in Denmark, and later in 1911 by Rous[7] in the U.S. Although not widely accepted at the time, we now know that these were landmark discoveries showing the relationship between viruses and cancer. A tissue culture assay for Rous sarcoma virus was developed by Temin and Rubin[8] in 1958. The formation of the provirus and its integration into the genome was described by Temin[9] in 1964. The Mendelian segregation of these proviral inserts was studied by Crittenden and Astrin.[10]

The role of the Bursa of Fabricius in antibody production was first reported in 1956.[11] The Bursa of Fabricius is the thymus-like gland of the chicken. The antibody-producing B-cells of both birds and mammals are so designated because of this discovery.

As we have seen, the domestic fowl has played a very important role in several landmark discoveries. The chicken provided us with some of the first genetic information discovered in the animal kingdom. The chick formed the fundamental basis for vertebrate embryology. The viral etiology of cancer was first discovered in the fowl and provided us with a basis for understanding retroviral biology. Finally, the role of the B-cells in the immune system were first described using the chicken.

III. LIFE CYCLE AND PRODUCTION REQUIREMENTS

The life cycle of an average strain of chickens takes approximately 6 months. Fertile eggs require a 21-d incubation period. The newly hatched chicks must be kept in brooders with supplemental heat for 3 to 4 weeks. The birds grow and feather rapidly and reach sexual maturity depending upon breed and type between 4 and 7 months. The life cycle can be entered at any point. Fertile eggs, newly hatched chicks, immature birds of either sex, or sexually mature adults are available commercially. If a breeding colony is to be established, it is usually easier to start with fertile eggs. Fertile eggs can be easily shipped, and if obtained from certified flocks, will not spread or introduce diseases into your animal care facility.

The freshly laid egg is really an embryo since it has spent at least 24 h in the hen's body. At the time of laying, the embryo consists of several thousand cells.[4,12] These embryos do not continue to develop if the temperature is kept below 20°C.

Storage conditions between 10 and 18°C with a relative humidity of 70 to 80% allows storage for up to 2 weeks.[13,14] Fertile eggs should never be stored in the refrigerator (4°C) because it is too cold and the embryos will be damaged. We have found it convenient to set eggs at 2-week intervals for hatching and for experiments which required embryos.

Eggs should be incubated under precise temperature and humidity conditions, particularly if the eggs are going to be hatched.[13,14] There are two basic types of incubators. Still-air incubators simply provide a chamber in which the temperature is controlled through the incubation period. The temperature of incubation for still-air incubators is 39.4°C. The relative humidity is kept between 50 and 60% during the first 18 d of incubation by providing a source of water in a flat, shallow pan. Forced-air incubators are operated at a lower temperature (37.5 to 37.7°C), with the humidity in the same range. With both types of incubators, the temperature is decreased during the last 2 to 3 d of the incubation period to 36 to 37°C while the humidity is raised to 70 to 75%. For good hatches, the eggs must be turned during the first 18 d of incubation. In the simpler incubators, this must be done by hand every 6 h. Most forced-air incubators have some kind of automated turning mechanism. Relative humidity is determined by comparing wet- and dry-bulb temperature readings. Still-air incubators are sufficient for embryonic work, and in fact other types of incubators, such as those used to grow bacteria, are usable.

The investigator can determine the precise time and temperature conditions which provide embryos at a particular stage of development. The stages of development of the chick have been standardized and are pictured and described in Hamilton's *Development of the Chick*.[4]

It is essential to clean and disinfect incubators thoroughly after the chicks and eggs have been removed. Fumigation with formaldehyde gas provided by a mixture of formalin and potassium permanganate is recommended.[13,14]

Still- and forced-air incubators are available from several sources. Unless a very large number of eggs are needed for embryonic or breeding work, commercial incubators are not usually needed. Several companies make smaller incubators for hobbyists, fanciers, and game bird breeders. They come with complete instructions for incubation and hygiene. Information about incubating and hatching chicks on a small scale can usually be obtained from your county extension agent, the poultry or animal science department of your land grant university, or from local fanciers and hobbyists.

Newly hatched chicks require elevated temperatures for the first 3 to 4 weeks of growth. The period of time during which the heat is supplied is called the brooding period, and the heat-supplying equipment are called brooders. The heat can simply be supplied by a heating lamp or with heating elements thermostatically controlled. The ideal temperature begins around 32°C and is dropped 3°C per week until room temperature is reached. The behavior of the chicks indicates temperature requirements as well or better than a thermometer. Chicks that are huddled together close to the heating source are too cold. Chicks that get as far away from the heating source as possible and begin panting are too hot. The ideal situation is indicated when the chicks are evenly dispersed throughout the brooding area. If the chicks need to be identified for genetic or experimental reasons, it is easy to wing-band them with bands that are available from several commercial sources.

Birds can be raised in floor pens or in cages with wire mesh bottoms. Unless facilities are available in the poultry or animal science department of your university for floor pen raising, it is more convenient to raise the birds in wire cages. Caging systems are more compatible with the usual laboratory animal care facilities. We have raised chickens for over 30 years with caging systems. Caging systems are available from a number of commercial sources and are relatively inexpensive when compared with rabbit, rat, and mouse cages because of the large number of caging systems that are sold commercially. We have found that individual cages for sexually mature birds reduce mortality and increase the efficiency of each bird. We use wire floored brooders and separate the sexes into group cages of 7 to 10 birds between 4 and 6 weeks of age.

One of the disadvantages of raising chickens is that they have a tendency to begin picking weaker and smaller birds and literally pick them to death. This is called cannibalism and makes debeaking a necessity. The birds are debeaked a week or two after they have become accustomed to their group cages. Debeaking involves removing a portion of the upper beak so that the birds cannot cannibalize each other.

The nutrition of the chicken has been thoroughly researched, and commercial feeds are available. The feeds vary in content depending upon the ages of the birds being fed. For example, starter feeds are designed to get the chicks growing rapidly. Laying rations are designed to optimize carbohydrate and protein for the laying hen. Some of the feeds contain antibiotics and coccidiostats. Coccidiosis is a protozoan infestation of the digestive tract which frequently occurs in birds. These additives are removed from laying rations because of the possibility of their transferral to eggs which are consumed by human beings. The contents of the various rations are clearly stated on the labels.

The birds mature in 4 to 6 months depending upon breed and size. The smaller breeds, like bantams, tend to mature more rapidly while the larger meat-type breeds mature more slowly. Growth, maturation, and egg laying are greatly influenced by photoperiods. It is easiest to control the photoperiod in windowless rooms which are frequently found in animal care facilities. A simple time clock controlling the light is sufficient. The desired photoperiod for growth is 12 h. The photoperiod for egg production is 14 to 16 h. If the facilities have windows, supplementary lighting must still be provided or else the birds will respond to decreasing day length and cease to produce eggs.

If birds are raised in floor pens, natural mating can take place. If the birds are raised in individual cages, then artificial insemination must be used. Since natural mating in wire-floored group pens is not satisfactory, artificial insemination is recommended. We have found artificial insemination to be convenient, quick, and effective. Size and behavioral differences between males and females are not a factor when artificial insemination is used. The basic technique of collecting semen from the males is described by North,[13] Johnson,[15] and Moreng and Avens.[16] The anatomy and physiology of the ejaculatory system in the male is described by Johnson.[15] We collect semen in small cups and aspirate it into 1-cc disposable plastic tuberculin syringes. The cloaca of laying hens is gently everted and 0.05 to 0.1 cc of semen or diluted semen is injected directly into the oviduct. This is done once a week to ensure good fertility. Fertile eggs may be obtained from an insemination for as long as 2 weeks.[15] Therefore, when changing hens in genetic matings, we allow at least 3 weeks between inseminations to ensure that the original rooster's semen is not fertilizing the eggs of the new mating.

We have supplied just 3 of many references on poultry production. The books by North[13] and Moreng and Avens[16] are for commercial production, while Banks[14] describes smaller-scale operations. Many other books and pamphlets about poultry raising are available in libraries and from county extension agents. The raising of domestic fowl is commercialized so equipment, facilities, feeds, disease control, and other requirements are highly developed. Information about facilities and equipment are readily available and competitively priced.

IV. ADVANTAGES AND DISADVANTAGES

The domestic fowl and related *Galliformes* possess many biological characteristics that make them especially suitable for biomedical research. These advantages, however, are coupled with some disadvantages. There are also some uniquely avian characteristics that make them especially suitable for specialized types of research. In this section, we shall discuss the advantages, disadvantages, and some applications of these unique features.

The vertebrate class *Aves* is the only class other than *Mammalia* that are homeothermic. This makes birds mammalian-like with regard to many of the fundamental processes of physiology, genetics, and biochemistry. Those of us working with birds would sometimes like to pass our experimental organism off as a "feathered mammal".

Chickens have a moderately short life cycle permitting production of two generations per year. This is not as short as the life cycle of the mouse (6 to 7 weeks). The Japanese quail *(Coturnix japonica),* however, has a life cycle of just 6 weeks, making it advantageous for certain research projects.

The highly commercialized poultry industry has developed housing, nutrition, and management practices that are very efficient and cost effective. In addition to poultry production, there is an interest in poultry raising as a hobby. Fanciers keep many different varieties of fowl in existence. These variations provide an extensive gene pool from which strains and stocks can be developed for specific research purposes. An International Registry of Poultry Genetic stocks is published every 3 years[17] and lists the sources of the various varieties, stocks, and strains.

Because a large number of chickens are raised in the U.S. each year, there is concern about disease control. As a result of this concern, disease-free flocks have been established. These specific pathogen-and/or antigen-free flocks provide a source of strictly controlled material particularly useful in virological and microbiological research.

One of the disadvantages of the domestic fowl is its chromosome complement. The haploid number is 39, with the 2N number being 78. Unfortunately, only 7 to 8 pairs of these chromosomes are large enough for distinctive morphological identification. The remainder are called microchromosomes and are difficult to see and distinguish from each other.[18] Methodology has been developed for dealing with this cytogenetic dilemma.[18] Because of the commercial emphasis in poultry breeding, mutations which were deleterious or not conducive to commerical applications were frequently discarded. This means that morphological and biochemical traits were not extensively mapped. However, a map of the traits of the domestic fowl is available.[19] As basic research interests increase and molecular genetic techniques develop, the genetic map will become more and more extensive. Progress has even been made in mapping and finding linkage groups on microchromosomes. Bloom and Bacon have found that the major histocompatibility complex of the chicken (B locus) is linked to the nucleolar organizer on a microchromosome.[20] Techniques have been developed to do gene dosage studies in birds aneuploid for this microchromosome.[21,22] Thus the disadvantages of chromosome morphology and the lack of genetic mapping information are being overcome by the diligent application of modern techniques.

Another disadvantage is the lack of sexual dimorphism at hatching. Sexing must be done by trained experts unless genetic crosses are appropriately set up so that morphological or color differences appear between males and females at hatching.[23] This genetic technique is called autosexing.

Transgenic experiments in mice are performed by injecting DNA into one-celled zygotes flushed from the oviducts of the female. Such experiments cannot be performed with freshly laid eggs, because as has already been mentioned, they contain embryos of several thousand cells.[4,12] One-celled zygotes occur only in the upper regions of the oviduct before the albumin and shell are placed around the yolk.

The most unique feature of birds is that they lay eggs. This development *in ovo* allowed us to obtain much of our basic embryological knowledge because the eggs could be opened and observed during development. This feature still provides material at various stages of development for experimentation. In mammals a study of developmental stages requires either uterine surgery or the killing of pregnant females. Because there is no gestation period and because chickens may lay over 200 eggs per year, it is possible to obtain a large number of progeny from a single mating.

Most mammals, other than man, are not visually oriented to their environment. Birds, however, are very visually oriented. Birds have the most highly developed visual systems of any animal.[24] In fact, the tawny owl *(Strix aluco)* has eyes similar to that of a human being.[25] Thus birds are uniquely suited for vision research.

Another unique feature of the class *Aves* is that they have nucleated red blood cells.[26] This archetypal characteristic has provided opportunities along several lines of research, including studies of the development of the hematopoietic system and studies of genetically inactive chromatin.[27]

Birds contain one third the DNA of mammals. Instead of the haploid genome being from 2 to 3×10^9 base pairs, it is somewhere between 6.7×10^8 and 1×10^9 base pairs. Some genes are therefore more simply organized and gene expression mechanisms less complex.[28] Molecular genetic analyses become simpler when there is less DNA to examine when looking for specific sequences.

Turkeys provide another unique experimental tool. Incubated, unfertilized eggs from a particular strain of turkeys will, in low frequency, produce embryos.[29-32] This parthenogenetic line of turkeys produces all male progeny because the sex-determining mechanism in birds is the reverse of that in mammals.[23] Naturally occurring parthenogenesis in a homeothermic organism provides a unique opportunity for studying developmental events which occur in the absence of fertilization.

V. CURRENT MODELS

In this section, we have selected some contemporary research projects which use chicken material alone or comparatively with mammalian material. The list of topics covered is not exhaustive. The examples given were selected to show a wide variety of research endeavors.

Several important studies investigating gene structure, organization, and expression have utilized chicken gene systems. Caplan et al.[33] have studied chromatin structure in the region of the adult beta-globin gene in repressed chicken red blood cells. Weintraub[27] reviews experiments in which chicken genes were transferred from one nucleus to another in both repressed and derepressed states and relates the results to chromatin structure. Kemper et al.[34] have examined protein-binding sites in the DNA of the chicken alpha-globin gene. These studies examine the function of DNase I hypersensitive regions in the chromatin.

Chambon and co-workers have extensively studied the structure and expression of the chicken ovalbumin gene as controlled by steroid hormones. Control regions have been mapped, and the relationship to steroid hormones and tissue specificity has been determined.[35,36] O'Malley and his research group have concomitantly investigated the structure and expression of the other egg-white genes in the fowl.[37]

Chick embryos have long been used in studies of morphogenesis. They are still used to provide up-to-date information about the role of somites,[38] collagen synthesis,[39] and dermal-epidermal interactions.[40] The movement of cell sheets and cellular aggregations is important during morphogenesis. Both chick and mouse embryos have been used to identify specific cell surface glycoproteins called cadherins responsible for selective cell-to-cell adhesions.[41]

The nuclear and cytoplasmic events occurring during differentiation have been investigated using chick embryo cells. One of the most extensively studied systems has been the determination and differentiation of myoblasts into completely differentiated muscle cells which produce their array of muscle-specific fibers. Holtzer and co-workers and former students are continuing to contribute significantly to our understanding of this process.[42-46]

Chick embryos are used in neurobiology research. The formation of axonal circuits in chick limb buds and embryonic retinas has been investigated.[47,48] The growth of nerve cell processes on muscles and in the cornea has also been studied.[49,50] The complete primary structure of the neural cell adhesion molecule for the chicken has been determined.[51] This molecule is important in the formation of nerve network patterns and the formation of nerve cell-muscle cell junctions.

The migrating neural crest cells of the embryo differentiate into many different adult cell types. Thus the neural crest of the chick has been studied because it is an excellent example of a multipotent cell type. Weston and Vogel have studied this system extensively with particular regard to the environmental control of the differentiation pathway.[52] Le-Douarin and co-workers have, through the use of quail-chick chimeras, gained insightful information as to the fate of neural crest cells and the mechanisms of differentiation.[53] Many of the experimental manipulations with the neural crest could not be performed on placental mammals, and thus our information about neural crest differentiation is enhanced by the use of chick embryos because of their development *in ovo*.

Comparisons of nucleated red blood cells from birds with the enucleated mammalian red blood cells by Lazarides has led us to a better understanding of the genetic control of the cytoskeletal architecture of these cells.[54] The nucleated chick red blood cell is convex and contains all three of the filament systems found in eukaryotic cells. The concave anucleate mamalian red blood cell has lost two of the three filament systems. Thus a comparison of the avian red blood cell with the mammalian red blood cell has shown some of the developmental steps used by mammalian red blood cells during nuclear elimination.

As has already been mentioned, studies of the immune system in the fowl have provided insights into its function in mammals and humans. Studies continue on the comparisons of B-cell development and the major histocompatibility locus (B locus) of the chicken with their mammalian counterparts. The differences as well as the similarities lead us to a better understanding of the ultimate function of these genes and their products.[55-57]

Up to this point, we have focused on how the domestic fowl provides information about normal functions related to mammals and humans. The chicken also provides some excellent models of diseases, both genetic and nongenetic.

Currently, the best model for the depigmenting disease called vitiligo is the "Smyth Chicken".[58] The *epi* chicken is an excellent model for studying the etiology and control of epilepsy.[59-61] The sex-linked paroxysmal *(px)* mutant of the fowl causes sound-induced seizures. This is again a model for studying epileptic seizures and for relating the abnormality to peripheral and brainstem auditory functions.[62] Hypertension and aortic rupture have a known genetic basis in turkeys.[63,64] Chicks 5 to 7 weeks old which are chimeric for quail spinal chord portions due to embryonic surgery performed at 2 d of incubation develop pathological signs similar to those found in individuals with multiple sclerosis and can serve as a model for this disease.[65] The next chapter illustrates how various pigment mutants in the fowl serve as models for different forms of human albinism.

Work with chickens and chick cells in culture has done much to elucidate the relationship between growth factors, cellular oncogenes, retroviruses, and cancer. Cellular oncogenes may be stimulated to overproduction and thus cause cancer-like transformations if a retrovirus has inserted close to the oncogene or, alternatively, if the oncogene has become integrated into the virus itself.[10,66] Some normal growth factors appear to be products of cellular oncogenes. Assays for growth factors are frequently done using chick embryo fibroblasts.[67] The function and regulation of the cellular oncogene, c-*myc*, have been determined in part using a chick system.[68] The expression and function of the viral oncogene v-*erbA* has also been determined using chick embryo fibroblasts.[69] The chick embryo fibroblast culture system is routinely used in oncogene research.

These are just some examples of disease entities in the fowl which serve as models for human diseases. Many other mutations and experimental systems have existed but have been lost or are not being used simply because they occur in birds and not in mammals. It is our opinion that much medically useful information would be obtained if more avian animal models of human diseases were studied. As has been pointed out previously, the advantages and uniqueness of the class *Aves* greatly exceed its disadvantages in biomedical research.

VI. FUTURE APPLICATIONS

Molecular genetics and genetic engineering techniques are being developed to utilize the domestic fowl further as a biomedical research animal. Womack's review[70] of genetic engineering applied to animal genetics completely ignored work done with the fowl. After reading this section, it will become evident that biotechnology as applied to poultry is a very active research area.

Animal biotechnology can provide three kinds of results applicable to human disease. First, fundamental regulatory information can be determined by molecular genetic manipulations in animals. Second, the feasibility of a technique for use in human beings can be determined in part by its success in animals. Third, animals may be engineered to produce certain pharmaceutical products which when consumed by people provide health benefits.

Three kinds of genetic engineering strategies will be discussed: insertional mutagenesis, site-directed mutagenesis, and gene transfer.

Retroviruses carrying selectable markers can be used as insertional mutagens to interrupt the functions of normal genes in cell cultures. The application of such methods to the differentiation of pigment cells has been attempted and is described in more detail in the next chapter.[71]

Murine retroviruses have been used to infect mouse zygotes.[72,73] In some cases, the viruses insert into the germ line and are passed on to succeeding generations. Since the proviral insertions interrupt the genes at the site of insertion, embryos and newborn mice are screened for developmental abnormalities. Using this technique to study a wide variety of developmentally interruptive insertions can help elucidate the genetic control of development. Retroviral insertions not only permit development to be studied but also allow the sequences and functions of the interrupted genes to be determined.

Avian retroviruses have also been developed for insertion into the germ line.[74-77] In some instances, birds with movable, multiple inserts acting somewhat like transposable elements were produced.[76,77] These individuals can be bred to produce novel additional inserts. Because chick embryos develop *in ovo*, it is possible to screen a large number of embryos and newly hatched chicks from such matings for abnormalities. Even early development can be studied because each egg can be opened and examined for fertility and the progress of development to the lethal stage. Thus, the chicken promises to provide a very suitable model for dissecting fundamental processes of development through insertional mutagenesis.

Thomas and Capecchi have developed vectors which cause site-directed mutagenesis due to the homology between the vector and the site of insertion.[78] Engineered avian leukosis viruses (ALV) have been designed to carry specific sequences and insert them into the genome.[79,80] These vectors could be designed for site-directed mutagenesis and for the delivery of specific genes which can be spliced into these vectors. The ability of such vectors to deliver and express the desired sequence of DNA that was spliced into them would suggest the feasibility of such systems for human gene therapy. For example, cis- and trans-acting upstream sequences, such as promoters and enhancers, and their effect upon the transferred DNA segment, could be provided by the chick system. The frequency of deleterious effects in the chick system would further indicate the feasibility for use in humans.

Gene transfer could be performed in the domestic fowl with the object to produce products of a pharmaceutical nature. It might be possible to transfer sequences designed to enhance the health of the consumer into the vitellogen (yolk) genes or albumin genes. It might also be possible to substitute some of the breast muscle genes for similar types of pharmaceutical products. Of course, the vectors would need to be engineered so that they underwent only one round of replication so that the birds themselves would not be viremic. Another application would be to consider the egg yolk, albumin, or breast muscle genes of the chicken to be potential synthesizing systems for pharmaceutical products. In this case, laying hens with substituted yolk or albumin genes would become "factories" for the production of substances such as insulin, factor VIII, and tissue plasminogen activator. Alternatively, the breast muscle genes of meat-type chickens might provide an advantageous source of such materials.

With imagination and diligent efforts, the domestic fowl could rise to its former position as a premier animal model for biomedical research.

VII. RESOURCES

Because the fowl has been extensively studied as a research animal and because of its commercial use, there are abundant resources available about nearly every respect of its biology. In this section a few basic works and sources to aid those interested in adopting the domestic fowl as their experimental model are listed.

ANATOMY AND PHYSIOLOGY
King, A. S. and McLelland, J., Eds., *Form and Function in Birds,* Vols. 1 — 3, Academic Press, London, 1985.[81]
Sturkie, P. D., Ed., *Avian Physiology,* Springer-Verlag, New York, 1986.[82]

GENETICS
Hutt, F. B., *Genetics of the Fowl,* McGraw-Hill, New York, 1949,[23] (Unfortunately no updated version is available.)
Somes, R. G., Jr., International Registry of Poultry Genetic Stocks, Storrs Agriculture Experiment Station Bull. No. 476, Storrs, CT, 1988.[17]
Somes, R. G. Jr., Linked loci of the chicken *Gallus gallus (G. domesticus), Genet. Maps,* 4, 422, 1987.[19]

Check with the nearest agricultural experiment station for information about genetic strains available in your area. Check with your county extension agent's office for information about local chicken fanciers, hobbyists, and shows.

Fertile eggs of various types which are specific pathogen and antigen free are available from SPAFAS, Inc., RFD No. 6, Norwich, CT 06360.

EMBRYOLOGY AND DEVELOPMENT
Romanoff, A. L. and Romanoff, A. J., *The Avian Egg,* John Wiley & Sons, New York, 1949.[83]
Hamilton, H. L., *Lillie's Development of the Chick,* Henry Holt, New York, 1952.[4]
Romanoff, A. L., *The Avian Embryo,* Macmillan, New York, 1960.[84]

SPECIALIZED REFERENCES

Lucas, A. M. and Jamroz, C., *Atlas of Avian Hematology,* Agric. Monogr. 25, U.S. Department of Agriculture, Washington, D.C., 1961.[26]

Lucas, A. M. and Stettenheim, P. R., *Avian Anatomy: Integument Parts I and II,* Agric. Handbook 362, U.S. Department of Agriculture, Washington, D.C., 1972.[85]

Kuenzel, W. J. and Masson, J. *A Stereotoxic Atlas of the Brain of the Chick (Gallus domesticus),* Johns Hopkins Press, Baltimore, 1987.[86]

Ookawa, T., Ed., *The Brain and Behavior of the Fowl,* Japanese Scientific Societies Press, Tokyo, 1983.[87]

Toivanen, A. and Toivanen, P., Eds., *Avian Immunology,* CRC Press, Boca Raton, FL, 1987.[88]

PRODUCTION AND MANAGEMENT

Banks, S., *The Complete Handbook of Poultry Keeping,* Van Nostrand Reinhold, New York, 1979.[14]

Moreng, Robert E. and Avens, John S., *Poultry Science and Production,* Reston Publishing, Reston, 120, 1985.[16]

North, M. O., *Commercial Chicken Production Manual,* 3rd ed., AVI Publishing, Westport, CT, 1984.[13]

The nearest arigultural experiment station or your county extension agent's office will have access to government publications on poultry rearing and can direct you to experienced individuals who can answer your questions.

ACKNOWLEDGMENTS

We thank Mim Sawtell for her careful preparation of this manuscript. This work was supported in part by grants GM18969 and BRSG #RR07055 from the National Institutes of Health and by a grant from Li-Cor, Inc., of Lincoln, NE.

REFERENCES

1. **Punnett, R. C.,** *Heredity in Poultry,* Macmillan, London, 1923.
2. **Bateson, W. and Saunders, E.,** II. Poultry, *Rep. Evol. Comm. R. Soc.,* 1, 87, 1902.
3. **Cuenot, L.,** La loi de Mendel et l'Heredite de la pigmentation chez les souris, *Arch. Zool. Exp. Gen.,* 10, 27, 1902.
4. **Hamilton, H. L.,** *Lillie's Development of the Chick,* 3rd ed., Henry Holt, New York, 1979.
5. **Ellerman, V. and Bang, O.,** Experimentelle leukamie bei huhnern, *Centralbl. J. Bakt. Abt. I* (orig.), 46, 595, 1908.
6. **Ellerman, V. and Bang, O.,** Experimentelle leukamie bei huhnern. II. *Z. Hyg. Infektionskr.,* 63, 231, 1909.
7. **Rous, P.,** A sarcoma of the fowl transmissible by an agent separable from the tumor cells, *J. Exp. Med.,* 13, 397, 1911.
8. **Temin, H. M. and Rubin, H.,** Characteristics of an assay for Rous sarcoma virus and Rous sarcoma cells in tissue culture, *Virology,* 6, 669, 1958.
9. **Temin, H. M.,** Nature of the provirus of Rous sarcoma, *Natl. Cancer Inst. Monogr.,* 17, 557, 1964.
10. **Crittenden, L. B. and Astrin, S. M.,** Genes, viruses, and avian leukosis, *Bioscience,* 31, 305, 1981.
11. **Glick, B., Chang, T. S., and Japp, R. G.,** The Bursa of Fabricius and antibody production, *Poult. Sci.,* 35, 224, 1956.
12. **Johnson, A. L.,** Reproduction in the female, in *Avian Physiology,* 4th ed., Sturkie, P. D., Ed., Springer-Verlag, New York, 1986, 403.
13. **North, M. O.,** *Commercial Chicken Production Manual,* 3rd ed., AVI Publishing, Westport, CT, 1984.
14. **Banks, S.,** *The Complete Handbook of Poultry-Keeping,* Van Nostrand Reinhold, New York, 1979.

15. **Johnson, A. L.,** Reproduction in the male, in *Avian Physiology,* 4th ed., Sturkie, P. D. Ed., Springer-Verlag, New York, 1986, 432.

16. **Moreng, R. E. and Avens, J. S.,** *Poultry Science and Production,* Reston Publishing, Reston, Va, 120, 1985.

17. **Somes, R. G., Jr.,** International Registry of Poultry Genetic Stocks, Storrs Agriculture Experiment Station Bull. No. 476, Storrs, CT, 1988.

18. **Shoffner, R. N., Krishan, A., Haiden, G. J., Bammi, R. K., and Otis J. S.,** Avian chromosome methodology, *Poult. Sci.,* 56, 334, 1967.

19. **Somes, R. G., Jr.,** Linked loci of the chicken, *Gallus Gallus (G. Domesticus), Genet. Maps,* 4, 422, 1987.

20. **Bloom, S. E. and Bacon, L. D.,** Linkage of the major histocompatibility *(B)* complex and the nucleolar organizer in the chicken, *J. Hered.,* 76, 146, 1985.

21. **Muscarella, D. E., Vogt, V. M., and Bloom, S. E.,** The ribosomal RNA gene cluster in aneuploid chickens: evidence for increased gene dosage and regulation of gene expression, *J. Cell Biol.,* 101, 1749, 1985.

22. **Bloom, S. E., Briles, W. E., Briles, R. W., Delany, M. E., and Dietert, R. R.,** Chromosomal localization of the major histocompatibility *(B)* complex (MHC) and its expression in chickens aneuploid for the major histocompatibility complex/ribosomal deoxyribonucleic acid microchromosome, *Poult. Sci.,* 66, 782, 1987.

23. **Hutt, F. B.,** *Genetics of the Fowl,* 1st ed., McGraw-Hill, New York, 1949.

24. **Meyer, D. B.,** The avian eye, in *Avian Physiology,* 4th ed., Sturkie, P. D., Ed., Springer-Verlag, New York, 1986, 38.

25. **Martin, G. R.,** Eye, in *Form and Function in Birds,* Vol. 3, King, A. S. and McLelland, J., Eds., Academic Press, Orlando, FL, 1985, 311.

26. **Lucas, A. M. and Jamroz, C.,** *Atlas of Avian Hematology,* Agric. Monogr. 25, U.S. Department of Agriculture, Washington, D.C., 1961.

27. **Weintraub, H.,** Assembly and propagation of repressed and derepressed chromosomal states, *Cell,* 42, 705, 1985.

28. **Landsman, D., and Bustin, M.,** Chicken non-histone chromosomal protein HMG-17 cDNA sequence, *Nucleic Acids Res.,* 15, 6750, 1987.

29. **Darcy, K. M., Buss, E. G., Bloom, S. E., and Olsen, M. W.,** A cytological study of early cell populations in developing parthenogenetic blastodiscs of the turkeys, *Genetics,* 69, 479, 1971.

30. **Olsen, M. W. and Buss, E. G.,** Segregation of two alleles for color of down in parthenogenetic and normal turkey embryos and poults, *Genetics,* 71, 69, 1972.

31. **DeFord, L. S., Buss, E. G., Todd, P., and Wood, J. C. S.,** Estimation of haploid cell content of parthenogenetic turkey embryos: a cytofluorometric study, *J. Exp. Zool.,* 210, 301, 1979.

32. **Harada, K. and Buss, E. G.,** The chromosomes of turkey embryos during early stages of parthenogenetic development, *Genetics,* 98, 335, 1981.

33. **Caplan, A., Kimura, T., Gould, H., and Allan, J.,** Perturbation of chromatin structure in the region of the adult beta-globin gene in chicken erythrocyte chromatin, *J. Mol. Biol.,* 193, 57, 1987.

34. **Kemper, B., Jackson, P. D., and Felsenfeld, G.,** Protein-binding sites within the 5′ DNase I-hypersensitive region of the chicken alpha D-globin gene, *Mol. Cell Biol.,* 7, 2059, 1987.

35. **Dierich, A., Gaub, M., LePennec, J., Astinotti, D., and Chambon, P.,** Cell-specificity of the chicken ovalbumin and conalbumin promoters, *EMBO J.,* 6, 2305, 1987.

36. **Gaub, M., Dierich, A., Astinotti, D., Touitou, I., and Chambon, P.,** The chicken ovalbumin promoter is under negative control which is relieved by steroid hormones, *EMBO J.,* 6, 2313, 1987.

37. **Scott, M. J., Huckaby, C. S., Kato, I., Kohr, W. J., Laskowski, M., Jr., Tsai, M. J., and O'Malley, B. W.,** Ovoinhibitor introns specify functional domains as in the related and linked ovomucoid gene, *J. Biol. Chem.,* 262, 5899, 1987.

38. **Bellairs, R., Ede, D. A., and Lash, J. W.,** *Somites in Developing Embryos,* Plenum Press, New York, 1986, 326.

39. **van der Rest, M. and Mayne, R.,** Type IX collagen, in *Structure and Function of Collagen Types,* Mayne, R. and Burgeson, R. E., Eds., Academic Press, New York, 1987, 195.

40. **Sengel, P.,** Epidermal-dermal interaction, in *Biolical of the Integument,* Gereiter-Hahn, J., Matoltsy, A. G., and Richards, K. S., Eds., Springer-Verlag, Berlin, 1986, 374.

41. **Takeichi, M.,** Cadherins: a molecular family essential for selective cell-to-cell adhesion and animal morphogenesis, *Trends Genet.,* 3, 213, 1987.

42. **Holtzer, H., Antin, P., Dlugosz A., Forry-Schaudies, S., Eshleman, J., and Nachmias, V.,** Using a co-carcinogen (TPA) and a carcinogen (EMS) to probe myofibrillogenesis, in *The Molecular Biology of Muscle Development,* UCLA Symp. Molecular and Cellular Biology, Emerson, C., Fischman, D., Nadal-Ginard, B. and Siddiqui, M. A. Q., Eds., Alan R. Liss, New York, 1986.

43. **Schafer, D. A., Miller, J. B., and Stockdale, F. E.,** Cell diversification within the myogenic lineage: *in vitro* hormone dependent and independent phases of adipocyte-mammary epithelial cell interaction, *Dev. Biol.,* 120, 245, 1987.

44. **Quinn, L. S. and Nameroff, M.,** Evidence for a myogenic stem cell, in *The Molecular Biology of Muscle Development,* UCLA Symp. Molecular and Cellular Biology, Emerson, C., Fischman, D., Nadal-Ginard, B., and Siddiqui, M. A. Q., Eds., Alan R. Liss, New York, 1986, 35.

45. **Obinata, T., Kawashima, M., Kitanai, S., Saitoh, Q., Masaki, T., Bader, D. M., and Fischman, D. A.,** Expression of C-protein isoforms during chicken striated muscle development and its dependence on innervation, in *The Molecular Biology of Muscle Development,* UCLA Symp. Molecular and Cellular Biology, Fischman, D. A., Emerson, C., Nadal-Ginard, B., and Siddiqui, M. A. Q., Eds., Alan R. Liss, New York, 1986, 292.

46. **Menko, A. S. and Boettiger, D.,** Occupation of the extracellular matrix receptor, integrin, is a control point for myogenic differentiation, *Cell,* 51, 51, 1987.

47. **Landmesser, L.,** Axonal guidance cues and the formation of neural circuits, *Trends Neurosci.,* 9, 489, 1986.

48. **Halfter, W., Deiss, S., and Schwarz, U.,** The formation of the axonal pattern in the avian embryonic retina, *J. Comp. Neurol.,* 232, 466, 1985.

49. **Bixby, J. L., Pratt, R., Lilien, J., and Reichardt, L. F.,** Neurite outgrowth on muscle cell surfaces involves extracellular matrix receptors as well as Ca^{2+}-dependent and -independent cell adhesion molecules, *Proc. Natl. Acad. Sci. U.S.A.,* 84, 2555, 1987.

50. **Sturges, S. A. and Conrad, G. W.,** Acetylcholinesterase activity in the cornea of the developing chick embryo, *Invest. Opthalmol. Visual Sci.,* 28, 850, 1987.

51. **Cunningham, B. A., Hemperly, J. J., Murray, B. A., Prediger, E. A., Brackenbury, R., and Edelman, G. M.,** Molecular genetics and complete primary structure of the neural cell adhesion molecule: cell surface modulation, alternative RNA splicing, and evolutionary relationships, *Science,* 236, 799, 1987.

52. **Weston, J. A. and Vogel, K.,** Environmental regulation of neural crest development, in *Advances in Gene Technology: The Molecular Biology of Development,* Proc. 19th Miami Winter Symp., Voellmy, R. et al., Eds., Cambridge University Press, Cambridge, 1987.

53. **LeDouarin, N. M.,** Ontogeny of the peripheral nervous system from the neural crest and the placodes. A developmental model studied on the basis of the quail-chick chimaera system, *Harvey Lect.,* Ser. 80, 137, 1986.

54. **Lazarides, E.,** From genes to structural morphogenesis: the genesis and epigenesis of a red blood cell, *Cell,* 51, 345, 1987.

55. **Weill, J. and Reynaud, C.,** The chicken B cell compartment, *Science,* 238, 1094, 1987.

56. **Pitcovski, J. Peterson, L., Lamont, S., and Warner, C.,** Identification and characterization of Class I MHC genes from the chicken B complex, *Fed. Proc., Fed. Am. Soc. Exp. Biol.,* 46, 945, 1987.

57. **Thompson, C. B. and Neiman, P. E.,** Somatic diversification of the chicken immunoglobulin light chain gene is limited to the rearranged variable gene segment, *Cell,* 48, 369, 1987.

58. **Smyth, J. R., Jr., Boissy, R. E., and Fite, K. V.,** The DAM chicken: a model for spontaneous postnatal cutaneous and ocular amelanosis, *J. Hered.,* 72, 150, 1981.

59. **Crawford, R. D.,** Genetics and behavior of the *epi* mutant chicken, in *The Brain and Behavior of the Fowl,* Ookawa, T., Ed., Japan Scientific Societies Press, Tokyo, 1983, 259.

60. **Crichlow, E. C.,** Electroencephalogram and brain biochemistry of the *epi* mutant chicken, in *The Brain and Behavior of the Fowl,* Ookawa, T., Ed., Japan Scientific Societies Press, Tokyo, 1983, 271.

61. **Johnson, D. D. and Davis, H. L.,** Drug responses and brain biochemistry of the *epi* mutant chicken, in *The Brain and Behavior of the Fowl,* Ookawa, T., Ed., Japan Scientific Societies Press, Tokyo, 1983, 281.

62. **Beck, M. M., Brown-Borg, H. M., and Jones, T. A.,** Peripheral and brainstem auditory function in paroxysmal *(px)* White Leghorn chicks, *Brain Res.,* 406, 93, 1987.

63. **Shoffner, R. N., Krista, L. M., Waibel, P. E., and Quarfoth, G. J.,** Strain crosses within and between lines of turkeys selected for high and low blood pressure, *Poult. Sci.,* 50(2), 342, 1971.

64. **Krista, L. M., Waibel, P. E., Sautter, J. H., and Shoffner, R. N.,** Aortic rupture, body weight, and blood pressure in the turkey as influenced by strain, dietary fat, *beta*-amino-propionitrile fumarate, and diethylstilbestrol, *Poult. Sci.,* 48, 1954, 1969.

65. **Kinutani, M., Coltey, M., and Le Douarin, N. M.,** Postnatal development of a demyelinating disease in avian spinal cord chimeras, *Cell,* 45, 307, 1986.

66. **Temin, H. M.,** Genetic mechanisms of oncogenesis, in *Carcinogenesis — A Comprehensive Survey,* Vol. 10, Huberman, E. and Barr, S. E., Eds., Raven Press, New York, 1985, 15.

67. **Smith, G. L.,** Multiplication-stimulating activity and the role of carrier proteins, in *Growth and Maturation Factors,* Guroff, G., Ed., John Wiley & Sons, New York, 1983, 293.

68. **Piechaczyk, M., Blanchard, J., and Jeanteur, P.,** C-*myc* gene regulation still holds its secret, *Trends Genet.,* 3, 47, 1987.

69. **Gandrillon, O., Jurdic, P., Benchaibi, M., Xiao, J. H., Ghysdael, J., and Samarut, J.,** Expression of the v-*erbA* oncogene in chicken embryo fibroblasts stimulates their proliferation *in vitro* and enhances tumor growth *in vivo, Cell,* 49, 687, 1987.

70. **Womack, J. E.,** Genetic engineering in agriculture: animal genetics and development, *Trends Genet.,* 3, 65, 1987.
71. **Oetting, W. S., Smith, G. L., and Brumbaugh, J. A.** Isolation of pigment genes using retroviral insertional mutagenesis, in *Advances in Pigment Cell Research,* Bagnara, J., Ed., Alan R. Liss, New York, 1988, 307.
72. **Gridley, T., Soriano, P., and Jaenisch, R.,** Insertional mutagenesis in mice, *Trends Genet.,* 3, 162, 1987.
73. **Copeland, N. G. and Jenkins, N. A.,** Eukaryotic transposable elements: identifying and studying genes important in mammalian development, *Birth Defects OAS,* 23(3), 123, 1987.
74. **Shuman, R. M. and Shoffner, R. M.,** Gene transfer by avian retroviruses, *Poult. Sci.,* 65, 1437, 1985.
75. **Temin, H. M.,** Retrovirus vectors for gene transfer: efficient integration into and expression of exogenous DNA in vertebrate cell genomes, in *Gene Transfer,* Kucherlapati, R., Ed., Plenum Press, New York, 1986, 149.
76. **Salter, D. W., Smith, E. J., Hughes, S. H., Wright, S. E., Fadly, A. M., Witter, R. L., and Crittenden, L. B.,** Gene insertion into the chicken germ line by retroviruses, *Poult. Sci.,* 65, 1445, 1986.
77. **Salter, D. W., Smith, E. J., Hughes, S. H., Wright, S. E., and Crittenden, L. B.,** Transgenic chickens: insertion of retroviral genes into the chicken germ line, *Virolology,* 157, 236, 1986.
78. **Thomas, K. R. and Capecchi, M. R.,** Site-directed mutagenesis by gene targeting in mouse embryo-derived stem cells, *Cell,* 51, 503, 1987.
79. **Hughes, S. H., Kosik, E., Fadly, A. M., Salter, D. W., and Crittenden, L. B.,** Design of retroviral vectors for the insertion of foreign deoxyribonucleic acid sequences into the avian germ line, *Poult. Sci.,* 65, 1459, 1986.
80. **Hughes, S. H., Greenhouse, J. J., Petropoulos, C. J., and Sutrave, P.,** Adaptor plasmids simplify the insertion of foreign DNA into helper independent retroviral vectors, *J. Virol.,* 61, 3004, 1987.
81. **King, A. S., and McLelland, J., Eds.,** *Form and Function in Birds,* Vols. 1-3, Academic Press, London, 1985.
82. **Sturkie, P. D., Ed.,** *Avian Physiology,* Springer-Verlag, New York, 1986.
83. **Romanoff, A. L. and Romanoff, A. J.,** *The Avian Egg,* John Wiley & Sons, New York, 1949.
84. **Romanoff, A. L.,** *The Avian Embryo,* Macmillan, New York, 1960.
85. **Lucas, A. M., and Stettenheim, P. R.,** *Avian Anatomy: Integument,* Parts I and II, U.S. Department of Agriculture Handbook 362, Washington, D. C., 1972.
86. **Kuenzel, W. J., and Masson, J.,** *A Stereotoxic Atlas of the Brain of the Chick (Gallas domesticus),* Johns Hopkins Press, Baltimore, 1987.
87. **Ookawa, T., Ed.,** *The Brain and Behavior of the Fowl,* Japanese Scientific Societies Press, Tokyo, 1983.
88. **Toivanen, A. and Toivanen, P., Eds.,** *Avian Immunology,* CRC Press, Boca Raton, FL, 1987.

Chapter 20

THE DOMESTIC FOWL AS AN ANIMAL MODEL FOR HUMAN ALBINISM

William S. Oetting and John A. Brumbaugh

TABLE OF CONTENTS

I. INTRODUCTION

The formation of pigment within the melanocyte provides an excellent model for developmental and genetic research.[1] The developmental biology of the melanocyte is complex, consisting of three phases: (1) the determination and differentiation of the melanoblast into the melanocyte, (2) the process of pigment granule (melanosome) formation or melanogenesis per se, and (3) the transfer and fate of the melanosomes in keratinocytes.[2,3] The melanocyte begins development in the neural crest as an unpigmented melanoblast. The melanoblast then migrates to its epidermal location, differentiating into a mature melanocyte as seen by pigment production. The large number of mutations isolated throughout the animal kingdom indicate the complexity of this process. Silvers, in his book *The Coat Colors of Mice,*[4] catalogs over 65 loci (with over 148 alleles) affecting the pigment system. The large number of loci required for pigment formation and deposition provides an opportunity to study the genetic control of development of a complex multigenic process.

Albinism has been described as an inborn error of metabolism since 1908, yet little is known about the specific mechanisms of the mutations causing the six or more types found in humans.[5] Oculocutaneous albinism (OCA) is a common genetic abnormality found throughout the world with a frequency of 1:17,000 in the U.S. and at least 1:2,000 in Equatorial Africa.[6] A reduction in melanin production in the skin can have devastating effects on the affected individual resulting in a marked sensitivity to ultraviolet radiation and a predisposition to skin cancer. Reduction of melanin in the eye during development is associated with life-long nystagmus, foveal hypoplasia with reduced visual acuity, and misrouting of the retinal-ganglion fibers at the chiasm with an abnormal optic radiation to the occipital cortex resulting in strabismus and loss of binocular vision.[7] The foveal hypoplasia and misrouting are pigment dependent and not determined by a specific pigment locus. Albinism can appear as the single genetic defect, or in association with other abnormalities such as Hermansky-Pudlak Syndrome (HPS),[8] Prader-Willi Syndrome (PWS),[9] and deafness.[10]

Understanding albinism in human beings is important, not only to determine the underlying causes of the condition itself, but to serve as a model to study other genetic diseases. Albinism provides a rich assortment of related genetic defects as compared with most other genetic disorders. Studying the interaction of these genes and how the mutations disrupt the developmental process of pigment formation will provide a better understanding of how other multigenic systems work.

The pigment system of the fowl provides a number of advantages as a model system for investigating albinism. Melanocyte specific pigment mutations have been isolated and well characterized. The oviparous nature of fowl provides easy access to embryos which supply a large number of melanoblasts for *in vitro* cell culture and biochemical analyses. The apparent homology between avian and human melanogenesis allows us to make direct comparisons between the two systems.

II. PIGMENT FORMATION

Two basic pathways are involved in the production of the pigment-containing organelle, the melanosome.[3,11] One pathway produces the pigment polymer, melanin, and the second produces the matrix of the immature pigment organelle, the premelanosome.

Two major pigments are found in both fowl and humans: eumelanin and pheomelanin. Melanins are formed by the catalytic activities of two enzymes (Figure 1) tyrosinase[12] and dopachrome oxidoreductase (DCOR).[13-15] The initial substrate for melanin is L-tyrosine and the final product is an amorphous polymer with a broad spectral absorption range. The addition of L-cysteine, or other sulfhydryl containing compounds, results in the yellow-red pigment, pheomelanin. Many of the products of the melanin pathway are highly reactive

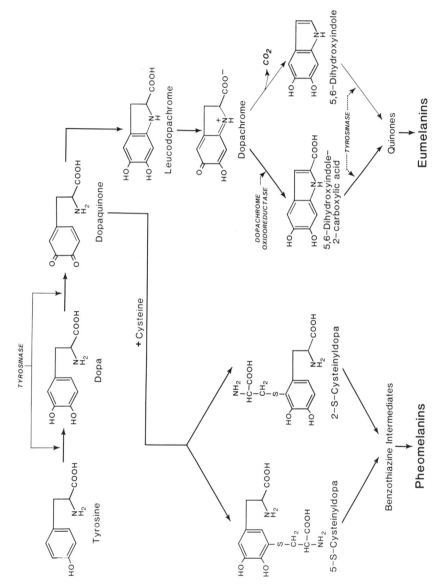

FIGURE 1. The melanin chemical pathway. Starting with the amino acid L-tyrosine, either pheomelanins (red, yellow) or eumelanins (black) are formed. The addition of L-cysteine results in the production of pheomelanin. (Courtesy of Dr. Richard King, Director, Clinical Genetic Services, University of Minnesota Health Center, Minneapolis, MN.)

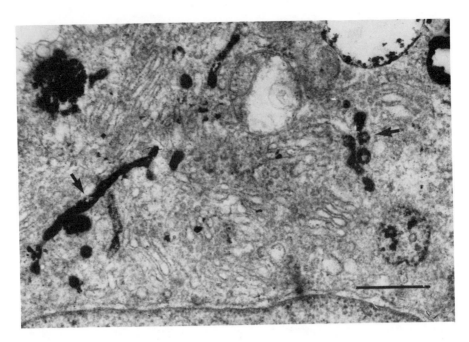

FIGURE 2.   Cytochemical dopa reaction of the Golgi region and coated vesicles. A thin section of a normally pigmented melanocyte was stained with L-dopa. Arrows point to electron opaque deposits indicating tyrosinase activity both in the Golgi region and within coated vesicles. Scale bar = 0.5 μm. (From Brumbaugh, J. A. and Oetting, W. S., *Bioscience*, 36, 381 — 387, 1986. With permission.)

molecules including oxygen radicals.[16,17] The synthesis of melanin therefore must be a tightly controlled reaction to protect the melanocyte from being damaged by its own by-products.

Tyrosinase is a bifunctional copper-containing enzyme that first catalyzes the oxidation of L-tyrosine to L-dihydroxyphenylalanine (DOPA), and then the dehydrogenation of DOPA to dopaquinone. Tyrosinase may also be involved in completing melanin formation at the distal end of the pathway.[18] Tyrosinase is produced in the rough endoplasmic reticulum (RER) and then transferred to the Golgi system, where it is glycosylated and packaged into coated vesicles.[11,19-21] Glycosylation has been shown to be important in both the intracellular maturation and translocation of tyrosinase.[21] As shown in Figure 2, cells stained with DOPA exhibit tyrosinase activity in the Golgi region and within small coated vesicles.

The second enzyme, dopachrome oxidoreductase (DCOR), is involved in converting the intermediate, dopachrome, to 5,6-dihydroxyindole-2-carboxylic acid (DHICA). This enzyme, like tyrosinase, is membrane bound. Both tyrosinase and DCOR activity have been found to be associated with melanosomes and may indicate a close association or physical coupling between the two enzymes during melanin synthesis.[22,23]

The premelanosome is a membrane-bound organelle that contains a protein matrix (Figure 3). This matrix is assembled in the smooth endoplasmic reticulum (SER) and is usually composed of regularly spaced, cross-linked fibers arranged in a zigzag fashion.[11,19,20,24,25] The matrix has a complex architecture that may consist of two or more interacting fibrous structures. As the matrix appears, the endoplasmic reticulum pinches off to form the membrane-enclosed premelanosome, which is usually rod-shaped or ellipsoid. The matrix of red pigment granules or pheomelanosomes is less organized than that found in eumelanosomes.[24,26] Cross-links between the fibers are few or absent, and the fibers themselves are shorter and show less zigzag folding. This results in a more spherical-shaped granule.

Pigment deposition actually occurs when the enzymatic and matrix components join. Coated vesicles released by the Golgi system carry tyrosinase, and possibly DCOR, to the

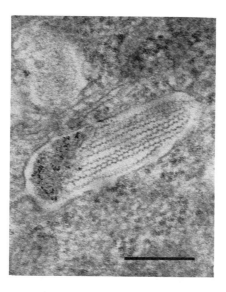

FIGURE 3.   Electron micrograph of a premelanosome. A thin section of a premelanosome before pigment deposition. Note the zigzag folding of the protein fibers and the cross-links. Scale bar = 0.25 μm. (From Brumbaugh, J. A. and Oetting, W. S., *Bioscience,* 36, 381 — 387, 1986. With permission.)

premelanosome and fuse with the premelanosomes' outer membrane. Like tyrosinase, the premelanosomal matrix proteins also contain carbohydrates, which may contribute to the tyrosinase-accepting function or *in vivo* melanizing function of tyrosinase upon fusion.[21] After fusion, the enzymes are activated and melanin synthesis begins. Melanin deposition continues on each matrix until, in normal genotypes, completely opaque black granules are formed with enzymatic activity gradually being lost.[27,28] A typical melanocyte (Figure 4) will contain numerous melanosomes which are eventually transferred from the tips of the melanocyte dendrites to the keratinocytes of the feather. This gives feathers their characteristic color.

The process of formation and ultrastructural features of melanosomes in both fowl and humans are essentially the same. The tyrosinases of human and fowl are antigentically similar and it is expected that strong homologies also exist between the matrix proteins of the premelanosomes of the fowl and man.

III. MUTANTS AVAILABLE

Because pigment can be easily seen, mutations have been isolated and studied for many decades.[29,30] It was both Cuénot,[31] using mouse pigment mutants, and Bateson and Saunders,[32] using pigment mutants of the fowl, who first showed that Mendelian principles applied to animals.

The domestic fowl, *Gallus gallus,* provides a rich assortment of mutations that affect melanogenesis.[33] Pigment mutations of the fowl consist of those that either affect pigment formation, or the pattern of pigment deposition, including dermal deposition, feather coloration, and retinal pigmentation. The mutations that we use in studying pigment biology are listed in Table 1. This table describes the pigmentation and tyrosinase activity found in both the newly hatched chick and cultured melanocytes for each locus. This list represents known mutations that directly affect melanin biosynthesis, but most likely does not represent all of the genes involved.

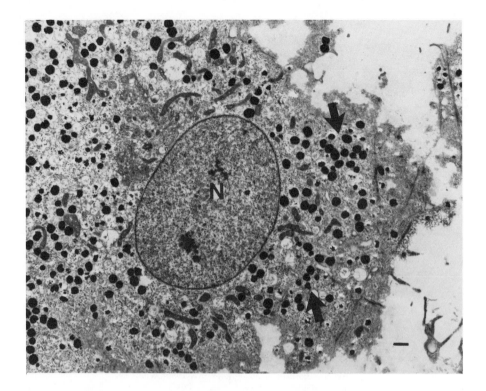

FIGURE 4.   Electron micrograph of a normally pigmented melanocyte. A low-magnification electron micrograph of a melanocyte showing numerous fully pigmented melanosomes. N = nucleus; scale bar = 1.0 μm. (From Brumbaugh, J. A. and Oetting, W. S., *Bioscience,* 36, 381 — 387, 1986. With permission.)

TABLE 1
Mutations of Melanin Biosynthesis

Locus name	Other alleles	Chick			Culture			Ref.
		Pigment	Tyase.	Pre.	Pigment	Tyase.	Pre.	
Normal (+)	—	Y	Y	Y	Y	Y	Y	—
Autosomal albino *(cᵃ)*	Y	N	N	Y	N	N	Y	45—47
Sex-linked albino *(sᵃˡ)*	Y	tr	Y	aber	tr	Y	aber	33, 38
Recessive wheaten *(eʸ)*	Y	tr	Y	aber	N	Y	aber	35—37
Pinkeye *(pk)*	N	dil	Y	Y	N	Y	tr	24, 40
Blue *(Bl)*	Y	tr	Y	aber	tr	Y	aber	43
Dominant white *(I)*	Y	tr	Y	tr	tr	Y	tr	38—42

Note: tyase — tyrosinase activity, pre. — premelanosome, Y — yes (present), N — no (absent), tr — trace, aber — aberrant, dil — diluted, gray.

None of the loci listed in Table 1 are linked, showing that each mutation affects a different gene responsible for pigment production. The action of these mutations also exhibits autonomy as shown by means of limb bud grafts and *in vitro* culture.[34] Therefore, the lack of pigment is due to mutations that affect the melanocyte itself and is not due to extracellular factors. The majority of the known pigment loci in the fowl are thought to be involved with the production of the premelanosome. These mutations produce aberrant melanosomes when observed with the electron microscope or normal melanosomes but in greatly reduced amounts.

Three loci, *E, S,* and *I* affect the synthesis and/or distribution of eumelanin and pheomelanin. The *E* locus is thought to be the major genetic controller of black-red melanin distribution in the chicken.[35] The most recessive allele, e^y, produces mostly pheomelanin-producing melanocytes, where the most dominant allele, *E*, produces exclusively eumelanin-producing melanocytes.[36,37] It was shown that fasting and thyroxin feeding experiments with the normal (black-red, e^+) adult changed the type of melanin synthesized.[36] It is thought there is an interaction between melanocytes and the ''zone of differentiation'' of the regenerating feather. Rapid movement through this zone favors eumelanin synthesis whereas slower growth favors pheomelanogenesis.

The *S* locus is a sex-linked locus that also affects red-black pigment deposition.[38] The dominant *S* mutant produces a normal eumelanin pattern, but no pigment (white) where pheomelanin would be expected. The recessive mutant s^{al} produces an imperfect form of albinism allowing pheomelanin synthesis to ''leak'' through, but no eumelanin synthesis. The *S* locus mutations seem to cause the formation of abnormal melanosomes.[33]

The *I* locus affects only eumelanogenesis.[38-41] Heterozygotes at this locus (I/i^+) show some leakage in the form of black flecking, indicating some somatic instability. There are two known mutant alleles at this locus, *I* and I^D.[42] Mutants at this locus produce a reduced number of pigment granules.[38]

The *Bl* mutation is incompletely dominant.[43] Homozygous *(Bl/Bl)* birds are white with spots of slate coloration, occasionally containing areas of normal (black) coloration. The heterozygotes *(Bl/bl⁺)* have slate colored plumage. The *Bl* mutation produces aberrant melanosomes as determined by electron microscopic observation. Though the *Bl* mutation appears to be unstable for melanocytes (somatic cells), there is no report of a revertant in germ line cells.[44]

The *pk* mutation is a recessive mutation producing grayish-brown feathers.[24,40] The number of premelanosomes produced are greatly reduced in number and usually incompletely melanized and poorly formed.

The recessive *C* locus mutations produce all white birds, equally affecting both eumelanin and pheomelanin production. There are presently three known mutant alleles of this locus.[45,46] The order of dominance, which follows the amount of pigment reduction, is as follows: *c* (recessive white), c^{re} (red eye), and c^a (autosomal albino), where c^{re} is incompletely dominant to c^a.

The *C* locus mutations produce normal premelanosomes but enzymatically inactive tyrosinase-like molecules.[47] The tyrosinase-like molecules produced by the mutant alleles have the same molecular weight and pI as native tyrosinase and will cross-react with anti-tyrosinase antibody. This suggests that the *C* locus is the structural gene for tyrosinase and is therefore the only pigment locus whose mode of action is known. Recently, the structural gene for tyrosinase has been isolated and linked to the *C* locus in the mouse and Type I (tyrosinase negative) OCA in humans.[48-50] The *C* locus mutations have been shown to reduce the growth rate of chicks.[51,52] This effect is thought to be due to the absence of melanin and not due to other regulatory roles of the *C* locus.[52] Apparently, mice do not show reduced growth rate when homozygous for the tyrosinase negative mutation which is homologous to the *C* locus in the fowl.[53] Since birds are visually oriented and mice are not, this difference may be related to the ability to find food and water.

One mutation that specifically affects the melanocyte but not melanogenesis is the pigment diluting mutation lavender *(lav)*. This mutation produces melanocytes which are defective in transferring melanosomes to keratinocytes.[54] This may be due to a microfilament related defect.[55]

Table 2 shows human albino loci that appear to be homologous to pigment specific mutations of the fowl.[6,56-61] With the exception of the homology between chicken autosomal albinism and human tyrosinase negative oculocutaneous albinism, these comparisons are

TABLE 2
Proposed Chicken-Human Homologous
Pigment Mutations

Chicken locus	Human homolog	Ref.
c^a	OCA (Ty-neg)	6
s^{al}	OCA (Ty-pos)	6
pk	Brown albinism	56, 57
I	Autosomal dominant OCA	58, 59
e^y	Red OCA (rufus)	60, 61

Note: OCA — oculocutaneous albinism, OCA (ty-neg) — tyrosinase-negative ouclocutaneous albinism, OCA (ty-pos) — tyrosinase-positive oculocutaneous albinism.

made by phenotypic considerations and not by the underlying mechanism of the mutations, which at this time are not known.

IV. CELL CULTURE

We have described a cell culture method for neural crest melanocytes.[62,63] This system produces cultures of melanocytes that retain the phenotype found in newly hatched chicks (Table 1). Though other culture systems for avian melanocytes exist, our system provides large numbers of melanocytes at high purity, useful for biochemical studies.[64,65]

Starting with embryos at stages 16 to 18, caudal somites are isolated, trypsinized, and then plated in a medium that has been preconditioned with Buffalo Rat liver (BRL-3A) cells.[62,66,67] BRL cells produce a number of growth-stimulating factors.[62,68,69] The medium is conditioned for 2 d, centrifuged and filtered to remove BRL-3A cells, and mixed (3:1) with fresh medium containing 10% calf serum. Resulting cell cultures have a purity greater than 98%, and 90 to 95% of genetically normal melanocytes produce copious quantities of pigment by day 17 of culture.[63] A typical experiment of 20 embryos will produce 8×10^7 melanocytes. Dividing cultures have been kept for over 6 months retaining both pigment production and normal morphology.

It has also been shown that the addition of the phorbol ester, 12-*O*-tetradecanoylphorbol-13-acetate (TPA) can act as an inhibitor of differentiated cellular functions. This has been demonstrated in chondrogenesis[70] and myogenesis.[71-74] The inhibition of melanogenesis in avian melanocytes has been reported by a number of investigators.[75,76] When melanocytes are grown in the presence of $10^{-7}\ M$ TPA, the cells will remain unpigmented throughout TPA treatment. The addition of TPA also produces a change in melanocyte morphology. Melanocytes are normally flat and dendritic in cell culture, but, upon TPA addition, become rounded or spindle shaped. This inhibition of pigmentation due to the presence of TPA is reversible if the TPA is removed from the growth medium.[77,78] The inhibition of pigment production has been confirmed at the electron microscopic level.[77,79] Both dopa-oxidase activity and premelanosomes were absent in TPA cultures, but then reappeared 24 h after TPA removal. After 48 h, pigment can be observed with the light microscope and cells return to their normal flat dendritic form. The resumption of pigment production and the return to normal morphology is cycloheximide sensitive. Inhibition of the differentiated functions(s) is thought to occur via interference with protein synthesis.[72,74,77] TPA appears to act as a "switch" which interrupts pigment production without stopping the growth of the cells.

V. HETEROKARYON STUDIES

The ability to culture mutant melanocytes provided us with the opportunity to categorize these mutants into complementation groups via heterokaryon analyses.[80-83]

Heterokaryons were used to study the human disease *xeroderma pigmentosum,* a biochemical defect in the ability to repair DNA.[84] At least seven genetically distinct forms of this disease were found using heterokaryon analyses. Heterokaryons were produced from different patients with the disease and examined for their ability to repair DNA. Cells that complemented each other (repaired their DNA) were considered to have different forms of the disease.

We have dissected the genetic control of pigmentation by combining both heterokaryon analyses with standard sexual crosses and compared heterozygous cells with heterokaryons. Pigment cells work well in single-cell analysis because the presence or absence of complementation, i.e., melanosome production, can be visually observed on a cell-to-cell basis.

For heterokaryon formation, cells of two mutant genotypes are mixed together and fused with polyethylene glycol (PEG).[85] This produces some cells with nuclei of different genotypes that share a common cytoplasm. Complementation results when the gene product from one nucleus is able to supply a necessary gene product which is deficient in the other nucleus.

In experiments, cells of one mutant were labeled with ^{3}H -thymidine for 24 h and then fused with unlabeled cells of a second mutant.[80] The fused cells were cultured for 48 h and examined for pigment production. The pigmented cells were then marked, and the culture plates prepared for autoradiography, allowed to expose, and then developed and reexamined with a light microscope. All cells found to contain pigment also contained two or more nuclei. In each case, at least one nucleus was radioactively labeled and one unlabeled. Nuclear ratios of more than 3:1 were rarely encountered, and usually, only one nucleus of each genotype was present per cell. In any case, dosage did not affect the results.

Heterokaryons were made between different combinations of mutants. From this, three complementation groups were found to exist. One group consisted of the *C* locus responsible for tyrosinase activity. The two other groups do not affect tyrosinase production and probably are responsible for the production or assembly of premelanosomes.

In these experiments two cases of noncomplementation were observed. The mutant *pk* was unable to complement with *Bl, e^y,* or *s^al* in heterokaryons, but did complement in cells from double heterozygotes. Also, dominant white *I* was able to complement with blue *(Bl)* in heterokaryons but not in cells from double heterozygotes. These results allow us to hypothesize that both *pk* and *I* are regulatory loci with nuclearly restricted gene products. Apparently the *pk* locus controls the *Bl, s^al,* and *e^y* loci. The role and position of the *I* locus in the regulatory scheme remains unknown. Heterokaryon analyses in *Neurospora* have indicated existence of nonallelic regulatory elements whose site of action is restricted to the nucleus.[82]

Using pigment as a marker of gene function, we have also shown extinction of pigment production by fusing mouse melanoma cells to chick embryo fibroblast cells of a pigmented genotype.[86] By observing tyrosinase activity in individual, nondividing control and fused cells, it was shown that extinction of gene function was not due to chromosome loss as might occur in dividing hybrids, but is the result of gene control interactions between the two parent cell nuclei. Extinction, in this case, was nuclearly controlled, but not nuclearly restricted. Fusion of enucleated fibroblast cytoplasts to the mouse melanoma cells failed to sustain extinction even in the absence of cell division.[87]

Fusions involving genetically inactive chick red blood cells, which are nucleated, with chick melanocytes did not produce extinction however. Restoration of pigment production was found instead.[88] Mutant melanocytes *(pk/pk)* that do not produce pigment while in culture were fused with inactive erythrocyte nuclei of the genotype *(Pk^+/Pk^+)* from pigment producing chick embryos. Restoration of pigment was found in five colonies.

Several fundamental developmental concepts have been derived from the various heterokaryon studies. The production of heterokaryons produced by the fusion of melanocytes of different genotypes has allowed us to divide pigment synthesis into several complementation groups. As a result of these fusions, two putative regulatory mutants have been identified. The fusion studies indicate that one locus controls other loci, indicating a hierarchical control of pigment cell differentiation. The comparison of the fusions between melanoma cells and intact fibroblasts with melanoma cells and enucleated fibroblasts has shown that the extinction or suppression of pigment synthesis by the fibroblast was under nuclear control, and that it depended upon a diffusible substance going from the fibroblast nucleus to the melanoma nucleus. The converse of extinction, i.e., gene activation, was shown when a developmentally inactive chick red blood cell nucleus was placed into mutant melanocyte cytoplasm. The inactive red blood cell nucleus was stimulated to initiate pigment production by the melanocyte cytoplasm. This indicated that the nucleus and/or cytoplasm of the melanocyte was able to activate previously inactivated genes.

VI. PROTEIN STUDIES

Although melanogenesis has been studied for some time, the number of proteins involved in pigment production is not known. An estimate of the minimum number of loci involved can be made by counting the mutants that have been isolated, but this cannot be used as a final determination.

The only characterized pigment protein in the fowl, as well as other vertebrate species, is tyrosinase.[47,89,90] The use of regenerating feathers provided excellent starting material for isolating tyrosinase activity.[90] Kilogram quantities of starting material have been used in large-scale isolation of this enzyme. Tyrosinase is a membrane-bound glycoprotein with a molecular weight of 72,000 Da. Unlike tyrosinase from mammalian sources, chicken tyrosinase is almost exclusively found in an insoluble (membrane-bound) form.[90,91] Upon isoelectric focusing (IEF), chicken tyrosinase resolves into ten bands with pIs between 5.06 and 3.97. These multiple forms will produce a single band upon SDS electrophoresis showing there is undetectable differences in the molecular weights of the various forms.[90] Incubation of tyrosinase with neuraminidase, which cleaves sialic acid residues from the molecule, caused an increase in the pI, but the multiple forms were retained. Apparently, other post-translational modifications produce the multiple forms of tyrosinase.

Characterization of the proteins of the premelanosome has not yet been realized. Previous reports describing these proteins have presented conflicting results.[92-94] Significant differences in both the number and molecular weights of premelanosomal proteins have been reported. We believe the major reason for these conflicting reports is that mature melanosomes have been used as the source for premelanosomal proteins. The mature melanosome is a subcellular particle very resistant to degradation. The melanin polymer is attached to the amino or sulfhydroxyl groups via quinone linkages making it unlikely that undegraded proteins can be removed from the mature melanosome.[95,96] It would seem advantageous to isolate proteins from premelanosomes before melanization has occurred.

The use of TPA as a reversible inhibitor has helped us to solve this problem.[63] As stated previously, cells growing in the presence of TPA are no longer producing tyrosinase or premelanosomes as shown by electron microscopy. Melanocyte cultures were grown in the presence of TPA between days 5 and 11 after initial plating. Cells were labeled with ^{3}H-leucine either on day 9, when cells were in TPA and pigment production inhibited, or day 13, 2 d after TPA removal. On day 13, the cells were producing both tyrosinase and premelanosomes, but pigment production was very slight and the premelanosomes were still susceptible to solubilization. The proteins of both cultures were then resolved by two-dimensional electrophoresis and the protein patterns compared. Any proteins synthesized in

343

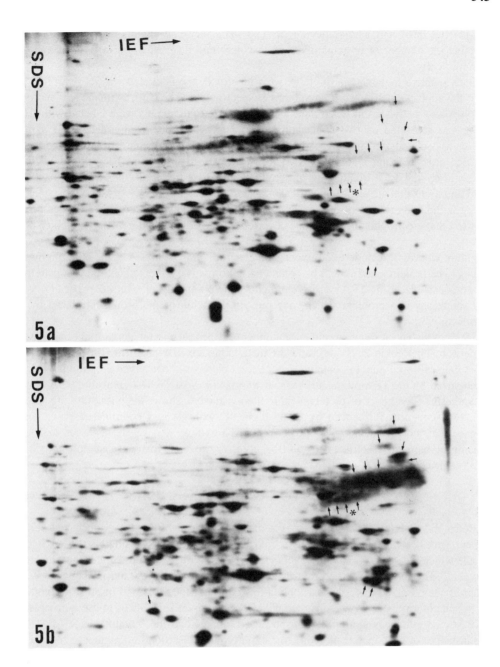

FIGURE 5. Two-dimensional gel electrophoretic fluorogram of ³H-leucine labeled melanocyte proteins during growth in TPA and after the removal of TPA from within the growth medium. Figure 5a shows proteins translated when cells are growing in TPA. Figure 5b shows proteins translated after TPA is removed. Arrows point to TPA-sensitive proteins. Asterisk indicates location of tyrosinase isozymes. (From Oetting, W. S., Langner, K., and Brumbaugh, J. A., *Differentiation*, 30, 40 — 46, 1985. With permission.)

the cells when TPA was removed but absent from cells labeled while still in TPA were said to be TPA-sensitive proteins and considered putative melanogenic proteins.

Figure 5 shows protein patterns of both TPA-inhibited (Figure 5a) and TPA-released (Figure 5b) cells. The asterisk indicates the location of tyrosinase, showing the multiple isozymes of the enzyme. Nine TPA-sensitive groups of proteins were observed, of which one was tyrosinase. The molecular weights ranged from 42,000 to 100,000 Da. These proteins were for the most part acidic, having pIs less than 5.1. This method was also used to determine the number of proteins involved in controlling differentiation in B16/C3 mouse melanoma cells.[97]

Once putative melanogenic proteins were identified, the next step was to determine if any of the spots were affected by mutations at specific loci. The *C* locus mutants, *c* and *c*[a], lack tyrosinase activity, but when melanocytes of these two mutants were labeled with ^{3}H-leucine and separated on two-dimensional gels, a row of spots co-migrated to the same position as isolated tyrosinase.[47] Additionally, anti-chicken tyrosinase mouse serum has identified cross-reacting material (CRM) for these two alleles. It has been proposed that both *c* and *c*[a] are CRM[+] mutants which produce tyrosinase-like molecules that are inactive due to mutations that are electrophoretically and antigenically "silent".[47] Altered enzymatic activity without alterations of electrophoretic mobility are not uncommon.[98-100] It has been estimated that electrophoretic techniques overlook 22 to 75% of the genetic variability per locus.

Future work will include two-dimensional electrophoretic analyses of the other pigment mutants. The protein patterns will be compared to normally pigmented cell protein patterns with emphasis placed on TPA sensitive proteins. The possibility of electrophoretically "silent" mutations is a problem which may prevent detection of the protein produced by a given locus.[101]

TPA has also been shown to induce a number of genes such as the genes for collagenase, stromelysin, interleukin 2, SV 40, and the metallothionein IIA gene (hMT IIA).[102,103] All of these genes were found to contain a common *cis* element, a conserved 9 bp motif, which is recognized by the TPA-modulated *trans*-acting factor AP-1.[102] The control of TPA upon the synthesis of a number of proteins within the melanocyte shows the possibility of shared regulatory sequences influenced by TPA. Pigment synthesis is a coordinated event, and common regulatory sequences would be expected for these genes. The isolation and sequencing of pigment specific genes will be necessary to look for consensus *cis* activator and/or enhancer elements.

VII. INSERTIONAL MUTAGENESIS

Retroviral insertional mutagenesis is a method that can be used to isolate and study pigment genes.[104] The Rous sarcoma retrovirus (RSV) contains a single-stranded RNA genome which is copied into a double-stranded DNA provirus which integrates into the host genome.[105] The insertion of the provirus appears to occur randomly and is stable.[106-108] It has also been observed that, at most, only a few copies of the virus will insert per genome, and frequently, only one copy is inserted.[105,109] These properties allow us to use retroviruses as insertional mutagens similar to transposable elements and the mutator phage Mu-1.[110] Two instances of gene inactivation due to natural proviral insertion into pigment genes have already been reported. Two mouse pigment genes, the dilute gene (*d*,[111]) and lethal yellow (*A*[y],[112,113]) were found to be due to the insertion of a murine leukemia virus (MuLV).

The use of spontaneously occurring insertional events is limited to those that can be found in nature. Utilizing artificially infected cell cultures allows for the screening of a large number of insertional events by looking for unpigmented or hypopigmented cells. When cells are made heterozygous for the desired gene, the probability of detecting an insertional event is greatly enhanced. Cells containing a desired insertion can be cloned and expanded, the genomic DNA isolated, and probed with sequences specific for the infecting virus. Regions of DNA flanking the virus would be part of the interrupted pigment gene. Such studies of pigmentation are possible at present only with chick embryo cells because of well-developed viral and culture methodologies.

A recombinant retroviral vector with the *Neo* gene substituted for the *src* gene has been constructed.[114] Cells heterozygous for both the s^{al} and c^a mutations $(C^+/c^a\ S^+/s^{al})$ were infected and then selected for *Neo* gene expression (drug resistance) using the antibiotic Geneticin® (G418). The double heterozygote increases the probability of observing an insertional event by providing a two-"hit" target. Heterozygous cells normally produce pigment, but if the provirus inserts into the single normal pigment gene at either locus, an amelanotic colony is produced.

From a total of over 60,000 colonies screened, 24 putative amelanotic or hypopigmented melanocyte colonies were selected from infected cultures. Figure 6 shows one such colony where three colonies are observed under phase lighting (Figure 6, top), but only two colonies are seen to be producing pigment when the phase ring is removed (Figure 6, bottom). Only two such colonies were identified by screening approximately 10,000 colonies in similar cultures of uninfected melanocytes. Colonies were ring-cloned, serially diluted, and seeded in multi-well plates. Most of the colonies, including the two derived from the uninfected cells, developed pigment at a later time in culture, but eight clones from the infected cells remained unpigmented and continued to proliferate.

The next step is to determine which of the two genes were mutated by the provirus. Utilizing heterokaryon analysis, the correct gene can be determined by fusing these cells with either of the two genotypes and looking for pigment production. For example, if the insertionally mutated cell is fused with cells of the genotype $c^a/c^a\ S^+/S^+$ and pigment is produced, then the insertion occurred within the S locus.

Once cell lines are cloned and expanded, the mutated gene can be isolated from a genomic library produced from these cells using the *Neo* gene as a probe, which is specific only for the infecting provirus and not endogenous viral sequences. In some instances *de novo* methylation of viral sequences occurs following insertion and interferes with cleavage by methylation sensitive restriction endonucleases. This does not appear to be the case with chicken cells. Chicken cells expressing avian retroviral DNA acquired by exogenous infection allow restriction enzyme cleavage of a number of potentially methylated restriction sites.[108,115] Thus, genomic library construction can take advantage of any single viral restriction site to obtain flanking DNA segments. The isolated pigment genes can then be used to screen cDNA libraries made from melanocyte poly(A) RNAs to isolate the transcription products of that pigment gene.

Sequencing both the cDNA and the genomic gene will provide an understanding of gene structure. Utilizing expression vectors, it will be possible to produce enough of this protein to make antibodies. Immunoelectron miscroscopy will then allow us to determine the location and function of the protein product within the melanocyte.

The isolation of two or more pigment genes will allow a search for specific regulating sequences (promoters, enhancers) for pigment genes. As previously discussed, we have identified a putative regulatory mutant (pinkeye) and a structural mutant (autosomal albino). Utilizing insertional mutagenesis, we are currently working to isolate these and other pigment genes. One advantage of insertional mutagenesis for gene isolation is that regulator genes which produce little or no gene product (RNA or protein) can be isolated. The cloned chicken pigment genes will be used as probes to monitor the coordinate control of gene transcription during differentiation. Using TPA inhibition of pigment production, followed by release, will allow us to study the timing of the expression of these genes. Comparing the actual sequences of normally functioning genes with their mutant alleles will allow the basis for malfunction and variation to be determined.[116]

Retroviruses are not only useful tools for mutagenesis, but will serve as a means of inserting genes into the genome of the chicken to study transgenic expression. Mouse cDNA for tyrosinase was inserted into a replication competent Avain Luekosis Virus proviral plasmid. A viral stock was produced and used to infect albino (c^a) melanocytes in culture. Dark pigment granules were produced in infected cells showing that the mouse tyrosinase

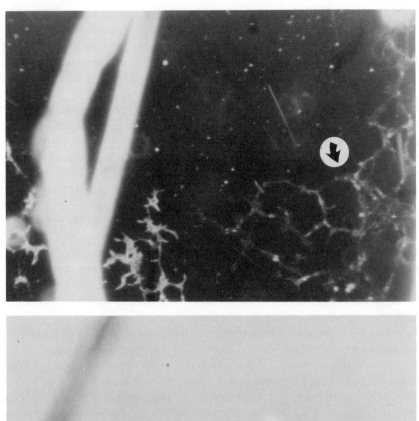

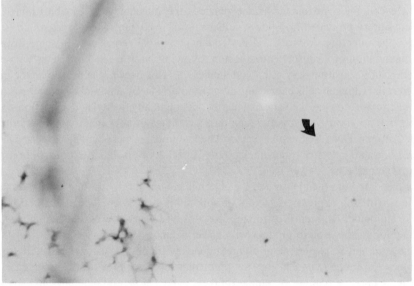

FIGURE 6.   Insertional mutagenesis of double heterozygous cells with a retroviral vector. Cells of the genotype s^{al}/S^+ c^a/C^+ and $s^{al}/-$ c^a/C^+ were infected with a retrovirus containing the *Neo* gene. Infected cells were selected with G418 and subcloned and allowed to grow into colonies. The grid line is the straight white line in the photograph. The bent white line is a portion of the identifying scratch on the bottom of the culture dish. (Top) Phase contrast — three melanocyte colonies are apparent. One to the left of the lines, one to the right, and one identified by the arrow. (Bottom) Same field as above but with the phase ring removed. The two left colonies are still visible because of the pigmentation. The colony identified by the arrow is no longer apparent. It is a "white" colony and therefore a putative insertional mutant at one of the pigment loci. (From Oetting, W. S., Smith, G. L., and Brumbaugh, J. A., in *Advances in Pigment Cell Research,* Bagnara, J., Ed., Alan R. Liss, New York, 1988, 307. With permission.)

"cured" the albino defect in the chick cells.[117] Tyrosinase could be developed as a marker for eukaryotes like β-galactosidase is for prokaryotes. Futher development of such a system will allow a wide range of genes to be inserted into the chicken genome.

VIII. SUMMARY

Pigment synthesis in the fowl consists of two major cellular pathways: the synthesis and transport of tyrosinase through the Golgi system, and the synthesis and assembly of pre-melanosomal matrix. The two components fuse with subsequent melanin synthesis and form a mature pigment granule.

Six mutations all causing partial or complete inhibition of pigment synthesis have been examined. The action of each mutation is autonomous, i.e., the pigment synthesizing machinery of the cells is defective.

A reproducible culture system, which provides pure cultures of melanocytes in moderately large quantities, has been developed. The tumor promoter, TPA, can be used to "switch" pigment synthesis "off" and "on" by its addition and removal, respectively, from the culture medium.

The culture system made it possible to do heterokaryon analyses with the various mutations and compare the heterokaryons with heterozygotes. As a result, two putative regulatory mutations were identified, namely *pk* and *I*. The ability to distinguish regulatory genes from structural genes allows strategies for determining gene functions to be cognitively developed.

Again the culture system, particularly with its TPA "switch", enabled us to identify at least nine proteins involved in pigment synthesis as determined by two-dimensional gel electrophoresis. These studies have not been completed for each genotype, but should reveal genetic differences.

Regulatory genes, by their very nature, produce little or no protein. They cannot be studied with conventional two-dimensional gel electrophoretic techniques. As a result, we have developed an insertional mutagenesis project involving the retrovirus, RSV. Sequences flanking the inserted viral genome should help us identify both regulatory and structural genes in the chick pigment cell system.

IX. APPLICATIONS TO HUMAN ALBINISM

One of the goals of studying avian pigmentation is to help advance the study and care of human albinism. At present, diagnosis of oculocutaneous albinism is based on imprecise clinical features and hairbulb tyrosinase activity. There is no specific care related to the primary defect for each type, largely because the primary defect is unknown. Until the different loci are isolated and linked to specific types of albinism, this will continue to remain the case. The lack of a highly proliferative, well-differentiated human melanocyte culture system has made it difficult to do biochemical and molecular studies directly on patient melanocytes. In the meantime, through the use of homologous avian pigment mutants, different types of albinism can be investigated. Homology of the different pigment loci has played an important role in isolating the human tyrosinase gene using an isolated mouse tyrosinase probe. It is expected that this homology will hold for the other pigment specific proteins as well.

Pigment genes isolated from the fowl, utilizing the different mutants and retroviral insertional mutagenesis, can be used to isolate their homologous counterparts in the human. The isolation of human probes to specific pigment loci will serve as markers for linkage studies with different types of albinism. With intralocus-specific probes available, family studies of restriction fragment length polymorphisms (RFLPs) will provide the best possible linkage data since little or no recombination would occur. This would ensure the most accurate diagnosis of carriers, noncarriers, and affected individuals. Accurate diagnosis is absolutely necessary for sound genetic counseling.

Isolation and sequencing of the avian pigment genes can be studied and compared for controlling sequences. Understanding the control of expression of the different pigment genes will be useful in understanding the control of expression of human pigment genes. Comparison of sequences between normal genes and mutant alleles will allow the basis of mutant function to be determined. This too may be helpful in understanding their analagous counterparts.

As we have shown, pigment production involves the control and integration of a multigenic system. Because homozygotes are viable, basic principles of genetic disease can be elucidated which are not possible with other genetically controlled processes. We hope that the study of avian pigmentation will not only provide information about pigment defects, but that they will also elucidate principles of developmental genetics applicable to other differentiating systems.

ACKNOWLEDGMENTS

We thank Mim Sawtell for her careful preparation of this manuscript. This work was supported in part by grants GM18969 and BRSG #RR07055 from the NIH and by a grant from Li-Cor, Inc., of Lincoln, NE.

REFERENCES

1. **Brumbaugh, J. A. and Oetting, W. S.,** What can we learn from chick embryo melanocytes?, *BioScience,* 35, 381, 1986.
2. **Jimbow, K., Quevedo, W. C., Fitzpatrick, T. B., and Szabo, G.,** Some aspects of melanin biology: 1950 — 1975, *J. Invest. Dermatol.,* 67, 72, 1976.
3. **Brumbaugh, J. A., Wilkins, L. M., and Moore, J. W.,** Genetic dissection of eumelanogenesis, in *Pigment Cell,* Vol 4, Klaus, S. N., Ed., S. Karger, Basel, 1979, 150.
4. **Silvers, W. K.,** *The Coat Colors of Mice,* Springer-Verlag, New York, 1979.
5. **Garrod, A. E.,** Inborn errors of metabolism, *Lancet,* July 4, 1, 1908.
6. **Witkop, C. J., Quevedo, W. C., and Fitzpatrick, T. B.,** Albinism and other disorders of pigment metabolism, in *The Metabolic Basis of Inherited Disease,* 5th ed., Stanbury, J. B., Wyngaarden, J. B., Fredrickson, D. S., Goldstein, J. L. and Brown, J. S., Eds, McGraw-Hill, New York, 1983, 301.
7. **Creel, D., O'Donnell, F. E., and Witkop, C. J.,** Visual system anomolies in human ocular albinos, *Science,* 201, 931, 1978.
8. **Hermansky, F. and Pudlak, P.,** Albinism associated with hemorrhagic diathesis and unusual pigmented reticular cells in the bone marrow: report of two cases with histochemical studies, *Blood,* 14, 162, 1959.
9. **Wiesner, G. L., Bendel, C. M., Olds, D. P., White, J. G., Arthur, D. C., Ball, P. W., and King, R. A.,** Hypopigmentation in the Prader-Willi syndrome, *Am. J. Hum. Genet.,* 40, 431, 1987.
10. **Reed, W. B., Stone, V. M., Boder, E., and Ziprkowski, L.,** Pigmentary disorders in association with congenital deafness, *Arch. Dermatol.,* 95, 176, 1967.
11. **Brumbaugh, J. A., Bowers, R. R., and Chatterjee, G. E.,** Genotype-substrate interaction altering Golgi development during melanogenesis, in *Pigment Cell,* Vol. 1, Riley, V., Ed., S. Karger, Basel, 1972, 47.
12. **Riley, P. A.,** The mechanisms of melanogenesis, *Symp. Zool. Soc. London,* 39, 77, 1977.
13. **Barber, J. I., Townsend, D., Olds, D. P., and King, R. A.,** Dopachrome oxidoreductase: a new enzyme in the pigment pathway, *J. Invest. Dermatol.,* 83, 145, 1984.
14. **Korner, A. and Pawelek, J.,** Dopachrome conversion: a possible control point in melanin biosynthesis, *J. Invest. Dermatol.,* 75, 192, 1980.
15. **Leonard, L., Townsend, D., and King, R. A.,** Dopachrome conversion in the eumelanin pathway, *Biochemistry,* 27, 6156, 1988.
16. **Grahm, D. G., Tiffany, S. M., and Vogel, S.,** The toxicity of melanin precursors, *J. Invest. Dermatol.,* 70, 113, 1978.

17. **Miranda, M., Bonfigli, A., Zarivi, O., Manilla, A., Cimini, A. M., and Arcadi, A.,** Restriction patterns of model DNA treated with 5,6-dihydroxyindole, a potent cytotoxic intermediate of melanin synthesis: effect of UV irradiation, *Mutagenesis,* 2, 45, 1987.

18. **Korner, A. and Pawelek, J.,** Mammalian tyrosinase catalyzes three reactions in the biosynthesis of melanin, *Science,* 217, 1163, 1982.

19. **Maul, G. G.,** Golgi-melanosome relationship in human melanoma *in vitro, J. Ultrastruc. Res.,* 26, 163, 1969.

20. **Maul, G. G. and Brumbaugh, J. A.,** On the possible function of coated vesicles in melanogenesis of the regenerating fowl feather *J. Cell Biol.,* 48, 41, 1971.

21. **Imokawa, G. and Mishima, Y.,** Importance of glycoproteins in the initiation of melanogenesis: an electron microscopic study of B-16 melanoma cells after release from inhibition of glycosylation, *J. Invest. Dermatol.,* 87, 319, 1986.

22. **Hearing, V. J., Korner, A. M., and Pawelek, J.,** New regulators of melanogenesis are associated with purified tyrosinase isozymes, *J. Invest. Dermatol.,* 79, 16, 1982.

23. **Lamoreaus, M. L.,** Dopachrome conversion and dopa oxidase activities in recessive yellow mice, *J. Hered.,* 77, 337, 1986.

24. **Brumbaugh, J. A.,** Ultrastructural differences between forming eumelanin and pheomelanin as revealed by the pink-eye mutation in the fowl, *Dev. Biol.,* 18, 392, 1968.

25. **Zimmerman, J., Winfrey, F., and Good, P.,** Observations on the structure of the eumelanosome matrix in melanosomes of the chick retinal pigment epithelium, *Anat. Rec.,* 200, 415, 1981.

26. **Jimbow, K., Ishida, O., Ito, S., Hori, Y., Witkop, C. J., and King, R. A.,** Combined chemical and electron microscopic studies of pheomelanosomes in human red hair, *J. Invest. Dermatol.,* 81, 506, 1983.

27. **Seji, J., Shimao, K., Birbeck, M. S. C., and Fitzpatrick, T. B.,** Subcellular localization of melanin biosynthesis, *Ann. N.Y. Acad, Sci.,* 100, 497, 1963.

28. **Menon, I. A. and Haberman, H. F.,** Formation of melanin-tyrosinase complex and its possible significance as a model for control of melanin synthesis, *Acta Dermatol.,* 58, 9, 1978.

29. **Bateson, W.,** The present state of knowledge of colour-heredity in mice and rats, *Proc. Zool. Soc. London,* 2, 71, 1903.

30. **Castle, W. E. and Allen, G. M.,** The heredity of albinism, *Proc. Am. Acad. Arts. Sci.,* 38, 603, 1903.

31. **Cuénot, L.,** La loi de Mendel et l'Hérédité de la pigmentation chez les souris, *Arch. Zool. Exp. Gén.,* 10, 27, 1902.

32. **Bateson, W. and Saunders, E. R.,** Part II, Poultry, *Rep. Evol. Committee R. Soc.,* 1, 87, 1902.

33. **Bowers, R. R.,** The melanocyte of the chicken: a review, in *Advances in Pigment Cell Research,* Bagnara, J., Ed., Alan R. Liss, New York, 1988, 49.

34. **Dorris, F.,** The production of pigment by chick neural crest in grafts to the 3-day limb bud, *J. Exp. Zool.,* 80, 315, 1939.

35. **Brumbaugh, J. A. and Hollander, W. F.,** A further study of the *E* pattern locus in the fowl, *Iowa State J. Sci.,* 40, 51, 1965.

36. **Brumbaugh, J. A.,** Differentiation of black-red melanin in the fowl: interaction of pattern genes and feather follicle mileu, *J. Exp. Zool.,* 166, 11, 1967.

37. **Brumbaugh, J. A. and Moore, J. W.,** The effects of the *E* alleles upon melanocyte differentiation in the fowl, *Can. J. Genet. Cytol.,* 11, 118, 1969.

38. **Brumbaugh, J. A.,** The ultrastructural effects of the *I* and *S* loci upon black-red melanin differentiation in the fowl, *Dev. Biol.,* 24, 392, 1971.

39. **Hamilton, H.,** A study of the physiological properties of melanophores with special reference to their roles in feather coloration, *Anat. Rec.,* 78, 525, 1940.

40. **Brumbaugh, J. A. and Lee, K. W.,** The gene action and function of two DOPA oxidase positive melanocyte mutants of the fowl, *Genetics,* 81, 333, 1975.

41. **Smith, L. T.,** On the allelism of blue plumage and dominant white, *Poult. Sci.,* 48, 346, 1969.

42. **Ziehl, M. A. and Hollander, W. F.,** Dun, a new plumage color mutant at the *I* locus in the fowl, *Iowa State J. Sci.,* 62, 337, 1987.

43. **Brumbaugh, J. A. and Lee, K. W.,** Types of genetic mechanisms controlling melanogenesis in the fowl, in *Pigment Cell,* Vol. 3, Riley, V., Ed., S. Karger, Basel, 1976, 165.

44. **Brumbaugh, J. A.,** unpublished data.

45. **Brumbaugh, J. A., Barger, T. W., and Oetting, W. S.,** A "new" allele at the *C* pigment locus in the fowl, *J. Hered.,* 74, 331, 1983.

46. **Smyth, J. R., Ring, N. M., and Brumbaugh, J. A.,** A fourth allele at the *C*-locus of the chicken, *Poult. Sci.,* 65(Suppl. 1), 129, 1986.

47. **Oetting, W. S., Churilla, A. M., Yamamoto, H., and Brumbaugh, J. A.,** *C* pigment locus mutants of the fowl produce enzymatically inactive tyrosinase-like molecules, *J. Exp. Zool.,* 235, 237, 1985.

48. **Yamamoto, H., Takeuchi, S., Kudo, T., Makino, K., Nakata, A., Shinoda, T., Takeuchi, T.,** Cloning and sequencing of mouse tyrosinase cDNA, *Jpn. J. Genet.,* 62:271, 1987.

49. **Kwon, B. S., Haq, A. K., Pomerantz, S. H., and Halaban, R.,** Isolation and sequence of a cDNA clone for human tyrosinase that maps at the mouse c-albino locus, *Proc. Natl. Acad. Sci. U.S.A.*, 84, 7473, 1987.

50. **Spritz, R. A., Strunk, K., Oetting, W. S., and King, R. A.,** unpublished data.

51. **Fox, W. and Smyth, J. R.,** The effects of recessive white and dominant white genotypes on early growth rate, *Poult. Sci.*, 64, 429, 1985.

52. **Pardue, S. L. and Smyth, J. R.,** Influence of *C* locus alleles on neonatal development in the domestic fowl, *Poult. Sci.*, 65, 426, 1986.

53. **Yanz, J. L., Herr, L. R., Townsend, D., and Witkop, C. J.,** The questionable relationship between cochlear pigmentation and noise-induced hearing loss, *Audiology*, 24, 260, 1985.

54. **Brumbaugh, J. A., Chatterjee, G., and Hollander, W. F.,** Adendritic melanocytes: a mutation in linkage group II of the fowl, *J. Hered.*, 63, 19, 1972.

55. **Mayerson, P. L. and Brumbaugh, J. A.,** Lavender, a chick melanocyte mutant with defective melanosome translocation: a possible role for 10 nm filaments and microfilaments but not microtubules, *J. Cell Sci.*, 51, 25, 1981.

56. **King, R. A., Creel, D., Cervenka, J., Okoro, A. W., and Witkop, C. J.,** Albinism in Nigeria with delineation of new recessive oculocutaneous type, *Clin. Genet.*, 17, 259, 1980.

57. **King, R. A., Lewis, R. A., Townsend, D., Zelickson, A., Olds, D. P., and Brumbaugh, J.,** Brown oculocutaneous albinism: clinical, opthalmological, and biochemical characterization, *Opthalmology*, 92, 1496, 1985.

58. **Frenk, E. and Calme, A.,** Hypopigmentation oculocutanée familiale à transmission dominante due à un trouble de la formation des mélanosomes, *Schweiz. Med. Wochenschr.*, 107, 1964, 1977.

59. **Witkop, C. J.,** Depigmentations of the general and oral tissues and their genetic foundations, *Ala. J. Med. Sci.*, 16, 331, 1979.

60. **Pearson, K., Nettleship, E., and Usher, C. H.,** A monograph on albinism in man, *Drapers Company Research Memoirs*, Biometric Series VIII, Dulav and Company, London, 1913, 329.

61. **King, R. A. and Lewis, R. A.,** Albinism, oculocutaneous, rufus type, in *Birth Defects Encyclopedia*, Muyse, M. L., Ed., Center for Birth Defects Information Services, Inc., Dover, in press.

62. **Giss, B., Antoniou, J., Smith, G., and Brumbaugh, J. A.,** A method for culturing chick melanocytes: the effect of BRL-3A cell conditioning and related additives, *In Vitro*, 18, 817, 1982.

63. **Oetting, W. S., Langner, K., and Brumbaugh, J. A.,** Detection of melanogenic proteins in cultured chick-embryo melanocytes, *Differentiation*, 30, 40, 1985.

64. **Bowers, R. R. and Gatlin, J. E.,** A simple method for the establishment of tissue culture melanocytes from regenerating fowl feathers, *In Vitro Cell Dev. Biol.*, 21, 39, 1985.

65. **Boissy, R. E. and Halaban, R.,** Establishment of proliferative, pure cultures of pigmented chicken melanocytes from neural tubes, *J. Invest. Dermatol.*, 84, 158, 1985.

66. **Hamburger, V. and Hamilton, H.,** A series of normal stages in the development of the chick embryo, *J. Morphol.*, 88, 49, 1951.

67. **Coon, H. G.,** Clonal culture of differentiated cells from mammals: rat liver cell culture, *Carnegie Inst. Washington Yearb.*, 67, 419, 1968.

68. **Straus, D. S., Coppock, D. L., and Pang, K. J.,** Low molecular weight mitogenic factor produced by BRL-3A cultured rat liver cells, *Biochem. Biophys. Res. Commun.*, 100, 1619, 1981.

69. **Massague, J., Kelly, B., and Mottola, C.,** Stimulation by insulin-like growth factors is required for cellular transformation by type B transforming growth factor, *J. Biol. Chem.*, 260, 4551, 1985.

70. **Lowe, M. E., Pacifici, M., and Holtzer, H.,** Effects of phorbol-12-myristate-13-acetate on the phenotype program of cultured chondroblasts and fibroblasts, *Cancer Res.*, 38, 2350, 1978.

71. **Cohen, R., Pacifici, M., Rubinstein, N., Biehl, J., and Holtzer, H.,** Effect of a tumor promoter on myogenesis, *Nature (London)*, 266, 538, 1977.

72. **Cossu, G., Pacifici, M., Adamo, S., Bouche, M., and Molinaro, M.,** TPA-induced inhibition of the expression of differentiative traits in cultured myotubes: dependence on protein synthesis, *Differentiation*, 21, 62, 1982.

73. **Croop, J., Toyama, Y., Dlugosz, A. A., and Holtzer, H.,** Selective effects of phorbol 12-myristate 13-acetate on myofibrils and 10-nm filaments, *Proc. Natl. Acad. Sci. U.S.A.*, 77, 5273, 1980.

74. **Croop, J. Dubyak, G., Toyama, Y., Dlugosz, A., Scarpa, A., and Holtzer, H.,** Effects of 12-O-tetradecanoyl-phorbol-13-acetate on myofibril integrity and Ca^{2+} content in developing myotubes, *Dev. Biol.*, 89, 460, 1982.

75. **Gimlius, B. and Weston, J. A.,** Analysis of developmentally homogeneous neural crest cell populations in vitro. II. A tumor-promoter (TPA) delays differentiation and promotes cell proliferation, *Dev. Biol.*, 82, 95, 1981.

76. **Sieber-Blum, M. and Sieber, F.,** Tumor-promoting phorbol esters promote melanogenesis and prevent expression of the adrenergic phenotype in quail neural crest cells, *Differentiation*, 20, 117, 1981.

77. **Langner, K., Oetting, W. S., Osborne, J., and Brumbaugh, J.,** A method for culturing chick embryo melanocytes which controls the gene products involved in melanogenesis, *Yale J. Biol. Med.,* 57, 384, 1984.

78. **Payette, R. Biehl, J., Toyama, Y., Holtzer, S., and Holtzer, H.,** Effects of 12-O-tetradecanoylphorbol-13-acetate on the differentiation of avian melanocytes, *Cancer Res.,* 40, 2465, 1980.

79. **Brumbaugh, J. A., Oetting, W. S., Langner, K. E., and Churilla, A. M.,** Molecular approaches to melanocyte differentiation: results and perspectives, in *Cutaneous Melanoma,* Veronesi, U., Cascinelli, N., and Santinami, M., Eds., Academic Press, New York, 1987, 145.

80. **Wilkins. L. M. and Brumbaugh, J. A.,** Complementation and noncomplementation in heterokaryons of three unlinked pigment mutants of the fowl, *Somatic. Cell Genet.,* 5, 427, 1979.

81. **Brumbaugh, J., Wilkins, L., and Schall, D.,** Intergenic complementation in chick melanocyte heterokaryons, *Exp. Cell Res.,* 111, 333, 1978.

82. **Wilkins, L. M., Brumbaugh, J. A., and Moore, J. W.,** Heterokaryon analysis of the genetic control of pigment synthesis in chick embryo melanocytes, *Genetics,* 102, 557, 1982.

83. **Antonious, J. and Brumbaugh, J.,** A regulator locus controlling melanocyte differentiation identified by somatic cell heterokaryon complementation analysis, *J. Cell Biol.,* 95, 449a, 1982.

84. **Bootsma, D. and Galjaard, H.,** Heterogeneity in genetic diseases studied in cultured cells, in *Models for the Study of Inborn Errors of Metabolism,* Hommes, F., Ed., Elsevier/North Holland Biomedical Press, Amsterdam, 1979, 74.

85. **Davidson, R. L. and Gerald, P. S.,** Induction of mammalian somatic cell hybridization by polyethylene glycol, in *Methods in Cell Biology,* Vol. 15, Prescott, D. M., Ed., Academic Press, New York, 1977, 325.

86. **Schwartz, M. S. and Brumbaugh, J. A.,** Extinction of cytochemical dopa oxidase activity in mouse melanoma X chick embryo fibroblast heterokaryons: a single cell analysis, in *Pigment Cell,* Seiji, M., Ed., University of Tokyo Press, Tokyo, 1981, 255.

87. **Schwartz, M. S. and Brumbaugh, J. A.,** Dopa oxidase expression in fibroblast X melanoma fusions: extinction in heterokaryons and maintenance in non-dividing cybeids, *Exp. Cell Res.,* 142, 155, 1982.

88. **Schall, D. G. and Brumbaugh, J. A.,** Restoration of pigment synthesis in mutant melanocytes after fusion with chick embryo erythrocytes, in *Cell Culture and its Applications,* Acton, R. T. and Lynn, J. D., Eds., Academic Press, New York, 1977, 491.

89. **Doezema, P.,** Tyrosinase from embryonic chick retinal pigment epithelium, *Comp. Biochem. Physiol.,* 46B, 509, 1973.

90. **Yamamoto, H. and Brumbaugh, J. A.,** Purification and isoelectric heterogeneity of chicken tyrosinase, *Biochim. Biophys. Acta,* 800, 282, 1984.

91. **Hearing, V. J., Nicholson, J. M., Montague, P. M., Ekel, T. M., and Tomecki, K. K.,** Mammalian tyrosinase: structural and functional interrelationship of isozymes, *Biochim. Biophys. Acta,* 522, 327, 1978.

92. **Doezema, P.,** Proteins from melanosomes of mouse and chick pigment cells, *J. Cell Physiol.,* 82, 65, 1973.

93. **Hearing, V. J. and Lutzner, M. A.,** Mammalian melanosomal proteins: characterization by polyacrylamide gel electrophoresis. *Yale J. Biol. Med.,* 46, 553, 1973.

94. **Zimmerman, J.,** Four new proteins of the eumelanosome matrix of the chick pigment epithelium, *J. Exp. Zool.,* 219, 1, 1982.

95. **Borovansky, J., Hach, P., and Duchov, J.,** Melanosome: an unusually resistant subcellular particle, *Cell Biol, Int. Rep.,* 1, 549, 1977.

96. **Seji, M.,** Melanogenesis, in *Ultrastructure of Normal and Abnormal Skin,* Zelickson, A. S., Ed., Lea and Febiger, Philadelphia, 1967, 183.

97. **Laskin, J. D., Piccinini, L., Engelhardt, D. L., and Weinstein, I. B.,** Specific protein production during melanogenesis in B16/C3 melanoma cells, *J. Cell. Physiol.,* 114, 68, 1983.

98. **Mohrenweiser, H. W. and Neel, J. V.,** Frequency of thermostability variants. Estimation of total rare variant frequency in human populations, *Proc. Natl. Acad. Sci. U.S.A.,* 78, 5729, 1981.

99. **Hamilton, H. B.,** Genetics and the atomic bombs in Hiroshima and Nagasaki, *Am. J. Med. Genet.,* 20, 541, 1985.

100. **Kelley, M. R., Mims, I. P., Farnet, C. M., Dicharry, S. A., and Lee W. R.,** Molecular analysis of x-ray-induced alcohol dehydrogenase *(Adh)* null mutations in *Drosophila melanogaster, Genetics,* 109, 365, 1985.

101. **Klose, J.,** Isoelectric focusing and electrophoresis combined as a method for defining new point mutations in the mouse, *Genetics,* 92, s13, 1979.

102. **Angel, P., Imagawa, M., Chiu, R., Stein, B., Imbra, R. J., Rahnsdorf, H. J., Jonat, C., Herrlich, P., and Karin, M.,** Phorbol ester-inducible genes contain a common *cis* element recognized by a TPA-modulated *trans*-acting factor, *Cell,* 49, 729, 1987.

103. **Lee, W., Mitchell, P., and Tjian, R.,** Purified transcription factor AP-1 interacts with TPA-inducible enhancer elements, *Cell,* 49, 741, 1987.

104. **Oetting, W. S., Smith, G. L., and Brumbaugh, J. A.,** Isolation of pigment genes using retroviral insertional mutagenesis, in *Advances in Pigment Cell Research,* Bagnara, J., Ed., Alan R. Liss, New York, 1988, 307.

105. **Bishop, J. M.,** Retroviruses, *Annu. Rev. Biochem.,* 47, 35, 1978.

106. **Varmus, H. E., Quintrell, N., and Ortiz, S.,** Retroviruses as mutagens: insertion and excision of a nontransforming provirus alter expression of a resident transforming provirus, *Cell,* 25, 23, 1981.

107. **Weinberg, R. A.,** Integrated genomes of animal viruses, *Annu. Rev. Biochem.,* 49, 197, 1980.

108. **Humphries, E. H., Allen, R., and Glover, C.,** Clonal analysis of the integration and expression of endogenous avian retro viral DNA acquired by exogenous viral infection, *J. Virol.,* 39, 584, 1981.

109. **Frankel, W., Potter, T. A., Rosenberg, N. Lenz, J., and Rajan, T. V.,** Retroviral insertional mutagenesis of a target allele in a heterozygous murine cell line, *Proc. Natl. Acad. Sci. U.S.A.,* 82, 6600, 1985.

110. **Bukhari, A. I., Shapiro, J. A., and Adhaya, S. L., Eds.,** *DNA Insertion Elements, Plasmids and Episomes,* Cold Spring Harbor Laboratory, Cold Spring Harbor, NY, 1977.

111. **Jenkins, N. A., Copeland, N. G., Taylor, B. A., and Lee, B. K.,** Dilute *(d)* coat colour mutation of DBA/2J mice is associated with the site of integration of anectotropic MuLV genome, *Nature (London),* 293, 370, 1981.

112. **Siracusa, L. D., Russell, L. B., Jenkins, N. A., and Copeland, N. G.,** Allelic variation within the *Emv-15* locus defines genomic sequences closely linked to the agouti locus on mouse chromosome 2, *Genetics,* 117, 85, 1987.

113. **Siracusa, L. D., Russell, L. B., Eicher, E. M., Corrow, D. J., Copeland, N. G., and Jenkins, N. A.,** Genetic organization of the *agouti* region of the mouse, *Genetics,* 117, 93, 1987.

114. **Hughes, S. H., Greenhouse, J. J., Petropoulos, C. J., and Sutrave, P.,** Adapter plasmids simplify the insertion of foreign DNA into helper-independent retroviral vectors, *J. Virol.,* 61, 3004, 1987.

115. **Hwang, L.-H. S. and Gilboa, E.,** Expression of genes introduced into cells by retroviral infection is more efficient than that of genes introduced into cells by DNA transfection, *J. Virol.,* 50, 417, 1984.

116. **Antonarakis, S. E., Irkin, S. H., Cheng, T. C., Scott, A. F., Sexton, J. P., Truska, S., Charache, S., and Kazazian, H. H.,** β-Thalassemia in American blacks: novel mutations in the "TATA" box and an acceptor splice site, *Proc. Natl. Acad. Sci. U.S.A.,* 81, 1154, 1984.

117. **Whitaker, B., Frew, T., Greenhouse, J., Hughes, S., Yamamoto, H., Takeuchi, T., and Brumbaugh, J.,** Expression of virally-transduced mouse tyrosinase in tyrosinase-negative chick embryo melanocytes in culture, *Proc. Pan Am Soc. Pigment Cell Res.,* in press.

Chapter 21

A MOLECULAR, EVOLUTIONARY PERSPECTIVE OF NONMAMMALIAN MODELS IN BIOLOGICAL RESEARCH

Mark W. Bitensky

TABLE OF CONTENTS

I. INTRODUCTION

At the outset, I would like to identify the intent of this writing, which is to convey a certain perspective about terrestrial biology. This perspective, while not provable with syllogistic rigor, is supported by experiment and example. In essence, I wish here to focus my attention upon an evolutionary point of view of the molecular biology of gene product optimization. Specifically, I shall focus on the idea that, working in intimate collaboration, the extreme plasticity of DNA, the potential for survival in emerging organisms, and the earth's environments which shape and nurture them participate in an ancient and a perennial process of progeny selection according to the recursive and exacting algorithms of Charles Darwin and molecular biology. This endless process of self-modification is bent upon optimization of the fit between organism and environmental niche. The recursive algorithms have been running (on our planet) for some 4 billion years and have produced an extraordinary array of living forms, some modest fraction of which are hinted at in the fossil record. I would especially emphasize, in the context of this volume of writings on nonmammalian models, that it is specifically the style and methods employed by the mammoth engine of evolution which make the utilization of such models so valuable in the biological sciences.

II. MURPHY'S LAW IN THE POSITIVE VOICE

In scrutinizing the splendid collection of biota which have tumbled from this evolutionary womb, it is possible to discern a pattern which can be expressed as a positive reformulation of Murphy's law: Anything that can possibly be made better will. I mean here to convey the remarkable facility that evolution has demonstrated in ferreting out the very best of all possible gene product optimizations. Implicit in my discourse is the fundamental assumption that the evolutionary process of species optimization (on a macroscale) is ultimately a derivative of gene product optimization (on a microscale). For example, the vertebrate rod is a true quantum detector. When fully dark adapted, it can mount a light response to a single photon.[1] Similar demonstrations of sensitivity optimization are encountered in the acoustical, olfactory, and chemodetection systems of a variety of mammals and insects.

III. GENE PRODUCT OPTIMIZATION

What in fact takes place in the process of gene product optimization over eons of evolutionary time? What sorts of features enjoy the benefits of exon shuffling[2] and other (more gradual) modes of peptide evolution? Parameters such as cooperativity among the proteins gathered into a regulatory ensemble, enhanced catalytic efficiency within the confines of a regulatory algorithm, and subcellular concentration within a functional domain — all can enhance the speed and efficiency of gene product function.

Thus, the collaboration of DNA plasticity, environmental vectors, and the existing infrastructure of gene-product form and function (which determine survivability) conspires to explore effectively a rather startling number of amino acid sequence combinations and permutations for their utility within the context of those functional algorithms which can materially influence the survival and population stability of a species.

Constrained only by the limits of the thermodynamic and combinatorial rules of physics and chemistry, evolution carries out its explorations primarily in the idiom of carbon, oxygen, nitrogen, and hydrogen (with help from sulfur, phosphorous, and a variety of metal ions).

While there is not yet a precise understanding of rates, it would appear that the process of evolutionary optimization will eventually and unerringly generate, within the confines of the harsh survival algorithms of Darwin and the ancient laws which govern the exploits of matter and energy, those clusters and ensembles of interacting gene products which most effectively accomplish the purpose at hand, whether it is sensory transduction, muscle contraction, or bioluminescence.

This relentless process of change and optimization is always aimed at a more perfect union between organism and environment. This is the first and last imperative. Aesthetics and theory have little importance here. What is operating is a pragmatic Darwinian empirical rule of competition and survival. If the ability to detect one photon or a single molecule of a pheromone confers some advantage, then no price is too dear for such quantum sensitivity, that is, no degree of modification of existing gene product structure or function can be considered excessive in meeting this goal. Great Nature, as she plays out her evolutionary epic, behaves with an obsessive zeal which would have astounded even Freud.

IV. A CONSERVATIVE AND PARSIMONIOUS HABIT

There is a second characteristic of this evolutionary saga that requires mention here. It concerns the fundamentally conservative nature of evolution's experimentations. While human social and aesthetic excursions are often tumultuous and disrupted by sudden changes in direction and values, evolution has a robust passion for continuity and for recycling those designs and motifs which have demonstrated utility in the past. Moreover, it seems invariably constrained to achieve improved performance by modifying existing components.

The reason for this seems apparent. It is far easier to modify an existing gene product to fit a new task than to start from scratch and build a new gene product from first principles: evolution is an ingenious and a gradual modifier — a tinkerer rather than a sudden inventor or creator of originals. Evolution can, of course, be most creative when, in the process of exon shuffling, it produces a totally novel arrangement of exons which may never before have resided in the same protein. While such a novel synthesis, in fact, provides totally new possibilities for gene product structure and function, it derives from, and consists of, functional elements which have been gradually introduced over very long periods.

Certainly the novel combination of previously unaffiliated exons may itself produce a necessity for refinements and adjustments in newly married, previously unacquainted exons. Such revisions would be directed at improving the compatibility of the newly joined exons to interact effectively in the context of the neophyte protein. The new gene product must adhere to the still unelucidated rules of protein folding[3] as it evolves toward a stable conformation which can retain the fused functions in the needed combination. Certainly novel combinations of exons which contained significant structural or conformational incongruities would probably be discarded even before there was sufficient time for attempts at revision.

The process of exon shuffling and the exploration of potential protein functions (vis-à-vis new combinatorial or mutational modifications) are analogous to tinkering with novel combinations of a large array of "off-the-shelf" solid-state components. My positive reformulation of Murphy's law is intended to capture the idea that if there is an as yet untried combination of native or revised exons which will do an old job better or do a completely new job, evolution will eventually try that combination and subsequently retain it because of the advantages conferred. If anything can possibly go right, it will, if given time enough!

I do not, by this statement, intend to suggest some mystical prescience on the part of the recursive algorithms of evolution. Rather, I wish to marvel at the efficiency of the evolutionary process in effectively exploring the functional potential lurking within the latent permutations and combinations of exons in the genome of any particular organism.

It would appear that this experimentation or tinkering with off-the-shelf components is always going on. The probability of achieving a useful new result should gradually diminish in time as the age of a particular stable environmental niche increases. This follows directly because, with the increasing age of a given stable environment (be it cave or stream, forest canopy or mountaintop), there has been time enough to permit more and more of the potentially useful gene product innovations to have been tried. Moreover, with sudden environmental shifts, very large numbers of new gene product possibilities become relevant to the changed conditions, and thus the process of refitting the organism into the changed or changing environment becomes witness to an increased pace of successful gene product innovation (see later discussion).

We may discern yet another feature within the conservative demeanor of evolution. This ancient process is, without question, the very first bona fide example of a no-frills flight. Evolution will not allow excess baggage, especially in the phenotypic compartment. A gene product which does not provide essential goods or services cannot long endure. One might venture the observation, in Darwinian idiom, that the production of unutilized products is expensive in energy and material, a profligacy which is antithetical to survival. Moreover, when two gene products exhibit a similar capacity to address a need or solve a problem, invariably the more parsimonious solution will prevail.

In addition to the conservative and parsimonious flavor of the collected works of Great Nature as seen through the unfoldings of the evolutionary epic, there is another idiosyncratic mode of procedure. This can only be characterized as a blatant opportunism which becomes discernible now and again, peeking out from behind evolution's impeccably conservative façade: again in the idiom of Murphy, anything that might do will be tried. For example, that venerable storage polymer of glucose — the glycogen macromolecule — is utilized in the role of lens material in some insect eyes!

V. INFORMATION SHARING

I would also introduce here a hypothesis concerning the scope of the availability of exons or even exon fragments as informational resources or components for use by the recursive algorithms of evolution. Although this hypothesis contains ambiguities in its temporal dimension, it is nevertheless both useful and relevant in the context of this writing. There is even some modest evidence in its favor preserved in the current genomic record. (The question is not whether the suggested phenomenon occurs at all, but rather how often and over what genetic distances.) Thus far the principal theme of these remarks is, simply stated, that evolutionary and environmental forces are remarkably adept at finding and/or optimizing the appropriate exons and exon combinations for the purpose of fine-tuning any organism into its environmental niche. Implicit in the way in which this concept is usually formulated is the assumption that the creative magic of evolution is worked upon the genome of the evolving organism. That is to say, the repertoire of chromosome segments available for recombining, the exons available for shuffling, and the choices of gene products available for single base modification are, in the ordinary sense, those that the organism calls its own. By and large, this view is valid with regard to any brief time span within the evolutionary history of a particular organism. However, there is a radical qualifier of this underlying assumption which deserves a hearing. I would submit that a more accurate perspective holds that the menu of components available in the perpetual honing of the adaptability and functionality of a genome to a particular environment is not restricted merely to the known DNA, which can be identified within the genome of the evolving species. I would instead suggest that the process has had and will have access to all of the DNA on the planet earth. This is a rather implausible contention which cannot have merit in the near term nor certainly for periods of months, years, and perhaps even centuries. The idea in the short-term is simply unsupportable. If one extends the period of consideration to major fractions of evolutionary time, however, one begins to perceive the fascinating possibility that if a giraffe has something that a whale really "needs", or vice versa, there is a distinct possibility that the required genetic information will become available across species or possibly even phyla. Of course, there are known mechanisms, such as viral infection, which have the capability of transferring chunks of DNA between species. Certainly the efficiency and the rapidity of such transfers become more modest as the genetic distance between donor and recipient increases. Yet when one contemplates very long periods of evolutionary time, one encounters the very solid possibility that an oganism which does not yet have a particular type of macromolecule and which would be profoundly benefited by its presence (or the presence of an ensemble of macromolecules) may, sooner or later, be the recipient of the needed information.

There are clearly genetic and other biological barriers to moving fragmentary sequences and/or entire exons across species, not only from giraffe to whale, but from whale to *Escherichia coli*. I speak not only of the obvious barriers to interbreeding but also of barriers with respect to the capacity of a putative vector to infect two unrelated hosts. Nevertheless, it is possible to envision gradients of compatibility, so that, in time, information can move slowly (perhaps even across all species and phyla) and particularly useful gene products are available in a very broad way. I would also point out that it is seldom easy to determine to what extent convergence is also operating here, as compared with lateral spread of information across distant species. It might eventually prove very interesting and perhaps surprising were it possible to develop some objective and accurate measure for the rates of interspecies gene product information transfer.

VI. EXTINCTIONS

This discussion brings us face to face with questions concerning rates of evolution. Certainly our conventional thought picture of evolution is that of a process which is terribly slow and which, in effect, is in harmony with what we call geological time intervals: that is, the time intervals required for the formation and destruction of continents. There is, however, more recent evidence that a large part of this slowness can be explained by the very success of evolution! In this view, the slowness results from the capacity of evolution to fill every nook and cranny on the surface of the planet efficiently and ingeniously, from the polar wastes to the steaming jungles of the equator, with an enormous variety of well-adapted life forms.

Thus, the three vectors of evolution (species survivability, terrestrial niches, and plasticity of DNA) have collaborated to fill (and are always collaborating to maintain in a filled condition) all terrestrial econiches. The apparent slowness of evolution may be nothing more than a reflection of the fact that "all seats are occupied". Before one can elaborate a new biological form, one must make available an unutilized or underutilized environmental space within which it can dwell. This is perhaps most dramatically illustrated by evidence from the fossil record which would indicate that following massive extinctions (perhaps associated with meteoric impacts or other unidentified cataclysms) the rates at which emptied econiches are refilled is prodigious and nowhere near as slow as our ordinary perception of the rate of evolution at equilibrium.[4] When econiches are filled, the evolutionary process is poised with enormous potential resident (as survival information) in the vast encyclopedia of terrestrial DNA. Evolution and its collaborators are waiting in the wings (with their myriads of untried exon combinations) to work their craft wherever and whenever new opportunities for niche fitting become available.

There is a molecular analogy to the blocking function performed by a species which occupies a particular ecological niche. An especially successful gene product can block the development of competitive gene products when it fills a particular functional niche with exceptional grace. For example, the photopigment family of rhodopsins and the related cone pigments[5] are encountered universally, even in organisms such as Chlamydomonas.[6] It would appear that the integral protein opsin, in collaboration with vitamin A aldehyde (as the 11-cis form of retinal), is so effective in serving the function of informing an organism about ambient photon fluxes that the appearance of competitive molecular forms simply has not occurred. This is to say that once the die is cast, that is, once a molecular niche is very well filled, it is likely that future uses for that type of molecule, for example, a photon-capturing application, will not be filled by initiating a totally new process of gene development. It is far more likely that the need will be filled by serial modification of one or more members of the existing family of rhodopsin photopigments. The efficiency and the parsimony in such a practice are apparent when one considers the cost of going back to the drawing board and the rather meager probability that something entirely different and as effective will emerge. This perspective argues for continual modification and adaptation to different and perhaps more subtle or specialized uses of an existing polypeptide as is the case for the human cone pigments which subserve the function of color vision.[5]

VII. THE INEVITABLE UTILITY OF NONMAMMALIAN MODELS

I would submit that biologists are entitled to have a certain confidence in their study of nonmammalian model systems. They are permitted confidence in the knowledge that the particular species (amphibian, reptile, or other) that is studied, has been optimized profoundly to its own environmental niche. Nevertheless, the themes that subserve major functions, whether they be muscle contraction, photon capture, or the clotting of blood, will be echoed and mirrored between human beings and nonmammalian models in ways that will not only help to elucidate the biochemistry and physiology of the human system but will also provide a tutorial on evolutionary techniques of optimization to specialized environmental domains. In the final analysis, the biologist can have genuine confidence that fundamental molecular themes of gene product function will be preserved, although varied somewhat, in order to adapt more perfectly to the lifestyle and needs of the model system. One should emphasize also that this extraordinarily conservative flavor of evolution applies not only to a particular gene product and its function but also to ensembles of gene products and their combined functions, as will be illustrated in my discussion of the GTP-binding protein algorithm of signal transduction.

This teleological or purposive view of evolution recommends an alternative approach to the shaping of biological experimentation. An investigator performs a gedanken experiment very much in the way that Einstein envisioned aspects of general and special relativity. In this approach, the investigator attempts to predict optimization features for a system with which he is already very familiar. Clearly the success of such an exercise will depend very largely on how extensively the infrastructure of the data is known and how well it provides a conceptual context that is grounded in reality. I am not proposing the creation of fantasy biota. Rather, if one has in-depth familiarity with the functions and components of a particular organelle or an enzyme cascade, one can begin to consider the putative advantages of having another layer of regulation or a particular topology or organizational scheme which would contribute to the more efficient functioning of that system. This approach becomes more powerful if one is able to demonstrate quantitatively an advantage (or an enhanced functionality) with respect to the postulated feature. It is, then, essential to devise an experiment that would straightaway demonstrate whether the feature predicted by the optimization exercise can be detected in the system under study. This form of conceptual boot-strapping seems justified by the vast and astonishing record of optimizations achieved by the evolutionary process.

VIII. THE METABOLIC CODE

Shortly before his death, Tomkins published a provocative paper in *Science* which spoke of a universal metabolic code.[7] The insights provided by the paper were compelling. In effect, Tomkins noted that 3′, 5′-cyclic AMP could be found in an extraordinary range of organisms: literally from *Escherichia coli* to the whale. It seemed to be a molecule almost ubiquitous and as fundamental as ATP. More significantly, Tomkins noted that the evolutionary process had conserved an abstract or symbolic meaning embedded within the function of this cyclic nucleotide. Specifically, he observed that in *E. coli* and *Dictyostelium discoidium,* as well as in vertebrates including reptiles and man, cyclic AMP was the molecular embodiment of hunger or glucose deprivation. Organisms as diverse as the coliform bacteria and the whale celebrated the status of hunger or caloric deprivation by elaborating a cyclic AMP signal. In coliform organisms, cyclic AMP was not only a herald of hunger but also a metabolic signal used to elaborate and deploy enzymatic tools which would allow the bacterium to utilize other sources of energy, such as fucose or mannose, in place of glucose.

The symbolic molecules (in this instance exemplified by cyclic nucleotides) could also elicit specialized functions within the cells where they were synthesized. Thus, while cyclic AMP (with remarkable consistency) symbolized the absence of glucose in many classes of single-celled and multicellular species, in later phases of evolution, cyclic AMP also began to symbolize a need for thyroid, melanocyte, or adrenal cortical function, or a need for renal-concentrating ability, but always within that specialized cell type capable of elaborating the specialized response. Moreover, these specializations appeared in multicellular organisms somehow to be derivative upon the primordial symbolism of hunger and/or the absence of glucose in the single-celled organism.

Study of the common slime molds (e.g., *D. discoidium*) revealed that a number of these organisms, when faced with substrate deprivation, will utilize cyclic AMP as a chemotactic signal which gathers together the single-cell ameboid forms into large multicellular aggregates, which evolve into a spore-producing body. This allows the progeny of *Dictyostelium* to survive a period of substrate deficiency and reemerge into adjacent (or the same) econiches under conditions favorable for continued growth and division. Since *Escherichia coli* release cyclic AMP, Dictyostelium can also migrate upstream in a gradient of cyclic AMP to find its favorite food.

In man the condition of hypoglycemia reliably elicits a cyclic AMP response to glucagon in hepatocytes and the subsequent degradation of glycogen, so that plasma glucose concentrations rise. How wonderful is this evolutionary process, which utilizes cyclic AMP in bacteria which must locate sugars in their environment (and cannot store them), and cyclic AMP in the mammalian liver as a tool for dismantling the giant glycogen polymers which serve as storage depots for glucose. Here the evolutionary process reveals a capacity to conserve the symbolic meaning as well as the functional classification for the product of a gene product! Moreover, in such complex animals as mammals, when one studies cells other than hepatocytes or adipocytes where fuel storage is a primary concern, cyclic AMP acquires a more generalized function: to elicit from a specialized cell that unique product or service altruistically elaborated for all other cells in the organism.

In Tomkins's distinction between symbolic and thermodynamic molecules, he emphasizes the different roles of symbolic molecules and ATP, the thermodynamic nucleotide which functions as a universally accepted currency for energy-dependent biological processes. In contrast, GTP which can participate in the commerce of "high energy" phosphates, can also find employment as a part-time symbolic molecule in its activation role exhibited in many signal transduction systems.[8] GTP-binding proteins are utilized in the signal-dependent stimulation of the synthesis of cyclic AMP, inositol bisphosphate, and the response of both rod and cone photoreceptors to photons. In such examples, Tomkins emphasizes the remarkably thematic conservatism of evolution in its deployment of symbolic molecules in homologous contexts with respect to their roles in announcing and reversing hunger and substrate availability. In the presence of an activated (liganded) receptor, GTP can bind to the alpha subunit of its heterotrimeric binding protein, thereby producing a regulatory signal. Its subsequent hydrolysis by the same subunit terminates that signal.

When 3', 5' cyclic AMP was initially discovered in the late 1960s by Rall and Sutherland,[9] there was not yet a glimmer of a suspicion that the molecule would find homologous functions in normal bowel flora as well as in cellular slime molds. It was only after its discovery in mammalian species that its presence throughout the animal kingdom was gradually revealed. Studies of the role of this nucleotide, in a variety of nonmammalian as well as nonvertebrate model systems, have produced extremely useful information in our ongoing efforts to understand biological regulatory systems.

In the context of the concept of inducible proteins, it is appropriate to add that an example alluded to previously (in the discussion of the metabolic code) has relevance here. Indeed, the proteins of the lac operon were the first in which it was shown that cyclic AMP was one of the elements required for the induction of a bacterial gene product.[10] In the absence of glucose, *E. coli* will synthesize cyclic AMP which is half of the signal for the production of alternative metabolic machinery. If lactose appears as well, then the enzymatic machinery for lactose importation and utilization is synthesized. The system remains consistent in its use of cyclic AMP to signal the absence of glucose, but it utilizes a double switching mechanism. It is sufficiently parsimonious that it will not begin to synthesize metabolic machinery for the metabolism of alternative sugars unless and until such a sugar appears for which the bacterium has the appropriate equipment. Thus the absence of glucose and the presence of lactose provide signals which allow expression of the lactose operon. The entire concept of inducible enzyme systems has been extensively studied in microorganisms and certainly such studies have significantly enhanced our approach to the understanding and further study of inducible proteins in mammalian systems, including such protective proteins as the ubiquitous metallothionein family of genes[11] and the even more complex family of microsomal oxidases.[12]

IX. THE CLONING REVOLUTION

A second most compelling and burgeoning area of study where nonmammalian systems have provided invaluable experimental leverage is the whole discipling of molecular biology and gene cloning. I must certainly include here the widespread use of bacteria, their plasmids and phage for cloning a variety of gene sequences and products, and their restriction enzymes, which have provided indispensable tools for the craft of molecular biology.[13] The great advantages of cloning have caused its practice to become almost commonplace for the production of large numbers of copies of specific mammalian DNA sequences and/or proteins.

It is also important to point out in this context, that the molecular biology community has acquired the tools which make is feasible to begin planning and carrying out the mapping and sequencing of the entire human genome. Inasmuch as this genome is estimated to contain something in the order of 3 billion base pairs, and an estimated 180,000 gene products, sequencing the entire genome is a task of the first magnitude. In the course of studying a variety of approaches to achieving this goal, many have naturally focused their attention upon the process of producing cloned fragments of human DNA as substrates for mapping, ordering, and sequencing activity. It became evident in the early phases of this enterprise that nesting strategies (that is, the ordering of smaller within larger DNA fragments) would provide substantial advantages for both mapping and sequencing efforts. Recently, pulsed-field gel electrophoresis developed by David Schwartz has provided an invaluable tool for the manipulation and separation of very large fragments of DNA.[14]

Of great interest here is a novel use of yeast to serve in the cloning of very large DNA fragments which would lend themselves quite well to such nesting strategies. The capability of constructing an extra yeast chromosome containing both the desired sequences and facilitating their replication in yeast is a feat which has now been accomplished in some laboratories.[15] In this example, a single-celled eukaryote is utilized as a factory in which to replicate very large human DNA fragments for the purposes of ordering smaller fragments and providing a framework within which the sequencing of human DNA can proceed rapidly. There is now evidence that a DNA sequence unique to the telomeric region of human chromosomes has thus far been found (by *in situ* hybridization) in the telomeres of all nine mammals so far studied, in both of the two bird species tested, and also in the one reptile. Thus the possibility is raised that all vertebrates may share common telomeric DNA sequences.[28] In the construction of yeast chromosomes a different but analogous yeast-specific telomeric sequence is utilized, as well as a yeast centromeric DNA sequence. Such yeast chromosome components will provide pragmatic advantages in the cloning and sequencing of very large (>5 megabase) fragments of human DNA.

Thus the advantage of being able to manipulate, clone, and analyze very large DNA fragments derives in part from the fact that they provide a context within which to order smaller fragments. Very large DNA fragments in the 1- to 10-megabase range thus simplify, in a profound way, the mapping and sequencing tasks.

X. CELL-ADHESION MOLECULES

D. discoidium has deservedly achieved notoriety for yet another feature which is as fascinating as its dramatic exemplification of the universal metabolic code. It was in the cellular slime molds that biologists first discovered a rather ingenious yet simple mechanism for cellular aggregation that is exploited by these and other organisms. At those specific times when environmental circumstances call for sporulation (for example, during an absence of moisture, warmth, or nutrients), slime molds flow together in an extraordinary migratory rush to form large clusters of organisms. To conclude the process of aggregation the amoebas produce a multicellular creature with the capacity for movement. The motile slug finally elaborates a graceful elongated stalk, crowned by a large fruiting body containing hundreds of spores.

Just before the time of sporulation, the slime molds begin to express a series of gene products which were previously absent from their phenotypic repertoire.[16] These gene products include a bifunctional stapling molecule which can literally bind the single-celled amoeboid organisms to each other, and an integral membrane anchor to secure the staples to the surface of the cells. Here we observe an inducible family of gene products which literally comprise a device for converting single, free-living amoeboid cells into a multicellular organism. The functional homologies between the currently fashionable cell adhesion molecules[17] and the slime mold staples is striking. Neurological cell-adhesion molecules are now credited with a fundamental role in the organization and coherence of the 10^{11} neurons in the human central nervous system. Moreover, cell-adhesion molecules of other persuasions appear to be performing similar functions in the embryogenesis of many other viscera included in the inventory of vertebrate organs. While the slime mold is widely regarded as a "lower" form which lives on a diet of fecal bacteria, it nevertheless depends for its reproduction upon molecular themes which are regarded as essential for the assembly of the most complex and ordered structure known in the cosmos, the human nervous system with all of its 10^{11} processing elements, each enjoying an average synaptic connectivity of 10^4.

XI. THE MOLECULAR ARCHITECTURE OF SIGNAL TRANSDUCTION

I would like to conclude this exposition with comments on the visual system. As a prelude to my discussion of vision, it is appropriate to return again to the cyclic nucleotides. It is in the program for the synthesis of cyclic AMP that we encounter an extraordinary molecular architecture which has received widespread attention in contemporary regulatory biology and even more widespread application by the machinations of evolution. I refer here to the three-panel architecture of signal transduction which consists of a receptor moiety, a heterotrimeric GTP-binding protein and an expressor moiety which elaborates, or in some way influences, the concentration of a "second messenger" molecule or ion.[18]

In addition to writing such epic masterpieces as *Dr. Faustus,* Goethe was an accomplished botanist who mathematically characterized the helical periodicity of distribution for leaves and stems about the canes of a rose. Goethe was a keen observer, well aware of the awesome bias which the human massively parallel processor can bring to an experimental context. He celebrated this awareness by coining one of his pithy aphorisms: "One can recognize only the familiar." In analyzing data it is far too easy to overlook an important but unexpected feature. The pioneers of cyclic nucleotide biochemistry[19] did not initially detect the critical role that the heterotrimeric GTP-binding protein plays in signal transduction.

The fact that there was a GTP co-factor required for signal transduction (Figure 1) was readily overlooked by Sutherland and Rall, for the very good reason that the available ATP preparations were almost invariably contaminated by GTP. In view of the formidable accomplishments of the pioneers, admiration of their work has never been diminished by this omission. In fact, one must admire the ingenuity of Rodbell and colleagues, who were finally able to identify the need for the initially elusive GTP co-factor.[20] The well-studied and widely utilized three-segment architecture of signal transduction is best described in terms of the specialized functions of each segment.[21]

The receptor panel is dedicated to the recognition of a signal which almost invariably arises from outside the cell. It is now known that the signal receptor moiety carries with it a committee of interactive partners, which includes a kinase, a phosphatase, and, in photoreceptors, a signal-quenching protein called arrestin. The second segment is composed of a heterotrimeric signal amplifier which is known as the GTP-binding protein. The number of GTP-activated alpha subunits which can be generated by a single liganded signal receptor varies from fewer than 10 in some hormone-sensitive cells to more than 1000 in vertebrate photoreceptors. The third (readout) segment or signal elaborator also has some amplification function but primarily modifies the concentration of a second messenger. The second messenger can be a cyclic nucleotide (cyclic AMP or cyclic GMP), an inositol bisphosphate, diacylglyceral, Ca^{++}, and possibly others as yet unidentified. The conversion of phosphatidyl inositol into inositol bisphosphate and diacylglycerol by phospholipose C (another GTP-dependent enzyme) results in the elaboration of two signals: diacylglycerol is an activator of protein kinase C; inositol bisphosphate appears to have the capacity to mobilize calcium from intracellular stores.

XII. OPTIMIZATION OF PHOTON CAPTURE AND EXTERNAL AWARENESS

It is indeed a remarkable fact that a transmembrane signaling apparatus which, by and large, appears dedicated to the transfer of chemically encoded information from the outside to the inside of a cell has also been exploited for use in the transduction of photons into a compilation of neural impulses which can be synthesized by the retina and occipital cortex into the visual fields. There are some rather fascinating evolutionary refinements in vision. The cone (which is responsible for color vision) operates at high signal intensity and has a more rapid response. The cone, like the rod, can show adaptation to varying levels of signal intensity. The rod, which operates at lower light intensities, nevertheless has a dynamic range of about 5 log units and, when fully dark adopted, can detect a single photon. Both rods and cones utilize opsin-like photopigments. In fact, such photosensitive macromolecules are utilized as photon detection devices throughout the animal kingdom, including the single-celled flagellated chlamydomonas which also has a chloroplast. A vast army of different

Extracellular Domain

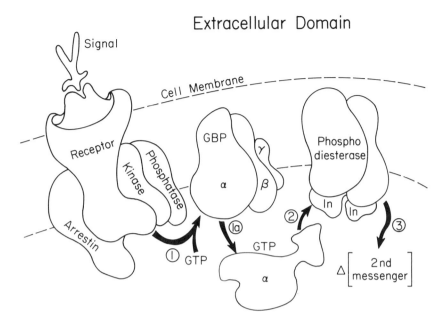

FIGURE 1.   The three-panel molecular architecture of signal transduction. The signal almost invariably arises from the extracellular domain and results finally in a change in the concentration of an intercellular second messenger (arrow 3). The principal logic of the three-panel architecture of signal transduction consists of a receptor panel, a GTP-binding protein panel, labeled GBP, and a third or an effector panel illustrated here by phosphodiesterase. The effector panel could be a phospholipase C, an adenylate cyclase, and other types of effector proteins. The receptor protein is known to be accompanied by a kinase and a phosphatase in the photoreceptor and cyclase systems. The kinase decreases the efficacy of the receptor only after it has acquired signal. The nonliganded receptor is not a substrate for the kinase. The phosphatase has not yet been well studied. The arrestin moiety has been encountered in the photoreceptor version of signal transduction where it contributes to the inhibition of liganded receptor (photoisomersized receptor). After the receptor has acquired signal, it can interact with the GTP-binding protein or second panel. This interaction (arrow 1) results in the binding of GTP to the α-subunit of the heterotrimeric GTP-binding protein. The α-subunit dissociates form the β- and γ-subunits and can go on to interact with the effector moiety (arrow 2). In the photoreceptor system, the α-subunit appears to enter the cytoplasmic domain as a feature of its interaction with the effector enzyme. In the cyclase system, it is quite likely that the GTP-bound α-subunit remains affiliated with the plasma membrane and interacts with the catalytic moiety by lateral diffusion in the plane of the membrane. The formation of many armed α-subunits per liganded receptor offers the opportunity for very great signal amplification. The armed (GTP-bound) α-subunit can then influence the activity of the effector enzyme. In the example shown, the rod phosphodiesterase inhibitor moieties change their interaction with the enzyme thus permitting the phosphodiesterase to become active. The putative role of such inhibitor moieties has not been established as a constant feature of the effector panel. The effector moiety changes the concentration of a second messenger. In this example, the second messenger is cyclic GMP, and the effector catalytic agent phosphodiesterase reduces its concentration. In other systems the second messenger can be cyclic AMP, calcium, or inositol bisphosphate whose concentrations are often increased in response to signal reception. However, for certain signals the concentration of cyclic AMP will fall, e.g., those signals which interact with the so-called inhibitory class of cyclase receptors.

invertebrate species house their opsin photopigment in a photoreceptor organelle called a rhabdome. Interestingly, in the invertebrate species, photon capture results in the depolarization of the plasma membrane of the rhabdome.

In the vertebrate subphylum, the capture of a photon is celebrated by hyperpolarization of the photoreceptor plasma membrane, be it rod or cone. There is now good evidence from patch clamping that both rods and cones utilize cyclic GMP as their second messenger and, moreover, that light can lower the concentration of cyclic GMP whose steady-state concentration maintains the patency of cations channels in the photoreceptor plasma membrane. Thus lowering of cyclic GMP as a consequence of light activation of photoreceptor phosphodiesterase will, in effect, reduce the patency of some fraction of the cation channels, thereby hyperpolarizing the cell membrane and resulting in the transmission of a visual data bit.

There have been extensive studies of the physiology and enzymology of both rods and cones. There remain many fascinating questions surrounding the mechanisms by which rods and cones have been functionally optimized to fill their discrete roles. It is clear for all photoreceptors thus far studied that opsin and isomers of Vitamin A aldehyde are the ubiquitous choice of evolution for photopigments from chlamydomonas to human beings. Moreover, a careful assessment of the molecular pocket in which the retinal resides in Schiff's base linkage to opsin in the photopigment of chlamydomonas has revealed an extraordinary degree of molecular conservation in the dimensions of this hydrophobic domain.[22] The hierarchy of binding affinities for a series of Vitamin A analogs and isomers is perfectly conserved between chlamydomonas opsin and bovine opsin.

There is also a related bacterial opsin which appears as the purple membrane of the halobacterium. This photopigment membrane protein is different than the opsin photopigment exploited for visual transduction. In the halobacterium, the opsin is a thermodynamic protein which utilizes the energy of photon capture for the creation of a proton gradient, that is, protons are expelled from the halobacterium as a consequence of the capture of a photon. The photon-generated proton gradient is exploited by the halobacterium for the synthesis of ATP.[23] Here, instead of serving as an information-containing signal that is descriptive of the external environment, photons are used to drive ATP synthesis. In man and chlamydomonas, opsins are used as symbol-deciphering macromolecules, and the photon has informational and not thermodynamic significance. There are some structural and molecular differences between bacterial opsin and rodopsin. These differences do not, however, obscure the fact that the proteins have a common ancestor and show a number of striking structural and chemical homologies, including the fact that they both make seven helical passes through the membrane as integral proteins and that Vitamin A aldehyde is used by both as the chromophore which is linked to opsin via a Schiff's base.

The fruit fly *Drosophila* is now enjoying a renaissance in studies of invertebrate visual transduction from the perspective of molecular biology.[24] Because of the availability of a wealth of visual and behavioral mutants, it has been possible to attempt characterization of the fly's vision in terms of a particular gene abnormality and thereby assign a function to each gene product in the vision cascade.

As is the case for other invertebrates, photon capture in *Drosophila* is celebrated by depolarization of the photoreceptor membrane. There is significant ambiguity concerning the identity of the visual second messengers in *Drosophila* since both cyclic GMP and inositol bisphosphate are found to be modified by photon capture. Which of the two has the singular responsibility for visual excitation and which the major responsibility for light adaptation is still under intense scrutiny.[25]

Recent studies by Whalen in my laboratory have utilized bovine and amphibian photoreceptors to study the fundamentals of rod visual transduction.[29] While both systems show striking similarities and use closely related gene products, there are some interesting differences. In the bovine system, there are two binding sites for the inhibitory subunits associated with the phosphodiesterase, and inhibition is relaxed when the alpha subunit of the GTP-binding protein is converted to an activator by illuminated rhodopsin and GTP.[26] In our comparisons of amphibian and bovine systems, the two exhibit subtle differences in their use of activator and inhibitory moieties for regulating the light-sensitive phosphodiesterase. It is not yet clear how such differences serve the distinct lifestyles of *Rana catesbiana* (primarily nocturnal) and various bovine species which are diurnal. We are inquiring whether the kinetic differences observed between these species (in the manner in which they activate their rod phosphodiesterase) might reflect the different ways in which they utilize their rods. The frog depends upon rod vision for twilight and for capturing insects in the dim light of early evening.

One other interesting difference between the frog and the cow is the size of the photoreceptor. The amphibian photoreceptor is substantially larger than that of the mammal, having a volume which is almost an order of magnitude greater. The corresponding receptor density per unit area of retina is much greater in the mammal. At the outset it seems certain that the neuronal circuitry and abundance in both retina and those brain regions which support visual data processing in the amphibian and mammal are strikingly different. Certainly the mammal possesses the neuronal fire power to support a far greater pixel density, as afforded by its much richer collection of small photoreceptors. It would appear more parsimonious to mount a smaller array of larger photoreceptors under circumstances where any greater number would not be justified in the face of the underlying neuronal circuitry. Thus, the size of a photoreceptor in general could reflect the underlying neuronal apparatus, where photoreceptor size would vary inversely with the neuronal number and complexity of the parallel processing capabilities of the host's nervous system. This may be too simple an analysis, however. The frog can only perceive moving objects, while mammals can continue to see an object whether it is in motion or stationary. It is also not yet clear what effect receptor size might have on photoreceptor response time, sensitivity, or the rates of dark and light adaptation. Moreover, metabolic differences between the poikilothermic amphibian systems and the mammals may also influence the size and packing density of photoreceptor cells.

Another fascinating evolutionary adaptation with respect to the deployment of retinal photoreceptors is found in raptors. Eagles and hawks can typically identify their prey over great distances. Here point-to-point discrimination is an essential feature of efficient predation, and the smearing of signal and pattern among adjacent photoreceptors would create unacceptable difficulties with respect to the location and capture of prey. Toward this end, raptors have developed a unique broad spacing of their retinal photoreceptors, which are placed further apart than the photoreceptors of other species of birds and other vertebrates. It has been demonstrated by Miller and collaborators that this strategy of photoreceptor deployment and other anatomical modifications of the raptor lens density allow the projection of a sharply resolved image upon the retina for very distant objects.[27] This example of the evolutionary focusing of gene product ensembles produces a result perfectly suitable to the predation pattern of raptors.

XIII. ENDNOTES

The examples and algorithms assembled here are intended to emphasize the conclusion that nonmammalian model systems are not merely convenient or useful because of simplicity, ease of handling, cost, or small size, but rather will often reflect the most luxurious examples of the evolutionary passion for gene product (and hence phenotypic) perfection. The richness of information derived from nonmammalian model systems goes well beyond the modest conveniences which may be encountered in some small animal model systems. When we take the trouble and time to study such model systems with appropriate care, we are very often rewarded with surprising insights into biological processes and especially a more profound understanding of the evolutionary process. Inasmuch as evolution is recognized as the primal shaping force of the gene content in all species, to understand the predilections and devices of evolution is to better understand all of biology.

ACKNOWLEDGMENTS

This work was supported by Institutional Basic Research Grant #81W. I wish to acknowledge the very excellent assistance of Alexandra Vigil in preparing the manuscript.

REFERENCES

1. **Hecht, S., Schlaer, S., and Pirenne, M. H.,** Energy, quanta and vision, *J. Gen. Physiol.,* 25, 819, 1952.
2. **Holland, S. K. and Blake, C. C. F.,** Proteins, exons and molecular evolution, *Bio. Syst.,* 20, 181, 1987.
3. **Geighton, T. E.,** The protein folding problem. *Science,* 240, 267, 1988.
4. **Sepkoski, J.,** Geologic implications of impacts of large asteriods and comets on earth, in *Mass Extinctions in the Phanerozoic Oceans: A Review,* Silver, L. T. and Schultz, P. H., Eds., Special Paper 190, Geological Society of America, Boulder, CO, 1982, 283.
5. **Nathans, J.,** Molecular biology of visual pigments, *Ann. Rev. Neurosci.,* 10, 163, 1987.
6. **Foster, K. W. and Smyth, R. D.,** Light antennas in phototactic algae, *Microbiol. Rev.,* 44, 572, 1980.
7. **Tomkins, G. M.,** The metabolic code, *Science,* 189, 760, 1975.
8. **Bitensky, M. W., Torney, D., Yamazaki, A., Whalen, M. M., and George, J. S.,** A model of the light dependent regulation of retinal rod phosphodiesterase, guanylate cyclase and the cation flux, in *Molecular Mechanisms of Neuronal Responsiveness,* Yigal, H., Ehrlich, R. H., Lenox, E. K., and William, O. B., Eds., Plenum Press, New York, 1987.
9. **Rall, T. W. and Sutherland, E. W.,** The regulatory role of adenosin-3', 5' -phosphate, *Cold Spring Harbor Symp. Quant. Biol.,* 26, 347, 1961.
10. **Daniel, J. and Danchin, A.,** 2-Ketoglutarate as a possible regulatory metabolite involved in cyclic AMP-dependent catabolite repression in *Escherichia coli* K12, *Biochemie,* 68, 303, 1986.
11. **Hildebrand, E. E. and Enger, M. D.,** Regulation of Cd^2/Zn^2 -stimulated metallothionein synthesis during induction, deinduction and superinduction, *Biochemistry,* 19, 5850, 1980.
12. **Hildebrand, C. E., Gonzalez, F. J., McBride, O. W., and Nebert, D. W.,** Human TCDD-inducible cytochrome P-450: localization on chromosome 15, *Nucleic Acids Res.,* 13, 2009, 1985.
13. **Nathans, O. and Smith, H. O.,** Restriction endonucleases in the analysis and restrictions of DNA molecules, *Ann. Rev. Biochem.,* 44, 273, 1975.
14. **Schwartz, D. C. and Cantor, C. R.,** Separation of chromosomes — size DNA's by pulsed field gradient gel electrophoresis, *Cell,* 37, 67, 1984.
15. **Burke, D. T., Carle, G. F., and Olson, M. V.,** Cloning of large segments of exogenous DNA into yeast by means of artificial chromosome vectors, *Science,* 236, 806, 1987.
16. **Gabius, H. J., Springer, W. R., and Barondes, S. H.,** Receptor for the cell binding site of discoidin I, *Cell,* 42, 449, 1985.
17. **Rutishauser, U., Acheson, A., Hall, A. K., Mann, D. M., and Sunshine, J.,** The neural cell adhesion molecule as a regulator of cell-cell interactions, *Science,* 240, 53, 1988.
18. **Shinozawa, T., Sen, I., Wheeler, G. L., and Bitensky, M. W.,** Predictive value of the analogy between hormone-sensitive adenylatec-cyclase and light sensitive photoreceptor cyclic GMP phosphodiesterase: a specific role for light sensitive GTP as a component in the activation sequence, *J. Supramol. Struct.,* 10, 185, 1979.
19. **Sutherland, E. W. and Rall, T. W.,** The properties of an adenine ribonucleotide produced with cellular particles, ATP, Mg^{++}, and epinephrine or glucagon, *J. Am. Chem. Soc.,* 79, 3608, 1957.
20. **Rodbell, M., Lin, M. C., Salomon, Y., Londos, C., Harwood, J. P., Martin, B. R., Rendell, M., and Berman, M.,** Role of adenine and guanine nucleotides in the activity and response of adenylate cyclase systems to hormones: evidence for multisite transition states, *Adv. Cyclic Nucleotide Res.,* 5, 3, 1975.
21. **Bitensky, M. W., Whalen, M., Torney, D. C., Tatsumi, M., and Yamazaki, A.,** A common algorithm for the transduction, amplification and cellular response to photons, hormones, and neurotransmitters, in *Mechanisms of Signal Transduction by Hormones and Growth Factors,* Myles, C. C. and Wallace, L. M., Eds., Alan R. Liss, New York, 1987, 3.
22. **Foster, K. W., Saranak, J., Patel, N., Zarilli, G., Okabe, M., Kline, T., and Nakanishi, K. A.,** Rhodopsin is the functional photoreceptor for photoaxis in the unicellular eukaryote chlamydomonas, *Nature (London),* 311, 756, 1984.
23. **Stoeckenius, W.,** The rhodopsin-like pigments of halobacteria light-energy and signal transducers in an archaebacterium, *Trends Biochem. Sci.,* 10, 483, 1985.

24. **Devary, O., Heichal, O., Blumenfeld, A., Cassel, D., Suss, E., Barash, S., Rubinstein, C. T., Minke, B. M., and Selinger, Z.,** Coupling of photoexcited rhodopsin to inositol phospholipid hydrolysis in fly photoreceptors, *Proc. Natl. Acad. Sci. U.S.A.,* 84, 6939, 1987.

25. **Deterre, P., Bigay, J., Forquet, F., Robert, M., and Chabre, M.,** Cyclic GMP phosphodiesterase of retinal rods is regulated by 2 inhibitory subunits, *Proc. Nat. Acad. Sci. U.S.A.,* 85, 2424, 1988.

26. **Snyder, A. W. and Miller, W. H.,** Telephoto lens system of falconiform eyes, *Nature (London),* 275, 127, 1978.

27. **Uchida, S., Wheeler, G. L., Yamazaki, A., and Bitensky, M. W.,** A GTP-protein activator of phosphodiesterase which forms in response to bleached rhodopsin, *J. Cyclic Nucleotide Res.,* 7, 95, 1981.

28. **Moyzis, R. and Meyne, J.,** personal communication.

29. **Whalen, M. and Bitensky, M.,** Comparison of the phosphodiesterase inhibitory subunit interactions of frog and bovine rod outer segments, *Biochem. J.,* 259, in press, 1989.

Chapter 22

ETHICAL AND REGULATORY CONSIDERATIONS IN THE USE OF COLD-BLOODED VERTEBRATES IN BIOMEDICAL RESEARCH

Rebecca Dresser

TABLE OF CONTENTS

I. INTRODUCTION

In recent years, the use of laboratory animals has come under increasing legal and ethical scrutiny. In late 1985, the two federal regulatory systems governing animal use in research, testing, and education were extensively revised. In addition, since the early 1970s, philosophers and other commentators have demonstrated a renewed interest in examining the morality of human attitudes and practices toward nonhuman animals.

How much of this heightened concern for animals extends to nonmammalian species? The overall focus has been on the primates, companion animals, and, to a lesser extent, farm animals, rabbits, and rodents, which are the animals most commonly used as models in science. In this chapter, I shall examine how the regulatory provisions and ethical principles governing the use of laboratory animals apply specifically to the species discussed in this volume.

II. FEDERAL LAW AND POLICY

Two federal regulatory systems affect the use of laboratory animals in the U.S. The Animal Welfare Act (AWA)[1] applies to research activities in both the public and private sectors. The first version of the statute was enacted in 1966, following public outcry over the theft of pet dogs for sale to research facilities. The law was amended in 1970, 1976, and 1985, to extend and supplement its coverage.[2]

The AWA requires the secretary of agriculture (secretary) to prescribe humane standards for the transportation, handling, care, and treatment of laboratory animals. Research facilities must register with the U.S. Department of Agriculture (USDA), must submit to inspections conducted by USDA personnel, and are subject to monetary penalties for violations of the rules. In 1985, new requirements for self-regulation were established, directing facilities to assemble an Institutional Animal Committee with community representation to oversee animal use at the facility.

However, none of the provisions apply to research on cold-blooded animals. The statute explicitly defines the word "animal" as "any live or dead dog, cat, monkey (nonhuman primate animal), guinea pig, hamster, rabbit, or other such warm-blooded animal, as the Secretary may determine is being used, or is intended for use, for research, testing, experimentation, or exhibition purposes. . . ."[3]

In addition, the secretary has exempted birds, rats, mice, and farm animals used in research from statutory coverage.[4] The secretary's action is now being legally contested, for many have argued that the law fails to give the secretary discretion to exempt from the law any warm-blooded animal used for scientific purposes.[5] But the Congress has thus far clearly excluded cold-blooded animals from coverage by the AWA, apparently based on a lack of knowledge or concern about the welfare of these laboratory animal species.

By contrast, however, the other major federal oversight system governing the use of laboratory animals applies to all vertebrate species. The funding policy of the U.S. Public Health Service (PHS) applies to animal use supported by the National Institutes of Health (NIH), Centers for Disease Control, Food and Drug Administration, Health Resources and Services Administration, and Alcohol, Drug Abuse, and Mental Health Administration.[6] Other funding entities, such as the National Science Foundation[7] and the American Heart Association,[8] apply the PHS policy or similar provisions to grant recipients.

In 1985, the PHS policy was extensively revised. The revised policy includes new substantive and procedural requirements for institutions and investigators. Each institution must assemble an Institutional Animal Care and Use Committee (IACUC) comprised of at least five members. Among the members must be "one Doctor of Veterinary Medicine with training or experience in laboratory animal medicine, one practicing scientist experienced in research involving animals, one member whose primary concerns are in a nonscientific area, and one individual who is not affiliated with the institution in any way other than as a member of the IACUC.''[6]

The IACUC has several responsibilities, including review of the institution's program for animal care and use, as well as inspection of every facility in which animals are housed for longer than 24 hours. The committee is authorized to suspend any activity that fails to comply with the PHS policy. Another major responsibility of the committee is protocol review. Proposals for the use of laboratory animals must be considered in light of the policy's substantive principles. The IACUC has the authority to approve, require modifications in, or withhold approval of the proposal's plan for animal care and use. In addition, if an ongoing project is determined to be in noncompliance with the policy, the committee may suspend it.

The PHS policy contains seven substantive principles governing laboratory animal care and use. Its requirements resemble those now contained in the AWA. Investigators performing painful or distressing procedures on animals are instructed to use anesthetics, analgesics, and tranquilizers, unless there is scientific justification for withholding such agents. Animals that would otherwise experience severe or chronic unrelieved pain or distress must be humanely killed unless this would interfere with a protocol's scientific validity. Euthanasia procedures must be consistent with the recommendations of the American Veterinary Medical Association Panel on Euthanasia[9] unless there are scientific reasons for using a different method.

The PHS policy also requires investigators and institutions to adhere to the recommendations of the NIH *Guide for the Care and Use of Laboratory Animals*[10] unless there are scientific reasons for a departure from the provisions. The *Guide* addresses institutional policies, laboratory animal husbandry, veterinary care, physical facilities, and other pertinent topics. It includes general standards for surgery and postsurgical care as well.

The NIH *Guide* also includes nine principles intended to govern research performed or sponsored by any U.S. government agency. Several of them closely resemble those included in the policy itself. Others, however, supplement the policy in important ways. According to one of these principles, ''Procedures involving animals should be designed and performed with due consideration of their relevance to human or animal health, the advancement of knowledge, or the good of society''. Another principle instructs investigators to select the appropriate species, quality, and minimum number of animals needed to obtain valid results and to consider available nonanimal research alternatives. A third principle states: ''Unless the contrary is established, investigators should consider that procedures that cause pain or distress in human beings may cause pain or distress in other animals''. The *Guide* itself recommends against the use of prolonged restraint or multiple survival surgery on a single animal except when such procedures are essential to research objectives.

Federal oversight of research institutions is provided by the NIH Office of Protection from Research Risks (OPRR). Each institution must submit annually to OPRR a detailed written assurance that describes the institution's level of compliance with the policy. Institutions also are subject to site visits by PHS officials at any time. Noncompliance with the policy could jeopardize PHS support of the institution.

III. ETHICAL CONSIDERATIONS

Government interest in the protection of laboratory animals is a recent development in the U.S. Until a short time ago, attitudes and behavior toward animals generally reflected religious and philosophical concepts suggesting that humans had few moral obligations toward nonhuman creatures. For example, one traditional view holds that humans appropriately have absolute authority over the rest of the natural world due to their God-given or evolutionary superiority.[11] According to another influential theory proposed by the French philosopher Descartes, nonhuman animals are like machines and thus lack the capacity to experience pain and other mental states.[12] As a result, humans need not follow any moral rules to avoid harming nonhuman creatures. A third position, put forth by Emmanuel Kant, is that humans have no direct duties toward animals but that cruelty to animals should be avoided if it encourages violence toward humans as well.[13]

The present federal policy governing the use of laboratory animals is, in part, attributable to the contemporary animal rights and welfare movement, which has challenged the traditional attitudes toward nonhuman animals. Over the past decade, increasing ethical concern for nonhuman animals has been translated into a political movement that is becoming increasingly activist at the grass roots, state, and federal levels. As a basis for their activities, members of this movement cite recent philosophical arguments supporting improved treatment for nonhuman animals.

Several contemporary philosophers have set forth arguments favoring an increase in the moral value humans assign to nonhuman species. Two of these positions have been influential in shaping an emerging ethical consensus regarding the appropriate guidelines for the use of laboratory animals. In addition, these views have to some extent been incorporated into the current federal regulatory policy. The humane-treatment and utilitarian positions on laboratory animal use assign moral significance to the interests of laboratory animals in avoiding harm and to the interests of humans and other animals in obtaining the benefits of continued use of laboratory animals.

In the remainder of this section, these two ethical perspectives and their relevance to laboratory animal use are discussed.

The humane-treatment view encompasses traditional animal welfare beliefs. The general position is that humans ought to have empathy for other creatures and thus should refrain from harming and seek to advance the well-being of animals under their care.[14] The humane-treatment position is evident in the federal policy governing laboratory animal use. The AWA and the PHS policy both emphasize the responsibility of the investigator and the institution to monitor laboratory animals for pain and distress and to administer pain- and distress-relieving agents when appropriate. Other federal provisions address the need to furnish laboratory animals with an environment that will meet their species-specific needs. In addition, the methods adopted to kill laboratory animals must be as painless as possible. Laboratory animal "quality of life" is the overall focus of the humane-treatment position and of the current federal regulatory policy.

A more complex aspect of the humane-treatment position concerns the acceptable justification for laboratory animal use. This issue is raised by the federal regulatory policy as well, for current rules require proposals for animal use to be evaluated for their "scientific necessity" and their "relevance to human or animal health, the advancement of knowledge, or the good of society."[2] According to the humane-treatment position, "unnecessary" harm must be avoided. The challenge then becomes to determine what animal harm qualifies as "necessary". As applied to laboratory animals, the goal is to reduce the harm animals experience, without reducing the important benefits that the use of laboratory animals provides to others. The traditional humane-treatment position, however, fails to set precise standards for the probability and magnitude of benefits to others that creates a legitimate necessity to harm laboratory animals. The current federal policy remains similarly vague on this point.

A few contemporary defenders of the humane-treatment view have attempted to address this problem. According to one such individual, the standard for necessary harm is met when the suffering of a laboratory animal is "likely to be offset by practical benefits to humans or animals, or else by a contribution to scientific knowledge which most informed persons would judge significantly proportionate to the amount of suffering in question".[15]

According to a second commentator, "necessity is a relation between a means (an action or policy) and an end (its objective)".[14] The investigator proposing a project in which laboratory animals will be harmed must demonstrate that no less harmful alternative procedure would produce equally valid and reliable scientific data. For instance, a lethal dose 50% (LD_{50}) test should not be performed if the less harmful limit test would provide sufficient information on toxicity of a substance. At a broader level, the question of necessity should include an examination of the probability that the proposal will meet its objective and an assessment of the importance of this objective to society.[14] For instance, is severe harm warranted in the LD_{50} when substantial evidence exists that the test produces unreliable results? Moreover, even if the test is sufficiently reliable, should it be performed to evaluate the toxicity of a new cosmetic that is nearly identical to one already on the market?

The second major ethical view that has influenced the development of guidelines on the use of laboratory animals is utilitarianism. The utilitarian theory mandates a similar examination of the justification for laboratory animal use, but it also attempts to compare the costs to the animals with the benefits bestowed on others. To be justified, proposals for the use of laboratory animals must contribute benefits to others that outweigh the harm inflicted on laboratory animals. The more harm that is imposed on animals, the greater must the probability and magnitude of the anticipated benefits.[16]

Utilitarians assign moral concern to the positive and negative experiences of all living creatures.[17] Suffering is bad and pleasure is good, no matter who experiences these states. According to the utilitarian, the pain experienced by a dog used in a cancer research project is just as morally significant as the added comfort the research project eventually contributes to the human cancer patient. First, the experiment should be designed to confer the least possible pain and distress on the dog. Second, the relative costs and benefits must be assessed. If the project confers little pain or distress on a few dogs, yet produces data significantly enhancing the lives of many human cancer patients, then the project is morally justified. Conversely, if large numbers of experimental dogs undergo severe pain and distress in a project that yields information that is accessible by other means, or lacks significant implications for cancer treatment, the project should not have been performed. The utilitarian analysis has been incorporated into various classification systems for evaluating proposals on animal use. Proposals that entail the greatest harm to laboratory animals are scrutinized most closely for their justification.[18]

At one level, the humane-treatment and utilitarian positions have clear implications for the use of animals in research, testing, and education. Every proposal for animal use must be designed to confer the least possible harm on the least possible number of animals. The more difficult questions arise in determining which proposals are morally justified according to the two ethical views. The difficulties in implementing this element of ethical analysis are attributable to three primary sources.

One is the problem of comparing human and nonhuman experiences. Given the current lack of precise data on how nonhuman animals experience many experimental interventions, it will often be impossible to assign comparative weights to the pain, distress, and well-being experienced by animal subjects and the human beneficiaries of research.[19]

The second difficulty stems from our inability to predict with accuracy the eventual results of many animal projects. Proposals that seemed unnecessary and trivial when they were undertaken have sometimes unexpectedly produced valuable knowledge, often in a field different from that intended, or several years later, after complementary investigations had been pursued. If we fail to allow seemingly unmeritorious or unnecessary projects to go forward, we shall protect more animals but also lose some knowledge.[20] The question is how much knowledge society is willing to delay or forgo in order to protect laboratory animals from being used.

The third troublesome issue concerns judgments on which benefits are necessary and important enough to society to warrant imposing harm on laboratory animals. For example, members of the animal welfare and rights movements often question whether the human need for new cosmetics and household products is sufficiently compelling to warrant painful and lethal safety testing on laboratory animals.[21] Questions also are raised regarding the true value to humans of a great deal of behavioral psychology research and educational projects using animals.[22] In addition, some argue that preventive and environmental health interventions would produce a greater reduction in human morbidity and mortality than the current research approach that consumes so many animals.[23] These are broad and complex issues with enormous social and political implications, which are unlikely to be resolved in the near future. As a result, at least in the near future, uncertainty is likely to characterize this aspect of ethical and regulatory analysis of the use of laboratory animals.

IV. SPECIAL ISSUES REGARDING THE USE OF COLD-BLOODED VERTEBRATES

Until now, this chapter has largely addressed regulatory and ethical issures bearing on the general topic of laboratory animal use. In contrast, this section will focus on the specific category of cold-blooded vertebrates. Are there special characteristics of this group that support treating its members differently from other vertebrate species? The literature on this topic is characterized by two competing views. One holds that all vertebrates should be governed by the same ethical principles; the other claims that differential treatment of cold- and warm-blooded vertebrates may be justified. Indeed, these disparate views are evident in current federal regulatory policy. The Animal Welfare Act excludes cold-blooded vertebrates from its coverage, while the PHS policy covers them.

The two disparate positions have roots in ethical theory as well. The humane-treatment view is based on the empathy humans have for nonhuman creatures. Different species elicit different psychological responses from humans depending on their historical relationships, physical characteristics, and mental capacities. "A gorilla will gather more sympathy than a trout, not so much because it is more intelligent as because it exhibits a range of needs and emotional responses to those needs that is missing altogether in the trout, in which evidence of pain can barely be detected."[14] This view probably underlies the idea that it is morally preferable to use "lower" rather than "higher" animals in scientific projects. Indeed, some commentators include the substitution of cold-blooded for warm-blooded animals among the alternative methods that are the focus of current efforts to refine and replace traditional animal models.[14]

Conversely, utilitarian theory supports the view that all species able to experience pain and other mental states are worthy of equal moral consideration. Those who argue for equal treatment of all vertebrates cite evolutionary and anatomical evidence that cold-blooded and warm-blooded vertebrates have similar capacities to experience pain.[24] Some scientists and other writers debate whether there is a significant difference between the ability of vertebrates and at least some invertebrate species to experience pain, thus calling into question the tendency to assign invertebrates a lower moral status than vertebrates. The following statement is included in the Society for Neuroscience Guidelines for the Use of Animals in Neuroscience Research. "As a general principle . . . ethical issues involved in the use of any species, whether vertebrate or invertebrate, are best considered in relation to the complexity of that species' nervous system and its apparent awareness of the environment, rather than the physical appearance or evolutionary proximity to humans."[25] Others have noted that some invertebrate species, such as the squid and octopus, have an "exquisitely developed" nervous system and sensitivity and that their use should receive the same level of ethical review given vertebrate use.[26]

Even utilitarians might support some differential treatment, however. It is difficult to discern a defensible moral basis for discriminating among animal species based on how cute and cuddly they are. It could be morally appropriate, however, to select animals for scientific use based on their capacities for more or less complex negative and positive experiences. For example, many of those who argue that it would be better to use a fish than a cat in a scientific project do so on the ground that the cat would be expected to experience greater harm from the confinement and interventions imposed as part of the project.[27]

To address this issue, it is necessary to examine the concept of harm. The harm that individuals experience is partly a function of their capacities for various positive and negative mental states. Although experiences such as physical pain and perhaps even fear may be found in both vertebrates and invertebrates, these species may still differ in their capacities to experience other states, such as anxiety and depression. Indeed, the documented presence of benzodiazepine receptors in all vertebrates except the cartilaginous fishes has led some to argue that vertebrates and not invertebrates are capable of experiencing anxiety.[19] The morally relevant question is whether certain species would experience various aspects of laboratory use more negatively than others would. If several species would be scientifically appropriate for a project, and some of them would suffer more than others would from exposure to specific experimental procedures, then it would be morally preferable to use a species that would suffer less than the others.

It is crucial, however, for such judgments to rest on demonstrable evidence, rather than on unexamined intuition. Assertions such as the following should be supported by empirical evidence: "I would argue that discrimination within the vertebrate kingdom is essential, for, while we have a duty to be considerate toward fish, amphibians and reptiles, in selecting animals for use in necessary laboratory experiments, scientists have a higher duty toward birds and mammals."[28] There is presently a compelling need for systematic investigation of the various states of harm different species are capable of experiencing. With the changing ethical and regulatory status of laboratory animal use, this question is becoming increasingly important. One must hope that these developments will generate support for further research on the capacities of the variety of nonhuman species used in scientific projects.

V. CONCLUSION

As this chapter indicates, the ethical and regulatory issues surrounding the use of laboratory animals are far from settled. Many questions still exist regarding how to assess the mental states of laboratory animals, how much moral value to assign to their interests, as compared with the interests of the potential human beneficiaries of animal use, and how to evaluate the scientific necessity for and overall merit of proposed animal projects. The answers to these questions are perhaps even less clear concerning cold-blooded vertebrates, whose moral status and mental awareness have not yet been the object of extensive study. The one thing that is certain is that a great deal of empirical and philosophical analysis will be needed to address these issues with the thoughtfulness and clarity they merit.

REFERENCES

1. U.S. Code, Vol. 7, Sec. 2131—2157(Suppl. 3), 1982 and 1985.
2. **Dresser, R. S.,** Assessing harm and justification in animal research: federal policy opens the laboratory door, *Rutgers Law Rev.,* 40, 723, 1988.
3. U.S. Code, Vol. 7, Sec. 2132 (g), 1982.
4. Code of Federal Regulations, Vol. 9, Sec. 1.1 (n), 1986.
5. **Cohen, H.,** Two questions concerning the Animal Welfare Act, Congressional Research Service, U.S. Congress, Publ. No. 85-927A, Washington, D.C., August 7, 1985.

6. Public Health Service, Policy on Humane Care and Use of Laboratory Animals, rev. ed., 1986.

7. National Science Foundation, *Natl. Sci. Found. Bull.*, 2, January 1986.

8. American Heart Association Task Force, Position of the American Heart Association on research animal use, *Arteriosclerosis*, 5, 310A, 1985.

9. American Veterinary Medical Association Panel on Euthanasia, Report of the AVMA Panel on Euthanasia, *J. Am. Vet. Med. Assoc.*, 188, 252, 1986.

10. U.S. Department of Health and Human Services, *Guide for the Care and Use of Laboratory Animals*, NIH Publ. No. 85-23, Bethesda, MD, rev. ed., 1985.

11. Thomas Aquinas, Differences between rational and other creatures, in *Animal Rights and Human Obligations*, Regan, T. and Singer, P., Eds., Prentice-Hall, Englewood Cliffs, NJ, 1976, 56.

12. **Descartes, R.**, Discourse on the method, in *Animal Rights and Human Obligations*, Regan, T. and Singer, P., Eds., Prentice-Hall, Englewood Cliffs, NJ, 1976, 60.

13. **Regan, T.**, *The Case for Animal Rights*, University of California, Berkeley, 1983, 174.

14. U.S. Congress Office of Technology Assessment, *Alternatives to Animal Use in Research, Testing, and Education*, Rep. No. OTA-BA-273, U.S. Government Printing Office, Washington, D.C., 1986, 37, 78, 80, 81.

15. **Fox, M.**, *The Case for Animal Experimentation: An Evolutionary and Ethical Perspective*, University of California, Berkeley, 1986, 168.

16. **Singer, P.**, *Practical Ethics*, Cambridge University Press, Cambridge, 1979, 58.

17. **Singer, P.**, *Animal Liberation: A New Ethics for Our Treatment of Animals*, Avon Books, New York, 1975, 7.

18. Scientists Center for Animal Welfare, Consensus recommendations on effective institutional animal care and use committees, in *Effective Animal Care and Use Committees*, Orlans, F. B., Simmons, R. C., and Dodds, W. J., Eds., American Association for Laboratory Animal Science, Cordova, TN, 1987, 11.

19. **Tannenbaum, J. and Rowan, A. N.**, Rethinking the morality of animal research, *Hastings Center Rep.*, 32, October 1985.

20. **Gallistel, C. R.**, Bell, Magendie, and the proposals to restrict the use of animals in neurobehavioral research, *Am. Psychol.*, 36, 357, 1981.

21. **Hermick, F. and Lehman, H.**, Unnecessary suffering: definition and evidence, *Int. J. Study Anim. Probl.*, 3, 131, 1982.

22. **Rollin, B.**, *Animal Rights and Human Morality*, Prometheus Books, Buffalo, NY, 1981, 105, 124.

23. **Carlsson, B.**, Ethical issues in animal experimentation: view of the animal rightist, *Acta Physiol. Scand.*, 128, 50, 1986.

24. **Drewett, R. F.**, Alternatives to the use of animals in behavioral research, *Alt. Lab. Anim.*, 14, 312, 1987.

25. Ad Hoc Committee on Animals in Research, Guidelines for the use of animals in neuroscience research, *Neurosci. Newslett.*, 14, 2, September/October 1983.

26. **Orlans, F. B.**, Review of experimental protocols: classifying animal harm and applying "refinements", in *Effective Animal Care and Use Committees*, Orlans, F. B., Simmons, R. C., and Dodds, W. J., Eds., American Association for Laboratory Animal Science, Cordova, TN, 1987, 50.

27. **Flemming, A. H.**, Animal suffering: how it matters, in *Effective Animal Care and Use Committees*, Orlans, F. B., Simmons, R. C., and Dodds, W. J., Eds., American Association for Laboratory Animal Science, Cordova, TN, 1987, 140.

28. Editorial, All animals are equal, but some . . . , *Alt. Lab. Anim.*, 14, 274, 1987.

INDEX

reproductive modes, 246
salamanders, 246—250
seasonal cycles, 258—259
semiaquatic species, 264—265
skin, 259
temperature, effects of, 258
temperature relations, 258—259
terrestrial species, 265—266
thermal acclimation, 258—259
thermoregulation, 258—259
toxins, 264
urinary bladder, 259—260
water relations, 259—261
water uptake and storage, 259—260
waterproofing, 261
Anaerobic metabolism, amphibians, 258, 261—263
Anatomical correspondences, reptiles, 298—299
Androgens, 283—284, 294
Anesthesia, 127, 167
Aneuploidy, 323
Animal genetics, 326
Animal models, 121—147, see also specific types
Animal pole, 100
Animal rights and welfare movement, 372
Animal Welfare Act, 370—372, 374
Animals in research, 149, see also specific types
Annelids, 21—46, see also Leeches
Anolis carolinensis, 290, 301—303
Anopheles quadrimaculatus, 67
Anorexia, 292
Anosmia, 291
Anoxia, 312
Ansa lenticularis, 298
Antennal gland, 49, 57—58
Antennal gland anatomy, 51—53
Anti-AChE, 79
Anti-ON$_1$/ON$_2$ antibody, 230, 232
Antibiotic proteins, 68
Antibiotics, 68—69, 322
Antibody population, 123
Antibody production, 320
Anticoagulant, 40, 42, 156
Antigenicity, 228
Antimalarial drugs, insects, 66—67
Antioxidant capacity, 315
Antioxidants, 18, 314
Anurans, 246—247, see also Amphibians
eggs, 254
fertilization, 253
general characteristics, 249—251
parental care, 254
polyploidy, 262—264
reproduction, 253—257
viviparity, 257
Aortic bodies, 179
Aortic chemoreceptors, 168
Apyrases, 40
Aquatic amphibians, 264—265
Arachnids, 154
Architecture of cells, 325
Arginine vasopressin, 281

Arginine vasotocin (AVT), 259, 284
Arrestin, 362—363
Arthropods, 62, see also Insects
Artificial insemination, 208, 322
Ascidian zygotes, 100—101
Ascidians, see Fertilization in ascidians
Ascorbic acid, 314
Asexual cycle, 15
Asphyxiation, 75, 81
Astrocytes, 230
Atherosclerosis, 113
ATP, see Adenosine triphosphate
ATP synthesis, 364
Atropine, 75, 82—84
Atropine sulfate, 77—78, 83
Attention structure, 298
August Krogh Principle, see Krogh Principle
Autoradiogram, 236, 238—239
Aves, 320, 326, see also Domestic fowl
Avian animal models of human diseases, 326
Avian leukosis viruses (ALV), 326—327
Avian retroviruses, 326
Axonal circuits, 325

B

B-cells, 320—321
B-glucanase, 114
B-lymphocytes, 68
Bantams, 322
Baroreceptor reflexes, 171
Baroreceptors, 171, 176, 178—179
Basal forebrain, reptiles, 290, 298—299
Basal ganglia, 298
Basic understanding of vertebrate systems, 122
Basic unity of physiology, 2
Bdellin, 40
Behavioral fever, 258
Behavioral sex, reptiles, 294—295
Binding of spermatozoa and eggs, 93
Biochemical basis, organophosphate poisoning, 75—76
Biological research, nonmammalian models, 353—367, see also Nonmammalian models
Bioluminescence assay, 67
Bioluminescence, insects, 67
Biomedical applications, marine animals, 149—159, see also Marine animals; and Marine biology
Biomedical education, 150—151
Biomedical research
cold-blooded vertebrates, ethical and regulatory considerations in use of, 369—376
domestic fowl, 319—331, see also Domestic fowl
insect models for, 61—72, see also Insects
medaka as tool in, 187—205, see also Medaka
reptile models for, 289—308, see also Reptile models, biological research
small fish, 185—214, see also Goldfish; Medaka
Birds, 311, see also Domestic fowl
Bisexual brain, 295
Black flies, 66

C

Oxidized glutathione, 313
Oxidizing situations, 313
Oxygen affinity, 313
Oxygen consumption, 112—113, 246, 261
Oxygen posthypoxic reinflux to tissues, buffering of, 313
Oxygen radicals, 112—113
Oxygen-reactive derivatives, threat of ill effects of, 314
Oxygen receptors, 168, 170
Oxygen reinflux, 312—313
Oxygen toxicity, 313
Oxygen transport, 261
Oxyluciferin, 67

P

Paedomorphosis, 247, 253
PAH, see *p*-Aminohippurate
Paleostriatum, 298
Palliative, 84
2-PAM, 75, 79, 82—84
Paralysis, 75, 77, 80—81
Paramecium, 14, 15
Paramecium caudatum, micronuclear transplantation, 16
Paraventricular hypothalamus, 299
Parental care, 253, 254, 301
Parkinson's disease, 41, 290, 299
Parkinsonism, see Parkinson's disease
Parthenogenetic line, 324
Pathogen production, insects, 65—66
Perceptual deficits, 298
Perifusion systems, amphibian model, 282
Peripheral blood cells, 123
Peripheral chemoreceptor, 179
Peroxidase, 112
Phagocytosis, 150
Pharmaceutical products, 326—327
Pharmacological basis, organophosphate poisoning, 76—77
Pheomelanin, 334—335, 339
Phorbol ester, 111
Phospholipase C, 110
Phosphorylated enzyme, 76
Phosphorylation, 76, 79, 228
Photic response, 302
Photinus pyralis, 67
Photomedicine, 301
Photon capture, 362—365
Photoperiod, 259, 322
Photoreactivation, 17
Photoreceptors, 359, 363—365
Phrenic nerve, 167—168
Phrenic nerve discharge, 169
Phylogenetic tree of animal kingdom, 151—152
Physiological basis, organophosphate poisoning, 74—75
Physiological control of secretion decapod crustacean bladder, 58
Physiological defense mechanism, 314

Physiological stress, reptiles, 290, 282—294
Physiology, insect, 62
Physostigmine, 83
Pigment formation, 334—337
Pigment loci, 338
Pigment mutants, 325
Pituitary gland, 280
Pituitary-gonadal axis, 282
Plasma concentrations of urea, 124
Plasmodium falciparum, 67
Plasmodium gallinaceum, 66
Plasmodium vivax, 66
Plasticity, 237
Platyfish-swordtail hybrid, 207—208
Play, 301
Pneumotaxic center, 163
Poecilia formosa, 216, 218
Poison glands, 264
Poisoning, see Toxicity testing
Polyclonal antibodies, 230
Polymerization, 313
Polymerized hemoglobins, 310, 312—313, 315
Polyploidy, 262—264
Polyspermy, 91, 95, 108, 109, 114
Pons, 167
Pontine area, 167
Pontine centers, 167
Pontine respiratory complex, 164—165, 167
Postnatal care, 301
Potential amphibian models, 282—284
Potential protein functions, 355
Power output, 258
PR, see Protective ratio
Preganglionic motor neurones, 167
Pregnancy test, 280
Premelanosome, 334, 336—337
Previtellogenic follicle, 297
Primitive streak, 129, 133
Proboscis, spastic paralysis of, 78
Production requirements, domestic fowl, 320—323
Progenesis, 253
Progesterone, 296—298
Programmed aging, 18
Prolactin, 282
Promoters, 327, 346
Prophylactic action, 67
Propranolol, 302
Prostaglandin E_1, 258
Prostaglandins, 281, 284
Protective mechanisms, 314, 315
Protective ratio (PR), 81
Protein expression, 234—235
Protein kinase C, 111
Protein studies, albinism, 342—344
Protein synthesis, 233—235
Proteinase inhibitors, 40
Proteins, 123
Proteins ON_1 to ON_4
　nerve injury, response to, 230—237
　sites of synthesis, 228—230
Proteoliasin, 114—115

Ureotelism, 260
Uricotelism, 260
Urinary bladder, 51—53, see also specific animal
    types
Urine, 7—8
Urine formation, 48
Urodela, 247, 249, 263
Utilitarianism, 373—374
UV damage, 196
UV irradiation, 17
UV light, 302
UV-A damage, 302

V

Vagal afferent fibers, 167
Vagal afferents, 163, 167
Vagal efferent fibers, 171
Vagal efferent output, 172
Vagal efferent supply, 178
Vagal input, 167
Vagal motoneurones, 172
Vagal motonucleus, 163, 178
Vagal motor column, 166, 172—173
Vagal preganglionic neurones, 178
Vagal respiratory neurones, 167
Vagal supply to heart, 172
Vagal Xth motor nuclei, 163
Vagal tone, 170—173, 178
Vagus, 172—173, 175
Vagus cranial nerve, 166
Vagus nerve, 170, 176
Vasoactive intestinal peptide (VIP), 123
Vasodilatation, 171
Vector, 326—327
Ventilation, nervous control of, 161—183
Ventilation rate, 176—177
Ventilatory cycle, 176
Ventilatory movements, 173
Ventral, bulbar respiratory nucleus, 163—165
Ventral tegmental area, 299
Vertebrate embryology, 320
Vertebrate hemoglobins, 310
Vertebrate systems, 122
Vertebrates, 162, 310—311, 374, 375, see also Cold-
    blooded vertebrates
Vesicles, 96, 114
Viability, 219—220
Viable mutant frequencies, 221
Viable mutants, 221—222
Vimentin, 227—228, 237

VIP, see Vasoactive intestinal peptide
Virological research, 323
Viruses, 320
Visceral alerting response, 171, 179
Visceral cardiac branch, 166, 172
Visible mutation, medaka, 197
Vision, 122, 362
Visual orientation to environment, 324
Visual system, 125, see also Goldfish, visual system
Vitamin B12, 18
Vitamin C, 17
Vitamin E, 17, 314
Vitelline, 115
Vitelline layer, 108, 113—114
Vitellogen genes, 327
Vitellogenesis, 281, 284, 296
Vitellogenic follicle, 297
Vitiligo, 325
Viviparity, 253, 257, 258
Voltage-gated ion channels, 96—97, 99
Volvox, 14
Vomeronasal chemical senses, 291

W

Water, 4—7
Water relations, 259—261
Waterproofing, 261
Whiptail lizard, 290, 295
Wing-banding, 322

X

X-ray sensitivity, medaka, 196
Xanthophores, 301
Xenodiagnosis, 65
Xiphophorus, 207—208

Y

Yeast, 360—361
Yellow plasm, 99
Yolk, 327
Yolk mass, 129
Yolk sac membrane, 129

Z

Zona pellucida, 90, 109, 116
Zona reaction, 109